DRILLING PRACTICES MANUAL

SECOND EDITION

SECOND EDITION

DRILLING PRACTICES MANUAL

Preston L. Moore

PennWell Books
PennWell Publishing Company
Tulsa, Oklahoma USA

Copyright © 1986 by
PennWell Publishing Company
1421 South Sheridan Road/P.O. Box 1260
Tulsa, Oklahoma 74101

Library of Congress cataloging in publication data

Moore, Preston L.
  Drilling practices manual.

  Includes bibliographical references and index.
  1. Oil well drilling—Handbooks, manuals, etc.
I. Title.
TN871.2.M562 1985   622′.3382   85–19086
ISBN 0–87814–292–4

Printed in the United States of America

1 2 3 4 5 90 89 88 87 86

CONTENTS

DEDICATION

This book is dedicated to my wife, Mary Jo Moore.
She has been my partner in marriage and business for 36 years.
Her encouragement and help were as responsible for this book
as the experience and knowledge that have been
inscribed herein.

NOMENCLATURE

AFE	authorization for expenditures
A_c	proportionality constant
A_p	pipe acceleration, fps^2
A	area of the piston or internal area of the liner, sq in.
A_n	area of all bit nozzles in use, sq in.
a	exponent between 0.4 (for hard formations) and 1.0 (for soft formations)
A_{pf}	area of contact between pipe and filter cake
B	bit costs, $
b_f	compressibility constant, dependent on solids type in the filter cake
bh	bottom hole
b	exponent between 1.0 and 2.0, depends on abrasiveness of fluid in contact with bearings
B_f	buoyancy factor
C_n/C_o	ratio of normal conductivity to observed conductivity, millimhos
CMC	carboxymethylcellulose
C_t	total drilling costs, $/ft
C_r	rig costs, $/hr
C_{dr}	drag coefficient, dimensionless
C	proportionality constant
C_d	drilling cost, $/ft
C_1, C_2, C_3, C_4	constants determined from field data
c	experimentally determined shale compactibility constant from 0.1 to 1.5
c_f	% of total fluid at the surface that is gas
cp	centipoise
D	labor + depreciation + rig supervision + rig insurance + miscellaneous rig expenses, $/hr
D_c	diameter of flow chamber, in.
D_n	normalized tooth wear
D_h	diameter of the hole, in.
D_i	inside diameter of pipe, in.

D_p	diameter of the pipe, in.
D_o	outer pipe inside diameter or hole size, in.
D_{op}	outside diameter of the pipe, in.
D_I	inner pipe outside diameter, in.
D_B	density of a sphere
D_b	bit diameter, in.
D_a	amplitude or diameter, in.
D_w	well depth, ft
d_B	diameter of a sphere
d	d exponent
d_c	d exponent corrected for mud weight and shale compactibility
D_l	internal diameter of the liner, in.
D_p'	equivalent diameter of cuttings, in.
D_g	depth at which gas entered the wellbore, ft
D_L	density of the liquid
d_f	height of filter cake
$D_2 - D_1$	horizontal distance between measuring points, ft
E	total distance east, ft
$E_2 - E_1$	east distance between measuring points, ft
e_1, e_2	mud weights at two different times, lb/gal
F_b	bit footage, ft
F_m	maximum permissible force, lb force
f_t	pressure losses in the straight-line portion of the curve, in turbulent flow, psi
F_i	impact force at the exit of the bit nozzles, lb force
F_{mo}	momentum force, lb force
f	constant
f_{pf}	coefficient of friction between pipe and filter cake
F_g	fracture gradient, psi/ft
fpm	feet per minute
f_f	friction between solids in the filter cake
F	total ft/bit
F_s	sticking force or total pulling force required to free stuck pipe
fps	feet per second
fph	feet per hour
gpm	gallons per minute

g_c	change from lb mass to lb force $= \dfrac{32.2\ \mathrm{lb_m} - \mathrm{ft}}{\mathrm{lb_f} - \mathrm{sec^2}}$
G	acceleration of gravity, ft/sec/sec
Hp_s	surface hydraulic horsepower
Hp_c	horsepower required in circulating system (excluding bit)
Hp_b	bit hydraulic horsepower
H	height of cement, ft
h_b	height of gas at the bottom of the hole, ft
K	proportionality constant
k_s	intercept of shear stress versus shear rate straight line on log paper
K'	constant
K''	proportionality constant
K_i	intercept of viscometer data on log paper
K_d	drillability constant
k	permeability of the filter cake
K'	consistency index, lb-sec/sq ft
l_f	length of flow chamber, ft
l	stroke length, in.
L	relationship between bit life and bearing life for bits
lb_f	pounds force
L_{dp}	length of drillpipe, ft
l_f	length of flow path, ft
l_t	tube length between pressure gauges, ft
M_p	pump maintenance, $/hr
M_{dp}	drillpipe maintenance, $/hr
M_{dc}	drill collar maintenance, $/hr
M_r	rig maintenance, $/hr
M	lb mass
M_c	mud costs, $/hr
MWD	measurements while drilling
M_{rt}	rig maintenance during round grips, $/hr
M_f	alkalinity of mud filtrate
M_{wl}	wireline maintenance, $/hr
$M_1 - M_2$	measured distance between measuring points, ft
N	rotary speed, rpm
n'	flow behavior index, dimensionless

n_d	deviation of a fluid from Newtonian behavior
N_2–M_1	north distance between measuring points, ft
N	total distance north, ft
N_{opt}	optimum rotary speed
N_{Re}	specified Reynolds number, dimensionless
n_c	slope of P_c vs Q_c on log paper
n	slope of F_i vs Q_c
n_s	slope of shear stress versus shear rate straight line on log paper
O	rig operating cost, \$/hr
p	pump pressure, psi
ppb	parts per billion
ppm	parts per million
P_t	total pressure loss in circulating system, standpipe pressure, psi
P_{lc}	pressure loss in circulating system (excluding the bit), psi
P_1	pressure of the formation, psi
P_{la}	pressure losses in laminar flow for the annulus, psi
P_d	shutin drillpipe pressure, psi
P_{dp}	pressure loss through drillpipe, psi
P_i	initial pressure, psi
P_{dc}	pressure loss through drill collars, psi
P_o	vapor pressure of water in oil-based mud, psi
P_{dca}	pressure loss through hole and drill collar annulus, psi
P_r	reduced pressure
P_{dpa}	pressure loss through hole and drillpipe annulus, psi
P_c	circulating pressure, psi
P_t	turbulent flow pressure losses inside pipe, psi
P_{bh}	bottom-hole pressure, psi
P_{ta}	annular pressure losses in turbulent flow, psi
P_h	hydrostatic pressure exerted by a column of mud and gas, psi
P_{su}	surge pressure, psi
P_{hs}	hydrational stress, psi
P_{m1}, P_{m2}	alkalinity measurements of mud filtrate
P_s or P_2	pressure at surface, psi
P_n/D_w	normal pressure gradient for the area, psi/ft
P_m	alkalinity of mud
P_{md}	pressure exerted by mud inside drillstring, psi

P_b	pressure drop or loss through bit, psi
P_{sc}	pressure loss through surface connections, psi
P_p	pore pressure, psi
P_T	total pressure desired, psi
P_p/D_w	pore pressure gradient, psi/ft
P_{pc}	pseudocritical pressure, psia
P_q	pressure at the circulation rate, psi
P_a	pressure on annulus gauge at the surface, psi
P_{at}	total pressure loss in annulus, psi
P_{ma}	pressure exerted by mud in the annulus, psi
P_{ff}	pressure exerted by mass of formation fluid, psi
P_{sd}	surface displacement pressure, psi
P	pressure at any point, psi
P_{mf}	alkalinity of mud filtrate
PV	plastic viscosity, poise or centipoise
P_v	vapor pressure of shale and water of hydration, psi
P_f	friction pressure, psi
P_{fc}	differential pressure across the filter cake
Q	flow rate, gpm
Q_c	circulation rate, gpm
Q_{cf}	pumping rate, cu ft/min
Q_b	pumping rate, bbl/min
Q_m	maximum circulation rate, gpm
Q_{max}	maximum pumping rate, gpm
r	pipe radius, ft
R	drilling rate, fph
RMS	rotor mud separator
R_g	natural gas constant (to convert units)
Re	Reynolds number, dimensionless
R_1, R_2	drilling rates corresponding to the respective mud weights, fph
r_s	resistance per unit weight of solids
rpm	cycles or revolutions per minute
R_t	ratio of unit tensile stress to minimum yield strength
R_p	particle Reynolds number
R_o/R_n	ratio of observed resistivity to normal resistivity, ohm-m
R_{mf}	resistivity of mud filtrate, ohms

R_w	resistivity of formation water, ohms
S/D_w	pressure gradient exerted by overburden rock, psi/ft
S	specific gravity of gas, dimensionless
S_v	solids content, vol %
spm	strokes per minute
T_i	initial temperature
t	total on-bottom time, hr
T	tension
T_{rt}	round trip time, hr
T_R	temperature, °R
T_1	temperature of the formation, °F
T_r	reduced temperature
T_2 or T_s	temperature at surface, °F or °R
T_{pc}	pseudocritical temperature, °R
T_a	average temperature, °R
T_{ab}	absolute temperature
T_b	bottom-hole temperature, °F
$v + 1$	filter-cake compressibility
v_c	fluid velocity when the flow pattern changes from laminar to turbulent, fpm
v_a	annular velocity, fpm
v	fluid velocity, fpm
V_f	total volume of filtrate collected
V_r	volume rate of gas flow, scf/day
V_v	vertical component of velocity, ft/min
V_n	nozzle velocity, fps
v_p	net upward velocity of cuttings, fpm
V_s	surface volume of gas, cu ft
V_{sc}	slip velocity of cuttings, fpm
V_1	volume of gas, cu ft
$100–V_1$	liquid volume, cu ft
V_b	bottom-hole volume of gas, cu ft
V_{pi}	pipe velocity, fpm
V_2	volume of gas at the surface, cu ft
V_d	displacement velocity, fpm
v	velocity, fps

V_2-V_1	vertical distance between measuring points, ft
∇	partial molar volume of pure water
V_B	terminal or settling velocity of spheres in a liquid
W	bit weight, lb/in.
w	solids content of mud per unit volume of filtrate
W_s	weight of shale + weight of water, lb/gal
W_c	weight of cement slurry, lb/gal
WOC	waiting on cement (to set)
W_m	weight of mud, lb/gal
W_{opt}	optimum bit weight
X	distance from the surface to the top of the gas bubble, ft
Y	yield point, dynes/sq cm or lb_f/100 sq ft
z_2	compressibility factor at the surface
z_a	average gas compressibility factor
z_1	compressibility factor under pressure in formation dimensionless
z_s	gas compressibility factor at surface
z_b	gas compressibility factor at bottom of hole
ΔP_s	hydrostatic pressure of mud − pore pressure of formation
ΔP_p	pressure loss over a given length of pipe, psi
ΔP	differential pressure across the fluid-end piston, psi
ΔP_d	pressure, dynes/sq cm
ρ	mud weight, lb/gal or slurry density, lb/gal
ρ_p	weight of cuttings, lb/gal
ρm	mud gradient, psi/ft
ρ_w	water weight, lb/gal
ρf	gas gradient, psi/ft
$\Delta t_n/\Delta t_o$	ration of normal travel time to observed travel time, μ sec
$\dfrac{\Delta v}{\Delta r}$	shear rate, sec^{-1}
ΔKE	changes in kinetic energy, ft-lb_f
Δv_1	changes in velocity, ft/sec
ΔV_f	volume of filtrate collected for a specific period
Δt	time filtrate is measured, sec
$\Delta V_o/\Delta V_n$	ration of observed velocity to normal velocity
$\Delta dC_n/\Delta dC_o$	ratio of normal modified d exponent to the observed modified d exponent

Δt_n	normal travel time, μ sec
Δt_o	observed travel time, μ sec
γ_s	shear rate, $\sec^{-1}$
γ	Poisson's ratio
γ_v	shear rate for a rotating viscometer
$\overline{\gamma}$ or γ_1	shear rate for a rotating viscometer, rpm
θ	mud properties, lb force/100 sq ft
μ_1	viscosity of liquid filtrate, cp
μ	viscosity of a liquid, cp
μ_p	viscosity of a liquid, poise
τ	shear stress, dynes/sq cm or lb_f/100 sq ft
θ_1	viscometer reading at a specific shear rate that equals the annular fluid velocity
μ_T	solids content effect on fluid properties in turbulent flow
θ_I	inclination angle at the first measuring point, degrees
θ_{II}	inclination angle at the second measuring point, degrees
$\overline{\theta}$	average direction angle, degrees
$\theta_{2\gamma}$	viscometer reading at 2.0 times the shear rate as shown for the denominator, lb_f/100 sq ft
θ_γ	viscometer reading at any shear rate, lb_f/100 sq ft

PREFACE

This book has been written to aid drilling engineers and drilling supervisors in drilling operations. I do not claim to have included all of the potential drilling technology. Portions of the book may be partially outdated by publication time. Also, there may be parts of the book that other experienced drilling personnel will think are incorrect. I, of course, believe that the contents are an accurate representation of the use of technology in drilling practices. One word of caution: in the 37 years I have been connected with drilling practices, I have changed my mind on several occasions concerning information I have published.

The units for equations in this book are English units. Translations are fairly easy. Giving metric units in addition to English units would have increased the book's length and perhaps have been confusing for those using this book.

Computers are becoming more common in field drilling operations; all of the equations in this book have been programmed for use on HP41-CV calculators. Technology is advancing rapidly. Measurements while drilling will probably be extended to actually measure some of the numbers such as annular pressure loss, which now are calculated. Actual measurements will improve the calculation's accuracy and will be a substantial step forward in well planning. Also, some major companies use satellite systems that can instantaneously display, in any office, drilling parameters from the field. These systems help substantially in explaining why certain problems occur and are a giant step forward in the promotion of safety.

Lastly, there is no portion of this book that could not be improved if time were available. Criticism is welcomed and when useful will be part of the foundation for future revisions.

PRESTON L. MOORE

INTRODUCTION

This book emphasizes drilling practices. Drilling equipment is discussed primarily in the context of practices. The general pattern of rotary drilling has changed very little since its inception in the Spindletop field of Texas in 1900.

TECHNOLOGY CONTINUALLY CHANGES

No doubt there is truth in the statement that we replow the same ground in the search for improvements. However, many times practices that appear to repeat previously discarded practices are used in current environments that enhance their chances for success. One example of this is the return to fixed-bladed bits.

The last fixed-bladed bits, called drag bits, were discarded in the 1950–60 decade and reinstated when diamond compact bits were introduced. The fixed-bladed bits were reintroduced because the cutting surfaces were improved by using manmade diamonds. Downhole motors and turbines are used more commonly now, and the reintroduction of oil-based muds improved lubrication and cleaning of the bits while drilling.

Changes in drilling equipment and practices that occurred in the last few years include the following:

- Improvements and changes in drilling bits
- Expanded use of turbines
- Increased utilization of oil-based muds
- Methods to measure drilling parameters while drilling, called MWD
- Expanded implementation of computers
- Introduction of power swivels on large drilling rigs

The changes in drill bits have been pronounced in the last two decades. Friction or journal bearing bits provided substantial improvements in bit life. Tungsten inserts provided substantial improvements in the life of cutting structures. Better lubrication makes it possible to turn roller-cone bits faster. Extended-nozzle bits have greatly improved bottom-hole cleaning. Bit nozzles designed to ensure clean cutting surfaces show promise.

The biggest bit change of all has been the introduction of fixed-bladed bits equipped with cutting structures made from manmade diamonds. The manmade diamonds initially called Stratapax® now are called diamond compacts. These diamond compacts come in many different shapes and sizes. Bit designs using the diamond compacts are changing. Presently there are diamond compact drag bits for soft formations and a standard diamond bit design using diamond compacts for cutting structures for medium-type formations. There is also a diamond compact bit shaped like a regular diamond bit that uses large manmade diamonds for hard formations.

A majority of the bits run probably will be fixed-bladed bits within five years if there are not further substantial improvements in roller-cone bits.

Motors have been used in directional drilling for many years, and operational turbines have been used for more than twenty years. There will be a resurgence in the use of turbines for routine drilling operations. The emergence of fixed-bladed bits enhances the value of turbines, and offshore operational costs make them more economically acceptable.

Oil-based muds, which have been used almost as long as water-based muds, have made a strong comeback. Several changes have made this comeback possible.

Drilling rates with oil-based muds were improved by using relaxed filtrate muds and diamond compact bits. Shale stability was enhanced by adding calcium chloride to the water phase of oil-based muds. Rental arrangements reduce the total cost of oil muds. Environmental objections have been minimized by using clear mineral oils.

Measurements while drilling were introduced many years ago, when the teledrift instrument for measuring inclination was introduced. Industry almost ignored the teledrift instrument because it did not measure hole direction also. Tools then were developed to measure hole inclination and direction while drilling, and the signals were transmitted to the surface through mud pulses, while circulating and not rotating. This technology is operational and available from several companies.

Additional information can be measured. Now a lithology log can be obtained while drilling. Claims for these measurements while drilling include a direct measurement of actual weight on the bit, annular pressure losses, and formation pore pressures.

These measurements while drilling are going to introduce expert drilling systems where potential drilling problems are analyzed and solutions suggested or automatically used.

Measurements while drilling re-open the door for expanded implementation of computers at the drilling rig and in the office. Satellite systems that transmit drilling data directly to the office will become more common. These systems increase the use of technology in drilling because more technical personnel will be involved in day to day operations. Operational personnel will use computers more because software is easy to use and more calculations will be necessary at the rig level.

The use of power swivels as a replacement for conventional rotary tables is an important advancement. Power swivels make torque measurement much easier and more accurate. Rotary speed control is improved, and auxiliary functions on the rig are improved with power swivels.

The changes and improvements in rotary drilling are substantial, yet the measurement of mud properties remains more or less a horse-and-buggy operation. Drilling mud is the heart of the entire drilling operation from the standpoint of drilling efficiency and safety. Yet reliance is still placed on funnel measurements of mud thickness. Flow behavior is described still by the use of yield point and plastic viscosity, both outdated parameters from the early days of two-speed viscometers.

Variable-speed viscometers are available and should be used. The thickness of water-based muds depends on temperature, and the thickness of oil-based muds depends on temperature and pressure. Often temperature is ignored; when measured, there is frequently no attempt to extend data to higher temperatures. No instruments at the rig measure the thickness or the weight of oil-based muds under pressure. Data show that pressure increases both the thickness and the weight of oil-based muds.

These changes in drilling mud as functions of temperature and pressure may affect significantly not only drilling efficiency but also safety. No one questions the need to know accurately the mud weight when drilling in abnormally high pore pressure areas. Actually, the foundation of much well control technology is based on mud weight and circulating mud thickness. If these numbers are unknown, there must be a certain amount of trial and error.

CHAPTER 1

RIG SELECTION

LARRY P. MOORE
MS&A Engineering Inc.

Rotary rig selection for the drilling of a well is one of the first tasks undertaken in the well planning process. Much of the remainder of the well plan hinges on potential limitations imposed by selecting the personnel and equipment for drilling. When choosing the rotary rig, the planner has two things in mind: (1) ensuring adequacy of the rig and (2) minimum cost. Both factors can be combined into one: cost effectiveness. In this discussion, a procedure for maximizing cost effectiveness of the rotary rig package is outlined.

WELL REQUIREMENTS

Before the rig can be selected, the requirements of the subject well must be determined. Total depth is the first thing that comes to mind, but anticipated hole sizes, hydraulic horsepower requirements, drillpipe, casing program, and potential hole problems should be considered.

Hole size and hydraulic horsepower requirements need to be known when deciding on mud pump equipment. For example, hydraulic horsepower (hhp) needed at the bit generally ranges from 3 to 5 hhp/sq in. of bit diameter. Assume that bit horsepower will be 65% of the available surface horsepower when bit horsepower is being maximized.

Drillpipe size is also a consideration. The larger the drillpipe diameter, the less circulating pressure losses through the drillpipe will affect the system, thus requiring less power and fuel to achieve the desired results. However, these benefits must be balanced against the increased cost and weight of the larger drillstring.

Hole problems known to exist in the area may also influence the rig

1

selection. For example, if severe sloughing shale is anticipated, then available surface horsepower requirements may be increased for more bottom-hole cleaning. If the likelihood of the pipe becoming stuck is increased because of lost circulation or deviation problems, then the safety factor for derrick loads may need to be increased.

DEPTH LIMITATION

After the requirements mentioned above are determined, establish your own minimum standards for the rig, commonly referred to as "depth rating." Unfortunately, there is no real standard for depth rating, so much of the term's meaning is lost. This necessitates an analysis by the planner to establish rating standards for the case at hand. Table 1–1 shows criteria for determining depth limitation.

Derrick and Mast

Derrick or mast ratings are made from the maximum hookload capacity. Know under what conditions these ratings were made when comparing them. For instance, a rating may be valid only with a certain type of crown/traveling-block line configuration. In addition, double check the credibility of the rating if it is attached to a non-namebrand mast. If there is some doubt, an API-certified inspector should verify all ratings. Other factors to consider with mast rating are leg loading, substructure rating, and maximum wind loading with drillpipe in the derrick.

In the context of rig selection, mast rating figures are needed to determine if the rig will be adequate under the maximum loading condition anticipated during drilling. This condition typically occurs while running the heaviest string of casing, so calculation of the maximum string weight is the first step in analyzing mast load.

To illustrate what occurs during mast loading, consider first the simplest of cases, demonstrated in Fig. 1–1. In this case, a 100-lb load is being lifted with a pulley. Total load on the derrick is 200 lb because the two lines are each exerting a downward force of 100 lb. Fig. 1–2 shows the familiar block and tackle system employed to lift the same 100 lb. The

■ **TABLE 1–1**   Criteria for determining depth limitation

- Derrick
- Drawworks
- Mud pumps
- Drillstring
- Mud system
- Blowout preventer
- Power plant

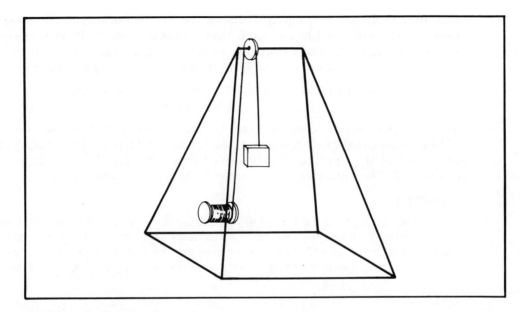

■ **FIG. 1–1** Simple pulley system

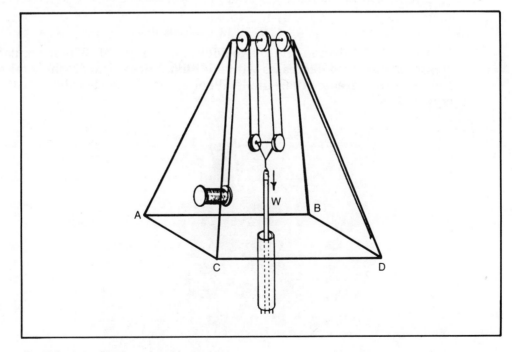

■ **FIG. 1–2** Block and tackle system

load here is distributed equally among the four lines strung between the crown and traveling blocks. Each line therefore has 25 lb of tension on it; the tension on the hoist line and the dead line is also 25 lb. This system enjoys a mechanical advantage of four to one according to the definition:

$$\text{Mechanical advantage} = \text{Weight lifted} \div \text{Force exerted}$$
$$= 100 \div 25 = 4 \qquad (1.1)$$

The total load on the derrick (see Fig. 1–2) is therefore the total number of lines multiplied by the tension in each line: $6 \times 25 = 150$ lb. A substantial reduction in the derrick load occurred from the Fig. 1–1 system to that of Fig. 1–2. The situation in Example 1.1 is directly analagous.

☐   **EXAMPLE 1.1**

The total weight of 9,000 ft of 9⅝-in. casing for a deep well is determined to be 400,000 lb. Since this will be the heaviest casing string run, the maximum mast load must be calculated. Assuming that 10 lines run between the crown and traveling blocks and neglecting buoyancy effects, calculate the maximum load.

SOLUTION:

The tension, T, will be distributed equally between the 10 lines. Therefore,

$$T = 400,000/10 = 40,000$$

The tension in the hoist and dead lines will also be 40,000, so the total load is

$$40,000 \times 12 = 480,000 \text{ lb}$$

Example 1.1 demonstrates two additional points. (1) The marginal decrease in mast load decreases with additional lines, and (2) the total mast load is always greater than the load being lifted. Fig. 1–3 shows a 12-line crown block.

■ **FIG. 1–3**   Top view of crown block

Derrick leg loading must also be considered in this analysis because the derrick is only as strong as its weakest leg. The typical hoisting arrangement is illustrated by Fig. 1–2. Notice that the hoist line is between legs A and B and that the dead line is fastened to leg D. Clearly, a more balanced load would be achieved if the dead line were fastened opposite the hoist line between legs C and D. But the pipe ramp, which is the avenue from which pipe off the racks is hauled to the rig floor, is always opposite the drawworks (or hoist) on a rig, so one of the legs opposite the hoist is the next best bet. See Fig. 1–4 for the locations of these components. Example 1.2 shows the effect derrick leg loading has on overall derrick loading.

☐ **EXAMPLE 1.2**

Calculate the equivalent mast load due to leg loading for the scenario of Example 1.1 using a load analysis.

SOLUTION:

Table 1–2 outlines the analysis procedure.

■ **Table 1–2**   Load analysis

	Centered Load	Hoist-line Load	Dead-line Load	Total Load
Leg A	2.5T*	0.5T	0	3.0T
Leg B	2.5T	0.5T	0	3.0T
Leg C	2.5T	0	0	2.5T
Leg D	2.5T	0	T	3.5T

* T is the tension in one line.

Referring again to Fig. 1–2, Table 1–2 shows the load distribution for each of the legs A, B, C, and D. The first category, the centered load, shows the minimum load on each leg due to the weight on the hook. The column labeled "Hoist-line Load" shows that loading due to tension on the hoist line is equally distributed between legs A and B. This is because the drawworks is located midway between these two legs. As the table demonstrates, loading from the dead line is borne entirely by leg D, as this line is fastened here. The category marked "Total Load" shows the load distribution for each leg. Leg D has the highest total, 3.5T, because of the unbalanced loading effect resulting from the dead line. Since the mast is only as strong as its weakest leg, the equivalent mast load is

$$4 \times 3.5T = 14T$$

The equivalent load in this case is

$$14/12 \times 480,000 = 560,000 \text{ lb}$$

Our total load due to the 400,000-lb casing is actually 40% higher at 560,000. However, this doesn't consider any overpull. If we assume an overpull of 25%

1 Mast-Derrick
2 Crown Block
3 Traveling Block
4 Drilling Line
5 Drawworks
6 Hook
7 Engines
8 Elevators
9 Swivel
10 Kelly
11 Kelly Bushing
12 Rotary Table, Master Bushing
13 Mud Pits
14 Mud Pumps
15 Shale Shaker

16 Reserve Pits
17 Casing
18 Drillpipe
19 Drill Collars
20 Drill Bit
21 Pipe Ramp
22 Rotary Hose
23 Doghouse
24 Blowout Preventer Stack
25 Cellar
26 Monkeyboard
27 Water Table
28 Mousehole
29 Mud-Gas Separator
30 Tongs

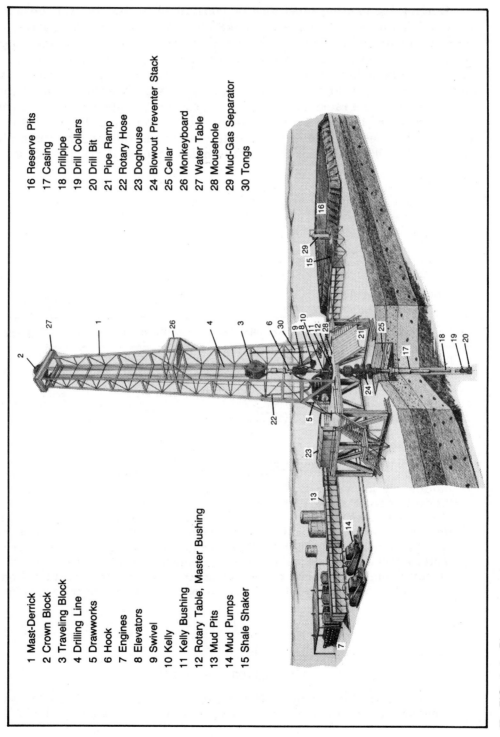

■ **FIG. 1–4** Rig components

of the string weight, or 100,000 lb, we must also go through the calculations of Examples 1.1 and 1.2. Therefore,

$$100,000 \times 12/10 \times 14/12 = 140,000 \text{ lb}$$

Now the total load is

$$560,000 + 140,000 \text{ lb} = 700,000 \text{ lb}$$

If, in addition, we incorporate a safety factor of 0.2, the required derrick capacity becomes 840,000 lb, over twice the string weight. Clearly, calculations of this type are essential to the rig selection process.

Drawworks

The applicability of the drawworks horsepower rating rests primarily on the time necessary to hoist a load rather than its actual ability to hoist the load. This relationship is expressed in Eq. 1.2.

$$\text{Time} = (\text{Force} \times \text{Distance}) \div (\text{Horsepower} \times 550) \qquad (1.2)$$

☐   **EXAMPLE 1.3**

For the case of the 9⅝-in. string described in Example 1.1, how long would it take to hoist one 40-ft joint with all the casing in the hole with a 500-horsepower (hp) and an 800-hp drawworks?

SOLUTION:

Remembering that the total mast load was 480,000 lb, then
500-hp drawworks:

$$\frac{480,000 \times 40 \text{ ft}}{500 \times 550} = 69.8 \text{ sec}$$

800-hp drawworks:

$$\frac{480,000 \times 40 \text{ ft}}{800 \times 550} = 43.6 \text{ sec}$$

The 800-hp drawworks would probably be preferable in this instance.

Fig. 1–5 gives hoisting time per stand versus number of stands in the hole for 400- and 800-hp drawworks. It assumes 90-ft stands with a weight of 1,935 lb each.

Another point worth mentioning about drawworks is that the power required to stop a load in free fall can be far greater than the drawworks' capacity. Sharp load increases caused by dropping a string of pipe, even for a short distance, should always be avoided.

Fig. 1–6 gives front and back views of a drawworks.

Mud Pump

The two types of reciprocating mud pumps on rigs today are duplex and triplex pumps. Figs. 1–7 and 1–8 show duplex and triplex mud pumps. Triplex pumps have largely replaced the duplex because of their substan-

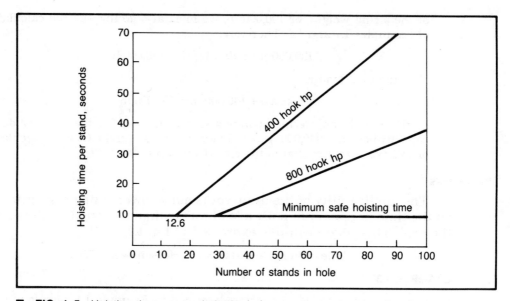

■ **FIG. 1–5** Hoisting time vs stands in the hole

tially lighter weight for equal horsepower ratings. A 6-in. liner for a 1,000-hp duplex pump weighs 394 lb, and the equivalent triplex liner only weighs 67 lb. Triplex pumps are also more appropriate for deep wells because they can operate at higher pressures with fewer fluctuations than the duplexes. Nevertheless, the double-acting duplex pump is still widely used.

The horsepower rating for these pumps is based on input horsepower. However, it is output horsepower that is relevant to the selection process. Output horsepower is determined by Eq. 1.3:

$$\text{Hydraulic horsepower} = PQ \div 1{,}714 \qquad (1.3)$$

where:

P = Pressure, psi
Q = Flow rate, gpm

Therefore, determining the maximum output horsepower entails fixing maximum values for pressure and flow rate. Maximum pressure is limited by the liner rating, which in turn is a function of the maximum permissible force on the power-end bearing. For example, the liner rating is essentially doubled when the size is decreased from 7 to 5 in. This is due to the relationship $F = PA$ and the fact that area, $A = \pi D^2/4$. The flow rate is a function of the liner size, stroke length, and pumping speed.

Much more relevant than theoretical maximum pressures and flow rates are the maximum sustained levels at which the pump has run. This information can be obtained only from the contractor and is crucial to determining the adequacy of the pumps and planning the hydraulics pro-

■ **FIG. 1–6** A modern drawworks (*courtesy National*)

gram. An example occurs in the case of a 1,500-hp-rated triplex mud pump. In order to prolong the life of the pump, the contractor runs the pump at a maximum sustained pressure of 3,000 psi and limits the pumping speed to 110 strokes per minute (spm) with a resulting flow rate of 365 gallons per minute (gpm). The resulting hydraulic horsepower is then

$$3,000 \times 365/1,714 = 639 \text{ hp}$$

Obviously, these parameters should be known before the rig is selected.

Drillstring

Of primary interest when examining drillpipe for rig selection purposes are the pipe dimensions and number of rotating hours since inspec-

■ **FIG. 1–7** Duplex pump (*courtesy Gaso*)

■ **Fig. 1–8** Triplex pump (*courtesy Gaso*)

tion. The larger I.D. pipe cuts down on pressure losses due to friction, thus enabling the mud pump to be run at lower pressures. This minimizes pump down time and conserves fuel. In addition, the lower circulating pressure losses improve the hydraulics program by making more hydraulic horsepower available at the bottom of the hole. The trade-off comes with the fact that large drillpipe is not as versatile when drilling small holes, and it is proportionately heavier.

Many rig contracts specify drillpipe inspection at the end of a certain number of operating hours. Some operators routinely inspect drillpipe before a well is spudded on a daywork contract. (For more information on daywork contracts, see the contracts section in this chapter.) These operators have probably been stung by poorly maintained drillpipe. Whether this is necessary on a routine basis is doubtful, but some discussion with the firm that performed the last inspection is advisable.

Mud System

The rig mud system (not including the main pumps) consists of the tanks, solids control equipment, stirrers, mixing system, and auxiliary equipment. Mud tanks in the active mud system typically include a discharge tank, a settling tank, and a suction tank, though some rig setups may add or subtract a tank. The most pertinent question about the mud system is "How much tank capacity is needed in the active system?" The procedure for determining required capacity involves calculating total pipe displacement at the projected total depth and multiplying the resulting volume by 2.5 (which is nothing more than a rule of thumb).

☐   **EXAMPLE 1.4**

Calculate the mud tank capacity necessary for the drilling of a 15,000-ft well given the following data:
Displacement of 4½-in. drillpipe = 0.0068 bbl/ft
Displacement of 6½-in. drill collars = 0.0336 bbl/ft
Length of drill collars = 720 ft
Displacement of tool joints = 10 bbl

SOLUTION:

$$14,280 \times 0.0068 = 97 \text{ bbl}$$
$$720 \times 0.0336 = 24 \text{ bbl}$$
$$(97 + 24 + 10) \times 2.5 = 328 \text{ bbl}$$

Therefore a tank capacity as close to 328 bbl as possible is desired. An amount greatly in excess of this number would increase treating costs and turnaround time. The use of centrifugal-pump-powered mud guns is the most effective way to ensure good mixing and stirring. Solids control equipment and techniques are discussed in chapter 6, but a definite determination should be made as to what solids control equipment is required. If the selected rig does not have it, then this equipment must be rented.

Auxiliary equipment for the mud system includes degassers, mixing tanks, a chemical barrel, charging pumps, and a pit-level indicator. The rig in question may have some, all, or more than the auxiliary mud equipment listed here. The important thing is to make note of these items on the rig inventory and to question the absence of equipment.

Blowout Preventer

The pressure rating of the required blowout-preventer equipment depends on the maximum bottom-hole pressure anticipated. Examination of rig pressure control equipment should not stop with the stack pressure rating, however. The checklist should include pressure ratings of lines, connections, and valves; availability of annular and rotating preventers, choke manifold, accumulator, remote and direct controls, and choke and kill lines. Make a sketch of the blowout-preventer system both to ensure the adequacy of the system and for reference.

Fig. 1–9 shows a diagram of an annular BOP, accumulator, manifold, and remote controls.

Power Package

The rig power-system concerns will be the number and sizes of engines and whether the power transmission is mechanical or electrical. The rig

■ **FIG. 1–9** Blowout preventer stack with spherical BOP at the top

prime mover must drive the drawworks, rotary table, mud pumps, auxiliary equipment, and, in some cases, rig lighting. Horsepower requirements of the prime mover depend to a great extent on calculations already made on the amount of power drawn by the hoist and mud pumps. Most newer rigs with prime movers over 1,000 hp are diesel-electric, i.e., the diesel engines run generators that convert the mechanical energy to electricity. The primary advantage of this system is that it allows more efficient power allocation between the rig components. It also eliminates the need for additional engines and generators to run rig lighting, and it is quieter.

Rigs with mechanical drives rely on a system of belts, chains, and sprockets to transmit power to rig components. This simple, direct method of power transmission is still used on many rigs. The main disadvantage of this system is that it often requires a separate engine for each rig component. This limitation is largely eliminated with the use of a mechanical compound that enables the engine outputs to be combined and distributed to rig components.

BID SOLICITATION

The preceding analysis not only serves the purpose of depth rating a drilling rig but also of determining specific rig requirements for a job. In this way specific minimum requirements can be enumerated when soliciting bids rather than asking for bids on, say, a 10,000-ft rig. Being as detailed as possible both on rig specifications and on what should be included in the bid (drillpipe, fuel, mobilization) eliminates confusion and makes bid comparisons possible. Quite often the rig best suited for a job is not selected because it cannot adequately be compared to other bids.

CONTRACTS

After the rig equipment specifications have been chosen, decide which type of contract is desired. The four types of contracts offered today are daywork, turnkey, footage, and a combination of these.

Under a *daywork contract,* the contractor supplies the rig and the crew and maintains them. The operator is responsible for the drilling operations and all associated activities. In return for the use of the rig and crew, the operator pays a set rate per day. This rate varies with the size of the rig and market conditions. In addition to the drilling day rate, a stand-by rate is specified. The stand-by rate goes into effect when drilling operations can't proceed and the rig is idle. This situation might occur if on-site personnel are waiting on orders from company management before they can resume drilling. Because the operator is in complete control of the well, he assumes the risks associated with getting it to bottom.

The contractor has control of the drilling operation in the case of the *turnkey contract.* Generally the contractor supplies everything necessary to get the well to total depth in return for a flat fee. The only variations in this agreement are the point at which the turnkey contract goes into effect and the point at which it is completed. A complete turnkey includes surveying and staking the well, building the location, moving the rig onto the location, drilling the well, running well logs, running and cementing the production string, and moving the rig off the location. Typically, the turnkey contract commences after the location has been built and continues until total depth is reached. Under these circumstances, the rig reverts to day rate at total depth until it is released. In the turnkey situation, all of the risks inherent in drilling the well are assumed by the contractor. This is obviously advantageous from the operator's point of view; however, the risk assumed by the contractor usually is reflected in the turnkey cost. In most cases, an operator with considerable experience in an area can drill more cheaply with a dayrate contract. When this experience is lacking, the turnkey may be the more cost effective choice.

A *footage contract* specifies a dollar amount per foot for the total depth of a well. In this sense it is similar to the turnkey because much of the uncertainty regarding drilling cost is removed. Under the terms of the footage contract, the operator is usually responsible for supplying the remainder of the intangible and tangible drilling necessities such as mud, bits, cement, rental tools, and casing. However, negotiable items on a footage contract might include rig move expenses and special rig equipment. A footage contract can revert to day rate under certain circumstances. For the drilling contractor's protection, a clause in the contract allows a maximum amount of standby time for such occurrences as lost circulation or fishing operations, after which the contractor charges a specified rate per day. The operator is given contractual assurance (as in the dayrate and turnkey contracts) that the wellbore deviation angle will not exceed a pre-established limit. In addition, the operator can assume control of the drilling operation upon notification of the contractor. Under these circumstances, drilling resumes with a dayrate contract. Because of these reversionary clauses, the operator and the contractor share the risks of drilling the well. Footage contracts are popular and are prevalent in some areas on shallow and medium-depth wells.

A *combination contract* usually consists of a turnkey to a certain depth and daywork thereafter to the total depth. This situation may arise when a contractor is unwilling to assume the risk for drilling a section of the hole or when an operator wants a turnkey but is unwilling to pay the cost of a turnkey over the entire hole. The combination contract is a compromise that shifts the risk between the two parties.

There are many drilling contractors in the marketplace. The financial stability of these contractors is of concern to the operator. To avoid the possibility of an incomplete job and/or liens placed against the well, some assurance of the financial viability of the contractors should be sought. These assurances obviously become more important if little is known about the firm. For example, if a contractor were forced into bankruptcy while drilling under a turnkey contract, drilling could be delayed indefinitely. With the possibility of lease expiration or drainage by offset wells, this clearly is something to be avoided.

RIG EFFICIENCY

A rig's efficiency is a function of the adequacy of rig equipment and the capabilities of the rig crew. Before selecting a rig, inspect the rig equipment (have an inventory on hand) for mechanical soundness. Spend a day observing the working operations of the rig so at least a qualitative assessment of the rig crew can be made.

Sheikholeslami, Miller, and Strong developed a method for quantitatively assessing rig efficiency.[2] This method consists of breaking down rig time into the categories of rotating time, trip time, rig lubrication, rig repair, connection time, nippling up blowout preventers, nippling down blowout preventers, slipping and cutting drilling line, rigging-up time, and rigging-down time. Several prospective rigs are chosen for comparison operating in similar geologic settings and at similar total depths. Times for each of the above components are tabulated for each rig. The fastest time in each category is labeled 100% efficiency. A ratio of the 100% efficient times (or fastest times) to the comparative rig times gives a rig efficiency factor for that category. The total of the fastest rig times divided by the total for the subject rig gives an efficiency factor for that rig. This approach has value both from the standpoint of rig selection and a contractor's evaluation of his rig fleet. A factor for rig efficiency can be directly applied when evaluating bids for daywork contracts. Dividing the dayrate fee by this factor yields an effective day rate. In this way, a bottom-line comparison of daywork bids that incorporates the factors which comprise rig efficiency can be made.

SUMMARY

The rig selection process involves first determining minimum rig equipment standards using criteria established from known well requirements. When these standards have been set, contract details need to be decided on so contracts can be compared.

Before selecting a rig, evaluate the working rig performance on site. Inspect the mechanical and personnel aspects of the rig. Subsequent to this, if a daywork contract is being evaluated, analyze the rig's efficiency in greater detail.

The outline of the rig selection process should be followed regardless of the contract type. In the case of turnkey contract in which the driller assumes the risk for getting the well to the total depth, a prudent operator checks the ability of the proposed rig to handle the job. The operator's responsibility is to produce a cost-efficient wellbore in good condition. If the rig chosen is inadequate for the job, the chances of obtaining a wellbore that will not inhibit completion and production operations are significantly reduced. For this reason, the rig selection process is crucial to planning the well.

■ **REFERENCES**

1. L'Espoir, John, "How to Determine Your Rig's Depth Limit," *Petroleum Engineer,* April 1984.
2. Sheikholeslami, B.A., Miller, J.D., and Strong, R.E., "A Practical Method for Evaluating Rig Performance," IADC/SPE 11364, February 1983.
3. "Procedure for Selecting Rotary Equipment," American Petroleum Institute, API Bulletin D10, August 1973.
4. Moore, Preston, and Cole, Frank, *Drilling Operations Manual,* chapter 5, The Petroleum Publishing Co., 1965.

CHAPTER 2

COST CONTROL

One of the first steps in drilling a well is to prepare an authorization for expenditures (AFE). A typical AFE form is shown in Fig. 2–1.

The AFE is an estimated well cost. There is no guarantee that the cost of the well to be drilled will even be close to the estimated costs on the AFE. Operators are obligated to provide themselves and potential investors a reasonable estimate of well costs. It is not uncommon to encounter problems during the drilling that extend drilling times and increase well costs. Chapter 3 includes many of the drilling problems that frequently occur.

Often two AFEs are completed: one for a dry hole and another for a producing well. As illustrated in Fig. 2–1, there are two main categories of charges. Intangible drilling/completion costs refer to money spent primarily for services. Tangible drilling/completion costs are mainly for equipment that becomes a permanent part of the well. The intangible and tangible costs are separated because they are taxed differently.

If, for some reason, during the drilling of a specific well the operator knows the total costs are going to exceed the estimated AFE, the operator generally issues a supplemental AFE for interest owner approval. On some wells there may be several supplemental AFE cost estimates.

COST CONTROL DURING DRILLING OPERATIONS

The primary objective in cost control during drilling operations is to minimize total well costs. Unfortunately, cost control is often equated to reckless drilling operations. Nothing could be farther from the truth. The operator who watches his costs closely is the one who emphasizes safety.

Prepared For: _____

_____

By: _____ Date: _____

LEASE _____ WELL NO. _____ AFE NO. _____

Legal Loc. _____ County _____ State_____

Objective _____ Drilling Contractor and Rig _____

INTANGIBLE DRILLING/COMPLETION COSTS

DESCRIPTION	$ DRILLING	$ COMPLETION	$ TOTAL
01 LOCATION — PERMITS, SURVEYS AND CONSTRUCTION		XXXXXXXXXXXX	
02 MIRU, RDMO		XXXXXXXXXXXX	
03 FOOTAGE ft. at $ /ft		XXXXXXXXXXXXX	
04 DAYWORK Days at $ /day		XXXXXXXXXXXX	
05 TUBULAR — CONDUCTOR 1 " @ $/ft.		XXXXXXXXXXXX	
05 TUBULAR — CONDUCTOR 2 " @ $/ft.		XXXXXXXXXXXX	
06 COMPLETION Days at $ /day	XXXXXXXXXXXX		
07 BITS, COREHEADS AND MILLS			
08 RENTALS — TOOLS AND EQUIPMENT			
09 FUEL, WATER, ELECTRICITY AND COMMUNICATIONS			
10 MISCELLANEOUS SERVICES			
11 TRUCKING			
12 LOGGING — ELECTRIC LINE, SLICK LINE, PERFS., ETC.			
13 LOGGING — GEOLOGICAL			
14 CEMENTING — SERVICES AND PRODUCTS			
15 DRILLING, COMPLETION AND PACKER FLUIDS			
16 STIMULATION — FLUSH, ACIDIZE, FRAC, ETC			
17 TESTING — DST, CORE AND FORMATION			
18 DISTRICT AND OVERHEAD EXPENSES			
19 WELLSITE SUPERVISION — GEOLOGISTS AND ENGINEERS			
20 CONTINGENCY — 10% OF TOTAL INTANGIBLE COSTS			
22 PURCHASES — TOOLS AND EQUIPMENT			
TOTAL INTANGIBLE DRILLING/COMPLETION COSTS			

TANGIBLE DRILLING/COMPLETION COSTS

TUBULARS - FOOTAGE	SIZE	WEIGHT	GRADE	UNIT COST			
—							
—							
—							
—							
—							
—							
—							
—							
—							
—							
—							
—							
—							
—							
—							
—							
—							
—							
—							
—							
51 CASING SETTING EQUIPMENT — SHOES, HANGERS, ETC							
52 PRODUCTION EQUIPMENT — SUBSURFACE				XXXXXXXXXXXX			
53 WELLHEAD AND XMAS TREE EQUIPMENT							
54 CONTROLLABLE PRODUCTION EQUIPMENT — SURFACE				XXXXXXXXXXXX			
55 NON-CONTROLLABLE PRODUCTION EQUIPMENT — SURFACE				XXXXXXXXXXXX			
56 LEASE INSTALLATION COSTS				XXXXXXXXXXXX			
TOTAL TANGIBLE DRILLING/COMPLETION COSTS							
TOTAL WELL COST							

■ **FIG. 2-1** Authorization for expenditure form

Cost control procedures have evolved since the year 1900. There have been no sudden changes or improvements in drilling practices. The primary objective in any cost control program is to maintain at least a daily estimate of total expenditures. Eq. 2.1 can be used to keep a daily record of costs or for any period considered appropriate by the operator.

$$C_t = \frac{B + C_r(t + T_r)}{F}$$

(2.1)

where:

C_t = total drilling costs, $/ft
B = bit costs, $
C_r = rig costs, $/hr
t = total on-bottom time, hr
T_r = round trip time, hr
F = total ft/bit

In this equation, C_t represents the total drilling costs in dollars per foot. The time interval included is left to the operator's discretion, so costs may be maintained on an hourly, daily, or per-bit basis. Bit costs, represented by B, are easy to estimate. The hourly rig operating costs, C_r, may be difficult to determine for the following reasons:

- Rig operating costs vary considerably, depending on daily expenditures for maintenance.
- Drilling practices, such as higher pump pressures, faster rotary speeds, and higher bit weights, used to reduce drilling costs, may increase rig operating costs.
- Overhead costs are not fixed. For example, in offshore operations the actual rig rental rate is frequently doubled to cover overhead and associated support operations.
- The hourly operating costs may include other costs such as drilling mud; however, these costs are highly variable and tough to calculate as average costs.
- Supplemental practices are often difficult to include, because they happen at specific times.

The rotating time in hours, t, includes all on-bottom time with the bit. Thus connection time is included in rotating time. Round trip time, T_r, is the total time necessary to pull and rerun a bit. This sometimes complicates calculations because, while the bit is out of the hole, some type of rig maintenance is commonly carried out. Any such maintenance is not included in the bit performance evaluation but is incorporated in total cost numbers. The total footage per bit, F, is easy to determine. A simple application of Eq. 2.1 follows:

EXAMPLE 2.1:

Determine the drilling costs in dollars per foot.

Well depth = 10,000 ft (when bit is pulled)
Bit costs = $2,000

Average rig costs = \$500/hr
Rotating time = 100 hr
Round trip time = 10 hr
Footage per bit = 1,000 ft

SOLUTION:

$$C_t = \frac{2,000 + 500\,(100 + 10)}{1,000} = \$57/ft$$

Note that an average rig operating cost of \$500/hr was used in Example 2.1, without considering the reasons rig operating costs differ. Each operator may handle his costs differently, and no specific procedure will be justified or condemned. Daily performance costs often do not add up to final well costs because the daily costs may emphasize comparisons, while final costs include all the money spent.

The primary objective in drilling oil and gas wells is to find hydrocarbons and then provide an environment for maximum hydrocarbon recovery. The economics of all phases of the operations are considered; however, opinions vary considerably on what to emphasize. The drilling man normally emphasizes the drilling operation. The formation evaluation man stresses practices for improved evaluation results. The completion man accentuates hydrocarbon recovery practices. If total well costs are to be minimized, all groups must compromise. Compromise is easy to discuss but difficult to accept. A geologist will often blame poor productivity from a promising hydrocarbon zone on the completion man. The completion man may find fault with the drilling procedures, and, of course, the drilling man may blame both groups for high drilling costs.

All operators want to drill minimum-cost wells and achieve the desired objectives. Their philosophies on how to drill the minimum-cost well may vary substantially. Sometimes the operator is overly influenced by the last problem he encountered; as a result, his drilling program for the next well reflects his desire to avoid the problem rather than to minimize costs. One operator may think a straight hole is more important than a fast hole. Another may be concerned more with pressure control than fast drilling with low mud weights.

Hole enlargement many times prompts the operator to increase well costs to prevent that problem. Specific hole problems are discussed in chapter 3. Hole problems are common and account for a percentage of deep-well drilling costs. While hole problems should be considered in planning a well, in general the cheapest wells are those that reach their objective in the shortest time.

Drilling with this philosophy of speed will result in more wells reaching their desired objective. Opponents of the fast drilling concept use specific examples to show where problems have occurred. In fact, it is possible to prove almost anything by selecting specific field examples. General trends offer the best evidence of drilling progress. The fast drilling concepts developed along the Louisiana coast during the late 1950s substantially reduce drilling times and costs in that area.

As in any operation, fast drilling must be tempered by other objectives or drilling problems encountered in a specific well. For example, drilling may have to be slowed in very soft surface hole sections to prevent the mud from becoming so heavy that circulation is lost. Gumbo clay may completely plug flow lines and stick the drillstring if drilling rates are not controlled in such sections. In pressure transition zones, the primary objective is to select a protective casing seat; this takes precedence over drilling fast. These exceptions illustrate that no philosophy or concept can cover all contingencies encountered in drilling.

Fast drilling promotes the idea of attacking hole problems and withdrawing slightly if necessary rather than establishing a program based on holding a position which to that point has never been very strong. Drilling in problem areas should never proceed from well to well using the same practices that resulted in hole problems. Sometimes this is done because it is easier to remain in the mainstream of activities than to make changes that may or may not be improvements and that are hard to explain if problems arise.

Certainly many factors must be considered in well costs rather than those related directly to drilling time. Casing programs are of prime importance in deep high-pressure wells. Primary cementing may be the key to reasonable completion costs. Formation evaluation methods can substantially affect well costs. Mud programs may minimize specific hole problems. These considerations and many others may be taken into account in planning a well. A study of factors that affect penetration rate and thus the total well time follows.

FACTORS THAT AFFECT PENETRATION RATE

Variables that affect penetration rate are enumerated below:

- Drill bit
- Bit weight
- Rotary speed
- Bottom-hole cleaning
- Mud properties

Fixed factors that affect drilling rate, such as rock hardness and type and formation pore pressure, cannot be changed; however, operator recognition is very important.

Bit Selection

Bit selection is based on past bit records, geologic predictions of lithology, and drilling costs in dollars per bit. Bit design and type constantly change, sometimes making selection difficult. Fixed bladed bits are becoming more popular, and regular rock bits frequently change. For these reasons there is a separate chapter (11) on bits.

Bit Weight and Rotary Speed

Increasing bit weight and rotary speed boosts drilling rate. However, these increases accelerate bit wear. Field tests show that drilling rate increases in direct proportion to bit weight. Fig. 2–2 shows typical field results.

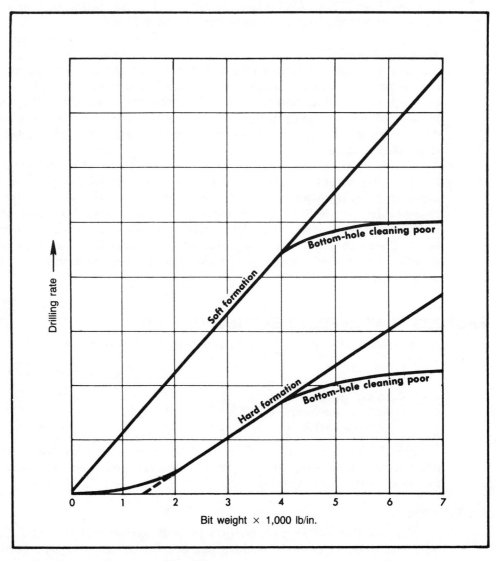

 **FIG. 2–2** Drilling rate vs bit weight

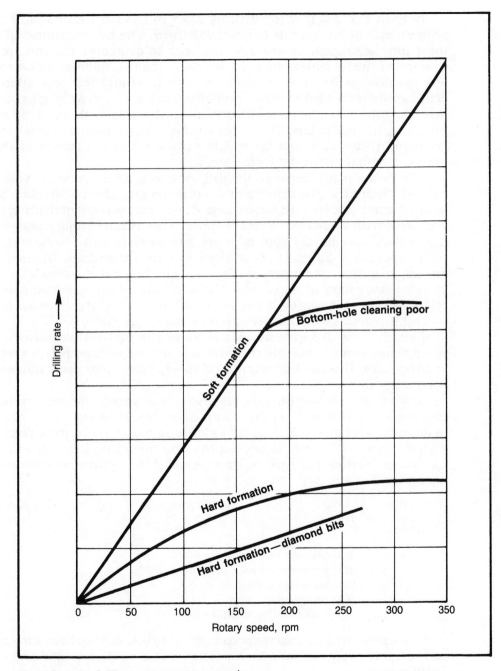

■ **FIG. 2-3** Drilling rate vs rotary speed

Note in Fig. 2–2 that the drilling rates in the hard formation are not proportional to bit weights below 2,000 lb/in. The bit, regardless of type, must impose enough pressure on the rock to overcome the compressive strength of the formation rock. In this case, about 1,500 lb/in. of bit weight was required to fracture the rock. In both hard and soft formations the drilling rate increased in direct proportion to bit weight as long as bottom-hole cleaning was adequate. This is generally true with any type of bit, including the insert bits. The upper limit of the proportional relationship between drilling rate and bit weight occurs when the matrix of the bit comes in contact with the formation.

The relationship between drilling rate and rotary speed is shown in Fig. 2–3. Notice that the drilling rate is directly proportional to rotary speed in soft formations. In hard formations, the rate of increase of drilling speed decreases with increases in rotary speed. This is the primary reason that high rotary speeds, 150 rpm or more, are used in soft formations. Low rotary speeds (40–75 rpm) are utilized in hard formations. Diamond bits are an exception, as shown on Fig. 2–3. Drilling rates increase in close proportion to rotary speed, provided bottom-hole cleaning is adequate. This relationship is the reason bottom-hole drives that rotate diamond bits at several hundred rpm's are preferred by many operators.

Figs. 2–2 and 2–3 give no indication of the best bit weight or rotary speed. They simply indicate the effects of bit weight and rotary speed on drilling rates. Use the bit weight and rotary speed that costs the least in dollars per foot.

The effort to select bit weights and rotary speeds for minimum cost in dollars per foot was termed "minimum-cost drilling." The increases in drilling rates due to higher bit weights or rotary speed were combined with the reduced bit life to predict the best operating limits for bits. Eq. 2.2 can be written for drilling rate versus bit weight and rotary speed from Figs. 2–2 and 2–3.

$$R = KWN^a \qquad (2.2)$$

where:

R = drilling rate, fph
K = proportionality constant
W = bit weight, lb/in.
N = rotary speed, rpm
a = exponent between 0.4 and 1.0

The exponent a on rotary speed is equal to 0.4 for very hard formations and to 1.0 for soft formations. An exact a can be determined from field drilling tests.

Eq. 2.2 is the instantaneous drilling rate at any point in time or can be used as an average penetration rate for a complete bit run. In fact, for tooth-type bits, there is a dulling trend and Eq. 2.2 is written as:

$$R = \frac{KWN^a}{1 + K'D_n} \tag{2.3}$$

where:

> K' = a constant
> D_n = normalized tooth wear

The constant K' and D_n can be determined from field drilling tests.

The relationship, L, between bit life and bearing life for bits is given in Eq. 2.4:

$$L = \frac{K''}{NW^b} \tag{2.4}$$

where:

> K'' = proportionality constant
> b = exponent between 1.0 and 2.0

The exponent b is a function of fluid type and varies between 1.0 and 2.0, depending on the abrasive characteristics of the fluid in contact with the bearings.

A plot of bit life vs. bit weight on log paper is the best method for obtaining b. A typical plot is shown in Fig. 2–4. Note that b is the slope of the straight line in Fig. 2–4. Field data should plot a reasonably straight line; however, this depends on the techniques used for pulling bits and for grading bits that are not worn out. Specific bit grading procedures are included in chapter 11. The primary methods used to determine when to pull bits include (1) a low rate of penetration compared with expected normal rates, (2) a sudden increase in rotary torque, and (3) an economic analysis based on cost per foot calculations.

Drilling rate is one method used for pulling bits since the inception of rotary drilling. Many times a combination of drilling rate and time is used. This procedure is imprecise and places a premium on bit-grading procedures.

Drilling torque is a common method for determining when to pull a bit. This procedure is subject to having a good torque indicator and to the operator's recognition that the increase in torque is due to locked bearings and not a change in formations. Fig. 2–5 shows how torque may be used.

The sudden increases in bit torque represent a warning signal with roller cone bits. With friction bearing bits the first indication may be the last before losing a cone in the hole. For this reason, always evaluate the length of time the bit has been used. Consider a situation in which the operator expected a friction bearing bit to last 100 hr. If the sudden torque increase occurs after 30 hr, the operator may not pull the bit until he sees a second indication of a sudden torque increase. If the sudden torque increase happens after 80 hr, the operator will probably pull the bit. If

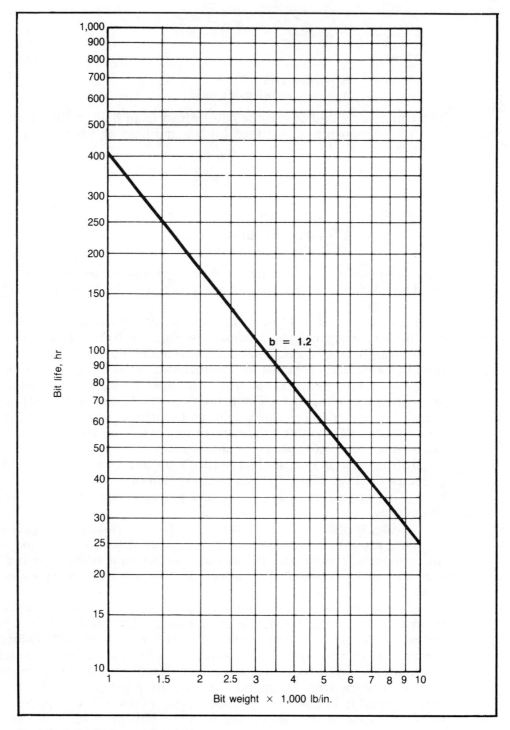

■ **FIG. 2–4** Bit life vs bit weight

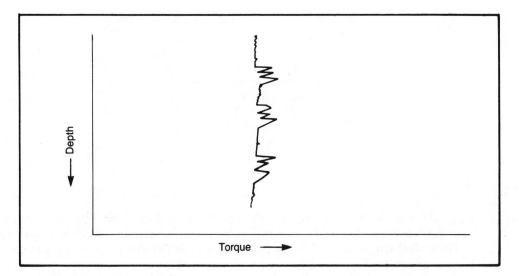

■ **FIG. 2–5** Torque vs depth

there is no sudden torque increase even after 120 hr of bit use, the operator may arbitrarily decide to pull the bit. Recovering a steel bit cone lost in the hole requires several hours, at least, and could take several days. For this reason, experience is helpful for knowing when to pull bits based on torque increases. A brave operator may decide not to pull the rock bit, regardless of the time in the hole, until he sees a rapid increase in bit torque.

No precise surveys are available, but an estimated 75% of rock bits are pulled green or before the bit is worn out. Because well costs may directly relate to the footage per bit, some effort to examine and update bit-pulling practices is almost always justified.

The third method used to determine when to pull bits is the economic analysis. This procedure involves a frequent determination of cost per foot while drilling. The bit is pulled when the cost per foots vs. drilling time reaches a minimum. This procedure is demonstrated in Example 2.2:

□   **EXAMPLE 2.2**

Determine when the bit should be pulled.

Well depth = 10,000 ft
Bit costs = \$600
Rig costs = \$200/hr
Round trip time = 1 hr/1,000 ft
Rate of penetration = 100 − 2t

SOLUTION:

Use Eq. 2.1 to develop the data shown in Table 2–1.

■ **TABLE 2–1**   Cost per foot data for Example 2.2

t, hr	R, fph	F, ft	Total Cost, $	C_t, $/ft
5	90	475	3,695	7.78
10	80	900	4,780	5.31
20	60	1,600	6,920	4.33
25	50	1,875	7,975	4.25
30	40	2,100	9,020	4.30
35	30	2,275	10,055	4.42
40	20	2,400	11,080	4.62

The data from Table 2–1 are plotted in Fig. 2–6. The bit should be pulled after about 25 hr. Actually, the precise pulling time is not critical. Note that the cost in dollars per foot was $4.30 after 30 hr. The economic procedure in Table 2–1 is primarily related to tooth-type bits where wear rate is predictable. This procedure also might be used with tungsten carbide insert bits when inserts are broken or pulled out of the matrix. The problem is that the wear rate with insert bits is unpredictable. From the standpoint of economics the insert bit should be pulled when the cost in dollars per foot begins to increase.

If bits are pulled on an economic basis it would be impossible to obtain meaningful bit wear information as shown in Fig. 2–4. Most grading sys-

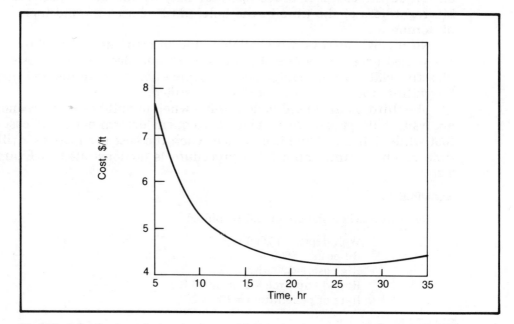

■ **FIG. 2–6**   Cost per foot vs hours run for Example 2.2 and data shown in Table 2–1

tems are based on an interpretation of bearing wear at some level before the bearings lock would not be accurate enough to use as shown in Fig. 2–4.

Some people strongly disagree with the assumption in Eq. 2.4 that bearing life in bits is inversely proportional to rotary speed. Certainly this assumption has limits. If the rotary speed reaches a level in which the bit bearings are destroyed by excessive heat, then obviously Eq. 2.4 does not apply. Also, if experience shows that bearing life is not proportionate to rotary speed, bit life (in hours) may be plotted against rotary speed (in rpm) on log paper much as bit weight vs. bit life plot (Fig. 2–4).

Optimized Drilling Costs

There are methods that predict the optimum bit weight and rotary speeds for drilling at a minimum cost in dollars per foot. Many techniques have been introduced. There are variable bit weight and rotary speed programs, variable bit weight and constant rotary speed programs, and constant optimum weight and rotary speed programs. Most of the methods depend heavily on the proper interpretation of bit wear.

All of the methods for optimizing bit weight or rotary speed are not discussed here. The basic procedure for optimizing begins with cost per foot, shown in Eq. 2.1. Next write expressions for drilling rate as a function of bit weight and rotary speed. Finally, express bit life as a function of bit weight and rotary speed. This was done in this chapter in Eqs. 2.1, 2.2, and 2.4. An optimum bit weight is determined using Eq. 2.5:

$$W_{opt} = \left[\frac{C_r K''}{(b-1)N(B + C_r T_r)} \right]^{1/b} \tag{2.5}$$

Eq. 2.5 is a simplified equation. It assumes that C_r is a constant number and that rotary speed does not change.

□ **EXAMPLE 2.3**

Determine the optimum bit weight.

> Well depth = 10,000 ft
> Bit costs = \$2,000
> Rig costs = \$2,000/hr
> Round trip time = 1 hr/1,000 ft
> Bit weight = 5,000 lb/in.
> Rotary speed = 75 rpm
> Bit wear, b = 1.5
> Bit life = 100 hr
> Average drilling rate = 10 fph

SOLUTION:

Using Eq. 2.4,

$$\frac{K''}{N} = (100 \times 5,000)^{1.5}$$

Using Eq. 2.5,

$$W_{opt} = \left[\frac{(2,000)\,(100)\,(5,000)^{1.5}}{(0.5)\,[2,000 + (2,000)\,(10)]}\right]^{1/1.5}$$

$$W_{opt} = \left[\frac{(2,000)\,(100)}{(0.5)\,(22,000)}\right]^{0.667} \times 5,000 = 34,606\ lb/in.$$

To show the effect of *b,* consider a change in the bit wear function from 1.5 to 2.0:

$$W_{opt} = \left[\frac{(2,000)\,(100)}{(1)\,(22,000)}\right]^{0.5} \times 5,000 = 15,076\ lb/in.$$

A change of 33% in the bit wear function, *b,* introduces more than a 50% reduction in the optimum bit weight. Note from Eq. 2.5 that, as the rig costs increase, the optimum bit weight also increases. Considering the results in Example 2.3, the optimum bit weight with a bit wear function of 1.5 or 2.0 is above the acceptable minimum for field practices. Field practices change as new bit designs are introduced. Bit weights of 7,000 lb/in. on insert friction bearing bits are not uncommon. The higher bit weights are generally imposed on hard formation bits. In any event, the effectiveness of any bit weight level depends on adequate bottom-hole cleaning. The final test, of course, is whether the cost in dollars per foot is reduced as bit weight is increased.

Eq. 2.5 can also be used to optimize rotary speed if bit weight is maintained at a constant level. However, if the wear term, *a,* in Eq. 2.2 is 1.0, then the limit on rotary speed is determined by rotary table power, the

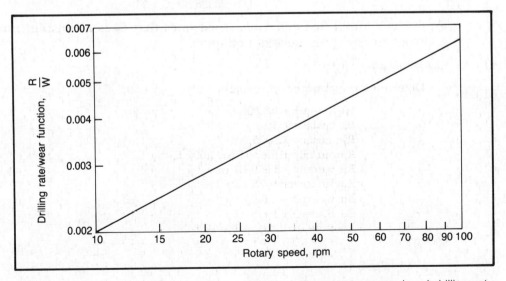

■ **FIG. 2–7** Method to determine the relationship between rotary speed and drilling rate for Example 2.4

drilling string torque, or some threshold limit on bit bearings. If a is less than 1.0 in Eq. 2.2, then the optimum rotary speed, N_{opt}, can be calculated from Eq. 2.6:

$$N_{opt} = \frac{a\,C_r\,K''}{(1-a)\,(B+C_rT_r)\,W^b}$$ (2.6)

The solution to Eq. 2.6 depends on the relationship between drilling rate and rotary speed. This relationship can be evaluated in field practices as shown in Fig. 2–7.

The slope of the line in Fig. 2–7 of R/W vs N on log paper is a. To illustrate, assume $a = 0.5$, and use the other data of Example 2.3 to determine the optimum rotary speed shown in the example below:

☐   **EXAMPLE 2.4**

Assume bit weight is 5,000 lb/in. and bit life is 100 hr. Using Eq. 2.4,

$$\frac{K''}{W^b} = (100)\,(75)$$

Using Eq. 2.6,

$$N_{opt} = \frac{(0.5)\,(2{,}000)\,(100)\,(75)}{(0.5)\,[2{,}000 + (2{,}000)\,(10)]}$$

$$N_{opt} = \frac{(2{,}000)\,(100)\,(75)}{22{,}000} = 682 \text{ rpm}$$

Example 2.4 gives an optimum rotary speed of 682 rpm. Remember that Eq. 2.4 assumes that the level of rotary speed is not a factor in bearing wear. Actually, at some rotary speed, the heat generated could cause rapid bearing failure. Determining a maximum rotary speed at which failure in field operations occurs is difficult; however, it is safe to assume that a 682-rpm rotary speed is above the level that most operators use in the field. Also, most bearing-type bits would be destroyed quickly at 682 rpm.

Rotary speeds with roller bearing bits in soft formations are usually above 100 rpm and, in many cases, above 200 rpm. The upper limit on rotary speed is generally set by the maximum safe speed relative to drill-string failure. Calculations show that there are rotary speeds at which the drillstring vibrations cause rapid drillstring failure. Downhole recording units confirm these critical rotary speeds. However, downhole recording units also reveal that any critical rotary speed shows at the surface through excessive vibrations. Thus, the operator need not calculate a critical rotary speed range. The only requirement is to change the rotary speed or bit weight when excessive vibrations at the surface are noted.

Rotary speeds with friction bearing insert bits were initially in the range of 45–60 rpm. Changes in lubricants and bearing design permit operators to increase rotary speeds to the 120-rpm level. The higher rotary speeds result in reduced drilling costs and substantially improved operations in wells with crooked hole problems that limit bit weights.

The optimized drilling equations can be extended by making the rig costs (in dollars per hour) a variable. One method for this is Eq. 2.7:

$$C_r = D + M_p + M_{dp} + M_{dc} + M_r + M_c \qquad (2.7)$$

where:

D = labor + depreciation + rig supervision + rig insurance + miscellaneous rig expenses, $/hr
M_p = pump maintenance, $/hr
M_{dp} = drillpipe maintenance, $/hr
M_{dc} = drill collar maintenance, $/hr
M_r = rig maintenance, $/hr
M_c = mud costs, $/hr

Evaluation of Terms in Eq. 2.7

In Eq. 2.7, D is a constant. Pump maintenance, M_p, is not a constant. As more bit weight or rotary speed is used to increase drilling rate, the pump pressure must be raised to increase bottom-hole cleaning. Eq. 2.8 can be used to evaluate the effect of pump pressure:

$$M_p + A_c e^{fp} \qquad (2.8)$$

where:

A_c = proportionality constant
f = constant
p = pump pressure, psi

The constants in the above equation are evaluated from pressure cost data in field operations.

The drillpipe maintenance, M_{dp}, is primarily related to rotary speed and can be expressed as shown in Eq. 2.9:

$$M_{dp} = CNL_{dp} \qquad (2.9)$$

where:

C = proportionality constant
N = rotary speed, rpm
L_{dp} = length of drillpipe, ft

The constant C can be evaluated in field operations. Since M_{dp} is zero at zero rotary speed, the cost at one rotary speed is sufficient. The drill collar maintenance, M_{dc}, is calculated from the same equation form as shown for drillpipe in Eq. 2.8.

Rig maintenance, M_r, is a function of rotary speed and round trips. Eq. 2.10 shows rig maintenance as a function of both rotary speed and expenses incurred during round trips.

$$M_r = C_1 + C_2N + C_3M_{rt} + C_4M_{wl} \qquad (2.10)$$

where:

C_1, C_2, C_3, C_4 = constants determined from field data
M_{rt} = rig maintenance during round trips, \$/hr
M_{wl} = wireline maintenance, \$/hr

Eqs. 2.7–2.10 are included for information. They have a small effect on drilling costs in dollars per foot. These formulas are used to develop new equations for optimum bit weight or rotary speed. Eqs. 2.7–2.10 may prove useful for predicting costs if field practices change.

Examples 2.3 and 2.4 show that the optimum bit weight and optimum rotary speeds are far above maximums used in field operations. The primary purpose of these examples is to emphasize that if only economics are considered in selecting bit weight and rotary speed, the operator will always be at the upper limits of both bit weight and rotary speed.

The optimized equations shown are for bits with cones that rotate on bearings. Fixed-blade bits, such as diamond and diamond compact bits, fall into a different category. The limit on bit weight is related to cleaning and cooling by the drilling fluid. Maximum rotary speeds are limited by the maximum rotational speeds of the drillstring or by the type of bottom-hole drive used, along with the available hydraulic horsepower available to activate the bottom-hole drive mechanism. Fixed-blade bits are being developed, and field practices with these bits are subject to daily change. The only certain facts are that bit weights with fixed-blade bits are less and rotary speeds are higher than with conventional rock bits.

Remember that many elements are involved in the selection of bit weight and rotary speed other than the cost per foot. One overlying factor in all calculations is the assumption that adequate bottom-hole cleaning is available. If the hydraulics program is inadequate, then the relationships shown in Eq. 2.2 are invalid. Also, crooked-hole problems or directional control must be evaluated when selecting bit weights. Rig and drillstring maintenance must be considered when selecting rotary speed. Thus there are no cookbook answers to the overall selection of bit weight and rotary speeds. Calculate the costs in dollars per foot whenever changes are made and make enough evaluations to eliminate the normal margins for error.

In general, the range of bit weights and rotary speeds used in 1984 are as follows:

Roller Bearing bits
Bit weight range: 4,000–7,500 lb/in
Rotary speed range: 100–180 rpm

Friction Bearing Bits
Bit weight range: 4,000–7,000 lb/in.
Rotary speed range: 50–120 rpm

Fixed-Blade Bits
Bit weight range: 1,500–3,000 lb/in.
Rotary speed range: 100–700 rpm

Upper levels of rotary speed can be achieved by downhole motors or turbines.

Bottom-Hole Cleaning

Chapter 8 describes the need for bottom-hole cleaning. There was a tendency in recent years to ignore bottom-hole cleaning in deep wells with abnormally high pore pressures. This was done because often a very fine balance is reached between a well kicking and lost circulation. Bottom-hole cleaning many times can be increased substantially with no adverse effect on hole conditions. The benefit of reduced drilling time should increase the possibilities of reaching the desired objective.

Remember that all of the drilling parameters used to predict pore pressure depend on good bottom-hole cleaning. The d exponent or drilling rate does not provide any accuracy if bottom-hole cleaning fails to remove the cuttings as they are generated.

Mud Property Effects

Mud property effects on drilling rate have been a research object for more than twenty years. At various times papers have proved that almost any mud property affects drilling rate.

Mud properties claimed to affect drilling rate include the following:

- Mud weight
- Solids content and type
- Mud viscosity, laminar and turbulent
- Water loss and spurt loss
- Liquid phase, water or oil

Mud Weight

All drilling personnel know that an increase in mud weight decreases drilling rate, if there is no change in pore pressure. The probable reason for this effect of mud weight on drilling rate is compaction of the formation just below the bit, as illustrated in Fig. 2–8.

If compaction is the main reason mud weight affects drilling rate, then apparently mud weight effects are subject to the kind of formation drilled. In soft shale formations, drilling rate is appreciably reduced if mud weight is increased with no increase in pore pressure. In hard formations, such as chert, dolomite, and dense limestone, mud-weight increase reduces penetration rate, but the effect is much less than that noted in the soft shales.

The approximate effect of mud weight on drilling rate can be calculated from an expression like Eq. 2.11.

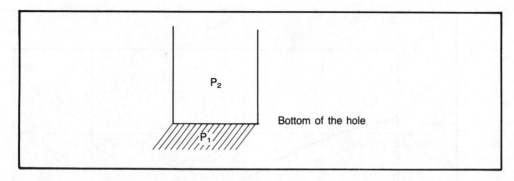

■ **FIG. 2–8**   Differential pressure compaction of the formation just below the bit

where:

$$\rho_1 \text{ and } \rho_2 = \text{mud weights at two different times, lb/gal}$$
$$R_1 \text{ and } R_2 = \text{drilling rates corresponding to the respective mud weights, fph}$$

$$\rho_2{}^c \log R_2 = \rho_1{}^c \log R_1 \qquad\qquad (2.11)$$

In soft formations c may be as high as 1.5 and in hard formations as low as 0.1. Consider the following example:

☐ **EXAMPLE 2.5**

If the mud weight is increased to 12.0 lb/gal, determine the drilling rate.

Formation type = shale
Drilling rate = 400 fph
Mud weight = 9.0 lb/gal

SOLUTION:

Assume c = 1.0

$$\text{Log } R_2 = \frac{9}{12} \log R_1 = \frac{9}{12} \log 400 = 1.95$$

$$R_2 = 90 \text{ fph}$$

Fig. 2–9 represents a field test in western Canada where the increase in pump pressure and the noted back pressure were imposed by a hand-adjusted choke in the discharge flow line. This test was run at a depth of 2,500 ft using a 9.0 lb/gal mud; thus the increase in back pressure of 250 psi represents an effective mud weight increase of 2.0 lb/gal. Note the decrease in drilling rate from about 95 to 30 fph, a common result at these mud-weight levels. Fig. 2–10 shows a laboratory test of drilling rate vs imposed hydrostatic pressure. Of significance in this test is the fact that the increase in hydrostatic pressure from 0 to 2,000 psi resulted in a decrease in penetration rate from 18 to 5 fph. The increase in hydrostatic

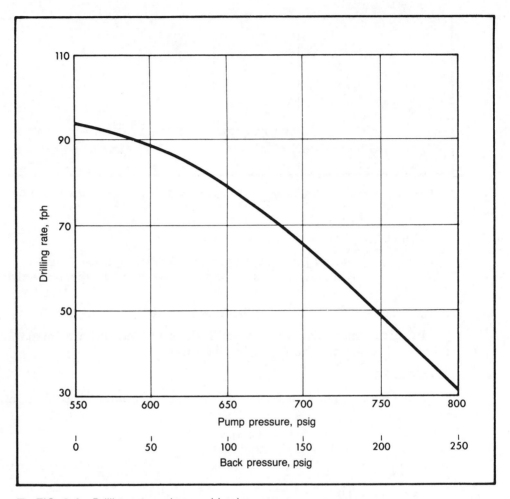

■ **FIG. 2–9** Drilling rate vs imposed back pressure

pressure from 2,000 to 4,000 psi resulted in a decrease of only 5.0 to 3.5 fph. Results shown in Fig. 2–10 are typical of actual field operations. The first unit of mud weight in excess of formation pore pressure reduces drilling rate more than each subsequent unit of mud weight increase.

Figure 2–11 shows the effect of mud weight on rotating time in two south Mississippi wells. Well A was drilled with 10.4-lb/gal mud and typified the normal practice in the field before drilling well B. Using a thinner mud weighing 9.6 lb/gal in well B resulted in a rotating time of 210 hr, less than half of the 500-plus rotating hours required to drill well A. More than just the mud weight was changed, and this test alone does not prove that all of the improvement was due to the reduction in mud weight. How-

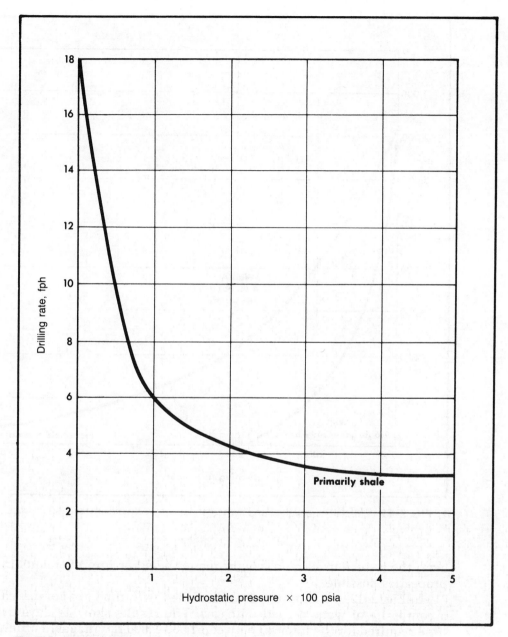

■ **FIG. 2–10** Drilling rate vs hydrostatic pressure

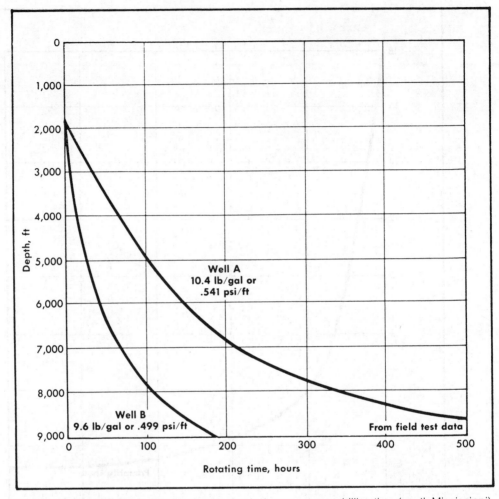

■ FIG. 2–11 Effect of mud weight and hydrostatic pressure on drilling time (south Mississippi)

ever, the reduction in mud weight made the other improvements in mud properties possible.

In Fig. 2–12, a composite of wells drilled with mud are compared with a composite of wells drilled with gas. The results show 14 days drilling time required for the interval between 1,500–9,800 ft with gas, as compared with 90 days for the same interval with mud.

Fig. 2–13 compares drilling with air and water in 2,600-ft wells in west Texas. The rotating time with air was about one-half the rotating time with fresh water. In both series of field tests (Figs. 2–12 and 2–13), the improved performance occurred primarily because the hydrostatic pressure was less than the formation pore pressure.

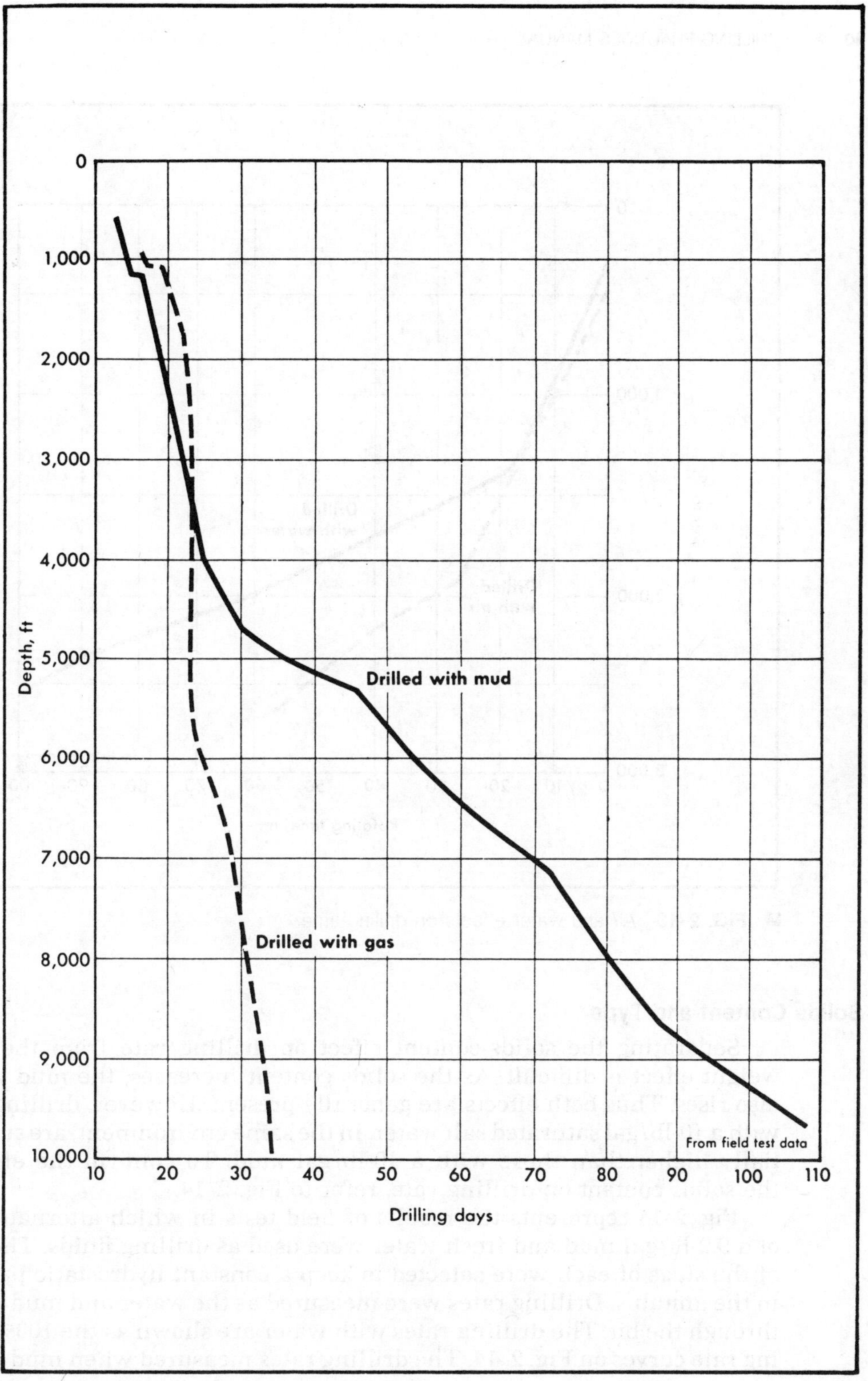

Depth, ft
0
1,000
2,000
3,000
4,000
5,000
6,000
7,000
8,000
9,000
10,000
10
20
30
40
50
60
70
80
90
100
110
Drilling days
Drilled with mud
Drilled with gas
From field test data

■ **FIG. 2–12** Gas and mud effects on drilling time

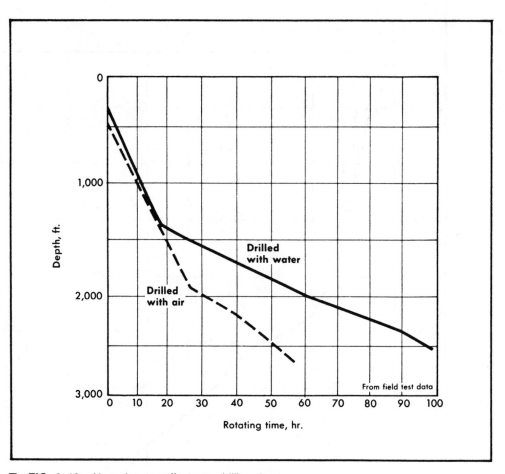

■ **FIG. 2–13** Air and water effects on drilling time

Solids Content and Type

Separating the solids-content effect on drilling rate from the mud-weight effect is difficult. As the solids content increases, the mud weight also rises. Thus both effects are generally present. However, drilling rates with a 10-lb/gal saturated salt water, in the same environment, are substantially higher than those with a 10-lb/gal mud. To confirm the effect of the solids content on drilling rate, refer to Fig. 2–14.

Fig. 2–14 represents the results of field tests in which alternate slugs of a 9.2-lb/gal mud and fresh water were used as drilling fluids. The sizes of the slugs of each were selected to keep a constant hydrostatic pressure in the annulus. Drilling rates were measured as the water and mud passed through the bit. The drilling rates with water are shown as the 100% drilling rate curves on Fig. 2–14. The drilling rates measured when mud passed

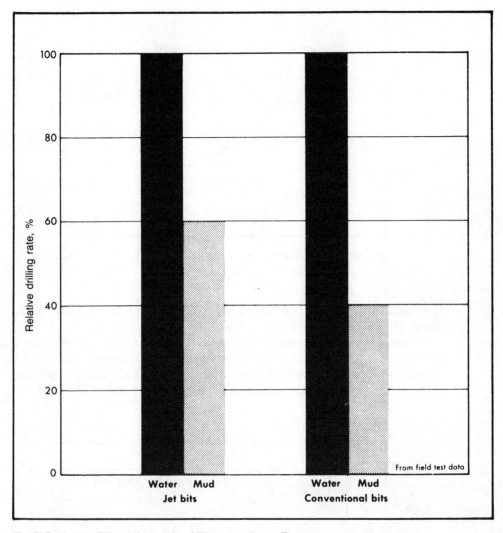

■ **FIG. 2–14** Effect of mud on drilling rate (west Texas)

through the bit are shown as 60% of those with water using jet bits and 40% of those with water using conventional bits. These tests were conducted primarily in west Texas and confirm that the solids content of the mud affects drilling rate. Further proof of the solids-content effect on drilling rate is given in Fig. 2–15.

In laboratory tests such as Fig. 2–15 there were no changes in hydrostatic pressure. Note that the drilling rates are 11 fph with 2% solids by volume and about 3 fph with a solids content of 12% by volume.

In addition to the solids content, the solids type and the state of solids

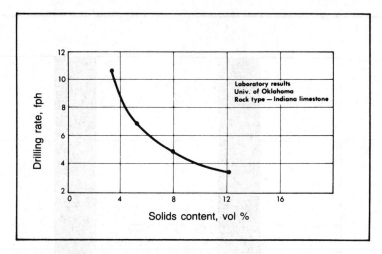

■ **FIG. 2–15** Quantitative effects of mud solids on drilling rate

dispersion also affect the drilling rate. Fig. 2–16 from work by Lummus shows the effect of solids dispersion on drilling rate.[1]

Notice, particularly below a solids content of 3% by volume, that the drilling was substantially faster with the nondispersed mud in Fig. 2–16. This same result is expected when comparing highly dispersible clay solids with nondispersible solids like sand and limestone.

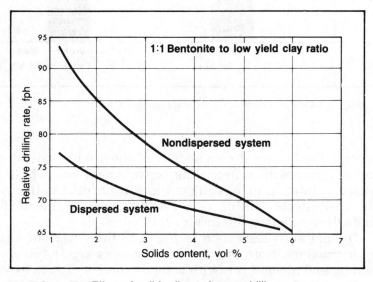

■ **FIG. 2–16** Effect of solids dispersion on drilling rate

Mud Viscosity, Laminar and Turbulent

The term viscosity was introduced to represent the thickness of a Newtonian fluid in laminar flow. In some way common usage expanded the original meaning to include the thickness of non-Newtonian fluids, such as muds. Later the concepts of thickness and viscosity became synonymous for drilling fluids, regardless of flow pattern; it is now common practice to refer to mud thickness as viscosity. In laminar flow, the mud thickness is defined by two terms, either plastic viscosity and yield point or from the power-law concept by n and k. These relationships are discussed in detail in chapter 5. In turbulent flow, thickness is a function of solids content, and the laminar flow viscosity terms have no meaning for thickness in turbulent flow.

The laminar flow viscous effects affect drilling rate only by the additional circulating-pressure loss imposed in the annulus as a mud thickens. This additional pressure increases the effective hydrostatic pressure in the annulus and affects drilling rate as shown in the section on mud weight effects.

The flow pattern of the drilling mud is turbulent as it passes through the bit, so the laminar flow viscous properties of the mud cannot be used to determine mud thickness. However, because plastic viscosity is a means of determining the relative solids content of a mud, the plastic viscosity can also be used to indicate the mud thickness in turbulent flow. A thin mud at high shear rates results in higher penetration rates than a thick mud if no other changes in mud properties occur.

Fig. 2–17, taken from Eckel, shows five muds with the same apparent laminar flow viscosity and with different ratios of yield point to plastic viscosity.[2] As the ratio of yield point to plastic viscosity (YP:PV) increases,

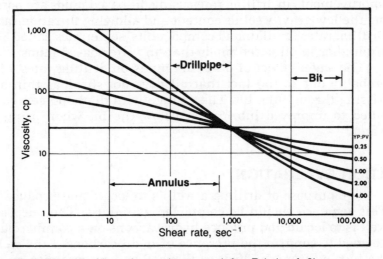

■ **FIG. 2–17** Viscosity vs shear rate (*after Eckel, ref. 2*)

so does the shear thinning characteristic of the mud. Note in Fig. 2–17 that the mud with a YP:PV of 4.0 sheared thinner than the other muds at high shear rates. Muds of this type usually are those with a low and flocculated solids content or with some polymer for thickening. As demonstrated in Fig. 2–17, all muds shear thin. The low-solids thick muds, in general, are the ones that thin the most at high shear rates.

Filtration Rate and Spurt Loss

The concept that drilling rates rise with increases in filtration rates was introduced by operating personnel. Drillers noticed that drilling rates decreased when the filtration rates were lowered. In all probability this drilling-rate reduction was more a result of the materials added to reduce filtration rate than of the filtration rate reduction itself.

When low-solids muds were introduced, it was not uncommon to have an API filtration rate of 10 cc and a high initial fluid loss, called *spurt loss*. The spurt loss was simply the fluid loss necessary to form a filter cake of solids. As a result, drilling rates with the low-solids mud with an API water loss of 10 cc may be substantially faster than a higher-solids mud having an API water loss of 20 cc. These drilling rates may or may not respond to changes in filtration rate, depending on other changes in the mud.

Fluid Phase

The fluid phase can have substantial impact on drilling rate. Air or gas is lightweight, and the effects of these two fluids were shown in Figs. 2–12 and 2–13. A continuing controversy is water versus oil. Drilling rates with a continuous oil-phase mud are generally lower than drilling rates using water-based muds, if rock bits or regular diamond bits are used. Improvements in drilling rates using oil-based muds are possible by reducing the low-gravity solids content and allowing filtration rates to increase. Drilling rates for diamond compact bits, sometimes called Stratapax® bits, are higher in oil-based muds than in water-based muds.

The stated effect of oil-based muds on drilling rates with rock bits is probably due to the fact that oil does not shear as thin as water when passing through the bit. The change with diamond compact bits is attributed to improved lubrication below the bit when using oil instead of water.

FORMATION EVALUATION

The purpose of drilling a well is to locate and produce hydrocarbons. This sentence should be completed as follows: The purpose of drilling a well is to locate and produce hydrocarbons on a commercial basis. It does no good to confirm the presence of hydrocarbons if the well cost exceeds the income from the hydrocarbons. Some compromise must be arranged when well costs become excessive, mainly because of extended formation evaluation programs.

One problem with the so-called compromise is that no department or individual ever admits that current programs can be trimmed without sacrificing critical information. For this reason, some of the primary formation evaluation methods, such as sample analysis, coring, logging, and testing, are discussed.

Sample analyses are a significant part of the search for hydrocarbons. Fluorescence from stained hydrocarbon samples may be the first indication that a potential productive horizon is present. In general, the drilling operation may be affected very little. There is some debate relative to the use of oil in the mud; this led to implementing essentially nonfluorescent oils.

Actually, this problem is minimized because ecology also frequently eliminates the use of oil, particularly in offshore exploratory operations. In any event, the elimination of oil emulsified in water-based muds may or may not affect drilling operations. This area is one in which little argument arises between those concerned with drilling and those concerned with formation evaluation.

Conventional, wireline, and sidewall coring represent desirable methods of evaluating potential productive horizons, particularly in exploratory wells. The only fault that should be emphasized relative to the frequency of coring is practices that continue extensive coring operations in development areas. Often coring in development areas is justified as the only means of gathering pertinent information about porosity, water saturation, and permeability. Actually, information may be obtained at a much lower cost from logs and pressure buildup or drawdown data.

While coring slows down drilling and increases cost, the coring operations would have little value if many special drilling precautions were not introduced. Most drilling programs include the statement: "Before commencement of coring, get the drilling mud in shape." Getting the mud in shape includes lowering water loss, either decreasing or increasing viscosity, and improving the mud's general appearance. These practices are habitual and benefit neither core recovery nor core condition.

Obviously fluid property changes have little bearing on core recovery. Specific cases arise in which special coring fluids are preferable. A desire to look closer at water saturation may prompt the use of oil-based drilling fluids. In addition, solids plugging of permeability may introduce a search for special muds. Many factors must be considered in the special cases; however, in general, oil-based fluids offer little improvement and frequently complicate evaluation.

Special fluids to prevent solids plugging of cores may be very expensive. Probably one of the best fluids for this purpose is bentonite mixed with fresh water, which is cheap and easy to prepare. Arguments against the bentonite-and-water mixture emphasize that bentonite is a small particle and is very detrimental. The truth is that bentonite is a small particle that adsorbs large quantities of water. It plasters on the core face in most instances, rather than entering the pore spaces. Thus, many expensive procedures associated with coring can be eliminated without losing any information.

Logging operations are extensive for both exploratory and developmental wells. Some logs are run specifically to determine formation pore pressure in wells in which abnormally high pore pressures have been detected. Thus, logs are for both formation evaluation and drilling operations.

Emphasis is placed on selecting only the logs required for evaluating the formation. This aspect of the business has improved substantially over the years, though some duplication still exists. Self-potential and gamma-ray logs may both be run to determine lithology when only one has utility. Several porosity logs may be run; different answers can confuse rather than clarify.

One bad habit that has developed in logging operations is the short trip, run back into the casing and back to the bottom before pulling the drillstring to log. This is a common, expensive practice in coastal operations and can result in prohibitive charges in offshore floating drilling. Several companies evaluated this practice; some studies observed operations for more than two years.

The result in most cases revealed that the short trip before logging was useless as insurance for getting the log to the bottom. In other words, the logs failed to go to the bottom with the same frequency, whether the short trip was made or not. Operators should check this practice carefully. Failures to get logs to the bottom can result in expensive, nonbeneficial prevention programs.

Logging as well as coring may introduce many changes in mud properties. The primary changes are directed toward reducing the invasion of filtrate into the formation. At times the composition of the liquid phase may be changed. Unfortunately, the quantity of filtrate invasion is equated to the level of the API water loss. While a reduction of 20 to 10 cc may reduce the short-term invasion, the total invasion over several days would probably be only slightly affected.

A reduction in the API water loss from 10 to 5 cc might actually increase the short-term invasion because of an increase in dynamic fluid loss. A more detailed explanation is given in chapter 5. The total invasion of filtrate and the depth of filtrate invasion could be related more to the pressure differential and the time of exposure than to the magnitude of the API water loss. This conclusion is an opinion only; however, it should be considered, and, when the opportunity presents itself, operators can make meaningful evaluations. These evaluations could be made by comparing the degree of flushing in the flushed zone and the depth of invasion with the time of exposure, pressure differentials, and measured filtration rates. Because so much emphasis is placed on keeping the filtration rate low—

to improve logging characteristics—this seems to be a necessary project for field operations research.

Formation testing may be carried out with wireline testers, conventional drillstem tests, or regular production tests. Reasons for testing include an evaluation of fluid type, formation pore pressures, and formation productivity. A conventional open-hole drillstem test is often run to determine productivity before spending money on a production casing string. However, it may also be run even though the operator has already decided to run casing. Operators often test formations shortly after they are exposed rather than wait until total depth is reached. Reasons given for this are to minimize exposure time for the formation of interest, to determine fluid contents, and to check pore pressures.

Problems with the practice are typified by a 12,000-ft well in Wyoming where 14 open-hole drillstem tests were run. The well records showed a total of 17 days of making open-hole drillstem tests. Actually, out of 154 days, 75 were required to run the tests and correct the subsequent hole problems. It would have been much cheaper to drill to total depth and set a small string of production casing.

Another conventional drillstem testing pattern that should be examined is the situation when samples, cores, and logs indicate the presence of producible hydrocarbons and a confirmation test is run before setting casing. If the confirmation test is to determine whether to set casing or not, then the test is probably justified. If casing is to be set regardless of the drillstem test results—common in some areas—then running the test is an expensive and futile exercise.

The drilling man is interested in finding hydrocarbons as well as drilling cheaply, and the formation evaluation man wants to drill cheaply as well as find hydrocarbons. The only conflict comes in the interpretation of requirements. While improved communication helps, the best method is to study procedures and practices on the basis of results versus dollars spent.

COMPLETION COSTS

The time and cost of completing wells increased substantially during the 1970s. Deeper wells make it harder to move pipe when cementing, and smaller clearances around production casing make it more difficult to place cement around the production pipe. Also, the ratio of exploratory-to-development wells has risen sharply. Operators are reluctant to leave a promising zone that fails to produce as expected. Their primary concerns are the quality of the cement job, the penetration of the perforations, and formation damage. All of these can be checked. Multiple curve bond logs give a good evaluation of a cement job. Communication indications can be checked with temperature and noise logs. Whether penetration has been successful is easy to determine. Formation damage is tougher to evaluate. Often more time is spent calculating skin effect and related problems than would be required to correct most of the damage.

Also, a strong emphasis is placed on moving off the drilling rig and moving in a smaller, less expensive completion rig, a practice that appears to have merit. However, the results depend on the expertise of the completion crews. High-pressure gas wells may be difficult to complete, and the completion personnel may be less familiar with the necessary requirements than are the drilling personnel. Too, time is lost on current production by moving the drilling rig and moving in the completion rig.

During completion operations, any problems are often blamed on drilling practices. For many years in the Rocky Mountains, practices that slowed drilling were promoted because of collapsed casing caused by running salt. Although the problem may never be completely solved, heavier casing opposite the salt section and more precautions in placing cement in the annulus through the salt section minimized it.

SUMMARY

The secret of cost control is to consider the total well costs. These include location, drilling, formation evaluation, casing, and completion costs. Cost control in drilling starts with the rig selection after the well is planned. Too often operators begin a deep, expensive well with an inadequate rig. Rigs should be examined carefully. Pump size is important, as are the normal operating procedures followed by the rig operator. Careful attention should be focused on the drillstring and the rig's operating efficiency on other wells. The success of any cost-control program depends on the rig.

Cost control or a minimum-cost well is sold as a service by many organizations. In most cases, some computerized operation is used. An accurate record of current and past operations is essential.

In general, all of the factors that affect drilling rate should be considered. Cost-per-foot determinations should be used on a continuous basis. All organizations may not agree with all the conclusions in this discussion. Any cost control program has to be tempered by potential hole problems.

An increase in emphasis and supervision usually results in reduced drilling costs. Expertise is essential in the increased supervision. The high-cost drilling operations create the necessity of having competent personnel at the rig during drilling.

Formation evaluation practices should be studied carefully. Extensive coring and logging programs do not necessarily increase the ratio of successful wells to dry holes. These programs do reduce the number of wells drilled. Formation testing programs should be analyzed. It may be cheaper to set small casing than conduct multiple open-hole drillstem tests. Also, be certain the test results are used to make the right decisions.

Completion practices may be affected by drilling practices; however, make sure that suggested changes that raise costs are not actually excuses for a dry hole.

A total cost-control program among geologists, engineers, and operating personnel is possible if they recognize the importance of being on the same team. Competition among these groups results in face-saving excuses that frequently are not related to the actual problem.

■ **REFERENCES**

1. Lummus, J.L., and Field, L.J., "Nondispersed Mud: A New Drilling Concept," *Petroleum Engineer,* March 1968.
2. Eckel, John R., "How Mud and Hydraulics Affect Drilling Rate," *The Oil & Gas Journal,* 17 June 1968.

CHAPTER 3

HOLE PROBLEMS

Drilling a small hole into underground formations that may or may not be well consolidated introduces the possibility of some type of hole problem. The potential for hole problems may be shown by the geology used in well planning or by past experience. The well plan should be designed to attack potential hole problems. Drilling programs based on caution and containment generally result in expensive wells that often fail to reach the desired objective.

In one sense drilling operations are unique because an operator is almost certain to encounter some type of hole problem in any area. Many hole problems are unexplained. Drilling practices may be standard, yet the hole sloughs where previously no similar problems existed. One obvious answer is that underground formations are not homogeneous; new problems may occur at any time in carefully planned wells. Unfortunately, hole problems in one or two wells frequently dictate practices for an area where no such problems existed previously. As a result well times are extended, the hole problems continue, costs increase, and the good old days are forgotten.

Hole problems covered in this chapter include lost circulation, drilling torque, pipe sticking, and hole sloughing. Other hole problems are discussed in various chapters.

LOST CIRCULATION

Lost circulation is the most common drilling problem. The range of lost-circulation problems begins in the shallow, unconsolidated sands and extends to the well-consolidated formations that are fractured by the hydrostatic pressure imposed by the drilling fluid.

Shallow, Unconsolidated Formations

In shallow, unconsolidated surface formations the drilling fluid may flow freely into the formation because of its high permeability. Drilling may continue without circulation or the mud may be thickened to slow the rate of loss.

Drilling without circulation may be dangerous in shallow, unconsolidated surface formations; before following this procedure the operator should be familiar with the area. Fluid lost into such formations may cause underground cavities by washing, and the result could be a surface cavity, the loss of a rig, and possible injury to rig personnel.

The most common method of handling lost circulation in shallow surface formations is to thicken the mud. This may be done in freshwater muds by adding flocculating agents such as lime or cement. In areas where only water is used, the water may be thickened with polymers. Also, lost-circulation material may be utilized. Since strength is not required, the best material is the cheapest bulk material available, such as cotton-seed hulls or sawdust.

Fast drilling rates in soft-surface hole formations may result in high annular mud weights. Fig. 3–1 shows the increase in annular mud weight experienced while drilling a 12¼-in. hole with a 9.0-lb/gal fluid.

In Fig. 3–1 the mud weight can exceed 12.0 lb/gal in the annulus with very high penetration rates and a relatively low circulation rate of 300 gpm. Increasing the circulation rate to 600 gpm reduces the mud weight substantially, and the higher circulation rate has little effect on circulating pressure losses. If trouble continues, the rate of penetration may have to be limited to control mud weight. Problems such as this occur frequently in surface formations of the U.S. Gulf Coast.

Below Surface Casing in Normal-Pressure Formations

Naming all of the causes of lost circulation in normal-pressure formations below surface casing would be difficult. Naturally occuring fractures are common. These fractured formations may or may not have subnormal pore pressures. Also, many unfractured formations fracture easily. The operator may not know the precise cause of lost circulation or even the exact location where circulation is lost.

The zone of loss may be determined by running a temperature log. If the temperature log is not definitive on the first run, pump an additional 50 bbl of fluid and rerun the temperature log. More expensive procedures are generally unnecessary; however, radioactive tracers may be used if the temperature log is not definitive.

A solution to lost circulation in this part of the hole is to drill without fluid returns to the surface. This practice requires large volumes of water

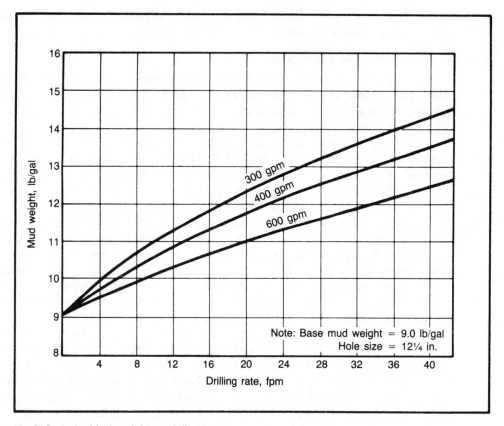

■ **FIG. 3–1** Mud weight vs drilling rate

and close supervision. In principle, the generated cuttings are removed from the bottom of the hole and deposited in the lost-circulation zone. The drilling operation must be watched closely for evidence of drillstring torque or drag. With close supervision the operator can detect potential hole problems and correct the operation before disaster strikes.

Specific problems in this part of the hole that may require special attention include seepage losses, a complete loss of circulation when cuttings must be obtained at the surface for formation evaluation purposes, and an area with a limited water supply.

Seepage losses are a common problem in hard-rock areas where naturally occurring formation fractures are present. For this type of problem, lost-circulation material often is carried in the mud until all the fractured zones have been drilled.

Frequently shale shakers are bypassed, and solids control becomes difficult. Bypassing the shale shaker often results in increased mud weight, and the operator is caught in a vicious cycle. The heavier mud weight raises the borehole pressure and increases the need for lost-circulation material. Probably one of the best ways to handle this problem is to keep the mud weight low, use lost-circulation material only when circulation is lost, and screen out the lost-circulation material when the fractured zones are penetrated.

The requirement for getting cuttings back to the surface or the lack of water may preclude drilling without fluid returns to the surface. Methods used to correct these problems include the following:

1. Medium and coarse lost-circulation material, batch-mixed to a concentration of 20–40 lb/bbl in 25–100 bbl of mud. The amount of mud utilized depends on the specific problem.
2. The use of solid plugs such as cement or diesel-oil bentonite. Usually a cement plug consists of a lightweight cement slurry spotted opposite the zone of loss in stages. A diesel-oil bentonite plug is difficult to use. All of the water in the tanks and the pipe must be removed; then 300–350 lb/bbl of bentonite is mixed into the diesel oil. This slurry is displaced and spotted in the zone of loss. The zone of loss must contain water to hydrate the bentonite, or water must be displaced down the annulus to hydrate the bentonite.
3. Time-setting plastics have also been used: a liquid plastic material is displaced into the formation, and the material forms a solid within the fracture.
4. A fourth method of getting circulation is to utilize an aerated fluid column or a viscous foam. This procedure has been implemented often in areas of Libya and in the Rocky Mountains of the U.S. The amount of air required to obtain a specific hydrostatic head is shown by Example 5.3 in chapter 5. Viscous foams can be used; however, some compromise must be made so the foams can be broken on their return to the surface. Aeration or foams substantially increase the costs of the operation and are generally last-resort methods.

Many lost-circulation problems have been cured simply by keeping the drilling-fluid weight close to the weight of water and by using very thin fluids. Over 50% of the initial lost-circulation problems in Libya were cured in this manner.

Lost Circulation While Using Weighted Muds

Lost circulation while using weighted muds generally occurs as a result of induced fractures caused by the hydrostatic pressure imposed by the drilling mud and associated activities such as fluid circulation and pipe movement. Lost circulation of this type may occur in normal-pressure formations above a pressure transition zone or below protective casing in the abnormal-pressure section of the hole. The pressure-transition section of the hole is that section at which the pore pressure begins to increase above normal to the point that protective casing is set to prevent fluid loss.

The loss of fluid in normal-pressure formations above the pressure-transition zone may be at any point from the bottom of the surface casing to the top of the pressure-transition zone. The danger in a case such as this is that, when circulation is lost, fluid begins to flow from a formation in the pressure transition zone. This danger places a premium on knowing the surface pressure that can be contained if the blowout preventers and surface choke lines are closed.

Procedures to be followed in closing in wells are discussed in chapter 14. As a precaution the operator should always test the cement job around the surface casing when abnormal pore pressures are expected. This may be done by drilling 15–20 ft below the surface casing shoe and applying pressure to the level of the total pressure expected at this point. If hole conditions at a later time make it appear the total pressure will exceed the test pressure, the pressure test should be repeated before increasing the mud weight above the previous test level.

A pressure test at the surface casing shoe does not indicate that a mud weight of the test magnitude can be used because a lower formation may fracture at a lower pressure. For this reason open-hole sections are sometimes pressure tested. Tests of open-hole sections should be performed with the drillstring in the casing to prevent sticking. Specific casing-seat testing procedures are given in chapter 14.

Along the coastal areas of Texas and Louisiana, the test pressure that can be contained below the surface casing may increase with open-hole exposure time. For example, finding the test pressure below surface casing to be equal to an equivalent mud weight of 14.0 lb/gal when testing immediately after the cement job is not unusual. Then, on a subsequent test after several days of drilling, the formation at the same point will hold a 15.5-lb/gal mud.

This increase in formation strength may be due to plugging the formation permeability with drill solids, which raises the pressure required to fracture the formation. Regardless of the cause, the phenomenon exists frequently. Tests several days after the cement job may give the operator more freedom in selecting the setting depth of protective casing.

In coastal area formations accidental loss of circulation in this part of the hole should not occur. In some areas, such as the deep Anadarko and Delaware basins where highly fractured low-pressure formations may be encountered in many zones below surface casing, the problem is more complicated, and testing programs are less definitive. However, pressure testing programs are still desirable in the Anadarko and Delaware basins.

If circulation is lost in the normal-pressure section of the hole in these areas, a batch treatment with lost-circulation material, or a solid plug, may be used. Both procedures were discussed in the previous section. However, when circulation is established, think about setting protective casing because the operator is faced with a potential underground blowout.

Consider next the use of weighted muds below the protective casing or protective liners. In the abnormal-pressure section of the hole, below protective casing take the following steps to avoid accidentally losing circulation:

1. Pressure test 15–20 ft below the casing shoe to the equivalent pressure level expected or to the leakoff pressure. Leakoff pressure is the pressure at which the formation begins to take fluid. A specific explanation of leakoff pressure is given in chapter 13.
2. If equipment limitations prevented reaching the leakoff pressure, retest after increasing the mud weight.
3. If the test pressure used equaled only the anticipated pressure requirements and subsequent drilling developments have raised these requirements, retest to leakoff. All of the testing should be performed with the drillstring inside the protective pipe.

If circulation is lost in the hole section below the protective pipe, two conditions may exist: (1) circulation is lost but there is no evidence of a well kick, and (2) circulation is lost and an underground blowout is taking place.

In the first case—circulation is lost and there is no evidence of a well kick—one of the first decisions that must be made is whether to pull the drillstring off the bottom of the hole into the protective pipe. This procedure is desirable because the drillstring may stick if left in the open hole. This is an undesirable procedure because, if the well kicks when trying to pull the pipe, controlling the well properly may be very difficult.

For these two reasons the question of pulling the pipe cannot be answered in a general discussion. This is a field decision, and drilling information, formation pressure data, formation types, control equipment, and crew excellence are some of the factors that must be considered. However, the operator should be aware of the potential problem and establish procedures that would help in making the decision if the problem occurs.

To cure the lost-circulation problem, the general procedure is to reduce the hydrostatic pressure either by reducing the mud weight or by thinning the mud. The use of lost-circulation material at this point without reducing the hydrostatic pressure may be more detrimental than beneficial. If circulation material were displaced into the zone of loss, it would act as a propping agent and most probably would prolong and perpetuate the problem.

In the past, the recommended practice was to wait a given number of hours; in recent years this procedure does not appear to be necessary. Some waiting time is built into the basic procedure for determining the problem. When circulation is lost, the *first* requirement is to determine if the well is kicking. *Second,* the operator must decide whether to pull the drillstring back into the protective pipe. *Third,* an attempt should be made to fill the annulus with water, measuring the quantity of water required to fill the annulus carefully. *Fourth,* determine from the mud and water levels the actual pressure the formation will support. *Fifth,* begin to displace the fluid slowly.

In this last case, lift the pipe as circulation is initiated to minimize any surge of additional pressure on the formation. The time required to fill the annulus with water, measure the quantity of water, and initiate circulation is all the waiting time necessary. Example 3.1 shows how to determine the mud weight that can be supported by the formation and also the mud weight that will control the subsurface pressure.

☐   **EXAMPLE 3.1:**

Well depth = 16,000 ft
Protective casing seat = 12,500 ft
Mud weight = 17.0 lb/gal
Drillpipe size = 4½ in.
Hole size, casing I.D. = 8.5 in.
Annulus volume = 0.05 bbl/ft
Water required to fill hole = 20 bbl

Determine:
The effective hydrostatic head and mud weight in lb/gal.

SOLUTION:

$$\frac{20}{0.05} = 400 \text{ ft of water}$$

Pressure imposed at total depth:

400 ft of water × 0.433 = 173 psi
15,600 ft of mud × 0.884 = 13,800 psi
Total pressure at 16,000 ft = 13,973 psi

$$\text{Effective mud weight} = \frac{13,973}{(16,000)(0.052)} = 16.8 \text{ lb/gal}$$

Pressure imposed at the casing seat:

400 ft of water × 0.433 = 173 psi
12,100 ft of mud × 0.884 = 10,700 psi
Total pressure at 12,500 ft = 10,873 psi

$$\text{Effective mud weight} = \frac{10,873}{(12,500)(0.052)} = 16.75 \text{ lb/gal}$$

There is a slight difference in the effective mud weight at the casing seat, which in this instance is in the vicinity of the assumed zone of loss, and at the bottom of the hole. Trouble may be experienced when trying to regain circulation. Determining all operating conditions such as the amount of additional hole required, the specific conditions that resulted in losing circulation, and the specific mud weight required to contain the high-pressure formations is hard to do. However, if difficulty is experienced in regaining circulation in the manner prescribed and particularly if gas or water cutting of the mud is noted during any fluid circulation, the operator should consider setting another protective casing string or liner.

The second condition in which the well kicked when circulation was lost is one of the most dangerous conditions in drilling. Surface control of the well kick has been lost. There is a general problem of losing the hole and of having a dangerous blowout.

The problem in a case such as this is to control first the lowermost problem in the hole. In hard-rock areas, the lost circulation problem may be above or below the kicking formation. In soft-rock areas the lost circulation is generally above the kicking formation.

In hard-rock areas in which the lost-circulation zone may be below the kicking formation, the usual solution is to use a cementacious mixture to cure the lost-circulation problem first. After curing the lost-circulation problem, maybe the mud weight can be increased enough to control the kicking formation. If the problem persists, then casing may have to be set.

The problem in hard-rock areas is complicated by alternating zones of high and low pore pressures. Pore pressures in hard-rock areas may alternate from high to low to high more than 15–20 times. It would be impossible to set enough casing strings to isolate each high and low pressure zone.

When maintaining circulation with a mud weighing enough to control completely a high pore pressure zone is difficult, drilling with a mud weight below that required for control is not uncommon. In this instance, the operator generally uses a rotating packoff device.

If the high pressure zone cannot be controlled within the pressure limitations of the surface packoff equipment or the lost-circulation zone, the blowout preventers and choke line may have to be implemented to bring the well under control. More freedom is available to the operator in hard-rock areas because the formations are stable and mud can be pumped at a rate that often exceeds the feed-in rate.

When the kick is below the lost-circulation zone, methods that may be used to control the kicking formation include:

1. Displacing a heavier mud in the open hole below the lost circulation zone and decreasing the mud weight above the lost circulation zone
2. Setting a barite plug in the open hole below the lost circulation zone
3. Setting a cement plug in the open hole below the lost circulation zone

Experience in an area helps determine what should be used. Again, the final decision should be made by a knowledgeable operator at the rig site.

Displacement of a mud heavier than that being used to drill the hole in the open-hole section below the zone of loss may be used when the productivity of the kicking formation is low. No effort is made to define "low." If there is substantial doubt, method 1 should probably not be attempted. Example 3.2 illustrates the basic procedure.

☐ **EXAMPLE 3.2**

Well depth = 18,000 ft
Last casing seat = 14,500 ft
Mud weight = 18.5 lb/gal when well kicks
Hole size = 6⅛ in.

Determine:
The feet of 22.0-lb/gal and 18.0-lb/gal mud required to equal a total column of 18.6-lb/gal mud.

SOLUTION:

Let X = Length of column of 22.0-lb/gal mud, ft
18,000 − X = Length of column of 18.0-lb/gal mud,
18,000 ft of 18.6-lb/gal mud = 18,000 × 0.967 = 17,406 psi
Thus:

$$1.144X + 0.936 (18,000 - X) = 17,406$$
$$1.144X + 16,848 - 0.936X = 19,406$$
$$0.208X = 558$$
$$X = 2,680 \text{ ft of 22.0-lb/gal mud}$$
$$18,000 - 2,680 = 15,320 \text{ ft of 18.0-lb/gal mud}$$
$$\text{Volume of 6.125-in. hole} = 0.0364 \text{ bbl/ft}$$
$$\text{Amount of 22.0-lb/gal mud} = (2,680)(0.0364) = 97.5 \text{ bbl}$$

This shows at least 97.5 bbl of 22.0-lb/gal mud would be required to fill 2,680 ft of 6⅛-in. hole. About 120 bbl of 22.0-lb/gal mud should be used because some hole enlargement could be expected. If the hole were exactly to gauge, 120 bbl of mud would fill 3,300 ft of hole, and this would still be below the casing seat at 14,500 ft.

After the determination of mud weight and volume requirements, the 120 bbl of 22.0-lb/gal mud should be displaced with 18.0-lb/gal mud until the height of the 22.0-lb/gal mud is about 100 ft higher inside the drillstring than outside. Then pull the drillstring back to the casing seat and continue to displace slowly with a thin 18.0-lb/gal mud. Do not attempt to pull the drillstring off of the bottom if the well is still kicking after displacing the 22.0-lb/gal mud. Proceed to set a barite or cement plug.

The second method for controlling the formation kick is setting a barite plug. Two procedures may be used to do this: (1) the barite plug may be displaced and the drillstring pulled out of the barite, or (2) the barite plug may be displaced and the drillstring left in place. Again, the procedure chosen depends on the formation productivity and experience in the area or in similar areas.

The procedure for mixing the barite plug is the same, regardless of the setting method used. Essentially, the procedure is as follows:

1. Raise the pH of the mix water to about 10.0.
2. Add 6 lb/gal of chrome lignosulfonate. Stir vigorously.
3. Add barite to a maximum weight of about 19.0 lb/gal.

Some barite at weights above 19.0 lb/gal may not settle readily when left quiescent, and it should be checked before adding a barite plug. In critical areas where the danger of lost circulation is associated with a potential well kick, if the barite does not settle readily, a special order of barite that will settle should be kept available. The potential problem is too great to be unprepared.

If the drillstring is to be pulled from the barite plug, the top of the barite inside the drillstring should be left about 100 ft above the barite top in the annulus. This helps ensure that the barite inside the pipe is cleared as the drillstring is pulled. If the barite plug is successful, the barite will settle into a solid plug and that portion of the hole is lost.

If the drillstring is to be left in the barite, the barite should be overdisplaced out of the drillstring by about 5 bbl. This procedure clears the bit and, if necessary permits a second barite plug or a cement plug.

The volume of the barite plug depends on the amount of hole available and the estimated bottom-hole pressure. If conditions permit, the slurry volume, before settling, should be enough to control the bottom-hole pressure.

The advantages of a barite plug over the cement plug are: (1) the plug weight is higher, and (2) more than one plug can be set if the first is unsuccessful. A cement plug is a last-resort tactic, which leaves the operator in a very poor position if it is unsuccessful because he can no longer circulate fluid. This is not meant to imply that a cement plug should never be used. If the formation fluid is gas, at least the first attempt to shut off the well kick should be with a barite plug. If the formation fluid is water and the flow is strong, the operator may use a cement plug. Specific recommendations on a general basis are not very usable; decisions of this type must be made at the rig site where all potential problems can be evaluated.

Assuming the well kick is controlled by setting a plug of some type, the next step depends on the method used to control the kick. If the placement of the 22.0-lb/gal mud were successful, a liner should be set if further operations in the well are considered desirable. If a solid plug were set and the kick controlled, the lost-circulation problem can be generally solved by reducing the mud weight. Further sidetracking operations are conducted in a normal manner. However, before reaching the high-pressure permeable sand that resulted in the problem initially, a liner must be set and cemented and the cememt job tested as discussed previously.

No mention is made of using lost-circulation material or setting a solid plug to cure the lost-circulation problem when it is below the protective pipe and the well kicks. There may be areas where using solid plugs solved the problem as defined. If so, there is no argument against the procedure. The ultimate objective is to cure the problem, and any solution proved successful should be considered. General suggestions in this discussion are based on industry experience but certainly cannot include every operating area.

As mentioned, specific rules are difficult to establish for controlling lost circulation. However, some procedures that have been tried repeatedly are acclaimed by some and discarded later.

First, should lost-circulation material be used on a precautionary basis?

Frequently fine lost circulation materials are used for seepage losses with low-weight thin muds. In some areas, coarse lost-circulation materials are utilized, and the shaker is bypassed. In general, the precautionary use of lost circulation is considered poor practice. Laboratory tests determined that heavy loads of lost-circulation material may result in a four- to six-fold increase in circulating pressures. Thus the precautionary use of lost-circulation material has, in some cases, a greater chance of causing the problem than eliminating it.

Using any lost-circulation material on a precautionary basis in weighted drilling muds is considered very poor practice, though it continues to flourish in some areas. From 1960 to 1970, using about 6 lb/bbl of a fine lost-circulation material on a precautionary basis in weighted muds was common in U.S. coastal areas. The practice has, in general, been discontinued in the coastal areas. However, using lost-circulation material on a precautionary basis in weighted muds is a normal practice in the Rocky Mountain areas of the U.S. Many claim improvement using these procedures. The practice is considered poor as it was many years ago in the coastal areas, and with time it will be discontinued.

The use of lost-circulation material in weighted muds before losing circulation is detrimental to the problem, not beneficial. Lost-circulation material is a solid; circulating pressure losses are increased as the solids content is raised, and any rise in circulating pressure increases the danger of lost circulation.

New lost circulation material continues to be introduced into the drilling business. New claims are made, and on specific occasions for any material can be successful. Laboratory tests of materials are unreliable, thus field testing becomes the only dependable procedure, and field tests are difficult to evaluate.

PIPE STICKING

Many reasons have been advanced for the drillstring sticking. In 1937, Hayward reported results from laboratory and field research on why drillpipe became stuck; included was a series of discussions by prominent industry personnel.[1] This paper was followed by another by Warren in 1940.[2]

Both of these papers concluded that pipe sticking was due to keyseating, an accumulation of cuttings around the pipe, or balling up of the bit. These papers also discussed the effect of sticking pipe in filter cake. The first paper on wall or differential pressure sticking was presented by Helmick and Longley in 1957.[3] This paper was followed by many others that also discussed the mechanics of wall sticking.[4,5,6,7,8,9,10]

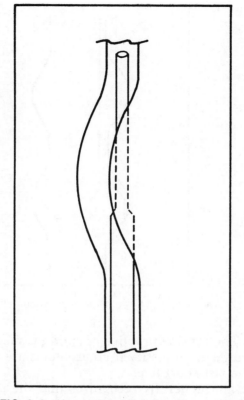

■ **FIG. 3–2** Keyseating drill collars in a crooked hole

Keyseating is still recognized as a primary reason for sticking pipe. The mechanics of keyseating involve wearing a groove in the wall of the hole with the drillstring and sticking larger pipe in the smaller groove when the drillstring is pulled.

Two conditions frequently contribute to potential keyseating problems. One is related to doglegs in the hole so that pipe is pulled into the wall with considerable lateral force while drilling. This condition is illustrated by Fig. 3–2. The second condition is when ledges that are close to gauge hole are left in areas of washed-out hole sections, as shown in Fig. 3–3.

Long-running bits also cause problems with keyseats. If a groove is being worn into the wall of the hole by the drillpipe, it will get deeper with time and increase the problem of pulling larger pipe through the smaller groove.

To offset long-running bits, making short trips with the bit is common practice. These short trips locate tight spots in the hole. The length of the short trip may be into the last casing string or simply through zones known to have caused past problems. The frequency of short trips depends on past experience and may also be dictated by angle and direction changes

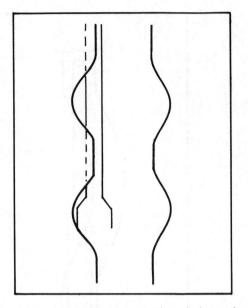

■ **FIG. 3–3**   Ledges in washed-out hole sections

in the hole. The routine practice of running short trips on a specific time schedule is difficult to justify if no problems are indicated during drilling or during the last short trip.

The operator generally recognizes that the pipe is stuck in a keyseat when the following sequence of events happens:

1. Some torque and drag are occurring.
2. The pipe sticks while in an upward motion, normally when pipe is being pulled.
3. After the pipe is stuck, there are no indicated restrictions on circulation.

Methods that may be used to free pipe in a keyseat include:

- Try to push the pipe back down the hole
- Jar on the drillstring
- Spot oil around the drillstring
- Locate the stuck point, back off, pull the drillstring, and wash over that portion of the drillstring stuck in the hole.

Drilling jars are usually run in the drillstring just above the drill collars in areas in which known keyseating problems occur. If these jars do not bump the drillstring out of the keyseat, then the free point is located and the pipe is backed off and pulled. Fishing jars are then run that exert substantially more jarring action on the drill collars than the regular drilling jars.

Hole cavings and cuttings that accumulate in cavities offer potential hazards to sticking pipe. The manner in which they might stick drillpipe is obvious. When pipe is stuck in this way, free fluid circulation is shut off or the pressure required for circulation is increased substantially.

Suggested methods for freeing pipe include:

1. If circulation is impossible, shut off the pump and release the pressure. Work the pipe slowly.
2. If circulation is possible but limited, circulate clear water to help remove the cuttings from around the pipe.
3. If step 2 is unsuccessful, spot oil around the pipe to reduce friction.

Drilling Torque

Drilling torque is common in all drilling operations. There are no straight holes; thus the drillstring lies against the hole wall while rotating. The problem of torque is often magnified by two extremes, one in which the operator attempts to drill a straight hole (less than 6° from vertical) and the high-angle directional well. The concept of keeping the hole very straight creates doglegs. High-angle directional wells increase the total drillstring weight against the wall of the hole.

When excessive torque occurs, mud treatment is emphasized. There are generally attempts to improve the filter cake lubricity by raising the mud's bentonite content or by adding materials such as walnut hulls or glass beads. The bentonite content can be checked by the methylene blue test and/or by a high-temperature, high-pressure filter test. Both tests are discussed in chapter 5. Adding walnut hulls or glass beads is simply a means of building a ball-bearing surface on the filter cake. When the walnut hulls or glass beads become completely imbedded in the filter cake, more must be added or else the torque will increase.

If the normal diagnosis of the mud shows a substantial bentonite content and adding walnut hulls or glass beads has no beneficial effects, then the operator looks for other causes. Doglegs, keyseats, and poor hole cleaning may also result in torque increases. The effects of doglegs and keyseats have been discussed. Poor hole cleaning may result in balled-up stabilizers and bits.

Sometimes the torque problem may be caused by the drilling assembly. Stabilizers may be digging into the wall or be unsuitable for specific drilling conditions. Also, long strings of drill collars are detrimental. When high torque is a problem, use only one stand of spiral drill collars and heavyweight drillpipe to put weight on the bit. Even if the operator is generally reluctant to use heavyweight pipe on a continuous basis to put weight on the bit, he should change the assembly as suggested to determine its effect on the torque problem.

Differential Pressure Sticking

The term *differential pressure sticking* was introduced in 1956 to identify a portion of drillstring sticking in the filter cake. Actually, operating personnel on drilling rigs had identified this problem as wall sticking many years before introducing the term differential pressure sticking. In 1937, Hayward recognized that pipe could become stuck in wall cake. He suggested that one method of solving the problem was to stop circulating in order to reduce hydrostatic pressure.[11] In 1938 and 1939, there are recorded cases in West Texas of backing off pipe above the stuck point, going into the hole with a packer, setting the packer, and releasing the hydrostatic pressure below the packer to free the pipe. Specific remarks on the service records indicate that the procedure was responsible for the pipe being blown free.

The primary advantage of calling wall sticking "differential pressure sticking" was that the latter term received more attention, and engineers began to emphasize the problem. The sticking force is described mathematically by Eq. 3.1:

$$F_s = \Delta P_s A_{pf} f_{pf} \tag{3.1}$$

where:

F_s = sticking force or total pulling force required to free stuck pipe

ΔP_s = hydrostatic pressure of mud minus pore pressure of formation

A_{pf} = area of contact between pipe and filter cake

f_{pf} = coefficient of friction between pipe and filter cake

The differential pressure is imposed by the magnitude of the hydrostatic pressure because formation pore pressures are at fixed levels. Thus, one method to minimize this effect is to drill with minimum mud weights. The problem of minimizing the differential pressure is often complicated by long sections of open hole where the formation pore pressures are substantially different. For this reason a given mud weight may be necessary to control the pore pressure in one open formation, which imposes a large pressure differential across another open formation.

Differential pressure sticking is recognized as a potential problem when the differential pressure reaches a given level in a specific area. For example, in one offshore area, the operator becomes concerned enough to initiate precautionary procedures when the differential pressure across open permeable zones reaches 1,200 psi.

The area of contact represents the total area of the pipe covered by filter cake across which the pressure differential is effective. Thus, the area of contact is affected by the following:

- Length of the permeable zone where the pipe contacts the filter cake
- Hole size and pipe size
- Pipe shape, whether externally flush or pipe with raised shoulders
- Thickness of the filter cake
- External stabilization of the pipe

The length of the permeable zone is a fixed parameter, so this factor cannot be changed by the operator. Even the hole and pipe sizes sometimes cannot be altered, or at least the reason the size combinations are used becomes more important than any special effort to minimize differential pressure sticking.

The pipe shape is a very important parameter and one that can be changed easily. Large, externally flush drill collars represent the ideal equipment to cause differential pressure sticking. In recent years special drill-collar configurations have been used. Some of these are (1) spiral collars with circulation grooves in the external surface of the drill collars, (2) square drill collars, (3) shouldered drill collars, and (4) heavyweight drillpipe.

The effect of special drill collars is to reduce the area of wall contact. The manner in which this is accomplished can be recognized easily. More discussion on special collars is given in chapter 9. The heavyweight drill-pipe has reduced the differential pressure sticking problem considerably, particularly in directional wells. This pipe, some of which is made by turning down the OD of regular drill collars, has upsets in the middle and on both ends of each 30-ft joint. This configuration reduces the area of contact. Other advantages to heavyweight pipe are that the reduction in stiffness lessens some of the dangers of keyseating and substantially reduces twist-offs at the connection point between the drill collars and the drillpipe.

The potential effect of the filter cake on contact area is shown in Fig. 3–4. The area of contact may be more than doubled by thickening filter cake. This has been the primary reason for controlling the high-temperature, high-pressure filtration rate. Actually, the rate of filtrate loss may be of minor concern to the operator; his primary interest is the filter-cake thickness.

Filter cakes during normal drilling generally reach an equilibrium thickness. This simply means that the rate of erosion by the circulating fluid equals the rate of deposition of new solids in the filter cake. This concept of cake erosion shows that short trips are unnecessary during long bit runs and are not required to remove filter cake before wireline logging.

When diamond bits were introduced in deep drilling in the U.S. coastal areas, the bit was usually pulled back into the last casing string every 24 hr, commonly called a *short trip*. The purposes of this short trip were to remove excessive filter cake and to ensure an open hole. Pipe-sticking problems were already frequent in the coastal areas. If the pipe stuck while using a diamond bit, the first question asked was, "How long was the bit in the hole?" This pressured operators who did not believe in short trips. However, studies of the problem showed that operators making short trips had more pipe-sticking problems that those who did not make the short trips.

Currently most operators do not make the short trips to open the hole or to remove filter cake. Watch carefully for potential keyseating problems.

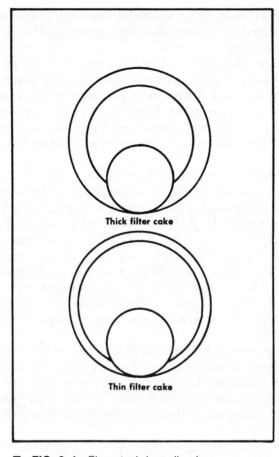

Thick filter cake

Thin filter cake

■ **FIG. 3–4** Pipe stuck in wall cake

If there is a severe dogleg in the hole, the drillpipe may create a groove in the wall that deepens with time. A deep groove may introduce such an extreme keyseating problem that drill collars and the bit cannot be removed from the hole. Problems such as this can be recognized because increased pulling weight occurs while the pipe is in motion. When such keyseating problems exist, it may be necessary to make short trips to ream the hole.

Short trips before logging are still common but are largely unnecessary. A complete study of this problem on a total of 50 wells, where short trips were made on about half of the wells, revealed the success in running logs to the bottom of the hole could not be related to whether the short trip was made. Apparently the practice continues because most operators remember only the cases in which logs did not go to the bottom of the

■ **TABLE 3–1** Friction factors between pipe and filter cake

Type of Mud	Additives	Friction Factor
Oil based	—	0.04
Unweighted water based	Good bentonite content	0.14
Weighted water based	Poor bentonite content	0.35

hole when the short trip was not made. Investigate all practices that require extra time because offshore wells, where costs exceed $50/min, place a real premium on conserving time.

External stabilization of drill collars is often employed to minimize crooked hole problems; however, this stabilization also minimizes the area of wall contact and reduces the tendency to stick the pipe.

The friction factor in Eq. 3.1 is never quantitatively defined in field operations. There is no accepted method for measuring the friction factor. In 1962, Annis reported some friction factor measurements, and for a short time a small tester of the type used by Annis was tried in operations.[12] Later the tester disappeared. In 1970, Mondshine published results concerning drilling-mud lubricity. Mondshine's work was related to lubricity and not to friction factors for differential pressure sticking.[13]

There are no accepted standards for friction factors. Some rule-of-thumb numbers are given in Table 3–1.

The numbers in Table 3–1 are not meant to be accurate quantitatively. They are relative numbers that show extremes and the general effect of the type of mud shown. Additives that are said to reduce friction factors are listed below:

- Oil emulsified in water
- Graphite (generally needs to be used with some oil)
- Detergents or soaps
- Special lubricants, generally a type of fatty acid

Oil emusified in water-based muds usually reduces the friction factor because oil droplets are in the filter cake. Graphite is a good lubricant but generally is effective only in the presence of some oil. Detergents or soaps feel slick to the touch. There are no field results that show these are effective in reducing pipe sticking. Actually, the detergents are mainly used to minimize bit balling. Special lubricants or the fatty acids reduce friction factors. These special lubricants are expensive, so it is difficult to evaluate field results economically.

The operator should watch for problems that may lead to differential pressure sticking. Example 3.3 illustrates a typical field problem.

□   **EXAMPLE 3.3**

Problem sand at 12,000 ft
Pore pressure of sand = 8,730 psi
Hydrostatic pressure of 14.8-lb/gal mud = 9,250 psi ·
Force required to pull collars after stopping opposite the sand = 60,000 lb
Maximum pull permitted on the drillstring = 100,000 lb

Determine:
The maximum permissible mud weight if no other changes are made.

SOLUTION:

$$F_s = \Delta P_s A_{pf} f_{pf}$$
$$60,000 = 520\ A_{pf} f_{pf}$$
$$A_{pf} f_{pf} = \frac{60,000}{520}$$
$$\Delta P_s = \frac{(100,000)(520)}{60,000} = 867\ \text{psi}$$

Permissible hydrostatic pressure = 8,730 + 867 = 9,597 psi

$$\text{Permissible mud weight} = \frac{9,567}{(12,000)(0.052)} = 15.4\ \text{lb/gal}$$

Also, the use of barite to increase mud weight will increase the friction factor. Thus, to even reach a mud weight of 15.4 lb/gal, the operator would have to change the drillstring, thin the filter cake, or add a lubricant.

Procedures for freeing pipe stuck in the filter cake include the following:

- Reducing the hydrostatic pressure
- Spotting oil around the stuck portion of the drillstring
- Washing over the stuck pipe

The hydrostatic pressure may be reduced by the methods listed:

- Reduce the mud weight
- Displace nitrogen into the hole
- Implement U-tube procedure
- Let a test packer in the hole above the stuck point

Reducing Hydrostatic Pressure

Reducing mud weight is an obvious solution. Using a mud weight that exceeds the requirements is frequently an alternative for the operator. Nitrogen is commonly pumped in hard-rock areas. The general procedure is to pump liquid nitrogen down the drillstring. The nitrogen goes to a gaseous state in the annulus and blows some of the mud out of the annulus. The volume of liquid nitrogen used is arbitrary. A controlling factor is, of course, the amount of hydrostatic pressure reduction the operator considers safe. Sometimes more than one batch of nitrogen is used. In one specific case, an operator used two batches of nitrogen, and the pipe remained stuck. He was considering that perhaps the pipe was stuck for some reason other than differential pressure sticking. A review of the wells in the area showed a sand section with thick wall cakes where the drillstring was stuck.

The conclusion was that the sticking problem had to be caused by differential pressure. The volume of nitrogen pumped on the third run was four times the amount used on each of the other two runs. The pipe came free. The operator risked a well kick to free his stuck pipe. The results were good, and, at least in this case, the right decision was made.

The U-tube method of reducing hydrostatic pressure may utilize unweighted or weighted muds. The general procedure is to pump a lowerweight fluid into the drillstring or annulus to lower the hydrostatic pressure in the annulus. Examples 3.4, 3.5, 3.6, and 3.7 show the U-tube procedure for unweighted and weighted muds.

☐   **EXAMPLE 3.4**

Assume: Pipe is differentially stuck at 10,000 ft
Mud weight = 9.6 lb/gal
Hole size = 8½ in.
9⅝-in. surface casing is set at 2,500 ft, 8.75-in. ID
Drillstring = 4½-in. OD, 3.82-in. ID
Drill collars = 600 ft of 6¼-in. OD, 2¾-in. ID

Annular capacity:
Drillpipe − hole annulus = 0.050 bbl/ft
Drill collar − hole annulus = 0.032 bbl/ft
Drillpipe − casing annulus = 0.055 bbl/ft

Internal capacity of drillstring:
Drillpipe = 0.014 bbl/ft
Drill collars = 0.0075 bbl/ft

Pressure data:
Estimated pore pressure = 4,330 psi
Hydrostatic pressure of 9.6 lb/gal mud = 5,000 psi
Desired reduction in hydrostatic pressure = 400 psi

Procedure:
First pump fresh water into drillstring. Let the mud U-tube back through the bit into the drillstring.

Determine:
The quantity of water required.

SOLUTION:
Pressure gradient of 9.6 lb/gal mud = 0.50 psi/ft
Required drop in mud level to reduce the hydrostatic pressure 400 psi

$$= \frac{400}{0.50} = 800 \text{ ft}$$

Determine the mud and water inside the drillstring to equal 4,600 psi.
Let X = water
10,000 − X = mud

Then:

$$0.433X + 0.5\,(10{,}000 - X) = 4{,}600$$
$$-0.067X = 400$$
$$X = 5{,}970 \text{ ft of water}$$
$$10{,}000 - 5{,}970 = 4{,}030 \text{ ft of mud}$$

Also, enough additional water must be used to drop the mud level in the annulus 800 ft.

$$\text{Additional water required} = 800\!\left(\frac{0.055}{0.014}\right) = 3{,}143 \text{ ft}$$

Total feet of water required = 5,970 + 3,143 = 9,113 ft
Barrels of water = (9,113)(0.014) = 128 bbl

☐   **EXAMPLE 3.5**

Assume: Same data as Example 3.4.

Procedure:
Pump water into the annulus. Let the water U-tube back until the mud level inside the drillstring is lowered 800 ft.

SOLUTION:
When the U-tubing action stops, the following conditions will exist:
5,970 ft of water
4,030 ft of mud

$$\text{Additional water required} = (800)\!\left(\frac{0.014}{0.055}\right) = 204 \text{ ft}$$

Total water required = 5,970 + 204 = 6,174 ft
Barrels of water required = (2,500)(0.055) + (3,674)(0.05) = 321 bbl

☐   **EXAMPLE 3.6**

Assume: Pipe is differentially stuck at 15,000 ft
Mud weight = 17.0 lb/gal
Hole size = 8½ in.
9⅝-in. protective casing set at 11,500 ft
Average ID of 9⅝-in. casing = 8.75 in.
Drillstring = 4½-in. OD
        3.82-in. ID
Drill collars = 7-in. OD, 600 ft
        2¾-in. ID
Annular capacity:
Drillpipe − hole annulus = 0.05 bbl/ft
Drill collar − hole annulus = 0.022 bbl/ft
Internal capacity of drillstring:
    Drillpipe = 0.014 bbl/ft
    Drill collars = 0.0075 bbl/ft
Pressure data:
    Estimated pore pressure = 12.645 psi
    Hydrostatic pressure of 17 lb/gal mud = 13,245 psi
    Desired reduction in hydrostatic pressure = 400 psi

Procedure:

Pump fresh water into the drillstring. Let the mud U-tube back through the bit into the drillstring until the mud pressure is reduced by 400 psi.

Determine:

The quantity of water required.

SOLUTION:

Pressure gradient of 17 lb/gal = 0.883 psi/ft

Required drop in mud level required to reduce the hydrostatic pressure

$$400 \text{ psi} = \frac{400 \text{ psi ft}}{0.883 \text{ psi}} = 453 \text{ ft}$$

Water and mud inside the drillstring equals a total hydrostatic pressure of 12,845 psi

Let X = Water with a pressure gradient of 0.433 $\frac{\text{psi}}{\text{ft}}$

15,000 − X = Mud

0.433X + (15,000 − X)0.883 − 12,845

0.433X + 13,245 − 0.883X = 12,845

−0.45X = −400

X = 889 ft

Thus, the final condition inside the drillstring requires 889 ft of water and 14,111 ft of mud. Enough additional water must be displaced into the drillstring to drop the mud height in the annulus by 453 ft.

$$\text{Additional water required} = 453 \left(\frac{0.05}{0.014} \right) = 1,618 \text{ ft}$$

$$\text{Total water required} = 889 + 1,618 = 2,507 \text{ ft}$$

$$\text{Barrels of water required} = \frac{(2,507 \text{ ft}) (0.014 \text{ bbl})}{\text{ft}} = 35 \text{ bbl}$$

☐   **EXAMPLE 3.7**

Assume same data as Example 3.6.

Problem: Reduce the hydrostatic pressure 400 psi by pumping water into the annulus.

SOLUTION:

Based on Example 3.6, a 400-psi drop in hydrostatic pressure requires 889 ft of water and 14,111 ft of mud in the annulus as a final condition. To accomplish this requires that the mud level inside the drillstring be at 453 ft. Thus, enough additional water has to be displaced into the annulus to drop the mud level inside the drillstring 453 ft.

$$\text{Additional water required} = (453) \left(\frac{0.014}{0.05} \right) = 127 \text{ ft in the annulus}$$

$$\text{Total feet of water in annulus} = 889 + 127 = 1,016 \text{ ft}$$

$$\text{Barrels of water required} = (1,016 \text{ ft}) \left(0.005 \frac{\text{bbl}}{\text{ft}} \right) = 51 \text{ bbl}$$

Obviously, if a back-pressure valve is being used, the U-tube procedure cannot be implemented. Also, there is always danger that the bit will be plugged while U-tubing. There are advantages and disadvantages to pumping into the drillstring or into the annulus. One advantage of pumping into the drillstring is that water can simply be U-tubed back onto the rig floor. This is a simple process that requires no additional equipment. Probably with an unweighted mud the best procedure is to pump into the drillstring. With a weighted mud, more controls are necessary. If water is pumped into the drillstring, the water inside the drill would have to be reversed out of the hole after freeing the drillstring. Also, with weighted muds it is always desirable to see a fluid level in the annulus in the event the well starts to flow.

The procedure of using a packer to release hydrostatic pressure is as follows:

1. Locate the stuck point.
2. Back off the pipe above the stuck point.
3. Pull the pipe free and rerun with drillstem-test type of packer on the bottom of the hole.
4. Screw back into the stuck portion of the drillstring.
5. Set the packer and release the pressure below the packer into the drillstring.

The amount of pressure reduction below the packer is controlled by the fluid weight inside the drillstring. This method requires a stable formation or casing where a packer can be set. This procedure is more dangerous than U-tubing, and it has no advantages.

Spotting Oil Around the Stuck Portion of the Drillstring

Using oil to free pipe that is stuck differentially is a common technique. Oil cannot penetrate the area between the drillstring and the filter cake readily. Thus, after oil is spotted, some waiting time is required. The actual process involves destroying enough of the filter cake to get oil between the remaining filter cake and the pipe. Generally, just plain crude oil in unweighted muds is inadequate. There needs to be an additive to help destroy the filter cake. With weighted spotting fluids, the oil must also contain an additive to destroy the filter cake. There are no standard procedures for spotting oil and waiting until the pipe comes loose. Suggested procedures follow:

1. Mix into the oil an ingredient that will help destroy the filter cake.
2. Displace the oil so there is 200 ft of oil above the stuck point.
3. Every 3 hrs displace the oil around the stuck pipe with a volume of new oil.
4. The oil volume required depends on waiting time. If you are going to wait only 24 hrs, then the volume of oil needed is 8 times the amount required to cover the stuck portion of the drillstring.
   Note: There are no magic numbers. Experience may dictate the necessary oil volumes and waiting time for a specific area.

Washing Over the Stuck Pipe

Fishing operations are covered in chapter 15.

SLOUGHING SHALE

Shale sloughs into the hole for two reasons:

1. Shale has a low matrix strength and compacts under the weight of the overlying rock.
2. Shale has a strong affinity for water, which further weakens the matrix and results in hydration or swelling of the shale.

Fig. 3–5 shows one theory of why shale is under stress and the reason shale sloughs into the hole.

Shale compacts because of its low matrix strength. Under normal-pressure conditions, the porosity in shale bed 3 is lower than the porosity in shale bed 2. This is because there is more weight on shale bed 3. When a well penetrates these shale sections, the pressure exerted by the mud column is less than the pressure exerted by the overburden rock. Thus, there is a natural tendency for shale to be pushed into the borehole. The normal pore pressure in the sand is due to a column of water extending to the surface. If for some reason communication with the surface is cut off for the sands during deposition, fluid in the pore spaces helps support the overburden, resulting in abnormally high pore pressures.

Shales naturally absorb large quantities of water. During the initial stages of deposition, shale is in a saturated water environment. As the

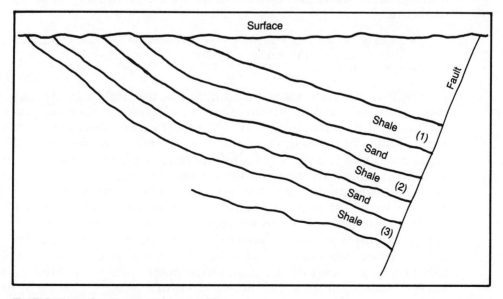

■ **FIG. 3–5** Sand and shale deposition

shale is buried, the weight of the overburden compacts the shale, squeezing out some of the water. When the borehole is opened and if a water-based mud is used, the shale absorbs water, which further weakens its matrix.

Young shales generally contain higher quantities of hydratable material than older shales.

Consequently, the geologically younger shales are more likely to contain highly hydratable material than the older shales.

Specific reasons for shale sloughing are given below:

- Hydration or swelling of clay materials in the shales
- Pressured shale sections
- Fractured shale sections in areas where bed dips are high

Hydratable shales are found in coastal areas where shale sections are geologically young. Extremely hydratable shale is called *gumbo*. Some of the gumbo sections in coastal areas may contain as much as 60% sodium montmorillonite clay (see chapter 5). Completely wetting the clay during drilling operations is difficult. Sometimes flow lines become plugged and actually lift the rotary drive bushing out of the rotary table.

There is no set solution for gumbo sections. Generally, if the gumbo section is not abnormally pressured, the drilling fluid should be clear water. When the water comes into contact with the gumbo, mud is formed in the same way mud is formed when bentonite is added to water to make surface mud. The only difference in drilling gumbo with clear water is that the mud is mixed at the bottom of the hole rather than at the surface.

If sloughing is associated with hydratable shales, the problem can be minimized by using inhibited drilling fluids. Commonly used inhibited muds are

- Calcium-based muds such as lime, gypsum, or calcium chloride
- Potassium chloride muds
- Oil-based muds

The first inhibited water-based mud was the high-pH lime mud. With only about 200 parts per million (ppm) calcium in solution, there was not much inhibition against clay swelling. However, the high-pH environment resulted in a very fluid mud with a high toleration for solids. Gypsum muds are run at a lower pH than lime muds with more soluble calcium. So the gyp muds are more inhibited against clay swelling but are less fluid than lime muds. The most inhibited of the calcium muds are those that are mixed using calcium chloride and run with a pH below 10.

Potassium chloride muds are used frequently to minimize clay swelling. Potassium is no more effective than calcium in minimizing clay swelling, but potassium muds are often more stable than calcium systems.

Clays do not hydrate in an oil environment. However, most oil-based muds contain some water. If the water salinity is not increased by adding salts, shales are frequently wet by osmotic pressure built up by a difference between the vapor pressure of the water in the formation and the vapor pressure of the water in the oil-based mud. The basic relationship is shown in Eq. 3.2:

$$P_{hs} = \frac{R_g\,T_{ab}}{\overline{V}}\,\ln\frac{P_v}{P_o}$$

where:

P_{hs} = hydrational stress, psi
R_g = natural gas constant (to convert units)
T_{ab} = absolute temperature
$\overline{V}$ = partial molar volume of pure water
P_v = vapor pressure of shale and water of hydration, psi
P_o = vapor pressure of water in oil-based mud, psi

If the vapor pressure, P_o, equals the vapor pressure, P_v, water is not moving into or out of the shale. Generally calcium chloride is added to the water phase of the oil-based mud to make certain the salinity of the water in the oil-based mud is high enough to prevent water from entering the shale from the mud. In practice, adding salt to the water phase of the oil-based mud has usually prevented the wetting of formation shales.

Pressured shales frequently slough into the hole unless the pressure exerted by the mud column is at least equal to the pore pressure of the shales. There are some exceptions. In geologically older shales, the matrix of the shale may be strong enough to allow the use of mud weight that exerts less pressure than the pore pressure. Thus, when using oil-based mud in which the water phase has been saturated with calcium chloride, some abnormally pressured shale sections often can be drilled under-balanced.

Highly fractured shale sections in areas with steeply dipping beds are a real problem. Control methods include the following:

- Filtration control in water-based muds
- Blown asphalt and gilsonite
- Keep the flow pattern of the mud laminar
- Oil-based muds

Filtration control may be of some value because the fractured shales contain micro-fractures, allowing filtrate but not whole mud to enter. Chapter 5 shows that there is no reason to control the API filtration rate to levels below 6 cc. Filtration control does not stop any shale from sloughing. By controlling the filtration rate, the degree of sloughing possibly is reduced.

Blown asphalt and gilsonite help plug fractures in fractured shales, and there is some indication that shale sloughing is often lessened when these materials are used. Laboratory tests show these materials imbedded in the fractures of cores from shale sections. By plugging the fractures, fluids cannot move into or out of the shale.

Fractured shales are subject to erosion caused by mud flow patterns, surge and swab pressures, and rotation of the drillstring. Turbulent flow patterns erode the hole and initiate sloughing problems. Surge and swab pressures cause unstable shale sections by moving whole mud into and out of fractures. A high rotary speed tends to knock shale off the wall and cause early sloughing problems. In past years, when bridges that prevented running wireline logs occurred in the hole, pipe was sometimes chained out of the hole to minimize hole erosion.

Sloughing shale is a drilling problem because of pipe sticking, bridges of shale in the hole, and an accumulation of shale on the bottom of the hole (called fill). The solution to the drilling problem is to keep the hole clean. Except for pressured shale, the primary secret to cleaning shale out of the hole is to thicken the mud to increase the mud's lifting capacity. Because of other problems, using viscous sweeps to clean the hole generally is desirable. The volume of the viscous sweep should cover 250–300 ft of the annulus. The number of viscous sweeps should be great enough to keep the hole clean. Lifting capacity is increased as mud is thickened, so the thickness of the viscous sweep depends on the cleaning requirements.

A standard procedure for controlling shale sloughing, such as the control of API water loss, never has been totally successful. As noted, filtration control may reduce the sloughing rate in fractured shales. However, caliper logs show that filtration control does not prevent shale sloughing. Certainly in swelling-type shales and pressured shales the control of filtration rate has no beneficial effects.

As noted, oil-based muds with calcium chloride in the water phase effectively reduce sloughing in fractured shales. If oil-based muds are used to drill fractured shale sections, these should be used before entering the shale, not after some of the shale has been drilled. Building viscosity in oil-based muds is more difficult than in water-based muds. Consequently, hole cleaning in enlarged sections of the hole may be harder with oil-based muds.

One method to minimize shale problems is to reduce time in the open hole. Shale is an unstable rock, and exposure time contributes to shale problems.

REFERENCES

1. J.T. Hayward, "Cause and Cure of Frozen Drill Pipe and Casing," *Drilling and Production Practices,* (1937), p. 8.
2. J.E. Warren, "Causes, Preventions, and Recovery of Stuck Drill Pipe," *Drilling and Production Practices* (1940), p. 30.
3. W.E. Helmick and A.J. Longley, "Pressure Differential Sticking of Drill Pipe," *Oil & Gas Journal* (June 17, 1957), p. 132.
4. H.D. Outmans, "Mechanics of Differential Pressure Sticking of Drill Collars," *Petroleum Transactions,* 213 (1958):265.

5. N.K. Tschirley and K.D. Tanner, "Wetting Agent Reduces Pipe Sticking," *Oil & Gas Journal,* (November 17, 1958), p. 165.
6. E.V. Tavan Jr., "Casing, Pipe Freed During Fishing Jobs with Oil-Base Mud," *Drilling,* February, 1959.
7. B.J. Sartain, "Drill-Stem Tester Frees Stuck Pipe," *Petroleum Engineer,* October 1960.
8. F.K. Fox, "New Pipe Configuration Reduces Wall Sticking, *World Oil,* December 1960.
9. Ed McGhee, "Gulf Coast Drillers Whip the Wall-Sticking Problem," *Oil and Gas Journal,* February, 1961.
10. E.L. Haden and G.R. Welch, "Techniques for Preventing Sticking of Drill Pipe, Drilling and Production Practice (1961).
11. Hayward, 1937.
12. M.R. Annis and P.H. Monoghan, "Differential Pressure Sticking—Laboratory Studies of Friction Between Steel and Mud Filter Cake," *Petroleum Transactions* (May 1962).
13. T.C. Mondshine, "Drilling-Mud Lubricity," *Oil & Gas Journal* (December 7, 1970).
14. T.C. Mondshine and J.D. Kercheville, "Successful Gumbo Shale Drilling," *Oil & Gas Journal* (March 28, 1966).

CHAPTER 4

WELL PLANNING

LARRY P. MOORE
MS&A Engineering Inc.

An essential first step in the promotion of good drilling practices is the planning of the well. A systematic approach to outlining procedures and practices must be followed in every case. What may seem like a relatively insignificant detail in the office can grow to enormous proportions in the field if confusion occurs during a critical moment. A comprehensive well plan is the first line of defense in controlling costs. Leaving anything to chance can result in needless additional expenditures.

The development of a sound, step-by-step procedure for writing a well plan is discussed in detail. While it is impossible to include every criterion applicable to each of the drilling environments encountered worldwide, what follows provides a minimum guideline for the drilling engineer. Data acquisition is the first step in well planning.

DATA ACQUISITION

The engineer called upon to derive a plan for drilling a prospect is typically given only minimal data. The exploration department usually supplies a location and the depth of the objective formation. Before one can proceed, he must seek additional information. This information comes primarily from offset wells. Table 4–1 gives the data that should be gathered. The items listed form the main pool of information necessary to prepare for drilling the proposed well.

Bit records can be obtained from bit companies and are a particularly useful data source. In addition to the bit type and the depths the bit was run in and pulled out, the record generally includes bit weight, rotary

■ **TABLE 4–1** Information from offset wells

Bit Records
Casing Depths
Electric and Mud Logs
Drillstem Tests
Mud Recaps
Drilling Times
Hole Problems

speed, penetration rate, deviation angles, nozzle sizes, pump pressure, and strokes per minute. Fig. 4–1 demonstrates a typical bit record.

Casing depths can be obtained from scout tickets. This information is a good starting point for designing the well's casing program.

Mud and electric logs are excellent information sources on the types of lithology to expect. Correlating bit records with lithology data from logs for bit selection optimization is a good idea. Fig. 4–2 shows bit data plotted on an electric log. Electric logs may also be used for pore pressure prediction and for the location of pressure transitional zones.

Drillstem test results indicate the presence of productive zones and give reliable bottom-hole pressure measurements.

Mud recaps can usually be obtained from the mud companies. These show mud properties at depth intervals and generally describe any hole problems. An example of a mud recap appears in Fig. 4–3.

Drilling time of offset wells is a key to predicting the drilling time of the subject well. This prediction is essential for an accurate cost estimate. Drilling times can be secured from scout tickets (see Fig. 4–4) or bit records.

A thorough study of all *hole problems* encountered on offset wells is very important to effectively plan any new well. This study should include a direct comparison of how these problems were dealt with in different situations. Thus, an understanding of what worked and why it worked can be gained. In this way problems can be anticipated and, if not entirely prevented, efficiently handled. Planning should include additional cost allowances in problem areas.

During the initial planning stages, question those who have control over the project. For instance, get a realistic estimate for the well's spud date. Costs often bear a direct relationship to weather. Extremely cold or wet seasons clearly should be avoided if possible. Inspect the proposed location to see if a relatively minor adjustment can have a great effect on building, maintaining, or restoring the drillsite.

Foreknowledge of the bottom-hole target area is also a necessary ingredient to planning, even if the proposed well is to have a "straight" wellbore. Close proximity to a lease line, fault, or water contact might impose severe target restraints that could slow drilling. In the same vein, if the well is

COUNTY	FIELD	STATE	SECTION	TOWNSHIP	RANGE	LOCATION	WELL NO.
LEA	W/C	N. MEX.	13	11-S	34-E	BOGLE FARMS	1

CONTRACTOR	RIG NO.	OPERATOR	TOOLPUSHER	SALESMAN
MORAN OIL PROD. & DRLG CORP.	3	EARL T. SMITH & ASSOCIATES	MILO PENFIELD	TOMMY EVERHART

SPUD	UNDER SURF.	UNDER INTER.	SET SAND ST.	REACHED T.D.	PUMP NO. 1	LINER	PUMP NO. 2	LINER	PUMP POWER	TYPE MUD
2-11-69	2-12-69	2-17-69			GD GXP	5½	NAT C-250		COMPOUND	

DRILL PIPE	TOOL JOINTS SIZE	TYPE	O.D.	DRILL COLLARS NUMBER	O.D.	I.D.	LENGTH	DRAWWORKS POWER
4½	4	H-90		28	8"		861	MID CONT. U-712-A 2-LRZ

CUMULATIVE $/FT.	NO.	SIZE	MAKE	TYPE	JET 32nd IN	SERIAL	DEPTH OUT	FEET	HOURS	FT/HR	ACCUM DRLG. HRS.	WT. 1000 LBS	RPM	VERT DEV.	PUMP PRESS	COST $/FT.	SPM 1	SPM 2	MUD WT	MUD VIS	MUD WL	DULL. COND. T	B	G	OTHER	FORMATION REMARKS
1.77	1	17-1/2	HUGHES	OSC3A		RT	396	396	6-1/2	61	6.5	ALL	100	1/2°	1800	1.77			SPUD MUD			N	D			RUN 13-3/8" CSG. 2-12-69
1.11	2	11-1/2	HUGHES	OSC3A	3-10	JB592	2135	1736	21-3/4	80	28	50	100	1/2°	1800	0.96	52		NATIVE			N	D			
1.40	3	11-1/2	HUGHES	OSCJ	3-10	HL403	3001	866	24	36	52	50	80	1-1/2°	1800	2.11	52					N	D			
1.64	4	11-1/2	HUGHES	OWV	3-10	HT329	3597	596	17-1/4	35	70	60	80	1°	1800	2.42	52					N	D			
1.89	5	11-1/2	HUGHES	OWV	3-10	FF923	4130	533	24-1/4	22	94	70	80	1/2°	1800	3.52	52					N	D			RUN 8-5/8" CSG. 2-17-69
2.07	6	7-7/8	HUGHES	XWR	2-10 1-9	HR792	4490	360	18	20	112	60	60	1/4°	1800	4.20	52		H2O			6	7			2-18-69 4131
2.23	7	7-7/8	HUGHES	XWR	2-10 1-9	HR784	4857	367	17-3/4	21	130	60	60	1°	1800	4.12	52		H2O			7	7			
2.35	8	7-7/8	HUGHES	XWR	3-10	HE753	5498	641	27	24	157	60	60	1/2°	1800	3.25	52		H2O			8	8			
2.43	9	7-7/8	HUGHES	XWR	3-10	FP598	5801	303	12-1/4	25	169	60	60	1/2°	1800	3.99	52		H2O			6	8			
2.49	10	7-7/8	HUGHES	XWR	3-10	HE924	5991	190	5-1/2	35	174	60	60		1800	4.26	52		H2O			7	7			
2.56	11	7-7/8	HUGHES	XWR	3-10	HY914	6497	506	20-1/2	25	195	60	60	1/4°	1800	3.40	52		H2O			7	8			
2.66	12	7-7/8	HUGHES	XWR	3-10	JW223	6790	293	15	20	210	60	60	1/4°	1800	4.79	52		H2O			6	8		CD	
2.75	13	7-7/8	HUGHES	XWR	3-10	RR#10	7062	272	14-1/4	19	224	60	60		1800	5.02	52		H2O			6	7			
2.84	14	7-7/8	HUGHES	XWR	3-10	HN909	7424	362	19-1/2	19	244	60	60	1/4°	1800	4.67	52		H2O			6	7			
2.94	15	7-7/8	HUGHES	XWR	3-10	JW217	7662	238	14-3/4	16	258	60	60	3/4°	1800	5.94	52		MUD UP			7	7			2-26-69 7570
3.04	16	7-7/8	HUGHES	XWR	3-10	HW128	7910	258	15-3/4	16	274	60	60		1800	5.75	52		98	40		8	8			
3.12	17	7-7/8	HUGHES	XWR	3-10	HW208	8228	318	18-1/2	17	293	60	60	1/2°	1800	5.21	52		98	40		8	8			
3.21	18	7-7/8	HUGHES	XWR	3-10	HS233	8372	144	10-3/4	13	303	60	60	3/4°	1800	8.30	52		98	40		6	8			
3.31	19	7-7/8	HUGHES	XWR	3-10	JW214	8597	225	18	13	321	60	60	2°	1800	7.28	52		98	37		6	7			
3.40	20	7-7/8	HUGHES	XWR	3-10	HS698	8835	238	16-1/4	15	338	60	60	2°	1800	6.47	52		98	37		8	8			
3.76	21	7-7/8	R	SCMJ	2-10 1-11	580	9267	432	55-1/4	8.3	393	35	40	1-3/4°	1900	11.09	56		98	37		3	6			3-4-69 8970
4.12	22	7-7/8	R	SCMJ	3-10	917	9733	466	62-1/4	7.5	455	40	40	1-3/4°	1900	11.22	56		98	36		2	6			3-7-69 9457
4.43	23	7-7/8	R	SCM	3-10	581	10,093	360	52	6.9	507	40	40	1°	1900	12.84	56		98	36		4	8			3-11-69 10093
4.52	24	7-7/8	HTC	WRT	3-10	AJ253	10,165	72	11-1/2	6.3	519	40	40		1900	17.98	56		94	47		7	7			
4.79	25	7-7/8	R	SCM	3-10	000	10,476	311	44-1/2	7.0	563	60	60		1800	13.45	56		97	44		4	6			
5.07	26	7-7/8	R	SCM	3-10	984	10,851	375	55	6.8	618	40	40	1°	1900	12.86	56		96	36		N	D			
5.31	27	7-7/8	R	SCM	3-10	597	11,165	314	45-3/4	6.9	664	40	40	1/2°	1900	13.63	56		96	32		8	8			
5.54	28	7-7/8	R	SCM	3-10	617	11,367	202	37	5.5	701	40	40	1/4°	1900	18.62	56		96	32		4	7			
5.63	29	7-7/8	HTC	XWR	3-10	JW224	11,415	48	9-3/4	4.9	711	60	60		1900	25.57	56		96	32		N	D			
5.92	30	7-7/8	R	FRC5J	3-11	462	11,612	197	31-3/4	6.2	742	40	40		1900	17.53	56		96	32		N	D			3-24-69 11571
6.00	31	7-7/8	HTC	WRJ	3-10	AV162	11,619	37	6-3/4	5.5	749	60	50	1/2°	1800	28.47	56		97	38		7	7			
6.23	32	7-7/8	HTC	X55R	3-10	JR082	11,765	116	34	3.4	783	40	40		1900	30.97	56		96	38		N	D			3-26-69 11697
6.32	33	7-7/8	HTC	XWR	3-10	KD488	11,817	52	12	4.3	795	60	60		1900	26.43	56		95	37		N	D			
6.53	34	7-7/8	R	SCM	3-10	RR	11,936	119	27-3/4	4.3	823	40	40		1900	20.08	56		95	37		8	8			
6.85	35	7-7/8	HTC	RG7XJ	3-10	JW123	12,114	178	58-1/4	3.1	881	40	40		2100	28.42	56		95	37		4	8		BTM	
7.10	36	7-7/8	HTC	RG7XJ	3-10	LM447	12,411	297	58-1/2	5.1	940	40	40		2100	17.11	56		96	39		3	8		BTM	
7.32	37	7-7/8	HTC	RG7XJ	3-10	ZD407	12,645	234	48-3/4	4.8	988	40	40		2100	19.25	56		96	38		4	8		BTM+	
7.50	38	7-7/8	HTC	RG7XJ	3-10	JR088	12,880	235	40	5.9	1028	40	40		2100	16.96	56		96	38		4	7		CR	
7.57	39	7-7/8	HTC	WRJ	3-10	AJ	12,940	60	12-1/4	4.9	1041	60	60		2100	23.72	56		96	39		6	5			
7.65	40	7-7/8	HTC	XWR	3-10	JE898	12,990	50	10-1/2	4.8	1051	60	60		2100	26.39	56		96	38		8	6		CD	
7.84	41	7-7/8	R	SCMJ	3-10	RR	13,098	108	29-3/4	3.6	1081	60	60		2100	31.28	56		95	41		4	8			4-11-69 13081
7.92	42	7-7/8	HTC	XWR	3-10	AN905	13,147	49	13	3.8	1094	60	60		2100	30.09	56		95	42		N	D			
8.02	43	7-7/8	HTC	XWR	3-10	HS243	13,220	73	19-1/4	3.8	1113	60	60		2100	25.36	56		96	42		8	8			
8.09	44	7-7/8	HTC	XWR	3-10	HC799	13,280	60	12	5.0	1125	60	60		2100	23.64	56		96	43		7	7			4-14-69 13223
8.15	45	7-7/8	HTC	XWR	3-10	LJ285	13,315	35	6-3/4	5.2	1132	60	60		2100	31.56	56		96	42		8	8			
8.23	46	7-7/8	HTC	XWR	3-10	LJ840	13,385	70	15-1/4	4.6	1147	60	60		2100	23.09	56		96	45		7	7		DST 1	
8.28	47	7-7/8	HTC	XWR	3-10	LJ288	13,396	11	1-3/4	11.2	1149	60	60		2100	73.35			95	46		1	1		DST 2	4-17-69

LOGGING 4-17-69 PLUGGED & ABANDON 4-20-69

FIG. 4-1 Typical bit record

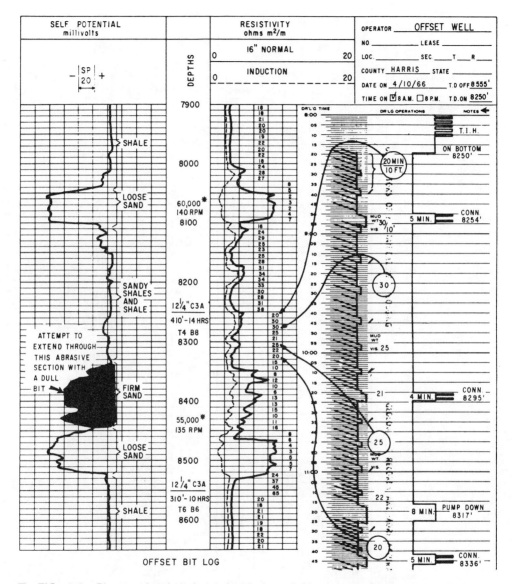

FIG. 4-2 Bit records plotted on electric logs (*after John Alterman, World Oil, March 1969*)

to be directionally drilled, extra time must be allowed. A rule of thumb is one additional day for each 4° of angle that will be built or dropped.

When planning hole and casing sizes, keep in mind the likelihood of drilling deeper than originally planned and/or the possibility of a dual completion. Depending on the answers to these questions, larger hole and casing sizes may be required.

COMPANY __APACHE OIL__ WELL __HOKEY #1__
CONTRACTOR __UNIT DRLG.__ FIELD __N.W. FT. SUPPLY__
LOCATION __35-25N-23W__ COUNTY __HARPER__ STATE __OKLAHOMA__

MUD COST ___5,745.97___
SALES TAX ___114.93___
DRAYAGE ___262.92___
TOTAL COST ___6,123.82___
COST PER FOOT ___.76___

MATERIALS USED / MUD AND CHEMICAL COST

ITEMS	UNITS	COST	ITEMS	UNITS	COST	ITEMS	UNITS	COST
Mil-Bar	Bag		Ligco	Bag		Uni-Cal	Bag	9
Milgel	Bag		Ligcon	Bag		C.S. Nulis	Bag	393
Super-Col	Bag	81 / 182.25	Lime	Bag	5 / 5.50	Kwik Seal	Bag	
Sal Water Gel	Bag	13 / 33.80	M-D	Can		Mil-Calc.Pl	Bag	4
Alisol S	DM	28 / 1,411.20	Mil CMC	DM		Mil-Fiber	Bag	
Ben-Ex	Bag		Mil-Flo	Bag		MilFlakes	Bag	
Bicarb	Bag		Mistrch	Bag		Milmica	Bag	
Caustic Soda	Bag	12 / 63.60	Pres.	DM		Drispac	DM	30
Flo-Sal	Bag	100 / 702.00	Soda Ash	Bag	32 / 165.12	Chek Loss	Can	2
LD-7	Can		Synergic	Can				

ENGINEER'S REPORT

	COST		
SURF PIPE 8-5/8" FT. 825 DATE 6-23-67	123.75	Intermediate Ft. Date	
Casing 5-1/2 7400 7-9-67	1,179.00		
Mud Up Depth At 3500 DR 600	61.25		
Total Depth 2nd Formation 7535	16.00		
Days on Mud AT 4 DR 9			
Date Completed 7-9-67			
D.S.T. Depth and Formation			
(1) 7308 Morrow	1,782.00		
(2) 7256-7272 Morrow	20.50		
(3) 7287 Chester			

LOST CIRCULATION ZONES:
(1) 2900-50 Bbl.
(3)
(4)
(5)
(6) Cores
(1)
(2)
(3)

DRILLING MUD RECORD

DATE 1967	DEPTH FT.	WEIGHT LB/GAL	SEC. QT	PLASTIC VISC. CPE	YIELD POINT	CPE	GELS IN.	GELS 10 MIN.	FILTRATE C.C.	CAKE 32ND	SAND %	% OIL	% SOLIDS	pH	SALT PPM	CALCIUM PPM	SULPHATES PPM	PF	LCM #/BBL	PRES. lbs/bbl	REMARK NUMBER	TREATMENT
6-22	1065	10.5	34			8-5/8 WOC					3.6			6.6	100,000	Hvy	Hvy	0	1			
6-23	Ran 82.5' DST																					
6-24	2233	9.8	29	3	2		0	0	NC	NIL	.5			9.0	95,000	Hvy	Hvy	0	0			
6-25	2888	9.9	30	4	3		0	1	NC	4	.6			8.7	80,000	Hvy	Hvy	0	0	AT	
6-26	3517	9.7	35	8	10		7	9	100	4	3.5	2.5	9	6.6	68,000	Hvy	Hvy	0	3	4	
6-27	4403	9.6	35	9	10		8	10	82	4	2.9	4.5	9.5	6.6	55,000	Hvy	Hvy	0	3	3	
6-28	5040	9.5	39	11	14		11	13	74	4	4	5	9.5	6.7	44,000	2210	Hvy	0	4	1	DR SA UC CS
6-29	5655	9.5	36	12.5	12		6	8	80	4	2	4	9	6.6	36,000	2000	Hvy	0	4	2	14 10 8 8
6-30	6100	9.3	38	13	18		8	9	86	4	3	3.5	7.5	8.5	35,000	300	Med	.15	2		2 2
7-1	6687	9.5	40	15	10		8	12	12	1	2.9	4	9.5	7.1	31,000	800	Med	0	4	2	3 4 2
7-2	6873	9.7	43	11	7		8	8	12	1	3	4	10	6.8	45,000	1200	Med	0	4	3	
7-3	7072	9.7	45	13	8		8	9	10	1	2.6	4	8	8.0	46,000	600	Med	0	4		2 5 4 7
7-4	7300	9.6	50	18	12		8	14	9.9	1	3	4	9	6.9	44,000	700	Med	0	4	3	2 3 3
7-5	7373	9.8	54	22	14		12	16	11.4	1	4	4	10		44,000					4	
7-6	7535 Ran Logs 3rd DST																				
7-7	7535	9.7	32	19	14		8	11			3	5		6.7	50,000	1000	Med	0	4		
7-8	7535 Ran DST #4 3rd 1310 down oil-drill pipe																				
7-9	7535 Ran 5-1/2" pipe																				

REMARKS: (1) DOWN 12 HR. MOTOR REPAIR
(2) RAN CONE OFF BIT – LOST ALMOST 30 HRS.
(3) RAN DST – NO PROBLEM
(4) RAN LOGS – HIT ONE BRIDGE ABOUT 1100 AFTER THAT NO PROBLEMS.
(5) RAN DST WITH 7" RUBBER. PULLED ONE OFF. RAN DST #2 OK. DST #3 – NO PROBLEMS.

TOPS:
TONKAWA 5612
MARMETON 6191
MORROW 7080
MORROW SD 7243
CHESTER 7287

FIG. 4-3 Drilling mud recap

```
STATE: OKLA.   COUNTY: COMANCHE
OPERATOR: WESTHEIMER-NEUSTADT CORP.
WELL: 1   FARM: RYAN
SEC. 23-3N-10W   LOC.: SE SE
POOL: WILDCAT ( 1 Mi S OF PROD )
ELEV: 1197 GR  1210 DF  1212 KB
CONT.: NICHOLS DRLG ( RT )
API NO.: 35-031-20204
FR: 1-20-69   SPUD: 2-27-69   COMP: 5-7-69
TD: 10,366   FORM: MC LISH
```

COMPLETION   RECORD

```
DRY AND ABANDONED
TD IN OLD HOLE 10,366
TD IN SIDE TRACK HOLE 9870

SIDE TRACK HOLE IS 308.18 NORTH AND
   278.26 WEST OF SURFACE HOLE LOC
```

CASING   RECORD

```
SIZE: 10-3/4 , DEPTH: 1427', SX: 700


LOG TOPS:_____(DF 1210 )
DETRITAL LM_____6973 (- 5763 )
HUNTON_____8102 (- 6892 )
SYLVAN_____8675 (- 7465 )
VIOLA_____8882 (- 7672 )
FAULT TO BROMIDE DENSE_____9560 (- 8350 )


1ST BROMIDE_____9652 (- 8442 )
FAULT TO TULIP CREEK_____9810 (- 8600 )
MO LISH_____10,155 (- 8945 )
TOTAL DEPTH_____10,366 (- 9156 )


SAMPLE TOPS:_____(DF 1210 )
UNCONFORMITY HUNTON_____8120 (- 6910 )
SYLVAN_____8658 (- 7448 )
VIOLA_____8910 (- 7700 )


(03-03) SPUD 2-27-69, 10-3/4" - 1427' - 700 SAX,
   DRLG 3012'
(03-10) DST #1 (GRANITE WASH) 3914-4040,
   OPEN 40 MIN, GTS/16 MIN, TSTM,
   REC 1320' GCMSW, ISIP 1620/40 MIN,
   IFP 1099, FFP 1427, FSIP 1554/40 MIN,
   DRLG 5246'
(03-17) DRLG 7147'
(03-21) DRLG 8000'
(03-31) DST #2 (DETRITAL LM) 7870-8008,
   OPEN 40 MIN, REC 100' MUD, 1HH 3709,
   ISIP 176/40 MIN, IFP 22, FFP 35,
   FSIP 136/40 MIN, FHH 3682,
   DST #3 (UNCONFORMITY HUNTON)
   8115-8210, OPEN 40 MIN, REC 3000' SWCM,
   5200' SW,  1 HH 3838, ISIP 3744/40 MIN,
   IFP 3294, FFP 3694, FSIP 3744/40 MIN,
   FHH 3872, DRLG 9203 LM
(04-07) DRLG 10,065 SH
(04-14) TD 10,363 - SCHL. 10,366
(04-21) PB 8087 - 160 SAX, DRLG CMT at 8210
(04-28) WS at 8210, DRLG 8901 LM
(05-05) at 9870 CIRC TO LOG
(05-07) TD WS HOLE 9870, SHCL. NO LOGS
   CALLED IN WS HOLE, WS HOLE IS
   308.18' NORTH AND 278.26 'WEST OF
   TOP HOLE LOC

   COMPLETED 5-7-69
   DRY AND ABANDONED
```

■ **FIG. 4–4** Scout ticket

GEOLOGICAL PROGNOSIS

The next step in well plan preparation involves securing the geological prognoses. Minimum requirements are shown in Table 4–2.

■ **TABLE 4–2** Geological prognosis

Formation Tops
Surface Casing Setting Depth
Objective Zones and Depths
Sample, Coring, and Test Depths
Logging Requirements

This geological information is a great help in refining the bit and mud programs.

Communication with the exploration department is vital at this point. Unplanned cores, tests, velocity surveys, and so on are sometimes unavoidable. Most of the time, however, it is quite feasible to plan in advance for the formation evaluation requirements. Good communication keeps surprises few.

Information supplied by exploration personnel assumes even more importance when the proposed well is a wildcat with little offset well data. In this case, geologic and geophysical data may be the primary tools for well plan construction. Of particular significance are bed dips, lithology, geologic ages of the formations, faults, and the presence of unusual formations such as salt.

From inference and analogy some conclusions may then be drawn concerning rock hardness, likelihood of deviation problems, and the existence of a situation conducive to lost circulation. Abnormal pressure and weak or unconsolidated beds may be implied. Clearly, considerable doubt will remain concerning the drilling environment likely to be encountered in unexplored territory. Caution and preparedness should always be paramount in the minds of operational personnel, but particularly when exploring virgin territory.

CONTRACTORS

Because of cost and overall effect, the drilling contractor is the single most important consideration in drilling a well. Typically, bids solicited are from qualified contractors operating in the area. (Chapter 1 details the qualifications.) At this stage in the well planning, decide which type of contract to solicit. Day work, footage, turnkey, or a combination may fulfill certain management requirements in different situations. Special provisions that need to be inserted in the contract on the company's behalf

■ **TABLE 4–3** Contractors and vendors

Mud	Mud Logging
Cement	Drillstem Testing
Drilling Contractor	Rathole Drilling
Subsurface Equipment	Dirtwork
Wellhead	Float Equipment
Casing Crew	Welders
Bits	Surveyors
Rental Tools	Transportation
Solids Control Equipment	Casing
Electric Logging	Coring
Tubular Inspection	Roustabouts

should be discussed at this juncture, e.g., lost circulation, deviation limits, *force majeure,* etc. (See chapter 1.)

After contacting the drilling contractors for bids, the next step is to solicit bids, estimates, and programs from the remaining contractors. This task can be time consuming. Phone contacts must be made in the planning stages and not when men and equipment are waiting on the location. Table 4–3 lists the vendors and contractors to be contacted.

The drilling contractor should be contacted first because rig selection is one of the first things that needs to be accomplished with planning a well. Other vendors who should be given priority are the mud, cementing, and bit companies as these contractors often supply valuable offset information useful in planning. The remaining vendors can be contacted in any order before the well plan is written.

Table 4–3 is not an all-inclusive list. It could easily be doubled by including specialized contractors such as diamond bit, directional, and fishing tool companies. Specialized or remote environments, e.g., jungles, arctic, or offshore, also require additional outside help. Mud, cement, and bit companies typically include a recommended program based on their offset data and experience in the area. These recommendations often vary substantially.

The planner must be sufficiently familiar with the area of interest and his company's policies and objectives to be able to integrate outside ideas with his own and come up with programs that fulfill all of the requirements. Obviously, these programs are flexible, to adapt to conditions actually encountered in the field. When writing these programs, be specific but leave room for on-site adjustments.

Designing the casing program is a prerequisite to the final cementing, mud, bit, and logging programs. The casing design includes choosing casing sizes, setting depths, and minimum cost plans. Casing size selection is based on the smallest size that accommodates all future casing strings. Considerations for the production string size include the anticipated tubing outer diameter and the feasibility of performing operations inside the string. Surface-casing setting depth depends on the depth of the freshwater table. Sometimes a second surface string must be set to case off a shallow lost-circulation zone.

Deeper wells frequently require one or more intermediate strings. The reasons for intermediate strings are usually abnormal pressure zones, low formation fracture gradients, lost circulation, or a combination of these. Intermediate setting depth is selected on the location and is often one of the most critical aspects of drilling the well.

Careful planning combined with thoroughly researched pore pressure and fracture gradient data is an essential element of drilling deep wells in problem areas. Casing string design is only mentioned here because it is intrinsic to well planning. The minimum cost string should be designed within the limits of predetermined safety factors for burst, collapse, and tension. When designing the string, consider also any special requirements such as internal coating or upgraded connections. Some operators in certain situations plan to have the casing inspected. This can be an expensive process, but, considering that as much as 10% of the joints inspected are rejected, it can easily pay for itself. Many wells have been lost because of bad casing.

AFE PREPARATION

Authorization for expenditure (AFE) preparation is one of the last steps in well planning. Preparing an AFE before the well plan is in its final written form is sometimes necessary because of time considerations. However, do not attempt to write the AFE until all the data have been gathered and interpreted. Rather than a best guess as to what the well will cost, an AFE should be an engineering summary reflecting costs.

Tangible Costs

Tangible, equipment, or capital costs are generally defined as expenditures for property that has salvage value. This includes tubulars, wellheads, salvagable subsurface equipment, and surface equipment. Calculation of tangible costs is a simple matter at this point because the tubulars, wellhead, and other equipment should be completely designed with firm bids in hand. Frequently, cost overruns occur because a radically different design is called for in the field. A typical example is the sudden realization that high-pressure connections are needed on a deep production string. Good planning heads off unpleasant surprises of this nature.

Intangible Costs

Intangible costs are expenditures for items that are not tangible. By far the most significant factor in estimating intangibles is drilling time. Interpretation of offset well data yields a depth-versus-days drilling time curve. In addition to daywork charges, some of the most prominent intangible costs are mud, rental tools, and bits. These costs are directly proportional to drilling time. Compare estimates compiled for these items with the drilling time curve to unveil any inconsistencies.

By the time the AFE is being written, contractors have been chosen from the list in Table 4–3. Estimates and, in some cases, firm bids have been received. Obviously, these estimates and bids aid greatly in constructing AFE costs. Check cost estimates received from contractors for inconsistencies and raise them if necessary.

Other intangible costs include rig moves, damages, overhead, taxes, legal fees, fuel, water, supervision, and plugging costs. Get reliable estimates for these expenses.

Many companies prefer to add an amount for contingencies, usually in the form of a percentage of total intangible costs. This is not a bad idea if the drilling is to take place in a problem area. The mistake sometimes made, though, is that the contingency charge encourages the planner to be less thorough in researching costs than he might otherwise be. Perhaps a special category should be created under the designation "hole problems," rather than adding a contingency expense on an AFE.

WELL PLAN FORMAT

When all the data have been assembled and the various contractors selected, the well plan can be written. One possible format is shown in Table 4–4.

The *introduction* should contain an overview of the most important facts associated with the proposed well: well name, location (supplemented visually with the plat), proposed depth, objective zone(s) and depth(s), and anticipated spud date. In addition, any significant difficulties likely to be encountered while drilling should be mentioned here.

One of the most effective ways of presenting a *summary* of the well plan is demonstrated in Fig. 4–5. In a quick, understandable manner, Fig.

■ **TABLE 4–4** Well plan format

1. Introduction
2. Location plat
3. Summary
4. Prognosis
5. Detailed technical data
6. Testing procedures
7. Emergency procedures
8. Vendor recommendations
9. Permits, plats, government regulations
10. Offset well data

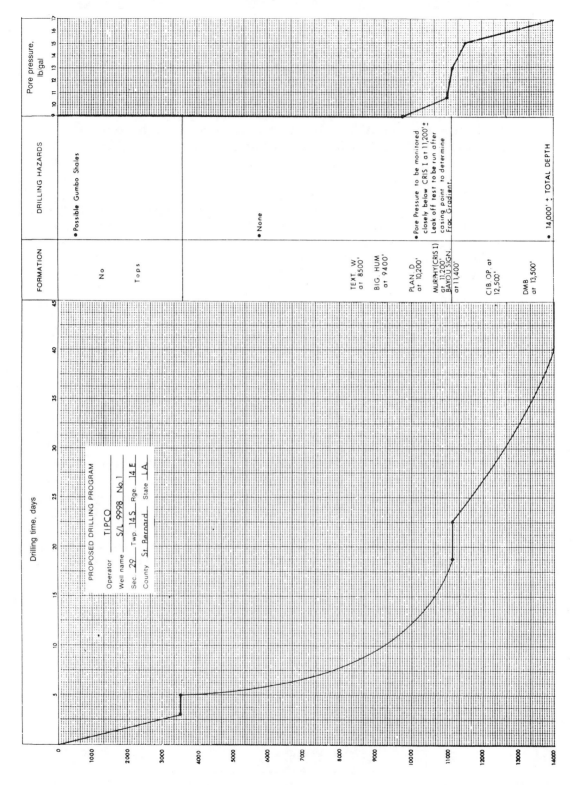

PROPOSED DRILLING PROGRAM

Operator ___TIPCO___

Well name ___S/L 9998 No 1___

Sec. _29_ Twp _14 S_ Rge _14 E_

County _St Bernard_ State _LA_

Pore pressure, lb/gal

DRILLING HAZARDS

- Possible Gumbo Shales

- None

- Pore Pressure to be monitored closely below CRIS I at 11,200' + Leak off test to be run after casing point to determine Frac Gradient.

- 14,000' + TOTAL DEPTH

FORMATION

No Tops

TEXT. W. at 8500'

BIG HUM at 9400'

PLAN D at 10,200'

MURPHY(CRIS 1) at 11,200'
BAYOU SIGN at 11,400'

CIB OP at 12,500'

DMB at 13,500'

Drilling time, days

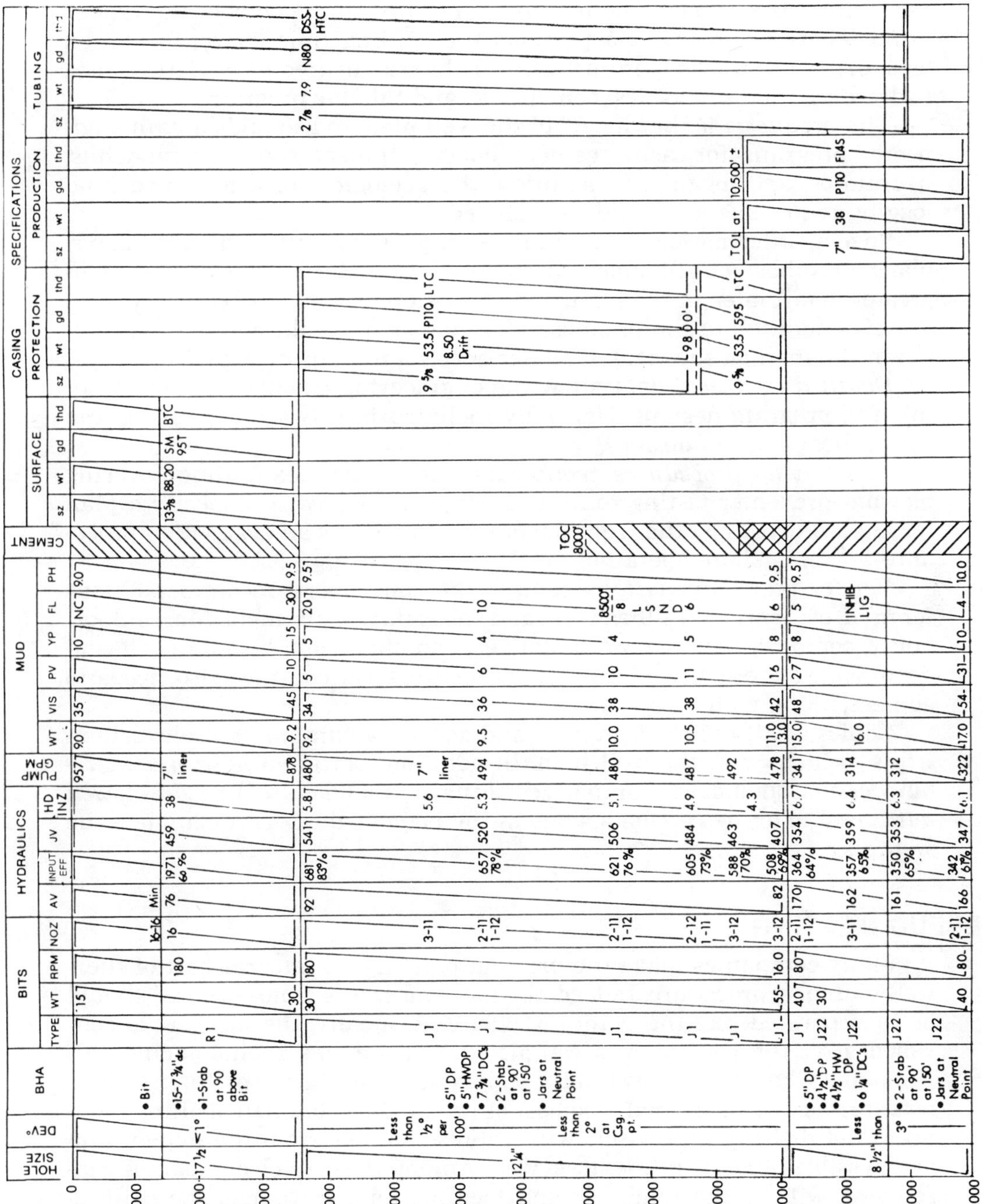

The chart is a well plan summary with a vertical depth axis (0 to 14,000 ft) and columns of data. Transcribed below as a table organized by depth.

DEPTH	HOLE SIZE	DEV°	BHA	BITS TYPE	BITS WT	BITS RPM	BITS NOZ	HYDRAULICS AV	HYDRAULICS INPUT EFF	HYDRAULICS JV	HYDRAULICS HD INZ	PUMP GPM	MUD WT	MUD VIS	MUD PV	MUD YP	MUD FL	MUD PH	CEMENT	SURFACE sz	SURFACE wt	SURFACE gd	SURFACE thd	PROTECTION sz	PROTECTION wt	PROTECTION gd	PROTECTION thd	PRODUCTION sz	PRODUCTION wt	PRODUCTION gd	PRODUCTION thd	TUBING sz	TUBING wt	TUBING gd	TUBING thd
0					15							957	9.0	35	5	10	NC	9.0																		
1,000			• Bit				16-16 Min				7" liner																									
2,000	17½	<1°	•15-7¾"dc / •1-Stab at 90 above Bit	R 1		180	16	76	1971 60.9%	459	38									13⅜	88.20	SM 95T	BTC												
3,000					30						8.78	9.2	45	10	15	30	9.5														2⅞	7.9	N80	DSS-HTC
4,000					30	180		92	681 83%	541	5.87	480	9.2	34	5	5	20	9.5																	
5,000	12¼"	Less than ½° per 100'																																	
6,000			•5" DP / •5" HWDP	J1			3-11				5.6	7" liner									9⅝	53.5 8.50 Drift	P110	LTC												
7,000			• 7¾"DC's / •2-Stab at 90' at 150' / • Jars at Neutral Point	J1			2-11 1-12		657 78%	520	5.3	494	9.5	36	6	4	10																		
8,000		Less than 2° at Csg. pt.		J1			2-11 1-12		621 76%	506	5.1	480	10.0	38	10	4	8500' 8 L S N D 6		TOC 8000'																
9,000				J1			2-12 1-11		605 73%	484	4.9	487	10.5	38	11	5										9800'										
10,000				J1			3-12		588 70%	463	4.3	492												9⅝	53.5	595	LTC	TOL at 10,500' ±							
11,000				J1	55	16.0	3-12	82	508 69%	407		478	11.0 13.0	42	16	8	6	9.5																	
12,000	8½"	Less than	•5" DP / •4½"DP / •4½"HW DP / •6¼"DC's	J1 / J22 / J22	40 30	80	2-11 1-12 / 3-11	170 162	364 64% / 357 65%	354 359	6.7 6.4	341 314	15.0 16.0	48	27	8	5 INHIB-LIG	9.5										7"	38	P110	FL4S				
13,000		3°	•2-Stab at 90' at 150' / • Jars at Neutral Point	J22 / J22			3-11 / 2-11 1-12	161 166	350 65% / 342 61%	353 347	6.3 6.1	312 322	17.0	54	31	10	4	10.0																	
14,000				40		80																													

FIG. 4–5 Well plan summary

4–5 summarizes most of the features contained in the body of the report. From left to right, Fig. 4–5 presents the drilling time curve, formation tops, drilling hazards, pore pressure, hole size, degree of deviation, bits, hydraulics, mud properties, cement fill, and tubular program.

The *prognosis* is the "meat" of the well plan and should contain engineered programs for casing, cement, float equipment, wellhead, mud, bits, hydraulics, and logging. In addition, the geological prognosis and mud logging requirements should be included.

Special information concerning testing procedures and contingency plans may be touched upon, although these subjects are discussed in greater detail later in the plan. A step-by-step drilling procedure is often a useful addition to the prognosis. Its inclusion should not, however, restrain the man on the rig, but rather act as a tool for his benefit.

Detailed technical data show the engineering calculations that went into the program designs. Normally included in this are the casing and hydraulics program designs.

The *testing procedures* section includes a detailed outline covering blowout preventer testing routine. *Emergency procedures* specify a plan for kick control and other contingency plans such as an evacuation procedure for an offshore operation.

Vendor recommendations contain the contractors' recommendations for mud, cementing, wellhead, bit, and subsurface equipment. More importantly, company names, representatives, and phone numbers for each vendor selected are listed. A copy of the drilling rig contract also needs to go here, along with a rig inventory.

Copies of all required *permits* and accompanying *plats* are excellent items to include in the well plan. In addition, pertinent *government requirements* should be summarized. Finally, inclusion of the *offset well data* used to arrive at conclusions in the body of the report complete the well plan.

PRESPUD MEETING

Many companies make the prespud meeting an integral part of their well planning procedure. Indeed, the meeting serves a purpose and should be incorporated into the planning checklist. Ideally, the meeting is convened after the well plan is completed. Participants should include all personnel who serve in any kind of supervisory role during drilling. Operator representatives normally consist of the drilling foremen, drilling engineer, and drilling supervisor, with representatives from the exploration and possibly land departments as well. Among the service companies represented will be the mud, cement, logging, and, of course, the drilling contractor. Special situations often dictate the presence of others such as rental equipment, directional, or subsurface tool people.

The primary purpose of the meeting is this: to go over the well plan and obtain an unqualified agreement from all in attendance on what their roles are and how these should be carried out. Reservations about any part of the procedure should be aired, discussed, and resolved. Anticipated problems should be discussed along with safety and emergency plans. Another key issue is the lines of communication among the participants. To avoid confusion, establish who contacts whom and when communication can be expected. Finally, the chain of command needs to be established so there is no doubt about who is responsible for what.

WELL HISTORY

The final act of well planning is actually performed after the well has been drilled. The well history is an evaluation of techniques and practices employed in obtaining the observed results. Every step of drilling the well should be described and analyzed.

Compare AFE costs with actual costs on an item by item basis. Explain over-expenditures and under-expenditures. This helps to head off inaccuracies on future cost estimates. For example, if some items were overspent because of inflation effects between the time the AFE was written and the time the well was drilled, compensate for this effect on the next AFE. There is no reason or excuse for repeating a mistake. This is true for cost estimates and, more importantly, for drilling practices. Those who don't learn from their mistakes are doomed to repeat them.

Note successful episodes and experiments. Good results obtained with, say, a new bit or experimental mud system are the types of things to include in a well history.

EXAMPLE WELL PLAN

The following (pages 91–106) is an example of the minimum that should be included in a well plan. Items 8, 9, and 10 from Table 4–4 are not shown in this example as these things are self-explanatory and, while necessary to a complete well plan, would be superfluous in this example.

Introduction

What follows is an engineered well plan for the drilling of State Lease 9998 No. 2, Lake Lery Field, St. Bernard Parish, Louisiana. The subject well tested the Lower Miocene Formation at an approximate depth of 14,000 ft. Total depth (T.D.) was +14,000 ft. The anticipated spud date was 20 May 1985, and T.D. was to be reached in 40 days.

No major hole problems were likely to be encountered in the top part of the hole, with the possible exception of some gumbo shales* in the top 3,000 ft. Geopressured formations were expected below the Cris I zone,

---

* Unconsolidated, sloughing shales

and intermediate casing was to be run in the transition zone. Pore pressures were to be closely monitored below the intermediate casing.

Summary

Fig. 4–5 is a visual summary of the well plan.

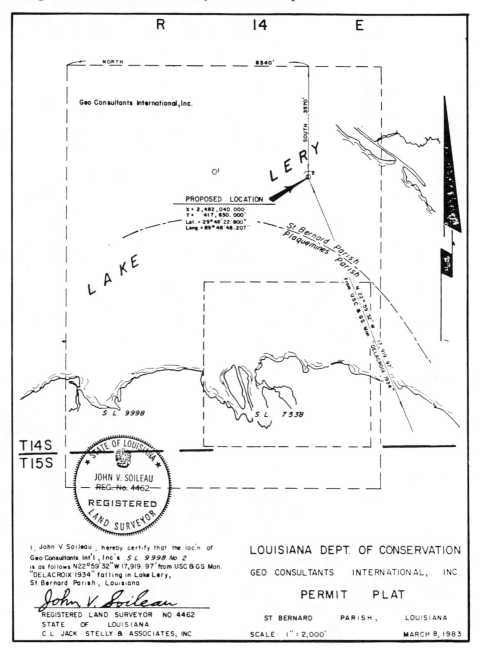
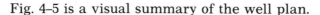

LEASE:	S.L. 9998
WELL NO.:	2
FIELD:	Lake Lery
LOCATION:	N 22° 59′ 32″ W 17,919.97′ from
	USC & GS Mon. "Delecroix" 1934
PARISH:	St. Bernard
STATE:	Louisiana
TOTAL DEPTH:	14,000′ +
OBJECTIVE:	Murphy (Chris I), Bayou Signette, CIB. OP., DMB, and
	Text W, BIG. HUM.

Logging	F.M. Tops		Depth	CSG/CMT	Bit	Mud
						Water Based *LSND* 0–3,500′ MW 9.0–9.2 VIS 35–40 WL 30
None				CIRC. CMT.	17½″	
			3,500′	13⅝″		*Water Based* *LSND* 3,500–8,500′ MW 9.2–10.0 VIS 34–38 NL 20–8
DIL/SLK/GR TDC/CNL/GR WST CST					12¼″	*INHIB-LIG* 8,500–10,500′ MW 10.0–11.0 VIS 38–42 WL 8–5
	Text. 'W'	8,500′				
	Big. Hum.	9,500′		TOC 8,500′		10,500–11,400′ MW 11.0–14.0 VIS 42–48 WL 5–4
	Plan 'D'	10,200′				
	Murphy (Chris I)	11,200′				
	Bayou Signette	11,400′	11,400′ +	9⅝′		*WATER BASED* INHIB-LIG 11,200–12,500′ MW16.0–17.0 VIS 48–54 WL 5–4
DIL/SLK/GR CNL/FDC/GR CST	CIB.OP.	12,500′		CIRC. CMT.	8½″	12,500–14,000′ MW 17.0–18.0 VIS 54–60 WL 5–4
	DMB	13,500′				
			14,000′	7″		

S/L 9998 Well No. 2 Drilling Prognosis

A. *Casing Design**

Section	Interval	Footage	Description
Surface:	0–3,500'	3,500'	13⅝", 88.20 lb SM95T, BTC
Protection:	0–11,400'	11,400'	
	0–9,400'	9,400'	9⅝", 53.5 lb, P110, LTC
	9,400–11,400'	2,000'	9⅝", 53.5 lb, S95, LTC
Production Liner:	10,500–14,000'	3,500'	7", 38.0 lb, P110, FL4S

B. *Float Equipment and Centralizers:*

Surface Casing: 13⅝" Weatherford-Lamb	Down jet float shoe, float collar, one (1) joint above float shoe, one (1) centralizer on middle of bottom joint with stop collar and one (1) centralizer on every other of next ten (10) joints. (Total 6)
Protection Casing: 9⅝" Weatherford-Lamb	Down jet float shoe and float collar, two (2) joints above float shoe, one (1) centralizer on collar between joints and one (1) centralizer on each of next six (6) joints with scratchers and wipers on bottom six (6) joints every ten (10) feet. Production intervals will be protected if required after log run.
Production Liner: 7" Brown Oil Tools	Type "V" set shoe with landing collar two (2) joints above shoe. Brown Oil Tools type "CMC" liner hanger.

C. *Cement*   B. J. Hughes, Inc.

All slurries to be lab tested for compatibility, compression strengths, and pumping times based on actual job conditions.

Surface: 13⅝" casing   17½" hole   100% excess circulate

Lead Slurry:	2,000 sx B.J. "Lite" + 0.4% D-6	(12.7 ppg, 1.84 cf/sx, 9.96 gal/sx)
Tail Slurry:	600 sx Class "N" + 0.2% R-1 + 0.4% D-6	(15.8 ppg, 1.15 cf/sx, 5.03 gal/sx)
Top Out Slurry:	100 sx Class "N" + 2% A-7 + 0.2% D-6	(15.8 ppg, 1.15 cf/sx, 5.03 gal/sx)

W O C   12 hr

Protection: 9⅝" casing   12¼" hole   50% excess   3,000' fill
20 bbl mud sweep at 14.0 ppg followed by

Lead Slurry:	875 sx 50:50:2 Pozmix (Class H:Fly Ash:Gel) + 0.5% R-1 + 0.4% D-6	(14.4 ppg, 1.22 cf/sx, 5.43 gal/sx)
Tail Slurry:	300 sx Class "H" + 0.4% R-1 + 0.4% D-6	(15.8 ppg, 1.15 cf/sx, 5.03 gal/sx)

W O C   12 hr

---

* See attachments for detail.

Production: 7″ liner 8½″ hole 75% calc. fill SQZ top.

20 bbl mud sweep + 2 gal PF-1 at 17.3 ppg followed by:
Slurry: 310 sx Class "H" + 35% D-8C + 0.9% D-22 +
21% W-5 + 3% KCR + 0.3% R-1
(Retarder to be determined)
(17.5 ppg, 1.43 cf/sx 5.05 gal/sx)
Squeeze: 15 bbl salt water followed by:
150 sx Class H + 18.1% W-5 + 0.8% D-19 +
0.2% R-1 + 3% KCI
(17.5 ppg, 1.13 cf/sx, 4.50 gal/sx)

D. *Wellhead equipment:* W-K-M (see attached)

Casing Head: 13⅝″ × 13⅝″ 5M psi with flanged outlet
Casing Spool: 13⅝″ 5M × 11 10M psi w/flange
Tree Assembly: 7¹⁄₁₆″ × 2⁹⁄₁₆″ with two 2⁹⁄₁₆″ gate valves,
single wing valve, single surface actua-
tor, and positive choke body.

E. *Mud Program:* Dowell Fluid Services (see attached)

Interval	Mud Wt.	PV	YP	VIS	Filtrate	Type
0– 3,500′	8.8– 9.0	7–12	5–10	33–59	N/C–18	LSND
3,500– 6,500′	9.0– 9.3	7–12	5–10	34–45	18–15	LSND
6,500– 8,500′	9.3–10.0	10–15	5–10	34–42	15–10	LSND
8,500– 9,500′	10.0–10.5	14–20	5–7	36–42	10–8	INHIB-LIG
9,500–10,500′	10.5–11.0	16–22	5–8	36–44	8–6	INHIB-LIG
10,500–11,000′	11.0–11.5	18–24	6–8	37–43	8–6	INHIB-LIG
11,000–11,400′	11.5–14.5	20–35	6–11	37–43	6–5	INHIB-LIG
11,400–12,500′	15.5–16.5	35–42	10–13	42–51	6–4	INHIB-LIG
12,500–14,000′	16.5–17.2	38–45	11–14	46–55	6–4	INHIB-LIG

F. *Bit program/hydraulics:** Hughes Tool Company (see attached)

(Pump pressure held constant at 3,000 psi)

Interval	Hole Size	Bit	Nozzles	W O B	RPM	Eff.	GPM
10– 3,500′	17½″	J1	16–16–16	15–30	180	60%	957–878
3,500– 6,000′	12¼″	J1	11–11–11	30–40	180	83%	480–469
6,000– 8,000′	12¼″	J1	12–11–11	30–45	180	78%	494–480
8,000– 9,500′	12¼″	J1	12–11–11	30–50	170	76%	480–466
9,500–10,100′	12¼″	J1	12–12–11	30–50	170	73%	487–473
10,100–10,600′	12¼″	J1	12–12–12	35–50	170	70%	492–478
10,600–11,000′	12¼″	J1	12–12–12	40–55	160	70%	478–440
11,000–11,400′	12¼″	J1	12–12–12	40–55	160	69%	440–420
11,400–11,800′	8½″	J1	11–11–12	40	80	64%	341
11,800–12,200′	8½″	J22	11–11–12	30–40	80	64%	327
12,200–12,800′	8½″	J22	11–11–11	30–40	80	65%	314
12,800–13,400′	8½″	J22	11–11–11	30–40	80	65%	312
13,400–14,000′	8½″	J22	11–11–12	30–40	80	61%	320

* Smith Tool Company equivalent bits approved

G. *Logging Requirements:*   Schlumberger Well Service

Surface: 17½″ hole   3,500′
              NONE

Protection: 12¼″ Hole   +11,400′
              DIL/SLK/GR from +11,400′ to surface pipe
              FDC/CNL/GR from +11,400′ to 4,500′ (optional)
              WST (Velocity survey)
              CST (Sidewall cores)

Production: 8½″ Hole   14,000′
              DIL/SLK/GR from   14,000′ to protection pipe
              FDC/CNL/GR from   14,000′ to protection pipe (optional)
              HRD (Dipmeter)
              CST (Sidewall cores)

H. *Geological Parameter:*

Formation	Estimated Tops	Estimated Pore Pressure
Text. W.	8,500′	9.0
Big. Hum.	9,400′	9.0
Plan. D.	10,200′	10.0
Murphy (Cris I)	11,200′	10.5
Bayou Signette	11,400′	11.0
CIB. OP.	12,500′	16.5
DMB	13,500′	17.0

I. *Mud Loggers:*   Exploration Logging of USA, Inc.

Place on at 7,500′   One (1) set Paleo sample every 30′ delivered to Paleo-Data, Inc., 6619 Fleur de Lis, New Orleans, LA 70124. GEMDAS Level XI with three-man crew to be utilized (see attached.)

J. *Other:*

1. *BOP drills* to be run on daily basis below 7,500′ and reported on daily drilling report.
2. *BOP tests* to be run weekly and at pipe depths.
   Test stack, manifold, and choke assembly to 7,500′ psi.
   Test annular preventor to 3,500 psi. Record on IADC report
3. *Wellhead equipment test pressures*
   a) 13⅝″ Casing spool: 5,000 psi
   b) 9⅝″ Casinghead: 10,000 psi
   c) 7⅟₁₆″ Tubinghead: 10,000 psi
4. *Contingency Plans and Policies*
   The following Contingency Plans and Policies are attached in Section IV to properly clarify the procedures and methods for handling various situations. These plans are as follows:
   a) Procedure for controlling a well kick
   b) Blowout preventor testing procedure
   c) Foul weather contingency plan

5. Security

TIGHT HOLE

All information concerning this well shall be considered *tight* and treated in the strictest confidential matter. That includes depths, mud weights, and geological markers. Mud loggers reports and log films will be obtained at the location and sent to the operator.

Drilling Procedure

1. MIRU   Glendel   Rig 8
2. Drive ±250′ of 20″ conductor pipe at 200 blows per foot to refusal.
3. Drill a 17½″ hole to 3,500′. Run and cement 3,500′ 13⅝″, 88.20 lb/ft, SM95T, BTC casing using a float shoe and float collar to be on when delivered (see detail). (Cement head must be given special notice.) Reciprocate the pipe 10′ through the cementing job. Pump slurry at 7–12 bpm or better.
   Casing Torque: Maximum 16,000 ft-lbs. Will torque to middle of triangle. T. H. Hill & Associates will have representative on location during all pipe runs.
   Cement as follows: (100% excess utilized in calculations)
   a) Mix and pump 2,000 sx B.J. "Lite" (12.7 ppg, 1.84 cf/sx 9.96 gal/sx)
   b) Mix and pump 600 sx Class H + 0.2% R-1 + 0.4% D-6′ (15.8 ppg, 1.15 cf/sx, 5.03 gal/sx)
   c) Drop the top plug and displace with ±515 bbl mud.
   d) Bump the plug with 1,000 psi above final pump pressure.
   e) Release the pressure to confirm float closure.
      Top out with 100 sx Class H + 2% A-7 + 0.2% D-6 (15.6 ppg, 1.18 cf/sx, 5.2 gal/sx)
   f) WOC 12 hr
   g) Set slips with 100% weight on slips.
4. Install a 13⅝″ 5M casing head. Test same to 5,000 psi. Install 13⅝″ 10M triple BOP. Test to 10,000 psi. Install 13⅝″ × 5M annular preventor. Test to 3,500 psi. Test casing to 2,000 psi (20 ppg equivalent). Hold each test for 30 min. Drill out 5′ of formation. Plot and run a leak off test to check fracture gradient.
5. Drill a 12¼″ hole to ±11,400′ with inclination surveys every 500′. See attached programs for mud properties, bit selections, and hydraulics. Install mud logger at 7,500′. Catch paleo samples every 30′ below surface casing. Casing depth to be determined from pore pressure plots and careful monitoring of drilling parameters.
6. Log well as per wellsite geologist. Schlumberger Well Service.
   Run DIL/SLK/GR from ±11,400′ to surface casing. Run CNL/FDC/SLK/GR from ±11,400′–7500′. Run WST (velocity survey) to surface. Run sidewall cores (CST) as required by wellsite geologist.
7. Run 9⅝″ protection casing as follows:

Internal	Length	Wt.	GR.	CPLG	Torque Min.–Max.	Turns
0– 9,400′	9,400′	53.5	P110	LTC	14,400–28,800	0.9, 3.5, 5.5
9,400–11,400′	2,000′	53.5	S95	LTC	14,100–28,200	0.9, 3.5, 5.5

Run down jet float shoe, two (2) joints casing and float collar. Bottom three (3) joints thread lok. Float shoe and float collar to be inspected and attached prior to delivery.

8. The following cement recommendation is based on obtaining cement from ±11,400' with 3,000' fill utilizing 50% excess. Actual volumes and retarder to be determined after mud compatibility tests and caliper log interpretations. Reciprocate the pipe 10 ft throughout the cement job. Pump all slurries at + 6 bpm.

    a) Mix and pump 20 bbl mud sweep spacer at 14.5 ppg.
    b) Mix and pump 875 sx 50:50:2 Pozmix + 18% salt + 0.5 R-1 + 0.4 D-6 (14.4 ppg, 1.22 cf/sx, 5.43 cf/sx)
    c) Mix and pump 300 sx Class H + 0.4% R-1 + 0.4% D-6 (15.8 ppg, 1.15 cf/sx, 5.03 gal/sx)
    d) Drop the top plug and displace with +800 bbl mud.
    e) Bump the plug with 1,000 psi above final pump pressure.
    f) Release the pressure to confirm float closure.
    g) WOC   24 hr

9. Drop the casing slips, set 100% of the pipe weight on the slips. Install 13⅝" 5M×11" 10M casing spool. Test same to 10,000 psi. Install 13⅝" 10M BOP stack. Test same to 10,000 psi including manifold and hydraulic choke assembly. Test annular to 3,500 psi. BOP stock as follows: 5" pipe rams, spool, variable (5" and 4½") pipe rams, blind rams, annular.

10. Drill an 8½" hole to +14,000' with inclination surveys every 500'. See attached program for mud properties, bit selection, and hydraulics.

11. Log well as per wellsite geogists. Schlumberger Well Services.
    Run DIL/SLK/GR from total depth to protection pipe. Run CNL/FDC/GR from total depth to protection pipe. Run HRD (dipmeter) from total depth to protection pipe. Run sidewall cores (CST) as per wellsite geologist.

12. Trip in hole. Circulate and condition mud to run production casing. Hole conditions dictate whether liner with tie back or full string will be run. POOH with drill pipe and drill collars.

13. Run 7" production liner as follows: (Brown Oil Tools "CMC" hanger)

Interval	Length	Wt.	GR	CPLG	Torque
10,500–14,000'	3,500'	38.0	P110	FL4S	7,500

Run BOT type "V" set shoe, two joints of casing, landing collar. Bottom three (3) joints thread lok.

14. The following cement recommendation is based on obtaining cement from ±14,000' to 75% caliper volume above shoe. Squeeze will be performed on top of liner. Run liner as per attached Brown Oil Tools recommendations. Actual volumes and retarders will be determined after mud compatibility tests and caliper log interpretation. Reciprocate the pipe 10' through the cement job. Pump all slurries at 4–5 bpm.

    a) Mix and pump 20 bbl mudsweep spacer at 17.3 ppg.
    b) Mix and pump 310 sx Class H + 35% D-8C + 21% W-5 + 3% KCI + 0.9% D-22 + 0.3% R-1 + 0.4% D-6 (17.8 ppg, 1.43 cf/sx, 5.05 gal/sx)
    c) Drop the top plug and displace with mud.

d) Bump the plug with 1,000 psi above final pump pressure. Release the pressure to confirm float closure.

e) Release liner: PU 400′ and reset squeeze packer.

f) Squeeze top of liner as follows: (pump at 3–5 bpm)

   1) Mix and pump 15 bbls saltwater spacer.

   2) Mix and pump 150 sx Class H + 18.1% W-5 + 0.8% D-19 + 0.2% R-11 + 3% KCl (17.5 ppg, 1.13 cf/sx, 4.50 gal/sx).

   3) Maximum squeeze 1,500 psi above breakdown pressure.

   4) Reverse clean. POOH.

15. Drill out to top of liner. Test liner top to 1,500 psi (20 ppg equivalent) POOH.

16. Drill out inside of liner and dress seal bore for packer tie back.

17. Run tie back packer with tie back sleeve. Test same to 1,500 psi (20 ppg equivalent).

18. Completion procedure to be determined after evaluation of open-hole logs.

S/L 9998 No. 2, Surface Casing Design, 13⅝″ O.D. Casing

3,500′ 13⅝″, 88.20 lb/ft, SM95T, BTC Casing (used)

Collapse

8.6 ppg water inside

9.5 ppg mud outside

$$(9.5 - 8.6) \times 0.052 \times 3,500' = 163.8 \text{ psi}$$
$$6,240/163.8 = 38.09 \text{ (High) SF}$$

If casing is empty, i.e., lost circulation:

$$9.5 \times 0.052 \times 3.500 = 1,482 \text{ psi}$$
$$6,240/1,482 = 4.21 \text{ SF}$$

Burst

14.5 ppg mud inside

8.6 ppg mud outside

$$(14.5 - 8.6) \times 0.052 \times 3,500' = 1,073.8 \text{ psi})$$
$$7,630/1,073.8 = 7.11 \text{ (High) SF}$$

or

14.5 ppg mud inside

air outside

$$14.5 \times 0.052 \times 3,500' = 2,639$$
$$7,630/2,639 = 2.89 \text{ SF}$$

State Lease 9998 No. 2, Protection Casing Design, 9⅝″ O.D. Casing

Section A (to be used as production casing also)

Depth: ±11,400′ (Depth to be determined by drilling parameters)

Assume: a) Collapse: air inside, 14.5 ppg outside

          b) Burst: water outside, 18.5 ppg inside

          c) Tension: no buoyancy

9⅝″, 53.5 lb/ft, S95, LTC Casing   Top at 9,400′

Collapse: $8,850/14.5 \times 0.052 \times 11,400 = 1.030$ SF

Burst: $(18.5 - 8.6) \times 0.052 \times 9,400' = 4,839$ psi
$9,410/4,839 = 1.94$ SF

Section B

9⅝″, 53.5 lb/ft, P110, LTC Casing   Top at surface
Depth: ±9,400′

Assume: a) Collapse: air inside, 14.5 ppg outside
b) Burst: air outside, surface pressure calculated as below.
c) Tension: no buoyancy

Collapse: $7,930 \times .982$ (corrected for tensile)$/0.052 \times 9,400 \times 14.5 = 1.099$ SF

Burst: 18.5 ppg $\times 0.052 \times 14,000 \times 0.773$ (assume SG = 0.65
corrected for weight of gas column at 14,000′) $= 10,410$ psi
$10,900/10,410 = 1.047$ SF

Tension: Joint Strength $1,422,000/53.5 \times 11,400 = 2.33$ SF
Body Yield $1,710,000/53.5 \times 11,400 = 2.80$ SF

State Lease 9998 No. 2, Production Liner, 7″ O.D. Casing

Depth: 14,000′

Assume: Collapse: air inside, 18.5 ppg mud outside
Burst: water outside, 18.5 ppg mud inside
Tension: no buoyancy
7″, 38.0 lb/ft, P110, FL4S Casing   Top at 10.500′

Collapse: $15,110/0.052 \times 14,000 \times 18.5 = 1.12$ SF
Burst: $14,850/0.052 \times 14,000 \times (18.5–8.6) = 2.06$ SF
Tension: $890,000/38.0 \times 3,500 = 6.69$ SF

Casing and Tubing Design

OPERATOR: __TIPCO__ LEASE & WELL NO.: __S/L 9998 No. 2__ DATE: __7/19/83__
FIELD: __Lake Lery__ PARISH: __St. Bernard__ STATE: __LA__ TOTAL DEPTH: __14,000__

		Liner:	″ Set @ ′ w/ lb/gal Mud
Conductor:	20 ″ Set @ ±250 ′ w/ driven lb/gal Mud		″ Set @ ′ w/ lb/gal Mud
Surface:	13⅜″ Set @ 3,500 ′ w/ 9.2 lb/gal Mud	Production:	″ Set @ ′ w/ lb/gal Mud
Intermediate:	″ Set @ ′ w/ lb/gal Mud	Tubing:	″ Set @ ′ w/ lb/gal Mud
	″ Set @ ′ w/ lb/gal Mud		″ Set @ ′ w/ lb/gal Mud

DEPTH AT BOTTOM OF SECTION (FEET)	LENGTH OF SECTION (FEET)	O.D.	LB./FT.	GRADE	CONN.	TENSION				COLLAPSE		BURST		BOX O.D.	DRIFT DIAMETER	CLEARANCE
						WEIGHT IN AIR (1000#)	DESIGN WT. IN AIR (1000#)	JOINT STRENGTH (1000#)	DESIGN FACTOR	MINIMUM PSI	DESIGN FACTOR	MAXIMUM PSI	DESIGN FACTOR			
3,500	3,500	13.625	88.20	SM95T	BTC	264.6	264.6	1,885	7.11	6340	4.21	7630	2.89	14.375	*12.250	4.167

1. TIPCO STOCK
* Special drift to 12.250″
Tension designed with no buoyancy.
Collapse designed with air inside.
Burst designed with air outside.

Designed By: _____

Casing and Tubing Design

OPERATOR: ___TIPCO___   LEASE & WELL NO.: ___S/L 9998 No. 2___   DATE: ___7/21/83___
FIELD: ___Lake Lery___   PARISH: ___St. Bernard___   STATE: ___LA___   TOTAL DEPTH: ___14,000___

		Liner:	___" Set @ ___ ' w/ ___ lb/gal Mud		

Conductor: ___" Set @ ___ ' w/ ___ lb/gal Mud   Liner: ___" Set @ ___ ' w/ ___ lb/gal Mud
Surface: ___" Set @ ___ ' w/ ___ lb/gal Mud   Production: ___" Set @ ___ ' w/ ___ lb/gal Mud
Intermediate: _9⅝_ " Set @ _±11,200_ ' w/ _14.0_ lb/gal Mud   Tubing: ___" Set @ ___ ' w/ ___ lb/gal Mud
___" Set @ ___ ' w/ ___ lb/gal Mud   ___" Set @ ___ ' w/ ___ lb/gal Mud

DEPTH AT BOTTOM (FEET)	LENGTH OF SECTION (FEET)	O.D.	LB./FT.	GRADE	CONN.	TENSION WEIGHT IN AIR (1000#)	DESIGN WT. IN AIR (1000#)	JOINT STRENGTH (1000#)	DESIGN FACTOR	COLLAPSE MINIMUM PSI	DESIGN FACTOR	BURST MAXIMUM PSI	DESIGN FACTOR	BOX O.D.	DRIFT DIAMETER	CLEARANCE
11,400	2,000'	9.625	53.5	595	LTC	107	107	1235	11.54	7,930	1.030	9,410	1.94	10.625	*8.50	1.625
9,400	9,400'	9.625	53.5	P110	LTC	609.9	609.9	1422	2.33	7,930	1.099	10,900	1.047	10.625	*8.50	1.625

* Special drift to 8.50"

Designed By: _____

Casing and Tubing Design

OPERATOR: ___TIPCO___   LEASE & WELL NO.: ___S/L 9998 No. 2___   DATE: ___7/29/83___
FIELD: ___Lake Lery___   PARISH: ___St. Bernard___   STATE: ___LA___   TOTAL DEPTH: ___14,000___

Conductor: ___" Set @ ___ ' w/ ___ lb/gal Mud   Liner: _7_ " Set @ _14,000_ ' w/ _18.5_ lb/gal Mud
Surface: ___" Set @ ___ ' w/ ___ lb/gal Mud   ___" Set @ ___ ' w/ ___ lb/gal Mud
Intermediate: ___" Set @ ___ ' w/ ___ lb/gal Mud   Production: ___" Set @ ___ ' w/ ___ lb/gal Mud
___" Set @ ___ ' w/ ___ lb/gal Mud   Tubing: ___" Set @ ___ ' w/ ___ lb/gal Mud

DEPTH AT BOTTOM (FEET)	LENGTH OF SECTION (FEET)	O.D.	LB./FT.	GRADE	CONN.	TENSION WEIGHT IN AIR (1000#)	DESIGN WT. IN AIR (1000#)	JOINT STRENGTH (1000#)	DESIGN FACTOR	COLLAPSE MINIMUM PSI	DESIGN FACTOR	BURST MAXIMUM PSI	DESIGN FACTOR	BOX O.D.	DRIFT DIAMETER	CLEARANCE
14,000	3,500	7.000	38.0	P110	FL4S	133	133	890,000	6.69	A 15,110	1.12	B 14,850	2.06	7.00	5.795	1.50

A. Collapse calculated with 18.5 lb/gal mud outside and air inside
B. Burst calculated with 18.5 lb/gal mud inside and water outside

Designed By: _____

Optimized Hydraulics Program Summary

COMPANY TIPCO	FIELD Lake Lery	COUNTY/PARISH St. Bernard	STATE Louisiana	DATE 7/19/83
CONTRACTOR Glendel Rig 8	LEASE & WELL NO. State Lease 9998 No. 2			DRILL PIPE SIZE WT., GRD 5", 19.50, "E" & S-135 4½"IF
PUMP #1 TYPE	MAX SPM / LINER SIZE / MAX PRESSURE	PUMP #2 TYPE	MAX SPM / LINER SIZE / MAX PRESSURE	DRILL COLLARS # OD ID 7¾" × 2¹³⁄₁₆" 4½"IF

HOLE SIZE	DEPTH FROM	DEPTH TO	JET SIZE 32nd's	STANDPIPE PRESSURE	GPM	GPM/IN. OF BIT DIAM. CROSS FLOW	ANNULAR VELOCITY	HP/in. at bit	% HP at BIT	JET VELOCITY	MUD WEIGHT	COMMENTS
17½	0	3,500	16-16-16	3,000	878	50	76	3.8	60	459	9.5	7" Liners @ 90%
							Set 13⅜" casing					
12¼	3,500	6,000	11-11-11	3,000	469	38	92	5.8	83	541	9.5	7" Liners @ 90%
12¼	6,000	8,000	12-11-11	3,000	480	39	94	5.6	78	520	9.7	7" Liners @ 90%
12¼	8,000	9,500	12-11-11	3,000	466	38	91	5.3	76	506	10.0	7" Liners @ 90%
12¼	9,500	10,100	12-12-11	3,000	473	39	93	5.1	73	484	10.5	7" Liners @ 90%
12¼	10,100	10,600	12-12-12	3,000	478	39	94	4.9	70	463	11.0	7" Liners @ 90%
12¼	10,600	11,200	12-12-12	3,000	420	34	82	4.3	69	407	14.0	7" Liners @ 90%
							Set 9⅝" casing					
8½	11,200	12,200	12-11-11	3,000	341	40	170	6.4	64	354	16.0	5" × 4½" DP
8½	12,200	12,800	12-11-11	3,000	314	37	162	6.3	65	359	17.0	6½" × 2 13/16"DC's
8½	12,800	13,400	11-11-11	3,000	312	37	161	6.2	65	353	17.0	6½" Liners @ 90%
8½	13,400	14,000	11-11-11	3,000	322	38	166	6.0	61	347	17.0	

Designed By: _____

Engineered Bit Program

HTC-1095-A

COUNTY St. Bernard			FIELD Lake Lery			STATE LA		LEGAL 14S-14E		COMPANY TIPCO		DATE
CONTRACTOR Glendel Rig 8						LEASE S/L 9998		WELL NO. 1		TOOL PUSHER		SALESMAN 504/872–0414 Doug Kuykendall
PUMP			PUMP POWER			MAKE RIG		RIG POWER		DRILL PIPE		DRILL COLLARS

RUN NO.	SIZE	TYPE	JET SIZE	DEPTH		FEET	HRS.	ACCUM. HOURS	WEIGHT 1000 POUNDS	RPM	PUMP PRESS.	SPM	DAYS	BIT COST	REMARKS
				FROM	TO										
1	17½	R1			3,500	3,500	45	45	15/30	180			3	4,940	
									Set 13⅜" csg.				5		
2	12¼	J1			6,000	2,500	30	75	30/40	180			7	2,430	
3	12¼	J1			8,000	2,000	25	100	30/45	180			8	2,430	
4	12¼	J1			9,500	1,500	25	125	30/50	170			10	2,430	
5	12¼	J1			10,100	600	25	150	30/50	170			11	2,430	
6	12¼	J1			10,600	500	25	175	35/50	170			13	2,430	
7	12¼	J1			11,000	400	25	200	40/55	160			15	2,430	
8	12¼	J1			11,200	200	15	215	40/55	160			16	2,430	
									Set 9⅝" csg.				20		
9	8½	J1			11,400	200	20	235	40	80			21	1,385	
10	8½	J22			12,200	800	70	305	30/40	80			24	4,291	
11	8½	J22			12,800	600	70	375	30/40	80			28	4,291	
12	8½	J22			13,400	600	70	445	30/40	80			32	4,291	
13	8½	J22			14,000	600	70	515	30/40	80			36	4,291	
														$40,497	

Blowout Preventer Testing Procedure

1. Run test plug in on a joint of drillpipe, set in seat, and back off from it. *Remove drillpipe joint.*

   Note: Plug to be machined so that there will be enough clearance between plug and pipe rams to clear tool joint when closed on joint of drillpipe made up in plug. The plug or drillpipe must be drilled so there is communication between inside of drillpipe and top of plug above seal surface.

2. Fill preventers with water.

3. Open all valves and chokes on choke lines and choke manifold to assure that water is flowing through each outlet. Let water run until clear.

4. Close adjustable choke and manifold valves on choke manifold, making sure they are full of water and have no trapped air. Do not close manifold valve immediately upstream or adjustable choke at this time.

5. Refill preventers if necessary.

6. Close blind rams with 1,500 psi.

7. Check closing line and preventer for leaks.

8. Pressure up to *working pressure* of preventer through test line. Hold for 10 min.

9. Check all valves, flanges, and seals that are under pressure for leaks and tighten if necessary. Check blind rams for leaks. Check test plug for leaks.

10. Close valves immediately upstream of adjustable choke. Open adjustable

choke. Hold pressure for 5 min. Close third valve from hole on choke line. Open all manifold valves on choke manifold. Hold pressure for 5 min.

11. Check valve for leaks.
12. Close second valve from hole on choke line. Open third valve out on choke line. Hold pressure for 5 min. Check valve for leaks.
13. Close inside valve on choke line. Open second valve from hole on choke line. Hold pressure for 5 min. Check valve for leaks.
14. Close inside valve on kill line side. Open inside valve on choke line side. Hold pressure for 5 min.
15. Check valve for leaks.
16. Close second valve out on kill line. Open inside valve on kill line. Hold pressure for 5 min.
17. Check valve for leaks.
18. Open second valve out on kill line. Close inside valves on kill line and choke line.
19. Disconnect test line; connect kill line.
20. Open blind rams with 1,500 psi.
21. Check opening line and preventer for leaks.
22. Screw into test plug with joint of drillpipe.
23. Purge all air from drillpipe and preventers.
24. Close drillpipe rams with 1,500 psi.
25. Check closing line and preventers for leaks.
26. With drillpipe full of water, attach test line to top of drillpipe.
27. Pressure up to *working pressure* of preventers through drillpipe. Hold for 10 min.
28. Check for leaks.
29. Release pressure.
30. Open drillpipe rams with 1,500 psi.
31. Check opening line and preventers for leaks.
32. Fill preventers.
33. Close 1-in. plug valve on closing line of annular preventer. Test to 1,500 psi. Check for leaks. Release pressure and open valve.
34. Close annular preventer with 1,500 psi. Keep accumulator pressure on closing side of annular preventer while testing.
35. Check closing line and annular preventer for leaks.
36. Purge air from drillpipe.
37. Pressure up to two-thirds of the rated *working pressure* of the annular preventer through drillpipe. Hold for 10 min.
38. Check for leaks.
39. Release pressure.
40. Open annular preventer with 1,500 psi, and remove test plug.
41. Open kelly cock.
42. Make up drillpipe safety valve on kelly.
43. Make up adapter sub on drillpipe safety valve.
44. Open standpipe and fill up line valves.
45. Fill kelly with water.
46. Close drillpipe safety valve.
47. Attach test line to adapter sub.
48. Open drillpipe safety valve and purge air from safety valve.

49. Close drillpipe safety valve.
50. Pressure up to *working pressure* of preventers and hold for 5 min.
51. Close kelly cock.
52. Open drillpipe safety valve.
53. Pressure up to *working pressure* of preventers and hold for 5 min.
54. Release pressure.
55. *Record test* on *Drilling Report.*

Statement of Policy Procedure for Controlling a Well Kick

The following procedures shall be adhered to in the event that the well should kick:

While Drilling:

1. When a primary warning sign of a kick has been observed, the kelly will be raised immediately until a tool joint is above the rotary table.
2. The mud pumps will be stopped.
3. The annular preventer will be closed.
4. The shutin drillpipe pressure, shutin casing pressure, pit gain, and time of day will be read and recorded.
5. The drillpipe will be checked for trapped pressure.
6. The exact mud weight necessary to kill the well will be calculated (no trip margin included).
7. The kill mud in the suction pit will be mixed.
8. While the kill mud is being mixed, a kill sheet will be prepared.
9. After the kill mud has been mixed, circulation will be initiated by adjusting the choke to hold the casing pressure at the shutin valve while the driller starts the mud pumps.
10. As the drillpipe is being displaced with the exact kill mud weight at a constant pump rate (kill rate), the choke will be used to adjust the pumping pressure according to the required pressures from the kill sheet.
11. The annulus will be displaced with the kill mud by maintaining the drillpipe pumping pressure constant at a constant pumping rate by using the choke to adjust the casing pressure as necessary.
12. After the kill mud has reached the flowline, the pumps will be shut down, the choke shut in, and the well will be checked for flow and/or pressure.
13. When the pressures on the drillpipe and casing are both zero, then the annular preventors can be opened, the mud circulated and conditioned, and a trip margin added.

While Tripping:

1. When a primary warning sign of a kick has been observed, the top tool joint will be set immediately on the slips.
2. A full-opening, fully opened safety valve will be installed in the drillpipe.
3. The safety valve and the annular preventer will be closed.
4. The kelly will be made up.
5. The safety valve will be opened.
6. The shutin drillpipe pressure, the shutin casing pressure, the pit gain, and time of day will be read and recorded.
7. If possible, the drillpipe will be stripped into the hole utilizing the pressure

method (cementing unit pump connected to choke line for accurate pressure readings and volumes).

8. When the drillpipe reaches total depth, the original mud weight will be circulated for the drillpipe volume. The kill weight mud will be calculated, and the well will be killed as above.

Well Control Data Sheet

Date _____

Name: _____ Rig _____ TD _____

I. *Pump Output*

	Type	Liner	Stroke	Maximum Pressure, psi	Displacement, bbl/Stroke
Pump No. 1	_____	_____	_____	_____	_____
Pump No. 2	_____	_____	_____	_____	_____

II. *Capacity and Displacement of Drill String*

		Cap	Displ		Cap	Displ

DP size × length _____" × _____' Cap & Displ _____ & _____ bbl/1,000' _____ & _____ bbl
DP size × length _____" × _____' Cap & Displ _____ & _____ bbl/1,000' _____ & _____ bbl
DC size × length _____" × _____' Cap & Displ _____ & _____ bbl/1,000' _____ & _____ bbl
Total Depth _____'   Totals—Cap & Displ _____ & _____ bbl

III. *Pump Strokes to Fill the Hole during a Trip*

$$\text{Pump Strokes to Fill the Hole during a Trip} = \frac{\text{DP Displ in bbl/1,000'} \times \text{length of stand}}{\text{Pump Displacement in bbl/stroke} \times 1000} = \underline{\hspace{2cm}}\text{stroke/stand}$$

IV. *Annulus Capacity*

Casing × DP _____" Length _____' Ann Cap _____ bbl/1,000' _____ bbl
Casing × DC _____" Length _____' Ann Cap _____ bbl/1,000' _____ bbl
Liner × DP _____" Length _____' Ann Cap _____ bbl/1,000' _____ bbl
Hole × DP _____" Length _____' Ann Cap _____ bbl/1,000' _____ bbl
Hole × DC _____" Length _____' Ann Cap _____ bbl/1,000' _____ bbl
Total Depth _____'   Total Annulus Capacity _____ bbl

V. *Pit Volume*   Record all volumes, circle active volumes

Pit No. 1   Size _____ × _____ × _____' Cap _____ bbl/in _____ bbl
Pit No. 2   Size _____ × _____ × _____' Cap _____ bbl/in _____ bbl
Pit No. 3   Size _____ × _____ × _____' Cap _____ bbl/in _____ bbl
Pit No. 4   Size _____ × _____ × _____' Cap _____ bbl/in _____ bbl
Sand Trap Size _____ × _____ × _____' Cap _____ bbl/in _____ bbl
Trip Pit   Size _____ × _____ × _____' Cap _____ bbl/in _____ bbl
Total Cap _____ bbl/in _____ bbl

VI. *Maximum Safe Surface Pressure*

Casing   Setting   Minimum Yield
Size _____ Weight _____ Grade _____ Depth _____ Pressure _____ psi
Working pressure of blowout preventers, wellhead, valves, and lines: _____ psi
Minimum frac grad: _____ lb/gal at _____ ft (casing shoe or weakest zone).
(Minimum frac grad − mud wt used) × depth of weakest zone × 0.052 =
                                    Max Initial Casing SIP _____ psi*

* After kick fluid is circulated into casing, the pressure may increase above this value without fracturing the formation.

VII. *Well Killing Volumes*

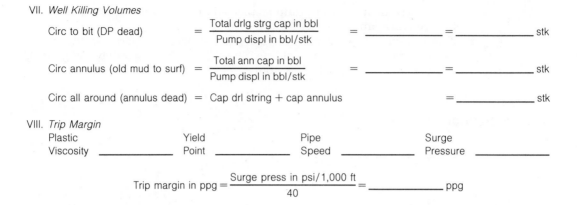

$$\text{Circ to bit (DP dead)} = \frac{\text{Total drlg strg cap in bbl}}{\text{Pump displ in bbl/stk}} = \underline{\hspace{2cm}} = \underline{\hspace{2cm}} \text{ stk}$$

$$\text{Circ annulus (old mud to surf)} = \frac{\text{Total ann cap in bbl}}{\text{Pump displ in bbl/stk}} = \underline{\hspace{2cm}} = \underline{\hspace{2cm}} \text{ stk}$$

$$\text{Circ all around (annulus dead)} = \text{Cap drl string + cap annulus} = \underline{\hspace{2cm}} \text{ stk}$$

VIII. *Trip Margin*

Plastic Viscosity _____   Yield Point _____   Pipe Speed _____   Surge Pressure _____

$$\text{Trip margin in ppg} = \frac{\text{Surge press in psi/1,000 ft}}{40} = \underline{\hspace{2cm}} \text{ ppg}$$

Vendor List

Drilling Contractor	Glendel	Abbeyville	318/898–1637
Mud and Chemicals	Dowell Fluid Srv.	Houma	504/876–6553
Fuel	TBD*		
Drive Hammers	Franks	Lafayette	318/233–0303
Conductor Pipe	Franks	Lafayette	318/223–1181
Casing Crews	Weatherford	Houma	504/851–0710
Tubular Goods	TIPCO STOCK	OKC	405/949-7700
Float Equipment	Weatherford	Houma	504/851–0710
Welding	TBD*		
Wellhead Equipment	W-K-M	Shreveport	318/222-3254
Bits	Hughes Tool Co.	Houma	504/872–0414
Bits	Smith Tool Co.	Belle Chasse	504/394-7260
Drilling Jars	Dailey Oil Tools	Houma	504/873–7780
Stabilizers	Weatherford	Houma	504/851–0710
Shock Sub	Hughes Tool Co.	Houma	405/745–2715
Degasser	Rig		
P-V-T Mud Monitors	EXL	Houston	713/495-9480
Mud Cleaner	SWECO	New Orleans	504/586–1504
Mud Logging	EXL	Houston	713/495-9480
Rental Equipment	Weatherford	Houma	504/851–0710
Cement and Service	B. J. Hughes	Venice	504/522-9210
Liner Running Equipment	Brown Oil Tools	Lafayette	318/632-6735
Open Hole Logging	Schlumberger	Belle Chasse	504/586-6666
Cased Hole Logging	CRC Wireline	Belle Chasse	504/394-3610
Well Testing	TBD*		
Packers	TBD*		
SC/SSSV	TBD*		
Adjustable Choke	SWACO	New Orleans	504/586–1504
Fishing Tools	RAM	Lafayette	318/856-6768
Dock Personnel	Delacroix Dock		
Dock Equipment	Amigo Dock	Hopedale	504/676-3755
Barges and Tugs	DuMar Marine	La Rose	504/566-1695
Crew Boats	Crewboats Inc.	Chalmette	504/277-8201
Manifold	Rig		

TBD: to be determined

The preceding example demonstrates the types of information that should go into the well plan. There can be rather wide variations on this theme, depending on well location and company policies.

Spare no effort when composing the well plan. It establishes the pattern that will be followed during the entire drilling process. It is the best tool engineering and supervisory personnel have in implementing safety and cost-control procedures.

■ **REFERENCES**

1. Kerr, J.W.: "Here's a Guide for Planning a Well Program," *Oil and Gas Journal (OGJ)*, 24 September 1968.
2. Adams, Neal and Frederick, Marsha: "How to Estimate Well Costs," *OGJ,* 6 and 13 December 1982.

CHAPTER 5

DRILLING MUDS

Drilling muds were introduced with rotary drilling in 1900. Initially the primary purpose of the mud was to remove cuttings continuously. With progress, sophistication was added, and more was expected from the drilling mud. Additives for almost any conceivable purpose were introduced, and what started out as a simple fluid became a complicated mixture of liquids, solids, and chemicals.

The drilling mud is discussed first from the standpoint of the functions it serves in rotary drilling; second, a brief résumé of the chemistry of the system is given. Then the basic composition of the mud is considered, followed by the specific controls used in field applications. Last, the trends in special mud compositions are reviewed. The primary functions of the mud follow:

- Lift formation cuttings to the surface
- Control subsurface pressures
- Lubricate the drillstring
- Clean the bottom of the hole
- Aid in formation evaluation
- Protect formation productivity
- Aid formation stability

LIFT FORMATION CUTTINGS

Cleaning the hole is an essential function of the mud. This function is also the most abused and misinterpreted. The drill solids generally have a specific gravity of 2.3–3.0; an average of 2.5 will be assumed. When these solids are heavier than the mud being used to drill the hole, they slip downward through the mud. Specific considerations of factors affecting the slip velocity of the cuttings are considered in chapter 7.

While the fluid is in viscous or laminar flow, the slip velocity of cuttings is affected directly by the thickness or shear characteristics of the mud. Thus, when the annular mud velocity is limited by pump volume or enlarged hole sections, it often is necessary to thicken the mud to reduce the slip velocity of the formation cuttings to keep the hole clean.

A water-based mud may be thickened by adding bentonite or drill solids, by flocculation of the solids, or by the use of polymers. An oil-based mud may be thickened by adding solids, in some cases by reducing the water in oil emulsifiers, and by adding polymers.

Sometimes the decision to increase the lifting capacity of the mud is complicated by the fact that any thickening of the mud may adversely affect other drilling conditions. For example, if the mud is thickened, circulating pressure losses increase and the danger of lost circulation increases. Small batches of viscous mud can be used to lift cuttings and to minimize the requirement for thickening all of the mud.

Water-based muds may be thickened cheaply by flocculating clay solids. However, when the solids flocculate, the filtration rate increases. The cost of reducing the filtration rate may make other thickening methods more desirable. Polymers thicken either water-based or oil-based muds. Polymers generally are considered the most expensive alternatives for thickening mud but may not be if all costs are considered.

What appears to be a simple decision on thickening the mud to increase lifting capacity is complicated by the resultant effects the thickening method may have on other objectives. The specific procedures must be determined by considering the actual problems at the time they occur.

CONTROL SUBSURFACE PRESSURES

Several problems are introduced in selecting mud weight to control subsurface pressures. Minimum mud weights are desirable. As mud weight decreases, drilling rates increase and lost-circulation problems are minimized. Actual mud-weight requirements in the normal range vary in different sections of the country. Fresh water frequently provides adequate weight to control subsurface pressures. In coastal areas a mud weight of 9.2 lb/gal may be required.

Abnormally high formation pressures require careful measurements of pore pressures to determine mud-weight requirements. Even a careful measurement of pore pressure may result in wide variations in mud weights used by companies drilling wells in the same areas because interpretations often differ. Therefore, emphasis should be dedicated to determining why the differences exist.

LUBRICATE THE DRILLSTRING

Lubrication and cooling have become important functions of the mud. The life of expensive equipment can be prolonged by adequate cooling and lubrication, and hole problems—such as torque, drag, and differential pressure sticking—are related directly to lubrication. Lubricants include bentonite, oil, detergents, graphite, asphalts, special surfactants, and walnut hulls. Bentonite acts as a lubricant only because it is slick when wet and reduces friction between the wall cake and the drillstring.

Oil has been the lubricant in drilling muds for many years. Limitations have been placed on the use of oil because of the associated disposal problems. Detergents have gained some attention as lubricants. Reported successful applications are regional, and field evaluations are always necessary. Graphite is a good lubricant but generally must be added with oil to be effective. It is used most frequently in areas where high rotating torques are a problem.

Asphalt has been a mud additive for several reasons. In oil-based muds it is often used as a low-gravity solids phase. It has also helped to stabilize shales by a claimed process of plugging off microfractures in the shale; less frequently it has been used as a lubricant. Special surfactants claimed to act as lubricants are introduced by many companies, but success is regional. Most surfactants are expensive, and their benefits should be analyzed carefully. Walnut hulls and similar products help reduce torque or drag; however, their use can become very expensive because in general they must be added in quantity every day.

CLEAN THE BOTTOM OF THE HOLE

Generally, bottom-hole cleaning is improved by having thin fluids at high shear rates through the bit. Therefore, viscous fluids can be potentially good fluids for bottom-hole cleaning if they have good shear thinning characteristics. A more detailed discussion of this phenomenon is included in chapter 2. Usually a fluid with a low solids content is best for bottom-hole cleaning.

AID IN FORMATION EVALUATION

Drilling fluids have been affected significantly by formation evaluation requirements. Viscosity has been increased to obtain better cuttings, filtration rate has been reduced to minimize fluid invasion, special fluids have been selected to improve logging characteristics, and muds have been changed to improve formation testing. Most of those precautions are taken more from habit than necessity. Oil-based muds make evaluating potential producing horizons difficult. Saltwater fluids make using a self-potential log to recognize permeable zones difficult.

Thick filter cakes may make it hard to obtain information from sidewall coring. Water or oil invasion certainly affects resistivity. In many cases the methods of evaluation are changed; in others the methods of measurement are not indicative of downhole conditions. Thus, the emphasis placed on fluid selection and treatment in this area needs to be reviewed in detail.

PROTECT FORMATION PRODUCTIVITY

Formation productivity is of major concern since noncommercial hydrocarbon zones are often blamed on formation damage introduced through the invasion of mud or filtrate. There is little doubt that keeping the downhole formation in its virgin state with no fluid entering the zone is desirable. In drilling, this generally cannot be done. In some areas of West Virginia and Kentucky, productive zones are drilled with air to keep liquid off of the formation. This practice has maintained formation productivity.

Also, in many areas the productive horizons are drilled using an oil-based mud to keep water out of the zone. This practice has been effective; however, in gas zones it may be more damaging than a saltwater fluid. Saltwater and high-calcium fluids also have been used to minimize formation damage. To some degree these fluids have been effective.

Even today, many companies reduce the filtration rate to low values, below 10 cc API, to minimize filtrate damage in the pay zone. This practice is normally ineffective for reducing filtrate invasion, and the biggest effect is the cloak of protection it provides for the practitioner. However, downhole filtration rates cannot be determined by static filtration tests at the surface, particularly at low values.

AID TO FORMATION STABILITY

Usually hole instability is associated with sloughing shales or salt sections. Water-based muds contribute to hole instability but may be treated to minimize their adverse effects. Common additives for minimizing shale sloughing in water-based muds are as follows.

- Calcium compounds
- Potassium compounds
- Asphaltic compounds

Oil-based muds, if treated with calcium chloride in the water phase, are also effective in providing hole stability.

If the shale section contains fluids in the pore spaces at higher than normal pressure, the weight of either water-based or oil-based muds may have to be increased.

Salt sections may be stabilized by drilling with either saturated saltwater or oil-based muds. Saturated saltwater muds must be saturated with the same type salt found in the salt formation. Also, as temperature increases, the solubility of salt in water increases. Thus, it is not uncommon to deposit salt at the surface where temperatures are lower and continue to leach salt downhole at higher temperatures. Salt is insoluble in oil, so no leaching occurs. In some cases, salt contains enough high-pressure liquids to flow into the hole if the pressure exerted by the mud weight is too low. In this case, with either water- or oil-based muds, the weight must be increased.

COMPOSITION OF THE MUD

Drilling mud consists of liquids and solids. Liquids may be mixed or varied, and different solids may be entrained in the mud. Often, chemicals are added. Table 5–1 shows the composition of mud.

Fresh water is the base of most muds. Fresh water is generally accessible, cheap, easy to control even when loaded with solids, and provides the best liquid for formation evaluation. Salt water has become more common because of its accessibility in offshore operations, which are expanding. Table 5–2 shows the salt required for different stages of saturation, beginning with fresh water. Table 5–3 shows the effect of temperature on the saturation level of sodium chloride.

Note that the salt required for saturation increases with temperature. Thus, salt often is deposited in surface tanks when a saturated saltwater mud cools. If salt beds are to be drilled, frequently the water is saturated before entering the salt sections. This minimizes the total hole enlargement, providing the mud is stirred vigorously at the surface as it cools. In some cases operators simply let the salt from the salt beds saturate the water.

There are many disadvantages to using salt. Mud costs are higher,

■ **TABLE 5–1** Liquid and solid composition of mud

	Solids	
Liquids	**Water Based**	**Oil Based**
Fresh water	Low gravity—specific	Low gravity
Salt water	gravity = 2.5	Amine-treated clays
Oil	Nonreactive, sand, chert,	Asphalt, gilsonite—
Mixtures of these fluids	limestone, some shales	specific gravity = 1.1
	Reactive solids, clays	
	High gravity	High gravity
	Barite—specific	Barite
	gravity = 4.2	Iron ore
	Iron ore—specific	
	gravity = 4.7–5.1	

■ **TABLE 5–2**   Salt conditions at 68°F

mg/l	Salt Added, lb/bbl	Salt, %	Weight of Solution, lb/gal
10,050	3.53	1	8.39
20,250	7.14	2	8.45
41,100	14.59	4	8.57
62,500	22.32	6	8.69
84,500	30.44	8	8.81
107,100	38.87	10	8.93
130,300	47.72	12	9.06
254,100	56.96	14	9.19
178,600	66.65	16	9.31
203,700	76.79	18	9.45
229,600	87.47	20	9.58
256,100	98.70	22	9.71
279,500	110.49	24	9.85
311,300	122.91	26	9.99

bentonite yield is reduced, formation evaluation methods are less effective, and potential corrosion problems may be increased. Mud costs go up because of the presence of the electrolytes such as salt in the mud, which reduces the benefits derived from most additives. Reducing the bentonite yield often results in poor filter cakes. Also, hole problems are frequently more severe.

Formation evaluation is affected primarily because of the reduced definition given by the self-potential log opposite permeable formations. The self-potential log is generated primarily by an electrochemical reaction between the mud filtrate and formation water, shown in Eq. 5.1.:

$$\text{Self-potential} = -K \log R_{mf}/R_w \qquad (5.1)$$

where:

K = proportionality constant
R_{mf} = resistivity of mud filtrate, ohms
R_w = resistivity of formation water, ohms

■ **TABLE 5–3**   Effect of temperature on sodium chloride

Temperature, °F	Salt to Saturate, lb/bbl
80	127
120	129
160	132
200	137

If the mud filtrate contains the same amount of salt as the formation water, R_{mf} equals R_w and the log of one is zero. Thus, no self-potential is generated. This concept has been used in reverse to obtain potential logs in areas where formation waters are close to fresh water. In this case salt is added purposely, and a positive self-potential log is obtained. Corrosion is more severe with salt water because the electrolytes are good conductors of electricity and because it is difficult to raise the pH.

When salt sections are to be drilled, the only alternative to salt water may be oil-based muds. If saltwater mud is to be used, an effort should be made to minimize costs. In high-weight muds—above 16.0 lb/gal—maintaining a high pH may be difficult and expensive. Thus, using a corrosion inhibitor that coats the pipe may be cheaper. If hole problems are severe, the operator may add prehydrated bentonite to the mud to improve the filter cake.

Some good effects may be realized from using saltwater muds. The swelling of clays is reduced with an increase in salt content. Thus, potentially productive formations containing swelling clays are damaged less by contact with filtrate. Also, many shales may heave or slough less in saltwater fluids than they will in freshwater fluids.

In application, saltwater muds should be considered special-purpose muds, that are basically more expensive to use than freshwater muds. Specific applications may require that saltwater muds be used, and the operator should consider carefully the best and cheapest method of application.

Oil as a base fluid has been used for almost as many years as water. Oil probably was implemented initially to protect potentially productive formations. Clays do not hydrate or swell in oil, and formation damage in oil zones is minimized. The next recognized advantage for oil was to minimize hole problems. Many recorded successful applications of oil for this purpose date back more than 30 years.

Two claims were made for reducing hole problems: (1) The shale zones would not enlarge due to sloughing and (2) the better lubrication would minimize torque and drag problems. Caliper logs until recently did not confirm the first claim, and the second claim was proven readily. However, when torque, drag, and wall-sticking problems were reduced, many operators also assumed the sloughing had stopped.

Because caliper logs were not run, the operators were unaware that sloughing had continued. It was discovered that hole enlargement continued when using oil-based muds because fresh water entrained in the oil-based mud was pulled out of the mud into the formation, thus wetting the clays. The resultant effect was that the clays continued to slough.

Mondshine suggested using calcium chloride in the emulsified water of an oil-based mud to prevent wetting formation clays.[1] He showed that if the water emulsified in the oil was fresh, it could be pulled out of the mud into the formation. Also, if the formation water was less salty than the water in the oil-based mud, water could be drawn out of the formation. In the water phase of oil muds, calcium chloride rather than sodium chloride—because of its greater solubility in water—proved successful in many field tests. Shale sections that had exhibited enlargement in previous drilling were kept near bit size.

Oil muds have been very effective in reducing torque, drag, and pipe-sticking problems. The primary application for this purpose has been in directional wells. Oil muds are more temperature stable than water-based muds. In deep wells where temperatures may exceed 350°F, oil muds are used frequently because they are more stable, provide better lubrication, and cost about one-third less to maintain than comparable-weight water-based muds.

A word of caution is necessary in changing from water-based to oil-based mud. In a deep well with several thousand feet of open hole, the operator is gambling when changing from a water-based to an oil-based mud. The open formations generally have adsorbed large quantities of water, and this water becomes a contaminant in the oil-based mud. Several days may be required to make the change, and the operator is taking a chance on losing the hole in the conversion process. If oil-based mud is to be used after using a water-based mud, the change should be inside a cased hole.

Oil as the continuous-phase liquid has some disadvantages:

- Oil costs more than water.
- Environmental pollution problems are increased.
- Drilling crews generally do not like to work around oil.
- Annular circulating pressures may be higher with oil-based muds.
- Gas kicks are more difficult to control because of the solubility of gas in oil.

Rental and buy-back arrangements have reduced the costs of oil-based muds. In addition, the daily maintenance requirements, particularly with weighted muds, are less with oil than with water.

Environmental pollution problems with oil are reduced by using non-polluting oils. A common oil liquid is number 2 diesel oil. Where pollution problems exist, such as in offshore areas, the diesel oil generally is replaced with mineral oils. This change minimizes the pollution problem; however, special oils cost more.

Generally, drilling crews dislike oil-based muds. Clothes do not last as long in oil, and working on the rig floor is more difficult. To compensate, crews usually are paid more when oil is being used.

The claim that annular circulating pressures may be higher using oil-based muds as compared with water-based muds irritates some oil-based mud proponents. Nevertheless, that is a fact.

This warning is given because lost-circulation problems are often more severe with oil-based than with water-based muds. This happens because oil-based muds are frequently thicker at the low annular shear rates than water-based muds. The annular effects are frequently unrecognized because the standpipe circulating pressures are sometimes less with oil-based muds. With variable-speed viscometers, the thickness of the mud can be checked quickly.

One additional word of caution: remember that pressure thickens an oil-based mud and has almost no effect on a water-based mud. Under normal conditions temperature thins both oil-based and water-based muds.

A mixture of oil and water is termed an *oil emulsion mud* if oil is added to a continuous water phase. It is called an *invert emulsion* if water is added to a continuous oil phase. Oil emulsion muds improve lubrication. Bit balling often is reduced when drilling in shales. Also, torque, drag, and pipe sticking problems may be lessened by adding 4–10% oil to a water-based mud. On some occasions, oil is added in larger quantities to water-based muds to lower the mud weight. Water may be used in oil-based muds to reduce cost and to better control the mud viscosity.

Solids

Solids for water-based muds, as shown in Table 5–4, are divided into two groups: low and high gravity. Low-gravity solids are further divided into nonreactive and reactive groups. As the term infers, nonreactive solids are those that do not react to a change in environment. The low-gravity nonreactive solids consist of sand, chert, limestone, dolomite, some shales, and mixtures of many minerals.

These solids are generally undesirable, and when larger than 15 microns in size, they may create an erosive environment that is detrimental to circulating equipment. The API classification for sand is any solid larger than 74 microns; however, many solids smaller than sand are detrimental to equipment. Table 5–4 shows solid sizes.

Reactive solids are clays, and the term "reactive" describes the action of these solids in water. Many definitions are given for clays, some of which are as follows:

■ **TABLE 5–4** Solid sizes

Microns	Inches	Shaker Screen Size
1,540	0.0606	12 × 12
1,230	0.0483	14 × 14
1,020	0.0403	16 × 16
920	0.0362	18 × 18
765	0.0303	20 × 20
210	0.00827	60 mesh
147	0.00579	100 mesh
74	0.00291	200 mesh
44	0.00173	325 mesh

- A solid with an equivalent diameter of less than 2 μ
- An electrically charged particle capable of adsorbing water
- A material that gives the appearance of swelling when water is adsorbed.

Two types of clays will be considered. The most common is sodium montmorillonite, commonly called either bentonite or gel. The other is attapulgite, or salt gel.

Actually, bentonite as marketed is not pure sodium montmorillonite. It is estimated that the best bentonite contains about 60–70% sodium montmorillonite. The other 40% might be calcium montmorillonite or other low-yielding clays, such as kaolonite. Sodium montmorillonite is a plate-like material which is often compared with the pages of a book. The plates are thin, and the total particle size may be less than 0.1 μ.

Hydration or swelling is accomplished primarily by the absorption of water to the surface of the clay. The amount of swelling that will occur observed by the measured increase in mud thickness or viscosity, depends on the available surface area and the total amount of water held on the clay. It has been theorized that a sodium montmorillonite plate would appear as shown in Fig. 5–1 if none of the edges had been broken.

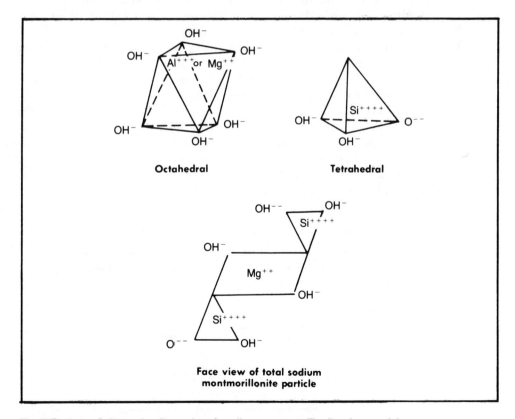

Octahedral

Tetrahedral

Face view of total sodium montmorillonite particle

■ **FIG. 5–1** Schematic diagrams of sodium montmorillonite clay particle

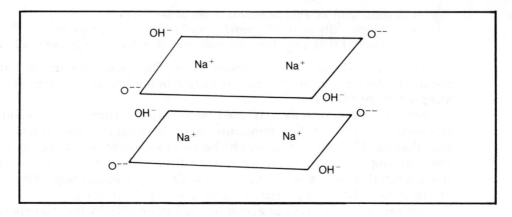

■ **FIG. 5–2** Schematic diagram showing sodium montmorillonite as plates with adsorbed sodium

Note that the basic clay structure does not contain sodium, which is an adsorbed cation as shown in Fig. 5–2. Many other cations may be adsorbed, such as aluminum, calcium, barium, potassium, hydrogen, and magnesium. Because these cations are adsorbed, they may be exchanged by changing the environment of the clays.

Clays with different adsorbed cations look the same, and the only way to differentiate is to establish standard tests. Therefore, bentonite must meet the API specifications listed in Table 5–5.

Clay yield in Table 5–5 is defined as the number of barrels of 15-cp mud that can be obtained using a ton (2,000 lb) of dry clay. In this case the 15-cp viscosity is determined by dividing the 600 reading by 2 as taken from a standard-size rotating viscometer. Note that the bentonite specifications require a clay yield of 91.8 bb/ton. The normal laboratory test simulates field units by using 22.5 g of bentonite in 350 cc of distilled water, which is comparable to 22.5 lb of clay/bbl of water.

■ **TABLE 5–5** API specifications for bentonite

Requirement	(API Standard 13A, March 1965)
Yield point, lb_f/100 sq ft	3 × plastic viscosity maximum
Filtrate	13.5 ml maximum
Wet screen analysis	
Residue on U.S. sieve no. 200	2.5% maximum
Moisture	10% maximum as shipped from point of manufacture
Yield	91.8 bbl of 15-cp mud/ton of dry bentonite

The plastic viscosity test is a fineness specification; yield point is a measure of clay activity, created primarily by the charged surfaces. The specification most overlooked is the one on water loss. Bentonite must have good filtration control characteristics; surprisingly, bentonite is the best material used in mud for controlling filtration rate.

These specification tests should be run in distilled water. As noted, the bentonite is reactive, and its hydration qualities are affected by changes in water composition. The yield of bentonite is reduced as the salt content, sodium chloride (NaCl), is increased. For example, there is no noticeable yield from bentonite in saturated salt water, and the yield of bentonite is reduced by more than 50% in sea water containing 50,000 ppm salt. Exact levels of clay yield are hard to obtain as a function of salt concentration because of other impurities in both the clay and water. The effect of salt on clay yield is known, but the exact causes are more difficult to explain.

Fig. 5–3 shows a hypothetical sketch of the clay plates in a saturated saltwater environment. All unsatisfied charges are satisfied with sodium cations. The bentonite particle becomes essentially inert. Failure of the bentonite particles to hydrate results in very little increase in mud thickness and a minimal reduction in filtration rate as the concentration of bentonite is increased. To be effective, the bentonite needs to be prehydrated in fresh water before it is added to saturated salt water. Substantial benefits are also obtained by prehydrating bentonite before it is added to sea water. The preferred method of prehydrating bentonite is as follows.

The bentonite is mixed in a tank of fresh water. The concentration varies, but about 40 ppb is all of the bentonite that can be added and still obtain a pumpable slurry. Adding about 2–4 ppb of lignosulfonate after hydrating prolongs the effects of the prehydration. The time required for complete hydration may vary from 6 to 24 hr. Usually, 6 hr provides adequate time if there is some reason to hurry.

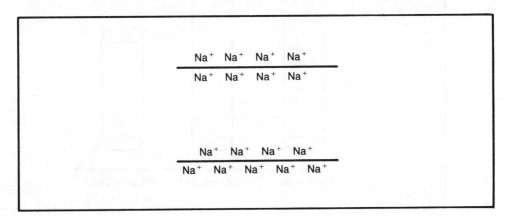

■ **FIG. 5–3** Bentonite clay in salt water

The yield of bentonite is also reduced when other cations—such as those of calcium, magnesium, and potassium—are present in the water. Calcium often is used in muds to prevent swelling formation clays. Muds of this type have been classified as lime, gyp, and calcium chloride systems. Actually, the lime-treated muds had only small quantities of calcium in solution.

In normal high-pH lime mud, only about 150 ppm calcium is in solution. In gyp muds, about 800 ppm calcium is allowed in solution. In the calcium chloride muds, as much as 3,500 ppm calcium is in solution. The effects of calcium on the swelling of bentonite is shown in Fig. 5–4.

The difference in soluble calcium in Fig. 5–4 is primarily due to the pH level of the mud. Calcium solubility is affected by pH, as shown in Fig. 5–5. Lime muds have a pH of about 12.0, and the amount of soluble calcium is a direct result of this pH level. Almost any mud may be converted to lime mud by raising the pH and adding calcium if it is not already present.

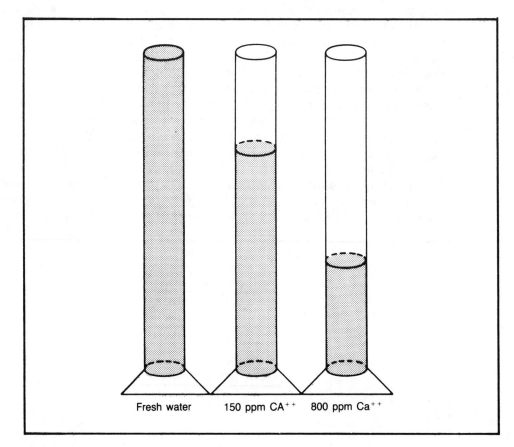

Fresh water   150 ppm CA^{++} 800 ppm Ca^{++}

■ **FIG. 5–4** Effect of calcium on clay swelling

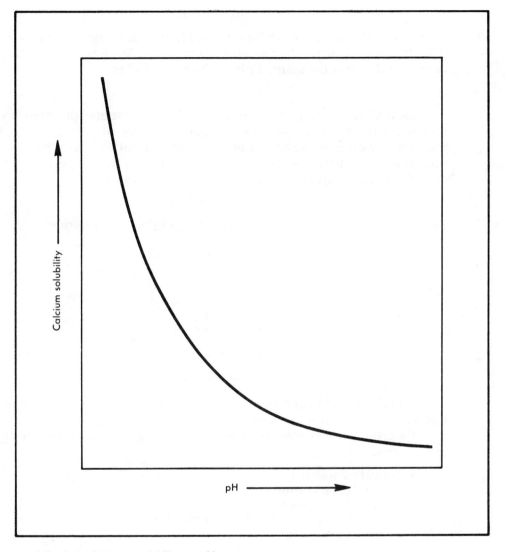

■ **FIG. 5–5** Calcium solubility vs pH

All chemical reactions are affected by their environment. In many cases muds have all types of different materials present, which substantially affects chemical reactions. One of the most important measurable characteristics of muds is pH. For this reason, pH will be defined and examined in detail.

pH

The pH value is defined as the negative logarithm of the hydrogen ion, H^+, content. This is shown as follows:

$$pH = -\log H^+ \tag{5.2}$$

Further significance can be attached to this definition when we realize that the product of the hydrogen ion concentration and the hydroxyl ion concentration is a constant. This is shown as follows:

$$H^+ \times OH^- = 1 \times 10^{-14} \tag{5.3}$$

The hydrogen ion represents the acidic component; the hydroxyl ion represents the basic or alkaline component. Thus, anything that reduces the concentration of the hydrogen ion results in an increase in pH. A neutral solution, typical of pure or distilled water, has the same concentration of hydrogen and hydroxyl ions. This is shown below:

$$H^+ = OH^- = 1 \times 10^{-7} \tag{5.4}$$

For this type of solution, the pH equals 7. Other combinations include:

• H^+	OH^-	pH
10^0	10^{-14}	0
10^{-4}	10^{-10}	4
10^{-7}	10^{-7}	7
10^{-9}	10^{-5}	9
10^{-11}	10^{-3}	11
10^{-14}	10^0	14

Thus, the pH of 14 represents the maximum concentration of hydroxyl ions and the minimum concentration of hydrogen ions.

The pH of water-based drilling muds may be increased for the following reasons:

- Control corrosion
- Activate organic thinners
- Stabilization temperature
- Raise toleration level for high solids content
- Reduce contamination

There is some dispute over the use of a high-pH mud to reduce corrosion. In areas where hydrogen sulfide gas is present, a pH above 10 will limit the corrosion effects of the hydrogen sulfide. If carbon dioxide is present, maintaining a high pH may be difficult because of the interaction between the hydroxyl ion and carbon dioxide, which is shown as follows:

$$CO_2 + OH^- = HCO_3^- \tag{5.5}$$

The formation of carbonic acid will lower the pH rapidly. Thus, it may be impossible to maintain a high-pH mud if carbon dioxide is also present. For this reason, when both hydrogen sulfide and carbon dioxide are present, normally oil-based muds are used. If an operator is caught unexpectedly in a situation in which both carbon dioxide and hydrogen sulfide enter a water-based mud, he should plan to use an amine type of coating agent immediately.

The need for a pH at the 9.0-plus level is generally a requirement when using organic mud thinners. If tannin products are used, the pH needs to be at the 11.0-plus level. One alternative is to premix the organic thinners in a chemical barrel with sodium hydroxide. A normal ratio in a premix chemical barrel is one-half caustic and one-half thinner, by weight. If the mud pH is lower than desired, use more caustic. If the mud pH is higher than desired, use less caustic.

The temperature stability and a higher toleration level for solids have been debatable issues. Water-based muds are more stable at high temperatures at a high pH (above 11.0). The higher toleration level for solids is probably related to the fact that the solubility of many contaminants is reduced as the pH is increased. For example, a 16.0-lb/gal mud may have a maximum solids toleration level of 36% by volume at a pH of 9.5. The same mud may tolerate a solids content of 40% by volume at a pH of 12.0.

Some claimed disadvantages to increasing pH above the neutral level are listed as follows:

- Sodium hydroxide or caustic is expensive.
- An alkaline solution is toxic and becomes more so as the pH is increased.
- It is difficult to maintain an elevated pH in a saltwater environment.
- Cement forms at high pH levels

Large quantities of caustic must be used to maintain the pH above 11.0 because the hydroxyl ions plate-out on solids and form other compounds with contaminants. For this reason, the cost may be high if a high-pH mud is to be utilized.

Sodium hydroxide is a very volatile and toxic material. If a premix tank for chemical additions is used, the caustic should be added to water already in the premix tank. Never put the caustic in a premix tank and pour water over the caustic. When water has been poured over caustic, an explosion may occur, causing serious injury. Also, at a pH level above 11.0, the mud may tend to destroy clothes and attack skin on contact.

In saturated saltwater muds, maintaining a high-pH mud is virtually impossible because the salt is a buffering agent. Thus, when there is a high salt content in the water, use an almost neutral pH and amine type of coating agents for corrosion control.

When high-pH lime muds were commonly used during 1950–60, there were claims that at high temperatures (above 250°F) the mud would solidify. Actual cases were reported in which the high-pH mud had formed a cement-like material. As pH was increased, calcium was less soluble, and the excess calcium hydroxide was increased. The reaction was as follows:

$$Ca^{++} + 2OH^- \rightarrow Ca(OH)_2$$
$$Ca(OH)_2 + \text{solids} + \text{temperature} = \text{cement}$$

There was no definite level at which cement was formed, but the problem was recognized. Most operators quit using high-weight, high-pH lime muds. The problem may still exist when the pH is increased to a 12.0 level if calcium is present in the water and the solids content of the mud is high.

Measurement of pH

The pH measurements in field operations are made by color strip charts or a pH meter or are estimated from alkalinity measurements. Color strip charts are rough estimates and probably suffice if the pH measurement is unimportant. If accuracy is needed, a pH meter should be used.

Alkalinity measurements are made by using hydrogen ions to neutralize the hydroxyl ions. The purpose of the alkalinity measurement is to measure the pH by the amount of hydrogen used. The problem with accuracy is that other materials in the mud unite with the hydrogen ion. This gives a false indication of the hydroxyl ion content. For example, the carbonate anion will unite with hydrogen as follows:

$$H^+ + CO_3^- \longrightarrow HCO_3^-$$

The bicarbonate anion, HCO_3^- is formed. Further additions of hydrogen ions may result in the following reaction:

$$HCO_3^- + H^+ \longrightarrow H_2O + CO_2$$

Thus the carbonate and bicarbonate ions unite with hydrogen, giving a false reading of the hydroxyl ion concentration.

Several alkalinity measurements have been made on drilling muds. Some of these are listed below:

$$P_{mf} = \text{alkalinity of mud filtrate}$$
$$P_m = \text{alkalinity of mud}$$
$$M_f = \text{alkalinity of mid filtrate}$$
$$P_{m1} \text{ and } P_{m2} = \text{alkalinity measurements of mud filtrate}$$

Rule-of-thumb estimates have been formed using these alkalinity measurements. The P_{mf} measurement is used often with the P_m measurement to estimate excess lime content, as shown below:

$$\text{Excess lime} = \frac{P_m - P_{mf}}{4} \qquad (5.6)$$

If the M_f alkalinity is equal to or more than 5.0 P_{mf}, assume the P_{mf} alkalinity is affected by carbonates rather than hydroxyl ions. The P_{m1} and P_{m2} measurements are made to provide a better estimate of the carbonate and bicarbonate ions. In actual fact, the carbonate ion does not contaminate the mud; it simply gives a false reading of the hydroxyl ion concentration. Thus, an ideal treatment, when carbonate is present, is to treat with calcium hydroxide. The effect is shown below:

$$Ca(OH)_2 \longrightarrow Ca^{++} + 2OH^-$$
$$CO_3^{--} + Ca^{++} \longrightarrow CaCO_3$$

Consequently, the calcium removes the carbonate as calcium carbonate and the hydroxyl ions remain in solution.

Sometimes a freshwater mud thickens when salt, calcium, or some similar contaminant is added either purposely or from formations being drilled. The mud thickens because of the flocculation of clays as shown in Fig. 5–6. As noted, the imbalance of charges on the edges of the clays results in an edge-to-edge attraction. When this happens, the mud thickens because, in addition to the adsorbed water, water is trapped between the flocs of clay as shown in Fig. 5–6. The tendency for freshwater clays to flocculate is reduced appreciably as the pH increases.

Salt Gel

Salt gel, an attapulgite clay, has been called a needle-like material. It does not occur in plates like bentonite, and yield or viscosity characteristics are obtained by vigorous agitation and cross-linking of particles. Actually, a salt-gel mud is similar to a normal flocculated mud. The particles are cross-linked, and water is trapped. There is essentially no water-loss control. Frequent additions of the salt gel are generally necessary to maintain a consistent mud thickness.

One question that arises is at what salt concentration should salt gel be used to replace bentonite. Bentonite is the preferred material always, even in cases when it must be prehydrated before being added to the saltwater fluid. Salt gel is more expensive, does not have the lubricating qualities of bentonite, provides a poor filter cake, and offers no aid to the control of filtration. It should be used only on a special application basis.

High-Gravity Solids

Barite is the high-gravity nonreactive solid used to increase mud weight. Barite is primarily barium sulfate ($BaSO_4$). API specifications for barite are shown in Table 5–6.

Note that the specific gravity minimum is 4.2. This is important. Barites with specific gravities as low as 3.9 have been detected in the field. Barite of this grade would make it almost impossible to run a 16.0-lb/gal mud or heavier without experiencing excessive viscosity problems. A low-grade barite increases chemical thinner requirements, and it may result in excessive circulating pressures in deep, high-temperature wells.

Other weighting materials in use are various grades of iron ore.

There are two types of iron ore utilized: illomenite, which has a specific gravity of about 4.7, and a material called itabarite, which has a specific gravity of 5.1. The primary complaint against using iron ore is the erosion

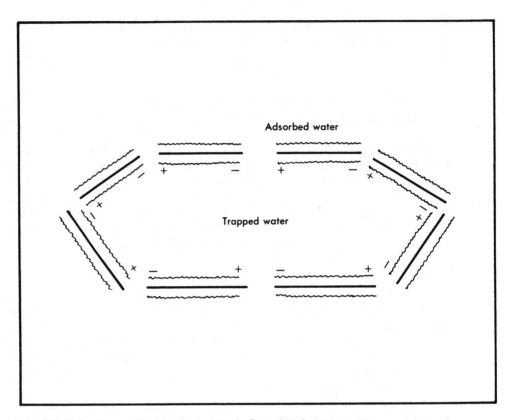

■ **FIG. 5–6** Clays with adsorbed water in flocculated state

of circulating equipment. Field and laboratory tests show that the erosional characteristics of iron ore can be reduced by a finer grind. Consequently, iron ore is ground finer than barite.

The higher specific gravity of iron ore offers the advantage of a lower solids content with weighted muds. Iron ores are used commonly in oil-based muds. There have been several reports of improved drilling rates when using iron ore as compared to barite.

Solids for oil-based muds are also divided into two groups, low- and high-gravity solids, as shown in Table 5–1. The low-gravity solids are generally clays that have been treated to oil-wet rather than water-wet surfaces. Often these clays are referred to as organophyllic clays.

In an oil environment, there are no electrical or chemical reactions as shown for the same type of clay in water-based muds. These clays in oil-based muds do the same thing as the clays in water-based muds: they provide the primary viscous characteristics of oil-based muds. In fact, in oil-based muds, if the operator wants to increase viscosity, the most common procedure is to add more clay. Viscous sweeps can also be used for

■ **TABLE 5-6** Barite physical and chemical requirements

Requirement	Numerical Value
Specific gravity	4.20 minimum
Soluble alkaline earth metals	
as calcium	250 ppm maximum
Wet screen analysis	
Residue on U.S. Sieve No. 200	3.0% maximum
Residue on U.S. Sieve No. 325	5.0% minimum

hole cleaning with oil-based muds by simply adding clay to the oil until the desired viscosity level is reached.

Other types of low-gravity solids may be used in oil muds. Some of the early oil-based muds used asphaltic materials as the primary source of low-gravity solids.

The high-gravity solids in oil-based muds are the same high-gravity solids used in water-based muds. Surfactants are added to the oil-based muds to preferentially oil-wet the surface of the weighting material and drill solids. Iron ore is used more frequently in oil-based muds for weighting material than in water-base muds.

CONTROL OF DRILLING MUD

Mud control can be divided into three basic categories: controlling mud weight, controlling viscosity and gel strength, and controlling water loss.

Controlling Mud Weight

Mud-weight control is almost synonymous with solids control. In recent years the emphasis on low-solids mud has been primarily an emphasis on low-weight muds. Mud-weight requirements should be based on that required to control formation pressures.

In coastal areas, normal formation pressures are generally in the range of 0.465 psi/ft. This equals a 9.0-lb/gal mud. Thus, to contain formation fluids during drilling operations, at least a 9.2-lb/gal mud is required. In inland areas it is not uncommon to find normal formation pressures as low as 0.40 psi/ft, which can be controlled adequately with fresh water. It may be necessary to use low-density fluids to maintain circulation. This has been true in the Rocky Mountain areas of Utah when formations are fractured and formation fluid pressures are less than an equivalent 5.0-lb/gal level.

The operator is encouraged to keep the mud weight just high enough to control formation pore pressures. As mud weight increases, drilling rates decrease, mud costs increase, and hole problems become more common.

In fresh water, a 9.2-lb/gal mud has about 7% low-gravity solids by volume, and a 10.0-lb/gal mud has about 13.5% low-gravity solids by volume. Fig. 5–7 is calculated using Eq. 5.7:

$$\rho = \rho_w(1 - X) + 20.8X \qquad (5.7)$$

where:

ρ = mud weight, lb/gal
ρ_w = water weight, lb/gal
X = solids fraction
20.8 = weight of low-gravity solids, lb/gal

The use of Eq. 5.7 is shown in Example 5.1.

☐ **EXAMPLE 5.1**

Assume:

Solids content = 0.07
All solids are low gravity
Liquid base is fresh water

$$\rho = (8.33)(1 - 0.07) + 20.8(0.07) = 9.2 \text{ lb/gal}$$

Mud weight can be used to determine the solids content in an unweighted mud. Methods for controlling the solids content are given in chapter 6.

While the equipment and techniques are available to control the mud weight going into the hole, the annular mud weight may become excessive. The increase in annular mud weight vs the rate of penetration is shown in Fig. 5–8, where it is assumed the input mud weighs 9.0 lb/gal.

This annular mud weight in Fig. 5–8 is based on the indicated drilling and circulation rates in a 12¼-in. hole and is calculated as shown in Example 5.2.

☐ **EXAMPLE 5.2**

12¼-in. hole volume = 0.15 bbl/ft
At 20 ft/min drilling rate, 3.0 bbl/min of cuttings are generated
Weight of cuttings = 875 ppb × 3 bbl/min = 2,625 lb/min

$$\text{Volume of cuttings} = \frac{2,625}{20.8} = 126.3 \text{ gpm}$$

Volume of input mud = 600 gpm
Weight of input mud = 600 gpm × 9 lb/gal = 5,400 lb/min

$$\text{Annular mud weight} = \frac{(5,400 + 2,625)}{(600 + 126)} = 11.0 \text{ lb/gal}$$

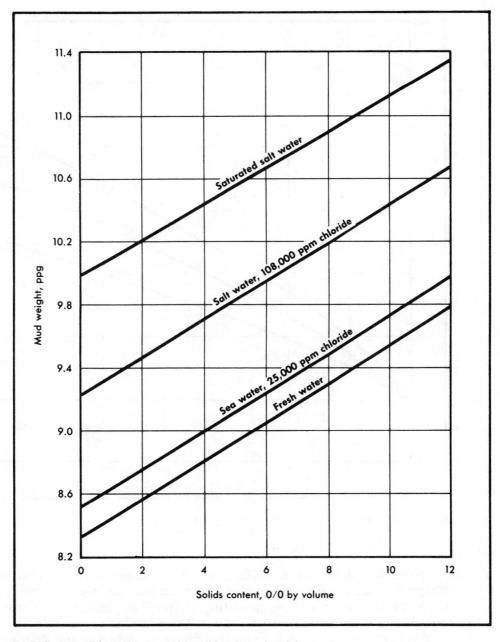

■ **FIG. 5–7** Effect of low-gravity solids on mud weight

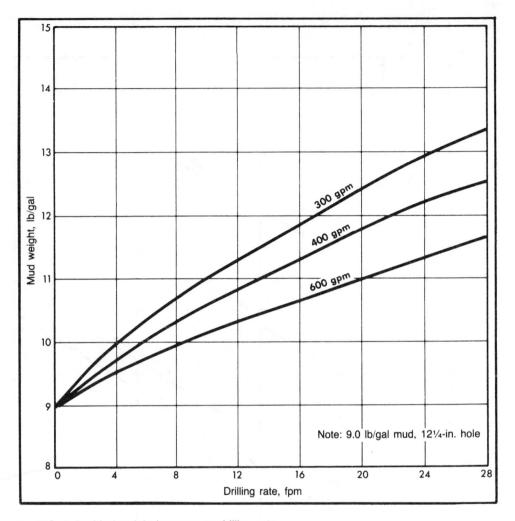

■ **FIG. 5–8** Mud weight increase vs drilling rate

Note that at the same drilling rate a reduction in circulation rate to 300 gpm would increase the mud weight to 12.4 lb/gal. Because these conditions are in a large surface hole, the friction-loss change due to altering circulation rate would be small. Thus, to minimize the hydrostatic head in the annulus, the operator has the choice of reducing the drilling rate or increasing the circulation rate.

Circulation is sometimes lost in large surface holes while drilling at high rates. Because the mud weights are predictable and the fracture gradients can be determined, the operator should not exceed permissible limits.

Mud Weight Less than Water Weight

Mud weight can be reduced to levels below a freshwater gradient by using oil or oil and fresh water or by aerating the fluid column. An average weight for oil is about 7.0 lb/gal; thus, a 50–50 mixture of only oil and water would weigh about 7.7 lb/gal. This assumes there are no solids. When even lighter mud weights are required, the fluid column can be aerated. Eq. 5.8 can be used to calculate air requirements for a given reduction in mud weight.

$$\frac{V_1}{100 - V_1} = \frac{(D_w \bar{\rho}_m - P_T) z_s T_s}{P_s z_a T_a \left(\ln \dfrac{P_T + P_s}{P_s} \right)} \tag{5.8}$$

where:

$$
\begin{aligned}
V_1 &= \text{gas volume, cu ft} \\
100 - V_1 &= \text{liquid volume, cu ft} \\
D_w &= \text{well depth, ft} \\
\bar{\rho}_m &= \text{mud gradient, psi/ft} \\
P_T &= \text{total pressure desired, psi} \\
z_s &= \text{compressibility factor at surface for gas} \\
T_s &= \text{temperature at surface, °R} \\
P_s &= \text{pressure at surface, psi} \\
z_a &= \text{average compressibility factor for gas} \\
T_a &= \text{average temperature, °R}
\end{aligned}
$$

This equation is used in Example 5.3.

□   **EXAMPLE 5.3**

Assume:

Well depth = 5,000 ft
Maximum permissible fluid pressure = 1,500 psi
Drilling fluid = water
$T_b = 140°F$
$z_b = 1.1$
$T_s = 60°F$
$z_s = 1.0$
$P_s = 14.7$ psia

SOLUTION:

$$D_w \rho_m = (5,000)(0.437) = 2,185 \text{ psia}$$
$$P_T = 1,500 \text{ psia}$$

$$\frac{V_1}{100 - V_1} = \frac{(2,185 - 1,500)(1)(520)}{(14.7)(1.05)(570) \ln \left(\dfrac{1,500 + 14.7}{14.7} \right)} = \frac{8.75 \text{ cu ft of air}}{\text{cu ft of mud}}$$

$$\frac{8.75 \text{ cu ft of air}}{\text{cu ft of mud}} \times 5.615 \text{ cu ft/bbl} = 49.2 \frac{\text{scf of air}}{\text{bbl of mud}}$$

In this case the water weight is reduced to an equivalent weight of 5.78 lb/gal. In some areas, if circulation is to be maintained, the fluid weight will have to be reduced to levels this low or lower.

Weighted Muds

Weighted muds are defined as those where barite has been added to increase mud weight. The measurement of solids content is important in weighted muds. With both low- and high-gravity solids in the mud, it is impossible to determine the solids content as a function of mud weight alone. The measurement of solids content is obtained by using a retort that generally contains no more than 15 cc of mud. A very small error may make a substantial difference in the measurements of solids content.

When conditions are critical, more precise laboratory methods for the determination of solids content should be used as a basis for checking field results.

Fig. 5–9 shows the effect of solids content on a 16.0-lb/gal water-based mud. As shown in Fig. 5–9, the composition of the drilling mud significantly affects the level of solids that can be tolerated in a weighted mud. The mud thickness or viscosity is affected by the following.

- Solids content
- Bentonite content
- pH
- Chemical treatment
- Contaminants present

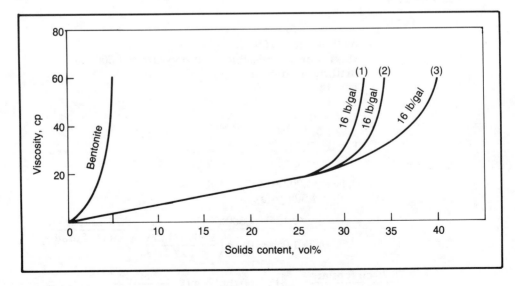

■ **FIG. 5–9**　Solids content vs viscosity

A close control of solids reduces the need to closely watch the other mud properties. Eq. 5.9 is a method to estimate the normal range in solids content for a weighted water-based mud.

$$S_v = 3.2(\rho_m - 6)$$ (5.9)

where:

S_v = solids content, vol%

Empirical data are unavailable to write an equation such as Eq. 5.9 for oil-based muds. In practice, solids do not become a critical factor in the control of oil-based muds.

A solids concentration of bentonite of less than 4% by volume will result in excessive thickening without any additional solids. Consider the solids problem in a 16-lb/gal mud. For a solids concentration of 32 vol%, about 25 vol% is barite and about 7 vol% is low-gravity solids. Only about 2% of the 7 vol% low-gravity solids can be expanded bentonite without excessive viscosity problems. On the other hand, if all low-gravity solids are low-yielding clays and inert materials, an additional 3–6% low-gravity solids can probably be tolerated with no viscosity problems.

In addition, very high concentrations of chemical thinners permit tolerating additional low-gravity solids. The decision on the solids concentration for a given weight mud should be based on the specific operating conditions. In a deep, high-temperature well where torque and drag have been problems, a good bentonite content is desired to reduce wall friction. However, by using a high bentonite content, the operator is lowering the solids toleration level of the mud. Thus, there may be a delicate balance between excessive downhole thickening and excessive torque and drag.

This sensitivity of mud to the type of low-gravity solids has led to a method for estimating the bentonite content in the mud, called the methylene blue test. This test, introduced by Jones, is based on the cation-exchange capacity, surface area, and state of dispersion of suspended clay particles.[2] Bentonite has a high cation-exchange capacity and a high surface area per unit of weight, and it is easily dispersed. For these reasons bentonite absorbs substantially more of the methylene blue solution than ordinary low-yielding clays.

Table 5–7 shows the adsorption capacities of the various bentonite samples and also of other low-yielding clays. The low-yielding clays have an adsorption capacity of less than one-sixth of the bentonite. On this basis it would require at least 6 vol% of low-gravity, low-yielding solids to equal 1 vol% of bentonite. Thus, the measurement of the equivalent bentonite content must also be accompanied by an estimate of the low-gravity solids content before it has any specific meaning. For example, 12–15 vol% low-gravity solids might be equivalent to 18–20 ppb of bentonite. Thus, with essentially no bentonite it would be possible to obtain a reasonable bentonite level.

■ **TABLE 5–7** Adsorption capacities of commercial bentonites for methylene blue

Sample	Adsorption Capacity, g dye/g clay	Calculation Factor*
1	0.271	2.91
2	0.269	2.93
3	0.253	3.12
4	0.267	2.95
5	0.287	2.75
6	0.306	2.58
7	0.284	2.78
8	0.246	3.20
9	0.257	3.07
10	0.280	2.81
11	0.276	2.86
12	0.249	3.17
13	0.292	2.70
14	0.365	2.16
15	0.265	2.97
16	0.271	2.91
17	0.252	3.13
18	0.294	2.68
19	0.270	2.92
Kaolinite	0.05	—
Illite	0.05	—
Attapulgite	0.12	6.5
Pennsylvanian shale	0.05	—

* For finding lb/bbl of bentonite in mud. Multiply factor by ml; 0.45% of methylene blue solution is needed to titrate 2 ml of mud.
From Jones, "New Test Measures Bentonite in Drilling Mud," *OGJ*, 1 June 1964.

In field practice, where weighted muds are often checked for equivalent bentonite content, the level of indicated bentonite may increase. Generally, this indicates an increasing low-gravity solids content that is bentonitic in nature. In high-weight muds there is an upper bentonite toleration level for the mud.

This simply means that excessive thickening may occur at bottom-hole pressure and temperature if this level if exceeded, as shown in Fig. 5–10. The upper limit on low-gravity solids at specific bottom-hole conditions will vary with mud weight, chemical treatment, and the type of low-gravity solids.

This discussion leads to the need for determining the quantity of low- and high-gravity solids in a given weight mud. The first requirement is an accurate measurement of solids content. After measuring the total solids content, determining the concentration of high- and low-gravity solids is

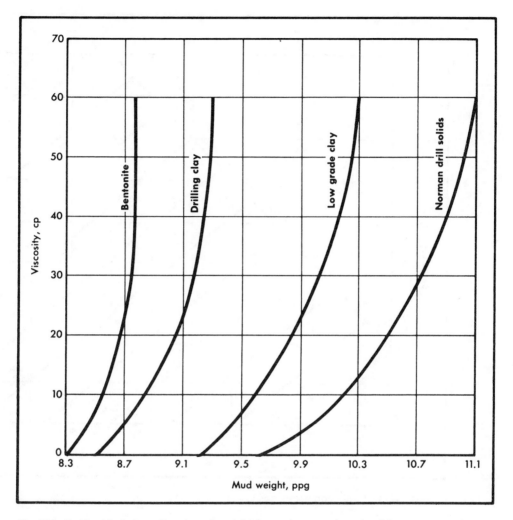

■ **FIG. 5–10** Mud viscosity vs mud weight for various grades of solids

simply a mathematical procedure. Fig. 5–11 can be used to estimate the amount of barite and low-gravity solids after the measurement of solids content. This chart was prepared from a simple mass balance as shown in Example 5.4.

□   **EXAMPLE 5.4**

Assume:

> Mud weight = 16 lb/gal
> Solids content = 32 vol%
> Specific gravity of low-gravity solids = 2.5
> Specific gravity of high-gravity solids = 4.3

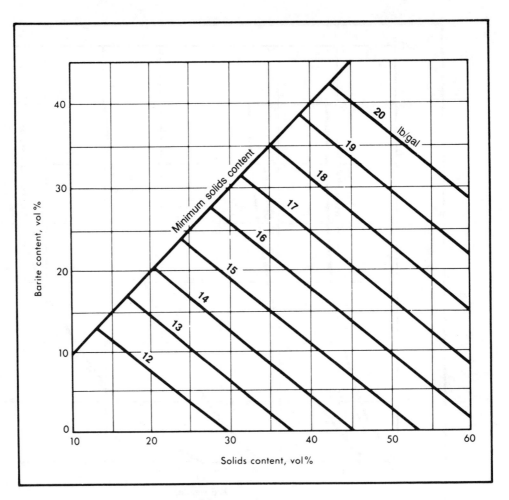

■ **FIG. 5–11** Barite content vs solids content

SOLUTION:

Let X = volume fraction of low gravity solids
Let 0.32 − X = volume fraction of high-gravity solids

Then:

$(X)20.8 + 35.8(0.32 − X) + (8.33)(0.68) = 16$
$X = 0.074$ (low-gravity solids)
$0.32 − X = 0.246$ (barite)

If other materials such as oil are in the mud, they can be added directly to the same mass balance. Some low-gravity solids weigh more and some less than 20.8 lb/gal; barite will still meet API specifications if it weights 35.0 lb/gal. These mass balances can be used to develop general equations for the barite required to obtain a given mud weight increase. Eqs. 5.10 and 5.11 have been developed for this purpose.

$$X = \frac{\rho_2 - \rho_1}{35.0 - \rho_1} \tag{5.10}$$

$$X' = \frac{\rho_2 - \rho_1}{35.0 - \rho_2} \tag{5.11}$$

where:

X = barite/bbl of final mud, bbl
X' = of barite/bbl of initial mud, bbl
ρ_1 = initial mud weight before adding more barite, lb/gal
ρ_2 = final mud weight after adding barite, lb/gal

Eq. 5.10 has more utility because it recognizes the limits on final mud volume. Example 5.5 illustrates the use of Eq. 5.10.

☐ **EXAMPLE 5.5**

Assume:

ρ_1 = 10 lb/gal
ρ_2 = 16 lb/gal
Mud volume = 1,000 bbl
Desired mud volume after increasing mud weight = 1,000 bbl

SOLUTION (using Eq. 5.10):

$$X = \frac{16 - 10}{35.0 - 10} = \frac{6}{25.0} = 0.24 \text{ bbl of barite/bbl of final mud}$$
Required barite = (1,000)(0.24) = 240 bbl

Therefore, 240 bbl of mud must be discarded before increasing mud weight. This 240 bbl of mud will then be replaced with 240 bbl of barite. Example 5.6 illustrates the use of Eq. 5.11.

☐ **EXAMPLE 5.6**

Assumptions:

Same as Example 5.4

SOLUTION:

$$X = \frac{16 - 10}{35.0 - 16} = \frac{6}{19.0} = 0.316 \text{ bbl of barite/bbl of initial mud}$$
Required barite = (1,000)(0.316) = 316 bbl
Total final volume = 1,000 + 316 = 1,316 bbl

Eq. 5.11 does not indicate how mud should be discarded for a final volume of 1,000 bbl. In fact, starting with 760 bbl of mud, Eq. 5.11 shows 240 bbl of barite (760 × 0.316 = 240) is needed to increase the mud weight to 16 lb/gal. Thus, only Eq. 5.10 should be used, even though many charts are available using Eq. 5.11. However, neither equation takes into consideration the solids content of the mud.

Special problems that may be encountered when making decisions to increase mud weight are as follows:

- At what mud weight should barite be used to obtain further increases in mud weight?
- Should oil be used in weighted water-based muds for lubrication?
- How high should the solids content be allowed in a specific weight mud before some means of reducing the solids content is used?
- How much bentonite should be added to weighted water-based muds?

The first problem is encountered most often in the pressure-transition zone of abnormal-pressure wells. To illustrate, a set of conditions has been assumed and the results are examined in Example 5.7.

☐   **EXAMPLE 5.7**

Assume:

1,000 bbl of mud
Mud weight = 10.0 lb/gal, unweighted
Solids content = 13.5 vol%

In five drilling days the mud weight will be increased to 16 lb/gal. Maximum desired solids content for 16-lb/gal mud = 32 vol%.

SOLUTION:

From Example 5.4, the 16-lb/gal mud with 32 vol% solids contains 7.4% of low-gravity solids. In 1,000 bbl this means 74 bbl can be low-gravity solids. In Example 5.5, where the mud weight was increased to 16 lb/gal, 240 bbl of mud was discarded. Starting with 760 bbl of 10-lb/gal mud means the operator would also be starting with (760 × 0.135) = 103 bbl of low-gravity solids. The net result would be that some of the weighted mud would have to be discarded before reaching the 16-lb/gal level.

A quick way to determine how much mud should be discarded can be obtained by determining the quantity of low-gravity solids that will be tolerated in a given final mud weight; divide this number by the solids content of the unweighted mud. In this example 74 bbl of low-gravity solids can be tolerated in 1,000 bbl of 16-lb/gal mud with a 32 vol% solids content. Dividing 74 by 0.135, the solids content of the 10-lb/gal unweighted mud gives 548 bbl of the 10-lb/gal mud that can be retained. If the calculation shown in Example 5.5 was followed, some of the weighted mud would have to be discarded before reaching the 16-lb/gal level.

These assumed conditions give a basis for determining the maximum mud weight before adding barite. Specifically, the following additions would be required:

548 bbl of original mud
246 bbl of barite
206 bbl of additional water

Adding the water to the retained mud shows the maximum permissible weight of the unweighted mud before adding barite. This is shown as follows:

	lb	gal
548 × 10 = 5,480		548
206 × 8.33 = 1,715		206
Total = 7,195		754

Maximum permissible mud weight before adding barite = 7,195/754 = 9.55 lb/gal.

This example illustrates one of the problems the operator faces in controlling costs. His course of action will have to be determined by the actual and anticipated mud-weight requirements. If the maximum mud weight will be 16.0 lb/gal and no unusual problems are anticipated, it may be unrealistic to emphasize a maximum solids content of 32 vol%. If, on the other hand, the mud weight may eventually exceed 17.0 lb/gal and several weeks of abnormal-pressure drilling are anticipated at temperatures above 300°F, he should have a minimum solids content before he begins to add barite.

On this basis specifying a maximum solids content of 31 vol% for the 16.0-lb/gal mud would not be unreasonable. With a maximum total solids of 31 vol%, the mud would contain the following:

> 412 bbl of original mud
> 254 bbl of barite
> 334 bbl of additional water

The mud weight before adding barite could not exceed 9.25 lb/gal. Considering the fact that the solids in the mud are primarily undesirable, the best approach might be to discard all of the old mud and begin drilling at the desired weight with freshwater bentonite and barite. Many other situations could be considered; however, this would serve no useful purpose. In actual operations, the cheapest and best method of mud-weight control can be determined by a few simple calculations such as those shown in these examples. The guesswork applied is often unnecessary and expensive.

The next problem is using oil for lubrication. In some cases the problem has been solved because of disposal problems. Oil emulsified in weighted muds takes up space, i.e., it reduces the free water content. This has the effect of increasing thickness and gel strengths. On the viscosity vs solids concentration curve shown in Fig. 5–10, the increase would move the equivalent solids to the right on any of the curves.

The problem may become critical in high-weight muds where the solids content is high by necessity. Oil additions are not recommended when mud weights exceed 17.0 lb/gal in deep high-temperature wells. Exceptions will exist when lubrication, by oil or from materials that must be added with oil, is necessary to minimize torque and drag.

Solids control and the use of bentonite in weighted muds are problems that continue to plague operations. In actual fact, definite limits cannot be set for either the total solids or bentonite content. Bentonite increases the thickness of mud, and in general when using weighted muds, one of the primary problems is keeping the mud thin.

On this basis, bentonite would be added only when barite begins to settle. However, bentonite helps improve the filter cake, and this helps reduce torque, drag, and differential pressure sticking. Therefore, bentonite would be added when improvements in lubrication are needed. Unfortunately, conflicting problems sometimes occur when thinning and better lubrication are needed at the same time.

Under such conditions the operator must compromise. The compromise may be based on economic considerations, or it may be based on which problem is the most pressing. Other special materials are available to improve lubrication, and muds may be thinned by raising pH, reducing the total solids content, and increasing chemical treatment.

Viscosity and Gel-Strength Control

Viscosity describes the thickness of muds in motion, and gel strength describes the thickness of muds that have been left quiescent for a period of time. In scientific terms, viscosity is a proportionality constant between shear stress and shear rate for Newtonian fluids in laminar flow. Thus, as a constant, shear rate would have no effect on viscosity. This is correct for true or Newtonian fluids such as water; it is not true for drilling muds. All drilling fluids shear thin. Therefore, the proportionality between shear stress and shear rate is reduced as shear rate is increased. As a result, the original scientific meaning of viscosity has been altered; for drilling muds it is used in a different context than for true fluids.

In drilling muds, viscosity has been adopted as the common expression for describing thickness. Unfortunately, certain thickness levels have been prescribed for drilling muds based on common usage rather than on current requirements. Consequently, this discussion will be dedicated first to defining objectives, second to the methods of measurement, and third to the methods of control.

Objectives

Mud thickness is controlled for the following direct reasons:

- To control circulating pressure losses in the annulus
- To provide adequate lifting capacity for removing formation solids
- To help control surge and swab pressures

Indirect considerations include the following:

- Drilling rates are higher with low-solids, thin fluids
- Muds may be thickened to minimize erosion in some unconsolidated shale formations because turbulent flow patterns with thin fluids may erode and excessively enlarge the hole

Methods of Measurement

The methods of measuring mud thickness are as follows:

1. *Marsh funnel.* This was the first method used to determine mud thickness. The measurement is made by comparing the time required for one quart of mud to run out of the funnel to the time required for one quart of water. The funnel is a calibrated instrument that is filled with 1,500 cc of fluid, and the fluid discharge is through a sized nozzle. One quart of water is supposed to be discharged in 26.5 sec. The relative time for the discharge of one quart of mud is an indication of mud thickness.

   There is no quantitative basis for using this number. For example, a funnel viscosity of 200 sec is not even proof that the mud is thicker than one with 100 sec when both fluids are in motion. Thus, the only benefit to be obtained from using funnel viscosity is to detect changes in mud properties that may be indicative of potential downhole problems.

2. *Rotating viscometer.* Several models of rotating viscometers are available. Field models are usually normal temperature types, which run at only two speeds, 300 and 600 rpm. The 300-rpm speed represents an approximate shear rate of 511 sec^{-1}, and the 600-rpm speed represents an approximate shear rate of 1,022 sec^{-1}. There are also six-speed model rotating viscometers, which are designed for speeds of 3, 6, 100, 200, 300, and 600 rpm.

   Some models of rotating viscometers have been built to run at variable speeds, and some run at speeds other than those listed. In addition to the normal-temperature rotating viscometers, several high-temperature rotating models are being used in laboratories. The high-temperature models permit mud evaluation while they are in motion at high temperature. Much of the work by Annis was done using a high-temperature rotating viscometer.[3] These high-temperature models provide a quick method of evaluating mud properties under environments similar to those encountered in the field.

   Maximum temperatures of 500°F and maximum pressures of 3,000 psi are the general rule for these viscometers. The pressure limitation is not believed critical for water-based muds where pressure is considered only as a factor to keep water from boiling. Pressure has a large effect on oil-based muds; to obtain accurate information, an instrument is required that measures at pressures equivalent to those encountered in the field, which are approaching 20,000 psi.

3. *Pipe viscometer.* The pipe viscometer is primarily a laboratory tool, although some field models have been used. This method of measurement meets the necessary criteria for accuracy; however, it is not easy to use in the field and fails the test for simplicity. Pressure drop is measured along a given length of pipe for mud at selected flow rates. By using heat exchangers, these pressure measurements can be obtained at any temperature and pressure permitted by the equipment in use.

The rotating viscometer is the common field instrument used to determine the effective mud thickness. However, the data obtained from the rotating viscometer must be converted to numbers that can be used on a quantitative basis. This requires either the use of equations that describe the flow pattern of the fluid or a plot of shear stress vs shear rate.

Eqs. 5.12–5.15 have all been used to describe the flow behavior of drilling mud.

Newtonian

$$\tau = -\mu\gamma_s \tag{5.12}$$

Bingham plastic

$$\tau = Y + (PV)(-\gamma_s) \tag{5.13}$$

Power law

$$\tau = k_s(-\gamma_s)^{n_s} \tag{5.14}$$

Power law with yield stress

$$\tau = Y' + k'_s(-\gamma_s)^{n'_s} \tag{5.15}$$

Note: No attempt has been made to make the units consistent in Eqs. 5.12–5.15; the equations are given for illustration only.

From Eq. 5.12, the viscosity, μ, is simply the proportionality constant between shear rate and shear stress and remains constant for all rates of shear. Eq. 5.13 shows that no single term describes thickness. Two terms are used for this purpose: the yield point, Y, and the plastic viscosity, PV.

From Eq. 5.14, thickness is described by k_s and n_s, and in Eq. 5.15, the description of fluid behavior has been further complicated by using three terms to describe thickness: Y', k'_s, and n'_s.

The primary criterion for any method of describing mud thickness is that it be reasonably accurate, have utility, and be easy to use.

Fig. 5–12 shows a shear stress, τ, vs shear rate, γ_s, diagram for a Newtonian fluid. The concept is easy to understand: μ is the slope of the straight line which goes through the origin. Fig. 5–13 shows a shear stress vs shear rate diagram for a typical drilling mud.

The two-speed rotating viscometer measures the shear stress in $lb_f/100$ sq ft at shear rates of 511 sec⁻¹ (300-rpm) and 1,022 sec⁻¹ (600 rpm). The slope of the assumed straight line between the two shear rates is called the plastic viscosity. The intersection of this straight line with the ordinate shown as the dashed line is called the yield point. Using plastic viscosity and yield point in this manner is known as Bingham plastic flow behavior.

The actual behavior of most muds in the shear-rate range below 511 sec⁻¹ is shown by the solid curving line. Thus, the assumption of Bingham plastic behavior is valid only if the mud is in laminar flow at a shear rate above 511 sec⁻¹, which is unlikely in field operations. Note that some

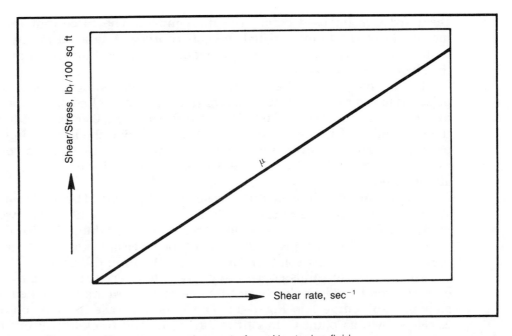

■ **FIG. 5–12** Shear stress vs shear rate for a Newtonian fluid

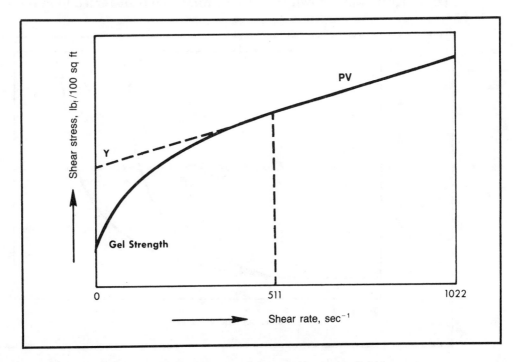

■ **FIG. 5–13** Shear stress vs shear rate for a non-Newtonian fluid

shear force is imposed before the fluid moves. The force that initiates movement of the fluid is called the *gel strength*. A tangent to any point on the curving line below 511 sec⁻¹ would show the fluid is thinning as the shear rate is increased.

The primary use of the plastic viscosity, which is measured in centipoises, is to show the effect of solids content on mud thickness. It is obtained by simply subtracting the 300 dial reading from the 600 dial reading on the viscometer. The instrument is calibrated to give the answer in centipoises. Actually, the difference should be multiplied by 0.937 to convert the units to centipoises. In field operations the conversion constant is generally ignored.

The magnitude of the plastic viscosity may be affected by solids content, size of the solids, and temperature. It is difficult to say that a given weight mud should have a certain plastic viscosity because the solid size is also a factor. Fig. 5–14, an abbreviated form of the viscosity vs solids concentration curve, shows how solid size may affect the mud thickness.

An increase in solids content at a given temperature may have little effect on total mud thickness until some critical range is reached, such as point C on curve 1 in Fig. 5–14. The same increase in solids on curve 2, also shown as point C, has little effect on mud thickness. From point A to point B the increase in solids has little effect on either curve 1 or 2. The performance shown for curve 2 might be transferred to curve 1 with

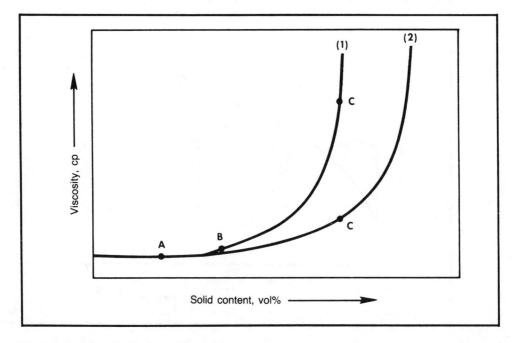

■ **FIG. 5–14** Viscosity vs solids content

no change in the total solids content other than a reduction in the size of solids.

The level of plastic viscosity for specific muds should be left an arbitrary term. In unweighted muds the solids content can be calculated, so there is no need to use plastic viscosity as a criteria for treatment. In weighted muds the plastic viscosity for given muds generally depends on solids volume, solids size, and temperature. Considering temperature constant for the moment, this means the plastic viscosity for a given weight mud may vary and still remain acceptable under certain drilling conditions.

This introduces the best way to use plastic viscosity when treating weighted muds. Measure the solids content as accurately as possible, and measure the plastic viscosity. If the solids content is in an acceptable range, so is the plastic viscosity, regardless of its magnitude. In critical wells measure the mud thickness frequently, watching for a change in plastic viscosity.

A sudden increase generally indicates a solids content increase. If undetected, this increase may result in a downhole disaster. Plastic viscosity is chosen as the parameter that should be measured often, rather than solids content, because it is easier and quicker to measure and may be obtained with less danger of an error in measurement.

The yield point is a pseudo number, as shown on Fig. 5–13. It is obtained by extrapolating the assumed straight line between the 300 and 600 dial readings on the viscometer. Thus, the number does not exist, and its utilization in quantitative calculations immediately introduces a known error. Yield point is determined quantitatively by subtracting the plastic viscosity from the 300 reading on the viscometer.

For field purposes, yield point is used as an indicator of the attractive forces between solids, or if there are no attractive forces, as an indicator of the deviation of the mud from Newtonian behavior. In general field practice, the yield point is used more as an indicator of mud thickness than the plastic viscosity. In unweighted muds the yield point is maintained at the level required for adequate hole cleaning. In weighted muds the desired level of the yield point is that required to support barite.

It might be argued that the gel strength, a measured value, would be more indicative of the solids support capacity of the mud, and this is probably true. The only argument in support of the yield point is that its magnitude is related qualitatively to the magnitude of the gel strength. The units of the yield point and gel strength are $lb_f/100$ sq ft. If either the yield point or gel strength is considered excessive, it may be lowered by reducing solids content or using chemical thinners.

The power law behavior shown by Eq. 5.14 more accurately represents drilling mud behavior than the Bingham plastic representation. However, using plastic viscosity and yield point has become so common in field operations that it is difficult to change. Thus, plastic viscosity and yield point are used often as treating mechanisms, and the n and k values from the power law equation are used for quantitative calculations concerning pressure losses and lifting capacity. Fig. 5–15 graphically represents the power law behavior shown in Eq. 5.14.

A more definitive equation for the straight line in Fig. 5–14 is shown by Eq. 5.16:

$$\log \tau = \log k + n \log \gamma \qquad (5.16)$$

The use and meaning of n and k become clearer in Eq. 5.7 for the Newtonian fluid is also expressed in log form, as shown in Eq. 5.17:

$$\log \tau = \log \mu + \log \gamma \qquad (5.17)$$

The only difference in Eqs. 5.16 and 5.17 is the value of the straight-line slope, n. In Eq. 5.16, n may be equal to or less than one. In Eq. 5.17 for Newtonian behavior, the slope of the line must always be one, and the intercept on log paper is shown as the viscosity. This analogy then shows that k for the power-law fluid is related directly to the viscous characteristics of the mud, and the line slope, $n,$ is related directly to the flow characteristics of the mud.

If n is one, the fluid is behaving as a Newtonian fluid. When n is less than one, the fluid behavior is non-Newtonian. Thus, both n and k are necessary to describe the thickness of the drilling mud. Equivalent

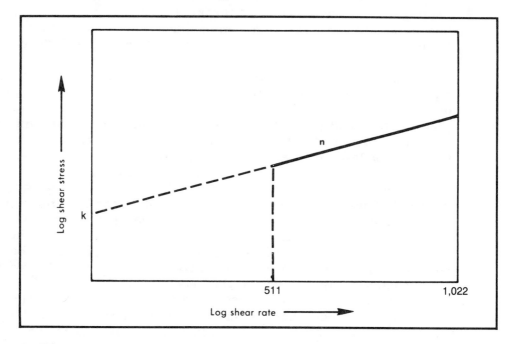

■ **FIG. 5–15**   Shear stress vs shear rate for a power law fluid

mud thickness for annular flow, using the power law equation, is shown in Eq. 5.18:

$$\mu = \left[\left(\frac{2.4\ \bar{v}_a}{D_h - D_{op}} \right) \left(\frac{2n+1}{3n} \right) \right]^n \frac{200k(D_h - D_{op})}{\bar{v}_a} \tag{5.18}$$

where:

μ = equivalent mud thickness, cp
$\bar{v}_a$ = average annular mud velocity, fpm
D_h = hole diameter, in.
D_p = outside pipe diameter, in.

The value of n can be determined using Eq. 5.19:

$$n = 3.32 \log \frac{\theta_{2\gamma}}{\theta_\gamma} \tag{5.19}$$

where:

$\theta_{2\gamma}$ = viscometer reading at 2.0 times the shear rate as shown for the denominator, $lb_f / 100$ sq ft
θ_γ = viscometer reading at any shear rate, $lb_f / 100$ sq ft

The value of k can be determined using Eq. 5.20:

$$k = \frac{\theta_\gamma}{\gamma^n} \tag{5.20}$$

The equivalent thickness for the power law fluid behavior is normally substantially less than that for the Bingham plastic fluid behavior. The power law relationship is considered the most accurate.

The power law with yield stress shown in Eq. 5.15 can be expressed in log form as shown in Eq. 5.21:

$$\log (\tau - Y') = \log k' + n' \log \tag{5.21}$$

The only difference in Eqs. 5.16 and 5.21 is shifting the straight line on log paper by the quantity of the yield stress, Y'. The 3-rpm reading on the viscometer is taken as the first estimate of yield stress. However, mathematical analysis using this approach becomes more complicated; rather than proceed in this direction, it is simpler to plot a curve of shear stress vs shear rate to determine equivalent mud thickness. As variable-speed viscometers become available in the field, a plot of shear stress vs shear rate will become a common method to determine equivalent mud thickness.

A simple method to determine equivalent mud thickness is enumerated as follows:

- Plot multispeed viscometer data on log paper or use a variable-speed viscometer.
- Determine the viscometer speed that equals the average annular velocity.
- Determine the viscometer reading at the viscometer speed that equals annular velocity.
- Calculate the equivalent thickness of the mud.

The viscometer speed that equals the average annular velocity can be determined using Eqs. 5.22, 5.23, or 5.24.

Viscometer speed in reciprocal seconds:

$$\gamma = \left(\frac{2.4\,\bar{v}_a}{D_h - D_p}\right)\left(\frac{2n+1}{3n}\right) \tag{5.22}$$

Viscometer speed in rpm, γ_1

$$\gamma_1 = \left(\frac{1.41\,\bar{v}_a}{D_h - D_{op}}\right)\left(\frac{2n+1}{3n}\right) \tag{5.23}$$

Viscometer speed in rpm assuming n = 0.7:

$$\gamma_1 = \frac{1.61\,\bar{v}_a}{D_h - D_{op}} \tag{5.24}$$

Note: The assumption of $n = 0.7$ introduces an error of less than 10% for all ranges of n.

Equivalent mud thickness may be calculated by Eqs. 5.25, 5.26, and 5.27.

Equivalent thickness using reciprocal seconds as the shear rate:

$$\mu = \frac{511\,\theta_\gamma}{\gamma}\ \frac{2n+1}{3n} \tag{5.25}$$

Equivalent thickness using rpm as the shear rate:

$$\mu = \frac{300\,\theta_{\gamma_1}}{\gamma_1}\ \frac{2n+1}{3n} \tag{5.26}$$

Equivalent thickness using rpm as the shear rate and assuming $n = 0.7$:

$$\mu = \frac{343\,\theta_{\gamma_1}}{\gamma_1} \tag{5.27}$$

Example 5.8 shows how the equivalent mud thickness can be determined using the various procedures available and also shows the magnitude of differences for this particular case.

□   **EXAMPLE 5.8**

Hole size = 8½ in.
Drillpipe size = 4½ in.
Mud weight = 10 lb/gal
Average annular velocity = 150 fpm
Viscometer readings:
$\theta_{600} = 100$
$\theta_{300} = \ \ 65$
$\theta_{200} = \ \ 50$
$\theta_{100} = \ \ 32$
$\theta_6 \ \ = 4.8$
$\theta_3 \ \ = 3.0$

SOLUTION:

Viscometer data for Example 5.8 are shown on Fig. 5–16. Use Eq. 5.19 to determine n:

$$n = 3.32 \log \frac{100}{65} = 0.62$$

Use Eq. 5.20 to determine k:

$$k = \frac{65}{511^{0.62}} = 1.35$$

Calculate equivalent thickness with Eq. 5.18:

$$\mu = \left[\frac{(2.4)(150)}{4} 1.2 \right]^{0.62} \frac{(200)(1.35)(4)}{150} = 132 \text{ cp}$$

Viscometer speed that equals average annular velocity is determined by using Eq. 5.23:

$$\gamma_1 = \frac{(1.41)(150)}{4} 1.2 = 63 \text{ rpm}$$

Viscometer reading at 63 rpm is given on Fig. 5–16:

$$\theta_{63} = 23 \text{ lb}_f / 100 \text{ sq ft}$$

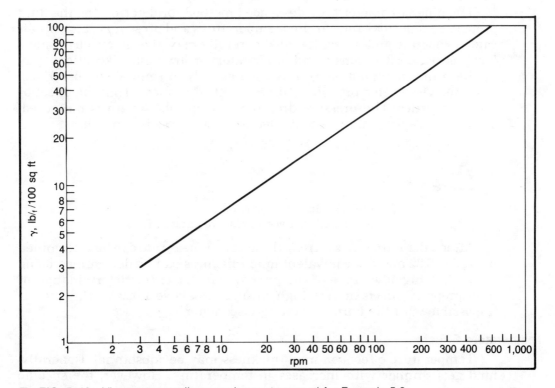

■ **FIG. 5–16** Viscometer readings vs viscometer speed for Example 5.8

Equivalent thickness is calculated by Eq. 5.26:

$$\mu = \frac{(300)(23)}{63}\,1.2 = 131 \text{ cp}$$

Viscometer speed that equals average annular velocity, assuming $n = 0.7$ is given by Eq. 5.24:

$$\gamma_1 = \frac{(1.61)(150)}{4} = 60 \text{ rpm}$$

Viscometer reading at 60 rpm is taken from Fig. 5–16:

$$\theta_{60} = 22.3 \text{ lb}_f/100 \text{ sq ft}$$

Equivalent thickness is determined by using Eq. 5.27:

$$\mu = \frac{(343)(22.3)}{60} = 127 \text{ cp}$$

The different procedures give about the same equivalent thickness. The simplifying assumption that $n = 0.7$ gives a slightly lower equivalent thickness. As the actual n drops to lower values, the assumption of $n = 0.7$ will produce larger errors, all on the low side.

Pipe Viscometer

The pipe viscometer provides a good method for determining the fluid properties at normal and elevated temperatures and pressures. It is complicated to operate and unless handled carefully may give erroneous results. For this reason it is considered a laboratory instrument. Exceptions may be the trailer-mounted data units used as aids in some field operations. The pipe viscometer provides a direct method of measuring the effect of fluid properties. The pressure drop for a given tube length is measured. This pressure drop may be converted to shear stress by Eq. 5.28:

$$\tau = \frac{300 \text{ PD}_i}{\ell_t} \qquad\qquad (5.28)$$

where:

D_i = internal pipe diameter, in.
ℓ_t = tube length between pressure gauges, ft

After calculating shear stress, the average shear rate can be determined using Eq. 5.22 and the equivalent mud thickness can be determined using Eq. 5.25. A big advantage of the pipe viscometer is the determination of fluid property effects at very high shear rates above 1,022 sec⁻¹, which is impossible with the normal rotating viscometer.

Temperature Effects

Temperature effects on mud thickness may be substantial. Generally, mud gets thinner with increases in temperature. However, the specific

■ **TABLE 5–8** Simulated Fann data

Temperature, °F	Reading at 1,022 sec⁻¹	Reading at 511 sec⁻¹	Plastic Viscosity, cp	Yield Point, lb$_f$/100 sq ft
68	136	70	66	4
72	111	62	49	13
120	83	47	36	11
160	62	34	27	7
220	40	25	15	10
320	32	22	10	12

effect may be decided by the type and total solids in the mud. For example, consider curve 1 in Fig. 5–14. If the solids concentration were at point B for this mud, any further dispersion of clays by high temperature might result in flocculation and severe thickening. If the mud were comparable to that shown on curve 2, an increase in temperature would probably result in thinning the mud.

Several investigators have conducted laboratory tests that describe the behavior of muds at elevated temperatures. Work performed by Bartlett is shown in Table 5–8.[4] Note that the plastic viscosity is reduced substantially with increases in temperature. The same table shows an erratic behavior for yield point as a function of temperature increases.

The reduction of plastic viscosity with increasing temperature is believed due to a thinning of the liquid phase of the mud. This belief is confirmed by Fig. 5–17, taken from work by Annis, which shows the normalized viscosity of water compared with the plastic viscosity of mud vs temperature.[5] The thickness of the water and mud followed the same reducing trend with increases in temperature until about 220°F was reached.

At this point the plastic viscosity of the mud did not decrease with further increases in temperature. One explanation is the additional dispersion of solids with temperature tends to increase the frictional effects of the solids at a rate comparable to the thinning of the liquid. The specific point of no further reduction in plastic viscosity depends on the type of mud. In Table 5–8 the plastic viscosity of the mud is 10 cp at 320°F and 15 cp at 220°F.

Muds may thin or thicken with increases in temperature. Evidence that proves this conclusion is shown in Figs. 5–18, 5–19, and 5–20 from Bartlett's work.[6] Note that in Fig. 5–18, the mud is thicker at 300°F below a shear rate of 200 sec⁻¹ than it is at the lower temperatures.

To confirm that the reaction with temperature depends on the type of mud, Fig. 5–19 shows a mud with 25 ppb of bentonite, 350 ppb of barite, and 9 ppb of lignosulfonate where thinning occurs with increases in temperature until a temperature of 220°F is reached. In Fig. 5–19, the mud

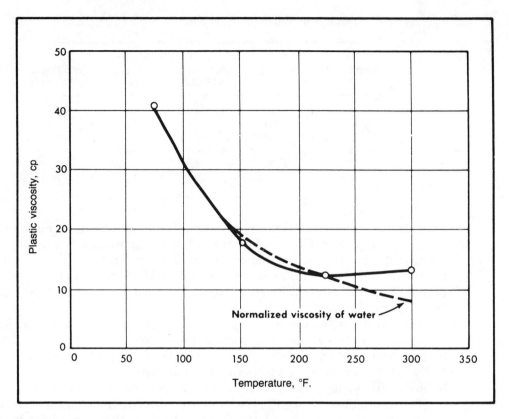

■ **FIG. 5–17** Effect of temperature on plastic viscosity of water-based muds

becomes substantially thicker at 320°F when the shear rate drops below 100 sec⁻¹.

The same mud composition but with 15 ppb of lignosulfonate is shown in Fig. 5–20, and the additional chemical treatment prevents the flocculation noted in Fig. 5–19. The mud thinned substantially with increases in temperature to 320°F. The pH of the mud in Figs. 5–19 and 5–20 is maintained at 9.5.

The effect of pH and the equivalent bentonite content is shown in Fig. 5–21. This figure shows the gel strength to be uncontrollable at an equivalent bentonite content of about 27.5 ppb and a temperature of 300°F for this particular mud. Of further significance in Fig. 5–20 is the reduction in gel strength noted when the pH is increased from 7.6 to 9.4.

These tests show that temperature has a substantial effect on the flow characteristics of drilling mud. Methods used in the field do not generally permit measuring fluid properties at temperatures above 160°F.

The quantitative effect of temperature on mud thickness can be estimated by measuring the thickness at three or more temperatures and plotting the data on log paper. In normal field operations, measuring mud

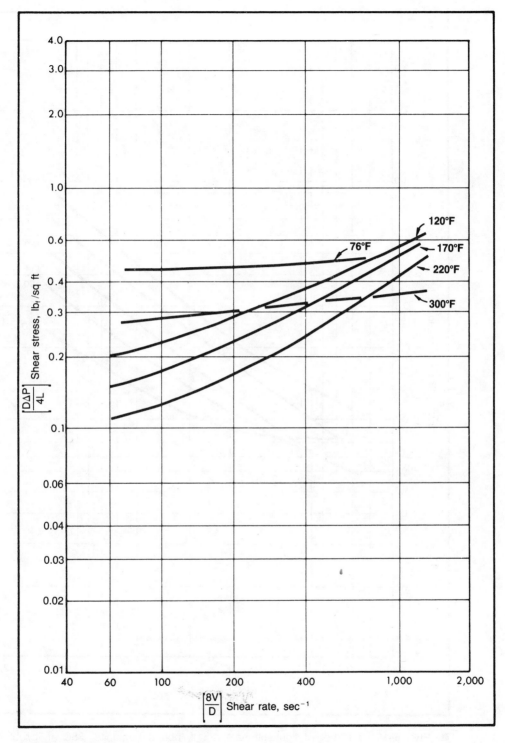

■ **FIG. 5–18** 21 lb/bbl of bentonite

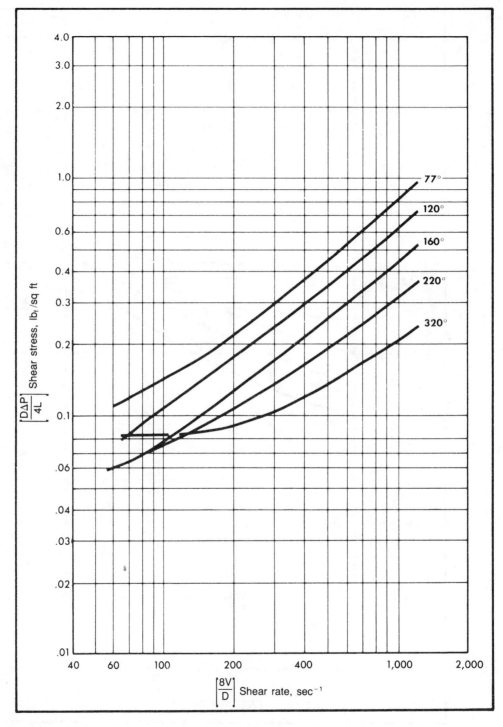

■ FIG. 5–19 9 lb/bbl of lignosulfonate, 25 lb/bbl of bentonite, 350 of lb/bbl barite, pH = 9.5

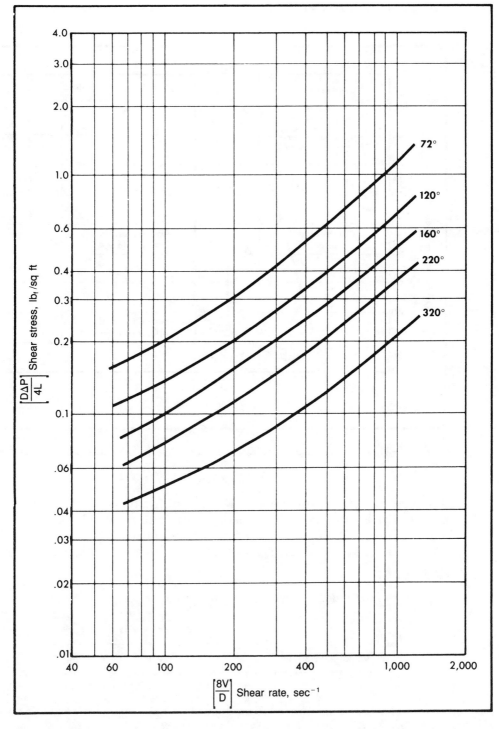

■ **FIG. 5–20** 15 lb/bbl of lignosulfonate, 25 lb/bbl of bentonite, 350 lb/bbl of barite, pH = 9.5

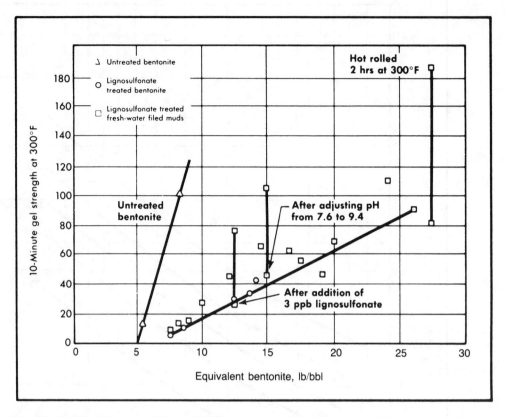

■ **FIG. 5–21** Effect of solids on 300°F gel strengths of field muds (*from Annis, courtesy SPE*)

thickness at temperatures above 150°F is difficult. Thus, if the mud thickness at 200°F must be determined, measure the mud thickness at lower temperatures, plot the data on log paper, and extrapolate to determine the thickness at 200°F. This procedure is shown in Example 5.9.

☐ **EXAMPLE 5.9**

Water-based mud, multispeed viscometer data:

			Mud Thickness		
Viscometer speed	80°F	120°F	160°F	200°F	250°F
3	3	2.3	1.9	1.6	1.4
6	5	3.6	2.9	2.4	2.0
100	24	17.5	13.5	11.0	9.2
200	38	26.5	20	16.5	13
300	50	34	26	21	17
600	70	49	37	30	24

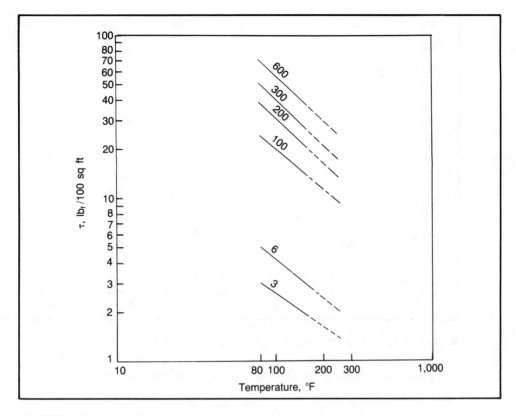

■ **FIG. 5–22** Mud thickness vs temperature

The viscometer readings at 80, 120, and 160°F were measured. Those readings at 200 and 250°F were obtained by extrapolating the readings as shown on Fig. 5–22. In field operations the data may not plot a straight line. There is certainly no magic to using a log plot. Any type of data plotting that results in a straight line for measured data potentially can be extrapolated to higher temperatures.

The data obtained by extrapolation is acceptable unless the mud flocculates at the higher temperatures. A potential indication of flocculation might be noted at low shear rates if the mud does not thin when the temperature is increased.

The data for 200°F are plotted in Fig. 5–23. The equivalent thickness of the mud can now be determined at 200°F. The use of these data are shown in Example 5.10.

EXAMPLE 5.10

> Well depth = 12,000 ft
> Annular velocity = 120 fpm
> D_h = 8½ in.
> D_{op} = 4½ in.

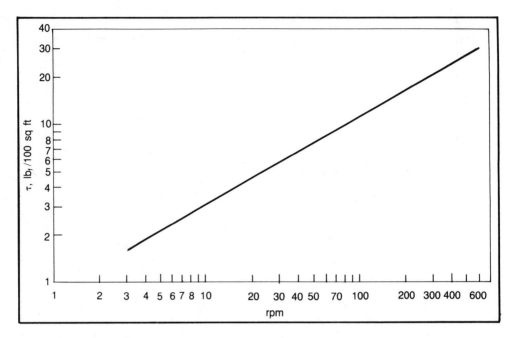

■ **FIG. 5–23** Viscometer data for 200°F from Example 5.9

Determine: Equivalent thickness of annular mud at 200°F

SOLUTION:

$$n = 3.32 \log \frac{30}{21} = 0.51$$

$$k = \frac{11}{170^{0.51}} = 0.801$$

$$\mu = \left[\frac{(2.4)(120)(2.02)}{(4)(1.53)} \right]^{0.51} \frac{(200)(0.801)(4)}{120} = 54.5 \text{ cp}$$

The equivalent thickness of the mud at 200°F is 54.5 cp. The thickness of the mud in the surface pits may be estimated by assuming the thickness to be equal to the 3.0-rpm speed at 80°F.

Use Eq. 5.27 to estimate thickness.

$$\mu = \frac{(343)(3)}{3} = 343 \text{ cp}$$

This comparison shows the mud in the surface pits to be more than six times as thick as the mud in the hole at 200°F.

The same procedure can estimate the thickness of an oil-based mud at elevated temperatures. However, pressure thickens an oil-based mud and has a minimal effect on a water-based mud. There are no field methods to estimate the actual effect of pressure. Laboratory tests have determined that pressure has a substantial effect on oil-based mud thickness. Also, the same tests show that the weight of the oil-based mud may be increased by as much as 1.0 lb/gal as pressure is increased. If the use of oil-based muds becomes routine, then more accurate field methods that utilize pressure changes will be needed to measure thickness and mud weight.

Methods Used to Control Viscosity

Controlling viscosity or mud thickness of course depends on the operator's objectives. Viscosity may be maintained at high levels to ensure adequate hole cleaning. Additives such as some polymers may be added to increase the mud thickness for adequate hole cleaning while keeping the mud weight and total solids content at low levels. With weighted muds some operators use polymers for increasing thickness in order to keep clay solids, which provide support for the barite, to a minimum.

In general if there are no specific hole requirements, the operator prefers to use the thinnest possible drilling fluid. Thinning is also emphasized when using weighted muds close to the fracture gradient of exposed formations.

When more viscosity is desired, it may be obtained by (1) adding bentonite, (2) flocculating the clay solids, and (3) adding polymers designed for the purpose. Bentonite is the primary viscosity builder in the mud system. If fresh water is being used, bentonite will increase viscosity rapidly with only small increases in mud weight. Flocculating clays in a freshwater mud is also a quick, cheap method of increasing viscosity. Flocculation can usually be done by adding lime or cement.

Polymers may be used as thickening agents and encapsulating materials to prevent dispersion of drill solids. Polymers are considered long-chain high-molecular-weight molecules. This definition leaves the door open to classify many materials as polymers. For the purposes of this discussion, polymers will be divided into three groups: (1) Polymers that thicken the mud by acting on the clay solids, (2) polymers that are colloidal in nature and essentially provide a substitute for bentonite, and (3) polymers that thicken the liquid phase and do not react with bentonite.

No attempt will be made to introduce all of the polymers that might fit any one of the above three categories, but typical examples of specific types will be discussed. However, in many cases the extent of a given polymer's utility is still being field tested; for any material to be successful, it must not only work but also be the best material economically. Specific applications might prove any one of the polymers successful, so the key is to keep accurate well costs. Remember that all results are good or bad based on comparisons with other alternatives.

A typical polymer that aggregates clay solids to increase mud thickness is formulated using a type of polyacrylate. This material aggregates bentonite, trapping water and increasing mud thickness. It also has the claimed advantage of aggregating undesirable clay solids, which are then easier to remove from the drilling mud. In low-solids unweighted muds the polymer, by increasing the viscosity-producing characteristics of bentonite, provides the necessary lifting capacity at a very low solids content, which is the same as saying at a very low mud weight. In low-solids weighted muds, the low-solids term applies to the low-gravity solids. The concept with a system of this type is to use a minimum of bentonite and increase the bentonite yield with the polymer in order to support barite. No chemical thinners should be used in either low-weight or weighted muds unless the operator intends to abandon the use of the polymer.

Polymers of the colloidal type that coats particles and provides additional lifting capacity are typified by sodium carboxymethylcellulose, CMC. Other polymers of this type are starches and gums. These materials help increase lifting capacity and also aid in improving filter cake characteristics. Sometimes combinations of these polymers are used, and it is difficult to evaluate all the combinations. They do not aggregate solids, as do the polyacrylates; however, they provide improved filter cakes. Many of these materials have been used in the drilling business since the 1920s. Improvements and different use guidelines have changed their utility.

The XC polymer, so called because it is produced by the bacterial action of genus *xanthemonas campestris* on sugar, increases viscosity independent of the solids content of the mud. The material goes into solution and is capable of thickening water regardless of the water's electrolyte content. One of the claimed advantages of the material is the thickening of the treated mud at low shear rates and the extreme thinning at high shear rates. Deily et al. presented precise information on this polymer and related information on its use, advantages, and limitations.[7]

Polymers of all types have gained more attention in recent years. The materials tend to shear thin at high shear rates, an advantage for drilling rate, and thicken at low shear rates, an advantage for lifting capacity. Most of the polymers have a temperature limitation, generally in the range of 300°F. Some are incompatible with particular mud additives, and some require certain pH levels. Cost is always a factor. The reference to cost means total well costs. An operator should never select a material based on the cost of that material alone or the mud cost alone. Some materials may increase mud costs yet reduce well costs.

These polymers accomplish different objectives, their costs differ, and the success or failure for any product is always influenced substantially by associated personnel. It is apparent from the widespread use of the polymers that operators believe they have gained benefits. The claimed benefits in the case of low-solids low-weight muds have in many cases been significant. The use of polymers in weighted muds is limited.

Viscosity control is generally considered in terms of reducing mud thickness. Methods used for control include mechanical aids, water dilution, and chemical thinners.

Mechanical aids are generally solids and viscosity control devices. They help control solids in the mud and also help control the type of solids, both of which are important in mud control.

A standard oilfield saying is, "If in doubt, add water." Many mud engineers have found this to be good advice; however, at times doubt exists when it should not. Methods are available to measure solids content, determine the type of solids, measure the pH, and determine the degree of contamination. Tests should be run and evaluated before the miscellaneous use of water. In weighted muds the materials cost is high, and each barrel of water must be weighted and treated.

Chemical thinners reduce the attraction solids have for each other. Table 5–9 includes some of the more common mud thinners.

Phosphates are inorganic materials, and they usually treat at any pH. Phosphates are also cheap; a normal mud treatment would be in the range of 0.2 ppb. The noted restrictions of 175°F and large quantities of calcium have limited the application of phosphates. Another factor that has limited the use of phosphates has been the trend toward nonthinned low-solids muds in the shallower sections of the hole.

Chrome lignosulfonates are the most common type of thinners. Many types are available, and they are sold under many trade names. These

■ **TABLE 5–9** Drilling mud thinners

Chemical	pH of 1% Solution	Limitations
Sodium acid pyrophosphate (sapp)	4.3	Decomposes and forms flocculating agent above 175°F; not effective in the presence of large quantities of calcium
Sodium tetraphosphate	8.0	Same as above
Chrome lignosulfonate	5.0	Material starts to decompose at temperatures above 300°F; pH needs to be at least 9.0
Lignite	3.2	Material starts to decompose at temperatures above 350°F; pH needs to be at least 9.0
Tannin	5.0	Not very effective if pH of mud is less than 11.0
Surfactants	—	Temperature stability above 300°F may be a problem; most are more expensive than other materials

thinners are organic in nature, and their base material comes from the lumber industry. Chrome is an added substance, and the method of manufacture may be the key to the effectiveness of these materials. Therefore, operators should run specification tests on chrome lignosulfonates if they are unsure of the source or if it is a new source of supply.

Some chrome lignosulfonates are similar in their degree of effectiveness; however, other materials called chrome lignosulfonate may not thin the mud. In Table 5–9 the chrome lignosulfonates begin to decompose at 300°F. This does not mean they cannot be used above this temperature. However, because of decomposition above 300°F, the operator will have to increase treatment to maintain a given concentration level. The higher the pH, the better these materials treat. At a pH less than 9.0, treatment with chrome lignosulfonate is not very effective.

Lignites are old mud additives. Chrome lignosulfonates were introduced between 1950 and 1960. Lignites were common mud additives at least ten years earlier. Lignite is a hydrocarbon product. While lignites are used as thinners, they are also used as water-loss control agents. They begin to decompose at about 350°F; however, these materials are used sometimes as the primary mud thinner to temperatures substantially above 400°F.

The tannins, typified by the trade name Quebracho®, were the most frequently used thinners until the introduction of the chrome lignosulfonates. To be effective, the tannins must be used in a high-pH environment. This was no problem because prior to the middle 1950s high-pH (12.0) lime muds were common and the tannins were very effective thinners. The reduction of mud pH to the level of 9.5–10.5 reduced the relative effectiveness of tannins to the chrome lignosulfonates, almost eliminating the use of tannins. If the mud pH is above 11.5, the tannins may be competitive or even better than some of the lignosulfonates.

Other types of organic thinners include iron lignosulfonates and combinations of tannins and lignosulfonates. Iron lignosulfonates are ineffective as mud thinners. Some combinations of tannins and lignosulfonates are mixed with caustic to create a desirable environment for making the organic materials soluble when added to water.

Surfactants are surface tension-reducing materials. They thin muds, help reduce water loss, and act as emulsifiers. The most common use of surfactants is as chemical emulsifiers. Usually, they have been uncompetitive economically with other thinners for routine mud treatment.

No chemical thinners are added to oil-based muds. Thinning is accomplished by adding oil. Sometimes, if water is present, the oil-based mud may be thinned by tightening the emulsion with surfactant emulsifier. Most oil-based muds are thickened by adding treated clays and may be allowed to thin by reducing clay additions.

Some ordinary rules of mud treatment with chemical thinners can be given. However, specific situations require specific techniques; thus, this is a general approach to mud treatment.

Mud treating decisions for weighted muds should be related to specific downhole objectives first and to the actual cost of treating second. Mud properties that affect decisions are total solids content, type of solids, pH, and known presence of contaminants. No chemical treatment is recommended for unweighted muds.

Consider first the solids content. We know that field retort measurements are approximations at best. However, this is the only method available in the field, and chances for errors in measurement emphasize the necessity to be accurate. While measuring the solids content accurately, determine the flow properties of the mud, using part of the same mud sample. Then measure these mud properties every 15 min for a complete cycle of the mud.

Muds are generally unhomogeneous in nature, and this procedure provides the operator with a specific look at the entire mud system. At the same time these tests are run, the same procedures should be used to determine pH, type of contamination, and bentonite content of the mud. Some knowledge is needed of the critical range of these mud properties.

Therefore, if everything measured at the surface is in an acceptable range, the properties will probably remain acceptable at downhole conditions where the temperature is much higher. Some indication of the effect of the higher downhole temperature can be determined by comparing the flow-line properties to the suction-pit properties. If the critical nature of the well justifies the effort, high-temperature pipe viscometer or high-temperature rotating viscometer tests should be run.

Flow properties of the mud should be checked frequently. If an increase in plastic viscosity is noted with no change in mud weight, this is an indication of an increase in solids content. If the increase in plastic viscosity is accompanied by an increase in yield point, the initial preferred treatment is a reduction in solids content, either by mechanical means or by water dilution. If, however, the yield point increases without an increase in plastic viscosity, then chemical treatment will probably be preferred. These procedures are based on the assumption that no pH change occurred. The n and k values could be used in the same manner. If n is low, chemical treatment may be required. If k is high, water may be needed.

When adding the organic chemical thinners, particularly chrome lignosulfonates and lignites, the preferred procedure is to mix these materials with caustic and water in a mixing tank or chemical barrel. These organic materials are not readily soluble; thus, when mixed in the presence of caustic and water, they are solubilized more readily than when added directly to the mud. As mentioned, pH is important in mud treatment, and one of its benefits is to increase the effectiveness of these chemical thinners.

The gel strength and yield point of the mud are related to pH. Generally, an increase in pH reduces gel strength and yield point. Conversely, a low pH results in increases in the properties. Thus, in many cases where gel strength is a problem, operators tend to continue increasing pH. Depending on the specific circumstances, a high pH (say, above 11.0) may result in large concentrations of excess lime $Ca(OH)_2$ and the formation of a cementaceous material if temperatures exceed 250°F. Needed, of course, would be a source of calcium, which may be obtained by drilling cement, from calcium in makeup water, from calcium-bearing formations, or from calcium added to reduce carbonate contamination.

The general assumption is that the adverse effects of contamination are caused by cations such as calcium and sodium. The carbonate anion may be equally troublesome, if not more so than the cations. This has resulted in developing new techniques to determine carbonate ion concentration. All of these methods are complicated by other mud additives and contaminants. If carbonate contamination is believed to be a problem, calcium may be added or the pH may be increased using caustic. In cases of severe problems, both procedures may be used at the same time.

Filtration Control—Water-Based Muds

The filtration rate is usually controlled for the following reasons: (1) to control the thickness and characteristics of the filter cake that is deposited on permeable formations and (2) to limit the total filtrate that enters underground formations.

There are basically two methods of measuring the rate of filtration: (1) static filtration rate tests, which are API approved and (2) dynamic filtration rate tests. The static tests include the standard API test of normal temperature (77°F) at 100 psi; the high-temperature tests generally run at 300°F and pressures of 100 and 500 psi. Static tests are meant to indicate fluid loss and filter cake buildup when the fluid is not moving. Dynamic tests represent fluid loss and filter cake buildup while drilling mud is being circulated.

Filter cake thickness and the friction between the filter cake and the drillstring relate to the problems of (1) differential pressure sticking, (2) torque and drag while drilling and tripping the drillstring, (3) running wireline tools and casing strings, and (4) sidewall coring. An increase in filter cake thickness enlarges the area of contact between the drillstring and the filter cake and, for a given friction coefficient, increases the danger of sticking the drillstring. It also increases the torque and drag on the drillstring.

Running wireline tools and casing is restricted by the reduction in hole size as the filter cake thickens. A thick filter cake may prevent formation fluid recovery on a wireline test. For these problems the operator is not concerned with the actual loss of filtrate. The filtrate loss is used to indicate the potential thickness of the filter cake.

Problems of differential pressure sticking, torque, and drag are related primarily to high-weight drilling muds in deep high-temperature environments. This is the primary reason the high-temperature static filtration rate test was introduced. Efforts were being made to simulate downhole conditions and standardize data and results to permit industry-wide usage. Basic correlations were developed quickly, and standard treating techniques emerged.

Two of these are (1) the rule-of-thumb maximum of a high-temperature water loss of 10 cc and (2) the concept of the compressible filter cake, determined by comparing the water loss at 300°F and 100 psi with that at 300°F and 500 psi. Results show the merit of these two techniques in some field operations; however, costs and problems often increase by simply depending on these rules.

To obtain a 10-cc static 300°F filter loss, the mud has either a high percentage of bentonite in the low-gravity solids, is highly treated with chemicals, or has a very high total solids content for the specific mud weight. Thus, other methods of measurement are needed to confirm what might actually occur downhole. The concept of the compressible filter cake is one such follow-up method. If the water loss shown by the 300°F and 500-psi test is the same or only slightly higher than the water loss obtained by the 300°F and 100-psi test, we assume the filter cake contains a high percentage of compressible solids such as bentonite. In fact, this particular test is a more accurate method to determine the effective bentonite content in the filter cake than the methylene blue test.

Eq. 5.29 provides a better understanding of the static filtration rate:

$$\Delta V_f = \frac{CP_{fc}^{1-b_f}\Delta t}{\mu_l rwV_f}$$

(5.29)

where:

ΔV_f = volume of filtrate collected for a specific time period
C = proportionality constant
P_{fc} = differential pressure across the filter cake
b_f = compressibility constant, dependent on solids type in the filter cake
Δt = time filtrate is measured, sec.
μ_l = viscosity of liquid filtrate, cp
r = resistance per unit weight of solids
w = solids content of mud per unit volume of filtrate
V_f = total volume of filtrate collected

If a high percentage of the low-gravity solids in the mud is compressible, b_f approaches 1.0 and the pressure does not increase the volume of filtrate for a specific time interval. The viscosity of the filtrate is reduced considerably with temperature and, as Fig. 5–16 shows, water is only one-fourth as thick at 225°F as it is at 75°F. Therefore, thinning the water would provide an increase of over 4:1 in the filtration rate at 300°F compared with 75°F. Further dispersion of clays at 300°F may reduce the ratio in some cases to about 3:1. However, if any high-temperature flocculation occurs, the 300°F water loss may be eight times the 75°F water loss.

This can be one check on whether the operator is close to the critical solids range, where a slight increase in solids can cause severe thickening. The r in Eq. 5.29 refers to the type of solids. The solids content is indicated by w. As shown, if the solids content increases and everything else stays the same, the water loss is reduced; however, this is at the expense of having a thicker filter cake.

Eq. 5.29 can be modified, and if all variables except time and volume are held constant, Eq. 5.30 can be used to determine filtrate volume as a function of time:

$$V_f = C' \sqrt{\Delta t} \qquad (5.30)$$

where:

$C' =$ proportionality constant that includes all of the variables held constant

Eq. 5.30 is the basis for measuring the 7½-min water loss and multiplying by 2 to obtain the 30-min water loss.

None of these methods for determining downhole static filter loss or filter cake thickness defines the problems of determining requirements. Potential requirements must be determined by the thickness of permeable formations open to the wellbore and the pressure differentials across these formations. If less than 100 ft of permeable formations are open to the wellbore, the operator may experience few problems.

If the pressure differential across the permeable formations is very low, (less than, say, 300 psi), again the operator may experience few problems. Controlling high-temperature water loss in the cases above may be a waste of money and may have a detrimental effect on the circulating mud properties. Never proceed blindly. Keep accurate records of the formations penetrated, formation pore pressures, and histories of past experience; treat the mud based on specific requirements of the well being drilled.

Another problem of thick filter cakes is running wireline tools, which causes trouble in shallow and deep wells. Low-solids muds in unweighted drilling fluids have minimized the problems of running wireline tools. In weighted muds, particularly in coastal areas, some operators make what is called a short trip before running wireline tools. If this short trip is made just before pulling pipe to run the logs, it is probably a waste of time, money, and effort. After several hours out of the hole, trips may need to be made with the bit and drillstring to wipe off filter cake. While circulating fluid, the filter cake reaches a maximum thickness, and the short trip just after circulating has stopped serves no purpose.

Operators are concerned with the total filtrate entering underground formations for the following reasons: (1) shale sloughing may be caused by filtrate invasion into the shales; (2) formation evaluation methods are affected by filtrate invasion; and (3) formation productivity is often affected by filtrate invasion.

The idea of wetting shales, which results in excessive sloughing, was first introduced as an engineering concept from 1920 to 1930. Since shales did slough as a result of getting wet, the concept was readily accepted. Actually, most shales are impermeable, and there is no possibility of wetting these shales by a normal filtration process. However, some argue that shales are full of microfractures and filtrate enters these fractures.

Filtration rate control is a normal practice when shales slough into the hole. There is no scientific evidence that shale sloughing is controlled by controlling water loss. Caliper logs show no particular relationship between hole enlargement in shale sections and water loss. Filtration control will not help reduce shale sloughing in hydratable shale zones. In older shales with naturally occurring fractures, the filtration rate control may help minimize sloughing. There is no indication shale sloughing can be stopped, ever, by controlling the filtration rate.

One problem with trying to control filtrate invasion is that the static filtration rate is not always a good indicator of the filtrate quantity that enters the formation. Filtrate invasion results from the following:

- Filtrate that enters the formation under static conditions
- Filtrate that enters the formation below the bit
- Filtrate that enters the formation across a dynamic filter cake

Filtrate that enters a formation under static conditions is probably less than 20% of the total filtrate invasion. Even in a 30-min API filtration test, the filtration rate near the end of the 30 min is very low. Filtrate that enters the formation ahead of the bit is high per unit of time; however, that time is short. Consequently, probably no more than 20–30% of the invasion occurs ahead of the bit. Probably 50% of the filtrate invasion occurs under dynamic conditions. The filter cake is kept thin by fluid circulation, and some of the filter cake is probably removed by pipe rotation.

There is no known correlation between filter loss ahead of the bit and static filtration rate. There is a qualitative correlation between static filtration rates and dynamic filtration rates across a filter cake. Dynamic filtration rates are measured by circulating the mud through or over the face of a core at a given temperature and pressure differential. Outman's proposed Eq. 5.31 determines dynamic filtration rate after the filter cake reaches an equilibrium thickness:[8]

$$\frac{\Delta V_f}{\Delta t} = \frac{\bar{k}}{\mu_l}\left[\frac{(\tau/f_f)^{(-v+1)}}{d_f(-v+1)}\right] \tag{5.31}$$

where:

$$\bar{k} = \text{permeability of the filter cake}$$
$$f_f = \text{friction between solids in the filter cake}$$
$$-v+1 = \text{filter cake compressibility}$$
$$d_f = \text{height of filter cake}$$

Note: Eq. 5.31 does not have consistent units. The equation describes behavior and should not be used to calcualte dynamic filtrate loss.

Note in Eq 5.31 that dynamic filtration rate increases with an increase in shear stress; thus, the quantitative magnitude of dynamic water loss depends on the flow rate, the viscous properties of the mud, the size of the annulus, and the characteristics of the cake. The f_f represents the friction between solids, which is an unmeasurable quantity; however, a material such as oil decreases the value of f_f, resulting in an increase in dynamic filtration rate. The d_f is the equilibrium thickness of the filter cake. Thus, the thinner the cake under circulating conditions, the higher the dynamic filtration rate for a given mud.

The $(-v + 1)$ is a measure of filter cake compressibility and Outman indicates the range of values to be 0.1–0.15. From this analysis we see that many factors affect the dynamic filtration rate that are not present when measuring the static filtration rate. This accounts for the lack of quantitative agreement between the static and dynamic filtration rates.

Laboratory tests have confirmed the difference between dynamic and static filtration rates. Figs. 5–24 and 5–25 show data obtained by Kreuger.[9] Fig. 5–24 shows tests using CMC, starch, and a metal lignosulfonate. Note that using CMC the minimum dynamic filter loss is recorded when the

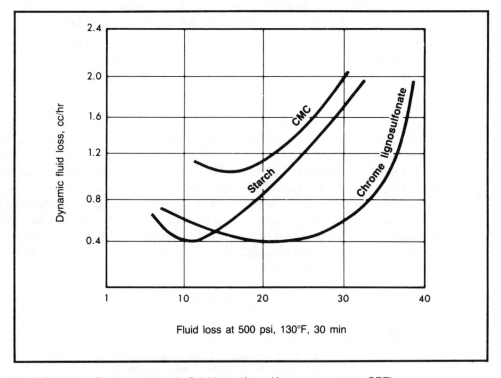

■ **FIG. 5–24** Static vs dynamic fluid loss (*from Krueger, courtesy SPE*)

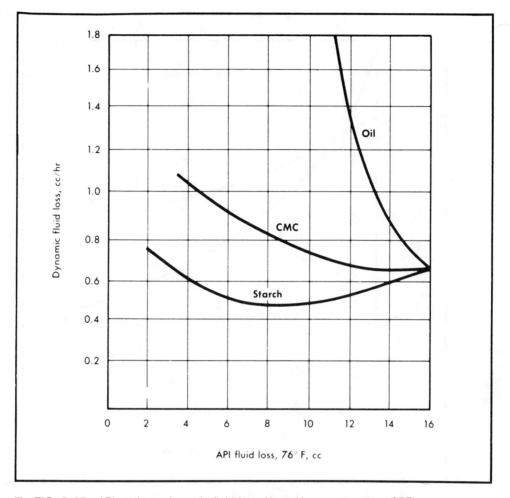

■ **FIG. 5–25**   API static vs dynamic fluid loss (*from Krueger, courtesy SPE*)

static filtration rate at 500 psi and 130°F is about 16 cc. For starch, the minimum dynamic filtration rate occurred at a static water loss of about 11 cc. The static water loss with the metal lignosulfonate was above 20 cc at the point where the dynamic water loss was a minimum.

Fig. 5–25 shows additional data, this time using starch, CMC, and oil. The comparisons are made between the dynamic and the standard low-temperature API static water loss. Further confirmation of these tests is given in Figs. 5–26, 5–27, and 5–28, which were obtained by Gray.[10] Fig. 5–26 compares the static and dynamic water loss using CMC. In this test the minimum dynamic water loss occurred when the static water loss was just above 6 cc. Fig. 5–27 shows a comparison between the static and dynamic water loss using a metal lignosulfonate. The minimum dynamic water loss was recorded when the static water loss was about 13.0 cc.

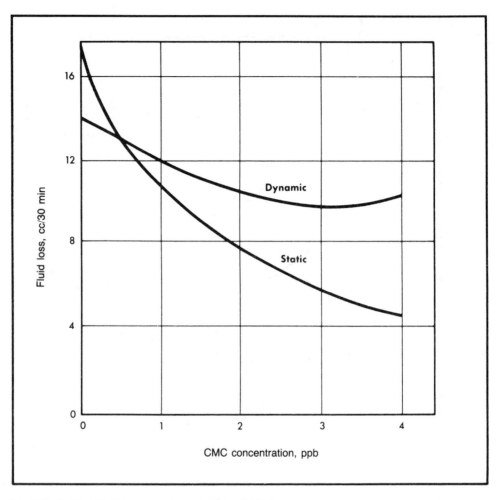

Fig. 5–28 compares static and dynamic water loss using oil. Adding oil increased the dynamic water loss and decreased the static water loss. There are no claims that this series of tests represents quantitative comparisons that can be applied in general field application. The tests show clearly that static filtration rate tests that are run in field operations provide only qualitative information relative to the total fluid that may enter underground formations.

Why does this happen? In the case of CMC, the material increases the shear characteristics of the mud. This characteristic of CMC means the shear stress is increased; thus, at some level of CMC concentration the increase in shear stress results in a thinning of the dynamic filter cake, which more often offsets the reduction in filter cake permeability offered by CMC. The same argument might be proposed for comparisons

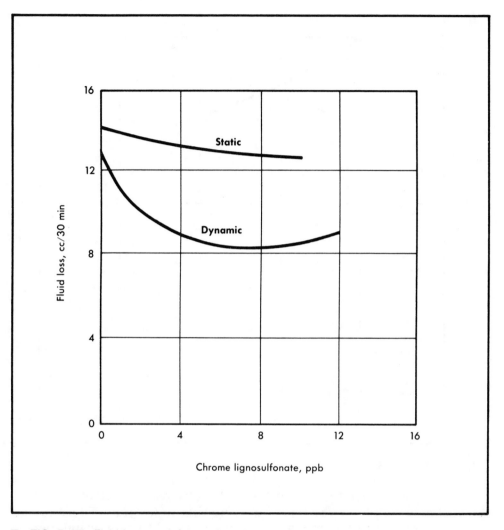

■ **FIG. 5–27** Fluid loss vs concentration of lignosulfonate

using starch, although as noted the starch used in these tests did not thicken
the mud as much as the CMC.

Different grades of CMC and starch are available; the quantitative reactions with other grades of these materials would be different from those
shown. The tests using oil are somewhat startling. In these tests the dynamic filtration rate was immediately increased by adding oil in emulsified
form. Referring back to Eq. 5.30, oil increases the shear stress and reduces
the friction between particles in the filter cake. Thus, oil thins the filter
cake, which would be beneficial from the standpoint of problems associated
with filter cake deposition but detrimental relative to the total fluid entering a permeable formation.

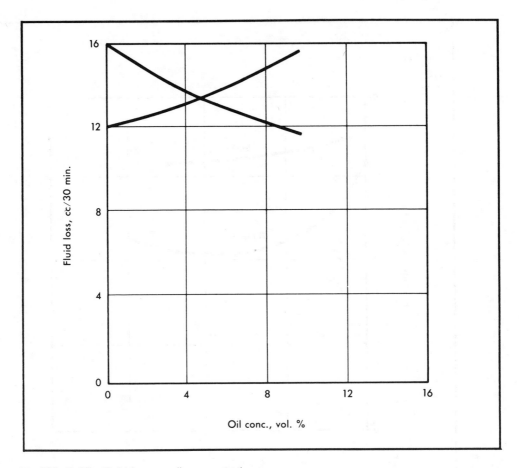

■ **FIG. 5–28** Fluid loss vs oil concentration

Ironically, oil has been used for many years to reduce the static water loss to minimize the total fluid entering underground formations, while in fact if filtrate were entering such zones, adding oil would increase invasion. Examples of this practice include the use of oil in the midcontinent U.S. to reduce water loss, often to values less than 4 cc, to prevent shale sloughing. If the primary purpose of the oil is to prevent wetting shales by filtrate invasion, the practice actually has the opposite effect if filtrate enters the shale.

Additives Used to Reduce Filtration Rate

Additives commonly used to control filtration rate are shown in Table 5–10. The most reliable method of reducing filtration rate is to maintain a good percentage of bentonite in a completely deflocculated mud. Often some other problem may prevent adding bentonite, particularly to weighted muds, and treating compromises must be accepted. For example,

■ **TABLE 5–10** Additives for controlling filtration rate

Bentonite	Provides a better basic filter cake
Metal lignosulfonates	Reduces filtration by deflocculating the mud; also increases the viscosity of the filtrate. Starts breaking down at 300°F; not very effective above 350°F
Lignite	Reduces filtration rate by deflocculating and plugging open spaces in filter cake. Starts breaking down at 300°F; effective to reduced degrees above 400°F
Sodium carboxymethyl-cellulose, CMC	Reduces filtration rate by coating solids and sometimes by minimizing flocculation. Starts breaking down at 300°F
Starch	Reduces filtration rate by coating solids. Will spoil in a freshwater environment where pH is less than 11.5, but a salt concentration of 250,000 ppm prevents spoiling in any pH water. Breaks down rapidly when temperatures exceed 275°F

in a deep high-temperature well where the total pressure exerted by the mud column is approaching the fracture gradient of an underground formation, the operator would be reluctant to add bentonite, which would probably thicken the mud and increase chances of losing circulation.

The most common agents used to control filtration rate in weighted muds are bentonite, metal lignosulfonates, and lignites. Bentonite is an excellent material for this purpose. Metal lignosulfonates are primarily deflocculants and are used often. When temperatures exceed 350°F, it is not uncommon to use primarily lignite because of its greater temperature stability. In unweighted muds, it is generally not necessary to control filtration rate; if it is controlled, bentonite, starch, and CMC are normal additives.

Drilling muds are always in a state of change. Generally, the best mud is one that is simple and easy to treat. There are exceptions because drilling conditions may require specific muds to minimize drilling problems and cost. The two problems most frequently encountered in drilling that may require special muds are hole enlargement due to shale sloughing and lubrication problems frequently associated with directional drilling.

Two methods are implemented to minimize shale sloughing with water-based muds: (1) using inhibitors to minimize clay hydration and (2) using polymers to minimize clay dispersion. The primary inhibitors used in water-based muds are calcium and potassium. Many polymers have been used to minimize dispersion.

The first inhibited water-based muds were formulated utilizing slaked lime, $Ca(OH)_2$. The two types of lime muds were (1) the high-pH (12.0) and high-lime-content (excess lime greater than 6.0 lb/bbl) muds and (2) the lower-pH (10.5 and above) and low-lime-content (3.0-lb/bbl range) muds. The inhibiting effects of these lime muds were limited because of the limited solubility of calcium in a high-pH environment. The high-pH, high-lime-content mud was very thin at high mud weights; this is the primary reason for its use. When it solidified at temperatures above 250°F, the lower-pH, lower-lime-content muds were introduced. These minimized the solidification problem but were not as thin.

In an effort to improve the inhibition effects of calcium and to further reduce the problems of high-temperature solidification, gypsum ($CaSO_4$) muds were introduced. The gypsum muds had a narrow pH range between 9.0 and 10.0. As a result, more calcium was in solution, and inhibition of clay hydration was improved. The primary problem was the higher gel strengths and the higher flow viscosity associated with the gypsum muds when compared with the lime muds.

About the same time gypsum muds were introduced, some calcium chloride muds were also introduced. The calcium chloride muds were sold under trade names and were not identified as calcium chloride types. Most of these calcium chloride muds were run at a pH of 11.0, which meant they also were subject to high-temperature solidification. However, these muds succeeded in reducing shale sloughing.

The next inhibited water-based muds were those treated with potassium compounds. Most of the potassium muds were formulated initially using potassium chloride (KCl); however, potassium hydroxide (KOH) is used more often now. Potassium muds are stable systems and may be improved by adding some lime. Combination potassium-lime muds frequently are mixed in conjunction with one of the polymers, which is used to further reduce dispersion of clays. The potassium muds also serve as completion fluids to minimize formation damage. When used as completion fluids, potassium hydroxide should always be used instead of potassium chloride.

The primary drilling problem with any inhibited water-based mud is the inhibition of clays that have been added to the mud. Common drilling problems are torque, drag, and differential pressure sticking. One way to minimize these problems is to add fully hydrated bentonite to the mud for filter cake improvement. In any inhibited water-based mud, the prehydration of bentonite is beneficial; however, the quality of the filter cake will be reduced because the bentonite will dehydrate. Thus, drilling problems associated with filter cake are almost always more severe when using inhibitors in water-based muds. For this reason, the operator must decide which problem is the most important when selecting the mud system.

Oil-based muds are no longer considered special mud systems. Many of the objections to oil-based muds have been overcome, and the advantages of oil-based muds have been expanded. One of the primary objections to

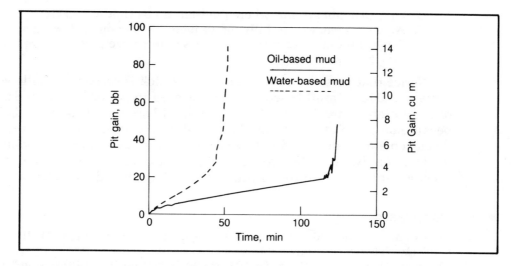

■ **FIG. 5–29**   Pit gain for a sample case (*from Thomas, et al., courtesy SPE*)

oil-based muds has been the pollution effects of oil in environmentally sensitive areas, particularly offshore. This objection has been overcome largely by using a clear mineral oil instead of number 2 diesel oil. Also, there is a continuing effort to improve the mineral oil.

Another problem has developed with oil-based muds that threatens their use, particularly in environmentally sensitive areas. Detecting some well kicks may be difficult because gas is soluble in oil. If the initial well kick is undetected, it is possible, during the normal displacement of mud, for gas volume to increase quickly. This sudden increase in gas volume, if not controlled immediately, may result in very high shutin surface pressures.

Several papers have been written on gas solubility in oil-based muds, and several examples given for potential kick control.

Fig. 5–29 shows the predicted pit gain vs time for gas that enters an oil-based mud and a water-based mud at 15,000 ft.[11] Fig. 5–29 is a specific example and does not represent any type of general solution.

We can conclude from this figure that if the gas kick is recognized early, control in the oil-based mud is not a serious problem. If the gas kick is not recognized and normal circulation continues, then the operator may encounter a sudden expansion, as illustrated in Fig. 5–29. This type of behavior may result in an almost sudden reduction in hydrostatic head and high shutin pressures at the surface. However, even with oil-based muds, some warning is given.

When using oil-based muds, the following steps should be taken:

1. Monitor pit levels carefully.
2. Watch standpipe pressures closely; beware of any sudden reduction in pump pressure.

3. Locate control systems where the well can be shut in quickly.
4. Watch flow shows carefully for an increase in annular flow rates.
5. Conduct training sessions on methods to recognize pore pressure changes for rig crews.

Notice that pressure control after a gas kick is recognized follows the same procedure in oil-based muds as it does in water-based muds. The only differences are (1) early detection in oil-based muds is more difficult because of the solubility of gas in oil and (2) the expansion of gas during displacement is more difficult to predict, again because of the solubility of gas in oil. None of the problems of gas kicks in oil-based muds should preclude the use of oil-based muds when they offer drilling advantages.

■ REFERENCES

1. T.C. Mondshine and J.D. Kercheville, "Successful Gumbo Shale Drilling," *Oil & Gas Journal* (28 March 1966).
2. Frank O. Jones Jr., "New Test Measures Bentonite in Drilling Mud," *Oil & Gas Journal* (1 June 1964).
3. Max R. Annis Jr., "High-Temperature Flow Properties of Water-Base Drilling Fluids," Paper SPE 1968, University of Texas (25 Jan. 1967).
4. L.E. Bartlett, "Effect of Temperature on the Flow Properties of Drilling Fluids," Paper SPE 1861, Fall meeting (1 Oct. 1967).
5. Annis, 1967.
6. Bartlett, 1967.
7. Deily, Lindblom, Patton, and Holman, "New Low-Solids Polymer Mud Designed to Cut Well Costs," *World Oil* (July 1967).
8. H.D. Outman, "Mechanics of Static and Dynamic Filtration in the Borehole," *SPE Transactions* (1963).
9. R.F. Krueger, "Evaluation of Drilling Fluid Filter Loss Additives under Dynamic Conditions," SPE Transactions (1963).
10. J. Gray, "A Dynamic Filtration Study of Drilling Fluids," master's thesis, University of Oklahoma (1965).
11. David C. Thomas, James F. Lea Jr., and E.A. Turek, "Gas Solubility in Oil-Based Drilling Fluids: Effect on Kick Detection," *SPE Journal,* (June 1984).

CHAPTER 6

DRILLING FLUID SOLIDS REMOVAL

GEORGE S. ORMSBY
Stonewall Associates

For many years drill solids were removed with seldom anything more than settling pits and vibrating shaker screens, few of which had openings less than one-sixteenth of an inch across. The mechanical processing of rotary drilling fluids entered a new phase with field-practical decanting centrifuges[1] in 1952, efficient 6-in. hydrocyclones[2] in 1954, much more efficient 4-in. hydrocyclones[3] in 1962, and very fine shale shaker screens[4] in 1966. Spinoffs and special equipment are being developed constantly.

This chapter reduces confusion and lays out available approaches to solids removal for better mud control at reduced mud cost.

SYSTEM MATERIAL BALANCE RELATED TO MUD COST

Anyone familiar with drillsite operations is aware that many active mud systems tend to increase in surface volume. The excess must be disposed of by some means after being hauled away or jetted out. Examine Fig. 6–1 with the following:

The subsurface mud system consists of the hole. This subsurface system increases at the rate of drilling new hole, plus caving or sloughing rate. Drilling the hole does not change the level in the surface tanks because every 100 bbl of hole drilled adds 100 bbl of cuttings and results in 100 bbl of hole to be filled with mud. The same is true of *caved* volume, and whether the formation might be bentonite or granite does not change this. The hydration of bentonite by water in the mud system neither shrinks nor expands the volume. It has no effect whatsoever on the system level.

Solids removal involves the removing of large amounts of liquid. Removal machines discard an additional 1 to 4 times as much liquid as the

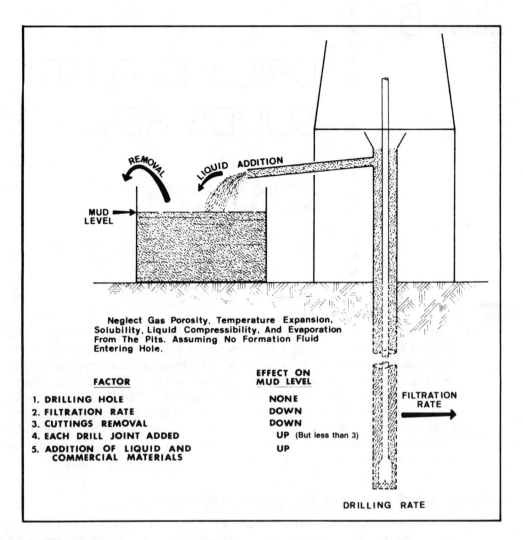

Neglect Gas Porosity, Temperature Expansion, Solubility, Liquid Compressibility, And Evaporation From The Pits. Assuming No Formation Fluid Entering Hole.

FACTOR	EFFECT ON MUD LEVEL
1. DRILLING HOLE	NONE
2. FILTRATION RATE	DOWN
3. CUTTINGS REMOVAL	DOWN
4. EACH DRILL JOINT ADDED	UP (But less than 3)
5. ADDITION OF LIQUID AND COMMERCIAL MATERIALS	UP

■ **FIG. 6–1** Mud system material balance

solids volume. Dilution and discard is even worse, with the liquid amounting to between 5 and 20 times the solids volume removed. To successfully drill a deep hole in the ground, all evidence indicates that at least an average of 85% of the hole volume drilled must be removed. No matter how it is done, for every 100 bbl of hole drilled, there will be between 200 and 1,100 bbl of discard, depending upon how we perform the removal. Obviously, the mud system level should be going down, not up!

Dynamic filtration to the formation varies from hundreds of gallons per minute to almost zero. Maximum rates occur when drilling massive sand and silt formations with very little or no effort to control filtration. Minimum filtration occurs drilling dense zones and while drilling permeable formations with low silt muds (explained later).

Solubility of the formations in the drilling fluid causes a decrease in the level, but the effect is too small to ever be detected in normal field operations. Evaporation is another minor volume-reducing effect except in geothermal drilling. In that application the loss can be significant, especially if mud cooling towers must be used.

The only significant item that can increase the mud system volume and level is drilling while the formation produces liquid. This is done regularly in a few areas, but it is a very special procedure for a special situation. All concerned are aware of exactly what they are doing (and why the pit level and system volume increase). If the formation begins producing unexpectedly during drilling operations, it is termed a "kick," and immediate steps are taken to stop it. So in a normal drilling operation the increase in the mud system level cannot be blamed on formation fluids.

Each time the kelly is drilled down, another joint of steel pipe is added, displacing that much mud. But the steel in the pipe is less than 10% of the volume of the hole drilled with the pipe. Considering that more than 85% of the cuttings must be removed and the liquid multiple that is removed with them, obviously the steel added does not discernibly raise the level.

Drilling a gas sand has a temporary, small increasing effect on pit level. After the sand is passed and all the gas broken out, there is a small decrease from the original level. Neither is significant if pore pressure is properly overbalanced. Temperature effects are too minor in a water-based mud system to see any changes in mud level from it.

Only one thing besides a kick can override the natural tendency of the mud level in the system to decrease—the volume of the material we add. Most of the volume added is the continuous phase of the mud, i.e., water in a water-based mud. This addition is shown in Fig. 6–1 as 5, addition of liquid and commercial solids materials. Water is a cost item, and though the scarcity varies more than most other materials, it is a major item in some areas.

Since buying material to increase pit level is made doubly expensive by the problem of disposing of the excess, there should be valid reasons for willfully adding so much to the problem and the cost. Before the advent of modern solids-removal equipment, the reason for "dumping," or "watering back," was to control mud properties. Today, it is usually a holdover in custom or the result of failure to understand (1) the problem and (2) the solids removal equipment available.

MUD TREATMENT SIMPLIFIED

The three basic procedures for treating liquid drilling fluids all relate to solids: (1) removing solids, (2) adding solids or their equivalent, and (3) treating solids chemically.

For specific controls, commercial clay and soluble solids are *added* to increase yield point, gel strengths, and plastic viscosity and to decrease filtration rate with minimum weight increase. Heavy commercial silt solids are *added* to increase the density of the slurry with minimum effect on the mud properties and solids volume.

Oil added to a water-based mud emulsifies, and the effect of the droplets is much like that of clay solids particles. The same is true of the water fraction of a continuous oil-phase mud.

Removing solids is necessary before a drilling fluid can be returned to the drill bit. Recirculated drilling fluids must be returned to as near the optimum drilling properties as economically practical. The more effective the mechanical solids removal, the less dilution and chemical treatment is required. If the total annular return fluid is being discarded after one pass by the bit, of course no separation effort is necessary. In most situations this latter is impossible for either economic or ecological considerations, or for both.

Treating solids—*mud chemistry* or *mud treating*—is the science and art of adding specific soluble materials to a drilling fluid to alter the behavior, directly or indirectly, of some specific solids in the slurry. The chemicals act on the clay particles, including hydratable shales, and not on the larger inert particles. Even though chemicals may be soluble, such as a salt, they do add to the true total solids.

DYNAMIC FILTRATION AND WATER DILUTION

Adding water does not necessarily dilute a drilling mud in a drilling situation. The water leaving the mud as filtrate in a dynamic fluid-loss situation can be very high. Downhole filtration has been reported as high as 400 gpm in long stretches of open hole in massive sand and silt sections and with very little selective solids control.[5] Failure to replace the filtrate loss would cause a decrease in volume and resultant increase in percent solids, weight, viscosity—all independent of drilling rate. However, if the filtrate loss is constantly replaced, the mud is constantly restored from the effects of filtration. This water is not causing a change in mud—it is only restoring, much like drinking water restores body fluids lost in perspiration. Whether this *pass-through water* amounts to most or a minor part of the total water requirement depends upon the ratio of the dynamic downhole filtration rate to the solids-removal rate.

Water added to replace solids removed from a system does change the system total solids. Water added over and above that necessary to balance filtration does dilute the total solids. Solids added by the bit do increase the total percent solids, but they do not increase the surface system level. If surface level is held constant, then total solids content is being altered only by the addition and removal of drilled and commercial solids regardless of the amount of water being added. All water being added is then replacing solids removed and liquid lost with the solids or restoring filtrate lost downhole.

PLASTIC VISCOSITY AND TOTAL SOLIDS CONTENT OF WEIGHTED MUDS

Frequently articles are circulated indicating that the plastic viscosity of a weighted drilling fluid of a given density varies with total solids content or with drill solids, or low-specific-gravity solids, content. Or it may be implied that a high plastic viscosity is a sure indication of the need to reduce low specific gravity solids, or drilled solids. Fig. 6–2 is a plot of over 350 random, bona fide field checks of drilling fluids in the coastal area of the Gulf of Mexico, in which the muds are classified low, medium, and high in low-specific-gravity solids according to the average specific gravity ($\overline{\text{ASG}}$) of the total solids. This purely arbitrary classification is shown at the bottom of the figure. These field checks represent many wells and operators, all service companies, and are typical for all of the various muds in use for about two years.

Although the highest plastic viscosity of 100 cp is for a "high" solids mud, the lowest plastic viscosities in any weight range are seldom for the "low" solids muds. Rather the "medium" solids muds are found consistently in the low viscosity range and in the high range as well. Although the "high" solids muds are the highest viscosity in each weight and are never the lowest viscosity in any weight, that statement must be tempered with the observation that the variations and overlap of viscosity in and between all classifications is more striking than their differences.

The wide range of viscosity, coupled with the lack of clear classification, indicates great differences in the character of the solids in drilling fluids. And so there are. The size and shape differences cause varying effects. The greatest size and shape variations are in the low-specific-gravity solids, or light solids, whether purchased or drilled. For this reason conclusions from laboratory tests on drill solids are often difficult to support in the field. Many diverse things can be "proved" in laboratory experiments involving drill solids, but field drilling conclusions are sometimes contradictory.

USING MUD STILLS TO DETERMINE TOTAL SOLIDS CONTENT

The first reaction of many persons upon seeing data similar to that of Fig. 6–2 is that the data must be inaccurate. The reaction has some validity. The field mud still, or retort, was developed in 1952 to obtain a rough measure of oil content. The condensate *gap* was recorded and assumed to represent "total solids." After some years, masses of these data were published with median lines as guides, ignoring the fact these figures were nothing more than inaccurate reflections of variations, with no discussion of what might cause the variations.

The API Committee on Standardization of Drilling Fluids Material became concerned enough about the inherent inaccuracy of mud stills,

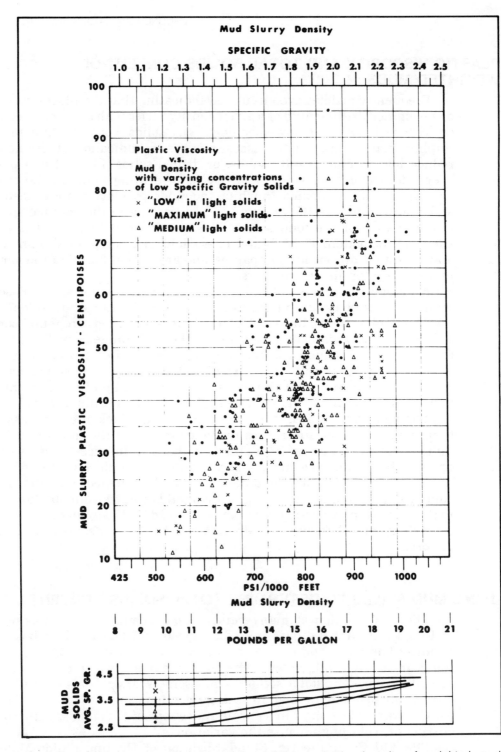

■ **FIG. 6–2** Effect of low-specific-gravity solids on plastic viscosity of weighted muds (*courtesy Pioneer Centrifuging Co.*)

as compared to the margins of solids control many operators are attempting to hold. They set up a task group on Determination of Liquids and Solids at the annual 1968 meeting. The later work of this group confirmed that dependable accuracy of field mud retorts does not justify the actions and expenditures many operators make based on solids tests.[6] Field readings of percent solids for years have varied from down two percentage points to up four percentage points from the true nonevaporite solids in a mud. Perhaps more significant, from a remedial point of view, is the fact that retorting affords no clue to the size or shape of the offending solids, or indeed that they are offending. Further committee work in the 1980s has improved the accuracy but not to the point most operators like to believe.

PARTICLE SIZE REFERENCES AND TERMINOLOGY

To appreciate any discussion of small particles, some sense of size is necessary. The micron is the most convenient unit to use in discussing small particles. Table 6–1 shows the micron size of common materials. The fingertip sensitivity can be used to determine the effectiveness of coarse solids removal by rubbing a test filter cake between the fingers. Since red blood corpuscles are very small, they make a good thinking reference when looking at tabulated sizes. Particle size distributions of the solids in drilling fluids show that most single particles are invisible.

It is sometimes inconvenient to refer to all sizes in exact micron sizes or micron ranges, so various disciplines have adopted some standard terminology for convenience. When an industry or group lacks a standard, individual investigators must use improvised or borrowed terms. Fig. 6–3 illustrates the terminology problem and compares usages often heard in the field.[7]

The word *clay* is used occasionally in this chapter to refer to particles

■ **TABLE 6–1** Micron relationships—familiar things

Material and/or Sense	Diameter of Size, microns
Human hair	30–200
Pollen	10–1,000
Portland cement dust	3–100
Milled flour	1–80
Talcum powder	5–50
Red blood corpuscles	7.5
Human eye resolution, normal	40 minimum
Cosmetic powder	35 maximum
Fingertip sensitivity	20 minimum
Between teeth sensitivity	6–8 minimum

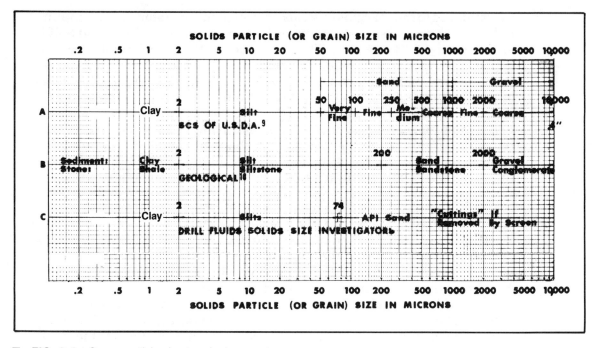

■ **FIG. 6–3**   Some particle size terminology and usage

less than 2 microns. Silt covers the 2 to 74-micron range. *Sand* and *API sand* are used interchangeably with each other and refer to particles larger than 74 microns, as determined by an API sand test.

These size usages do not change with the nature of particles, whether drilled limestone, quartz, dolomite, basalt, chert, commercial bentonite, barites, or steel from a milled fish.

A source of confusion in the application of solids-removal equipment has been the use of the word *clay* by most mud-treating authorities in reference to all commercial solids and all drilled solids in the system, i.e., to all solids except commercial weighting material. It is still found in the following erroneously expressed mud-solids equation:[8]

$$\frac{100\%}{\text{specific gravity of solids}} = \frac{\text{clay, wt\%}}{2.6} + \frac{\text{barite, wt\%}}{4.3} \qquad (6.1)$$

This is one of the most basic equations in drilling fluid solids work, and the equations to be presented in this chapter are derived from it, but the terminology is extremely unfortunate. The relationship expressed here concerns the specific gravities of solids materials, their weight percentages in a mixture of these solids, and the average specific gravity of the mixtures. The confusion arises because the term *clay,* used erroneously here in this mathematical expression as general for materials of 2.6 specific gravity, is recognized and used by solids investigators and by earth-related sciences as a size range, whether or not the size ranges are in exact agreement. Refer again to Fig. 6–3.

The damage from this unfortunate usage has come in this way: It is correct to say "decanting centrifuges can separate liquid and clay from the larger solids particles in the drilling muds processed through the machines," using the term "clay" properly in the size sense. It is completely erroneous to say "decanting centrifuges can separate low-specific-gravity solids (or drilled solids) from barites in the drilling muds processed through the machines." Yet the erroneous statement has been written and still is widely accepted because of the ambiguous use of the word "clay" in recognized mud treatment works.

"Clay minerals" is a term specifically denoting smectites. This chapter will not discuss this special group of clays in a manner requiring further use of the term.

THE RELATION OF SOLIDS PARTICLES SPECIFIC SURFACE AREA TO MUD PROPERTIES

The effect solids particle size alone can have on mud properties is illustrated in Fig. 6–4 from Ritchey.[9] Although another API barite might exhibit some variation, and "3 microns or less" barites in another test would not follow the same line as those in Fig. 6–4, a similar basic difference would exist. The difference in results when particle size is varied in a mud slurry is a matter primarily of surface area, for the surface area adsorbs, or ties up, water. Obviously, if there is more surface area, more water is adsorbed. Table 6–2 shows how surface area increases as particle size decreases, even though the shape and quantity (net volume and weight) remain exactly the same for 1 lb of glass balls.

Table 6–2 also shows that, if shape is changed, the surface area is affected. The sphere shape permits the least area possible for any given volume. Particles of crushed quartz having the same volume as quartz balls have more area per unit of weight and so would adsorb more water per pound of solids.

A layer of moisture adsorbed on the surfaces of solids as they become wetted contains neither chlorides nor clays. Dry solids added to the mud system, in the process of becoming wetted, literally take a portion of the available free liquid and make it unavailable to other solids or to reduce flow resistance. If plastic viscosity and/or yield point are already measurable, this adsorption has the effect of concentrating the solids, including bentonite, already in the system. The result can be the same as adding bentonite—an increase in viscosity and gels—even though the solids added may not be efficient viscosifiers and the increase in properties may not be desired or expected.

This surface area relationship also explains why it may be necessary in weighted muds to remove fine solids to reduce viscosity and at the same

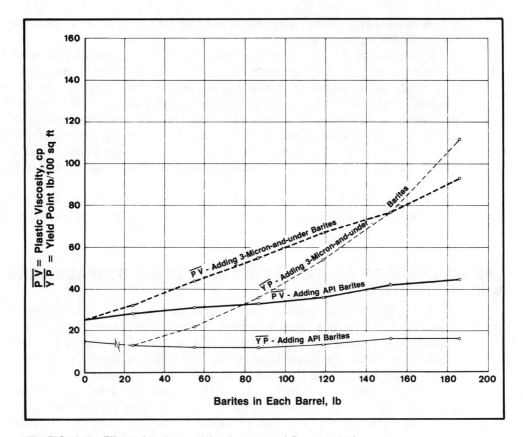

■ **FIG. 6–4** Effect of barite particle size on mud flow properties

■ **TABLE 6–2** Effect of particle size and shape on surface area

Particle Diameter or Equivalent Diameter, microns	Type of Particles	Sq ft/lb	Sq m/kg
50.0	Glass spheres	234.5	48.0
5.0	Glass spheres	2,345	480.4
1.0	Glass spheres	11,725	2,402
1.0	Crushed quartz	17,160	3,515
0.5	Glass spheres	23,450	4,804
0.5	Crushed quartz	34,320	7,030
0.1	Glass spheres	117,250	24,018
0.1	Crushed quartz	171,600	35,151

time add bentonite or some other high quality commercial product to control fluid loss. Viscosity consideration often does not permit adding any more of the colloidal solids necessary to control filtration unless total solids surface area is first reduced by removing a portion of the existing clays, most of which may be of poor quality.

The earth particles that reduce most readily to clay size are the soft flaky minerals, so that most clays are of a flatter shape than crushed quartz or ground barites. As purchased, bentonite claystone may be finely ground and sized (see Fig. 6–5).

When added to water, bentonite subdivides as it hydrates, resulting in tremendous increases in surface area and in its effect on viscosity and gels of a slurry. Before approximately 1975, API specification "Wyoming" (sodium-based) bentonite required approximately 2 hr to approach its full yield. This high quality material not only had to be from high grade ore.

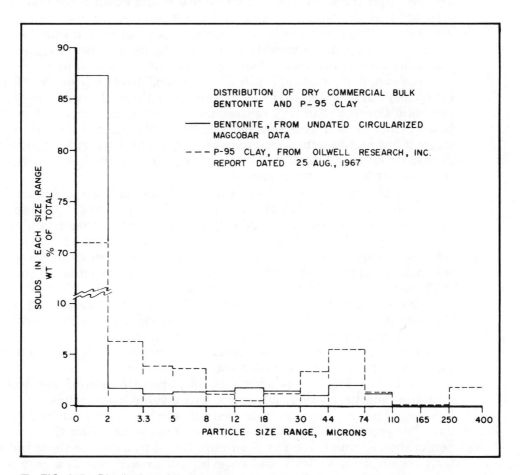

■ **FIG. 6–5** Distribution of dry commercial bulk bentonite and P-95 clay

it required some months of surface aging to mature before being milled and sold.

During the drilling boom starting in the middle 1970s, a shortage of surface stocks developed and the aging became impossible. The addition of small amounts of powerful polymer began, in order to meet the viscosity requirements of the API specification. Eventually the standards were lowered, more polymer was added, the standards were lowered twice more, and a situation developed in which a typical sack of bentonite produced an instant viscosity increase when added to water, lost that viscosity if circulated through a hot drill hole, but then matured in the system with a high viscosity increase occurring 7 to 21 days after its addition.

In the field, operators and mud engineers alike were unaware of the change in the product. The immediate increase from the polymer when "bentonite" was added was attributed to the bentonite. When the viscosity dropped after the mud was circulated one round down a hot hole, it was assumed that the bentonite had "faded out," and more was added—each circulation. When the excessive bentonite content in the mud began to mature rapidly and inexorably increased the mud's viscosity, the cause was a mystery. Several deep holes were lost to this phenomenon.

Curves A, B, C, and D of Fig. 6–6 are adapted directly from the "Typical Clay Yield Curve" diagram found in the 12th edition of *Principles of Drilling Fluid Control.*[9] Curve A represents the high-grade, matured, sodium-based bentonite still available in 1969. Note that when hydrated in fresh water it has more surface area than can be wet by a water-bentonite ratio of 95:5 or 19:1 by volume (5% solids). This did not include polymer, but this product is no longer available.

Curve B obviously represents a material having less specific area and is therefore less efficient at producing viscosity in a slurry. Although these curves are not part of any standard, they do assist in understanding the effect of more or less surface area per unit of volume. Curve B is commonly referred to as the curve for attapulgite and sepiolite in fresh water. It is said also to represent the API bentonite (including polymer) of the early 1980s as well as calcium-based bentonite.

Curves C and D represent clays of less viscosity-producing ability. Curve C is typical of kaolinite and curve D typical of illite. Neither of these is noted for its ability to control filtration rate.

Of all the clays discussed above, only bentonite, whether sodium or calcium based, controls filtration rate dependably.

Curve E of Fig. 6–6 is drawn as an approximation of the probable minimum plastic viscosity possible at any given percent solids in a water-based mud having filtration rate controlled to less than 15 cc by API standard low-pressure, low-temperature test. This is based on the field data points shown, which are the minimum found in over one thousand field tests screened for validity and covering all mud types. The only points below the curve are of muds mixed quickly to control a kick, with little or no filtration control.

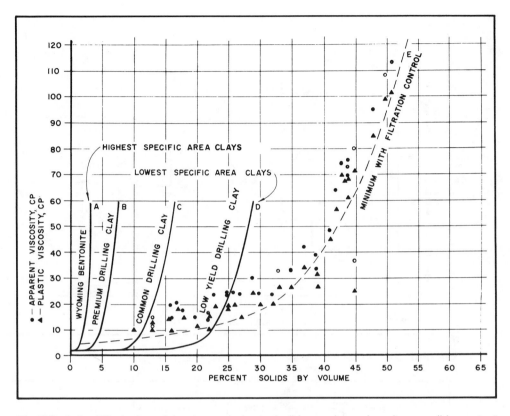

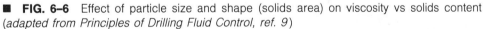

■ **FIG. 6–6** Effect of particle size and shape (solids area) on viscosity vs solids content (*adapted from Principles of Drilling Fluid Control, ref. 9*)

If the plastic viscosity of a water-based mud is significantly higher than curve E, its plastic viscosity can be reduced by reducing the clay content; if the viscosity is near curve E, the plastic viscosity probably can be reduced only by reducing total solids. This latter may or may not be possible, according to the slurry density necessary, and the amount and size of low-specific-gravity solids present. The size distribution of the light solids present determines the economics of reducing total solids.

The particle-size distribution of barite milled to meet API specifications are somewhat similar over the world, as might be expected. A maximum of 3% by weight is allowed to be of API sand size. The difference in the amounts of barites in the clay size is sometimes noticeable and affects viscosity problems. As already discussed, this is due to the specific surface area phenomenon.

Typical variations in barite particle-size distribution from suppliers of API-grade barites are shown in Fig. 6–7. In the past, barite suppliers have added phosphates, tannins, lignins, etc., at the bulk plants or the mill to counteract the viscosity-increasing effect of the water adsorption by the clay barite fraction. In 1985 the API barite specifications were re-

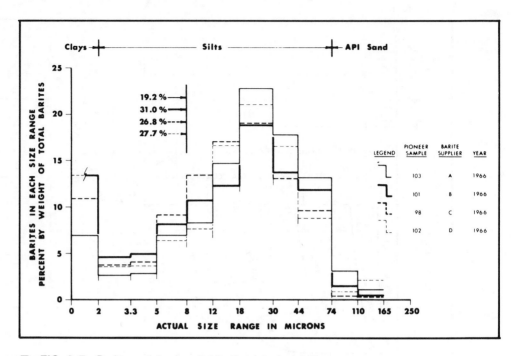

■ **FIG. 6–7** Barite particle-size distribution—typical API

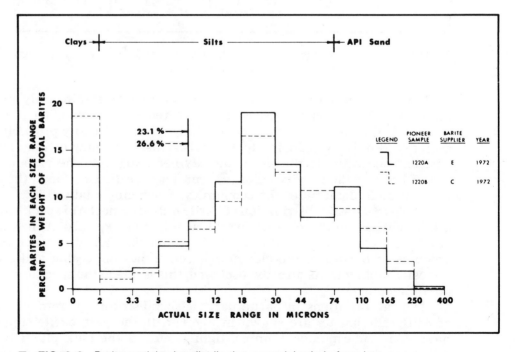

■ **FIG. 6–8** Barite particle-size distribution—special grinds from Iran

vised to prohibit the inclusion of nonbarite ore, to eliminate the requirement of at least 5% by weight larger than 44 microns, and to limit the amount of material under 6-micron particle size to a maximum of 30% by weight. Fig. 6–8 shows special particle-size distributions ground for use to build some of the highest drilling mud weights in the world, in Iran.

STOKES' LAW AND SOLIDS SEPARATION

Stokes' law gives the mathematical relationship of the factors governing the settling velocity of spheres in a liquid stated in its simplest form as Eq. 6–2:

$$V_B = \frac{d_B{}^2(D_B - D_L)G}{18\,\mu}$$

(6.2)

where:

V_B = terminal or settling velocity
d_B = diameter of the sphere
D_B = density of the sphere
D_L = density of the liquid
G = acceleration of gravity, ft/sec/sec
μ = viscosity of the liquid, cp

Using Eq. 6–2, or a modification of it, the settling velocity of any ball in any liquid can be calculated or predicted. Likewise, we can let any object settle in a liquid, measure its rate of fall, and calculate what size ball would settle at that rate. This gives us an equivalent spherical diameter. This method is commonly used for sizing small irregular particles, such as drill solids and barites, that are below a size easily screened. It is particularly valid for drilling-fluids work because settling occurs in the hole or in a mud pit and it is an essential part of the separation process in centrifuges, hydrocyclones, and settling traps.

Two solids particles of known different specific gravities, or densities, will settle at the same rate if their sizes, or equivalent spherical diameters, are of a certain mathematical relationship. This can be shown quite simply for two particles c and b if we assume their settling velocities are the same, or $V_c = V_b$. Then

$$\frac{d_c{}^2(D_c - D_L)G}{18\,\mu} = \frac{d_b{}^2(D_b - D_L)G}{18\,\mu}$$

$$d_c{}^2(D_c - D_L) = d_b{}^2(D_b - D_L)$$

(6.3)

$$d_c = d_b\sqrt{\frac{D_b - D_L}{D_c - D_L}} \quad \text{or} \quad d_b = d_c\sqrt{\frac{D_c - D_L}{D_b - D_L}}$$

If the normal range of values for barite density, light (or drill) solids density, and the liquid phase of the mud system are substituted in Eq.

6.3, the result always indicates that the light solids particle will be approximately 1½ times the diameter of the barite solids particle when the two settle together. A 10-micron barite particle and a 15-micron light solids particle cannot be separated by any settling device, including those using centrifugal force.

Although Stokes' law as presented here is for Newtonian fluids, the 1½ to 1 relationship is as valid in non-Newtonian muds as any rule of thumb yet devised, and predictions using it agree well with laboratory and field results.

Fig. 6–9 presents unusual data in that high-density solids were removed from a weighted oil mud by flotation, and then particle-size distributions were run independently on each type of solids.[10] The actual sizes are shown in Fig. 6–9. The sample is from mud that has passed through a 12 × 12 shale shaker screen, and the graph shows that a finer screen might have removed some more drill solids and perhaps without cutting deeply into the barites (assuming a fine screen could function properly in the situation).

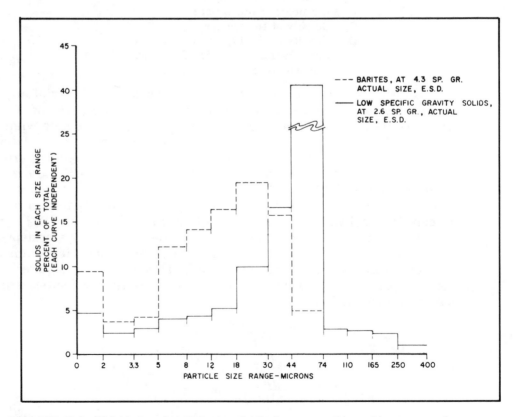

■ **FIG. 6–9** Weighted mud particle-size distribution, separation problem as seen by screen, actual equivalent spherical diameters

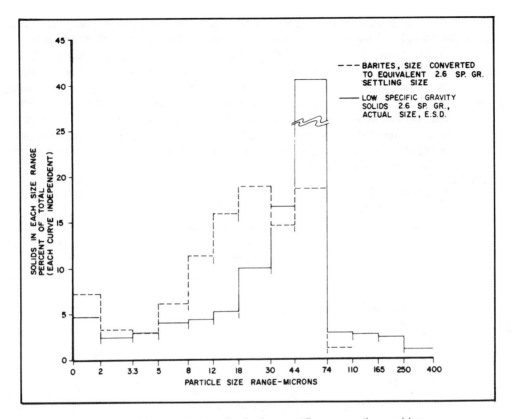

■ **FIG. 6–10**   Stokes' law equivalent distribution—settling separation problem

Fig. 6–10 shows the problem as it refers to any settling-type separation device. The barite distribution has been multiplied by approximately 1.5, to the equivalent settling size of the low-specific-gravity solids. This makes it obvious that any removal of light solids, at either the clay end or at the sand end, by any settling device, will also take some barites. This is not necessarily bad, and the removal may be very desirable, but the sacrifice of barites must be expected and understood.

THE CONTRAST BETWEEN WEIGHTED AND UNWEIGHTED MUDS

Water as a drilling fluid does not qualify as a mud. If there are no hole stability problems that prevent its being the most economical drilling fluid; if neither the geologist, paleontologist, nor production supervisor has valid objections; and if enough is available, water is seldom if ever surpassed. When the formation requires, or a supervisor demands, filtrate control or viscosity or gels in the drilling fluid, a mud is built. Or if the fluid density required is too high for liquid alone, mud properties are required to suspend weight material.

The first and most basic difference between an unweighted mud and a weighted mud is that money is spent on the unweighted mud to keep slurry density *down,* and money is spent on weighted mud to keep slurry density *up.* A more usable criterion is that if the mud contains suspended material added to increase or maintain slurry density, it is a *weighted mud;* otherwise it is an unweighted mud. In the search for simplicity, we sometimes miss that, as the change is made from an unweighted to a weighted mud system, the philosophy, or logic, of solids treatment also must be changed because the problems almost completely change. This is shown in Table 6–3.

■ **TABLE 6–3** The difference in mud solids problems between weighted and unweighted water base muds

Item	Weighted Mud	Unweighted Mud
Costliest portion	—Weight material	—Liquid, solubles, and clays
Size of:		
Ideal solids	—Mostly silt	—Mostly clays
Solids from bit	—Clay, silt, sand	—Clay, silt, sand
Specific Gravity of:		
Ideal solids	—High	—Low
Drilled solids	—Low	—Low
Source of:		
Needed solids	—Commercial barite silt	—Commercial clays
	—Commercial colloids	—Drilled clay and some silt
Detrimental solids	—Drilled silt and sand	—Drilled silt (excess) and sand
	—Drilled clay	
	—Barite sand	
	—Clays mechanically degraded from larger particles.	
Detrimental effects of:		
Drilled clays	—$\overline{PV}$ increase (major)	—Density (minor)
Barite clays	—$\overline{PV}$ increase (major)	—Not applicable
Size degradation of all solids	—$\overline{PV}$ increase (major)	—No problem with good removal system
Drilled silt over 44 microns	—$\overline{PV}$ increase (minor)	—Density (major)
	—Source of degraded clays	—Abrasion (major)
		—Filter cake character (major)
		—Density (major)
Drilled sand	—Abrasion (variable)	—Abrasion (major)
	—Filter-cake character	—Filter-cake character (major)
	—Source of degraded clays	
Barite silt	—$\overline{PV}$ Increase (minor)	—Not applicable
	—Source of degraded clays	
Barite sand	—Abrasion (major)	—Not applicable
	—Filter-cake character	
	—Source of degraded clays	

Factors Governing Total Solids Control in Weighted Muds

In weighted muds overriding factors cause total solids to vary, sometimes more than purposeful effort by the operator. If the reasons for the variations are not understood, the blame may fall on the type of mud, supervision, removal equipment, etc., when it actually may be caused by another one or none of those things. Table 6–4 lists the governing factors.

Contrary to popular belief, total solids do not always and consistently tend to run high in weighted muds. In drilling a medium-hard formation with toothed bits and with good solids removal, the necessity to mix new mud for the volume of the high percentage of large cuttings removed maintains a minimum solids content. When we understand that the most economical and trouble-free method of drilling an extremely hard or soft formation with weighted mud at best involves some higher total solids content than drilling a medium-hard formation, we can concentrate on the optimum solids removal with equipment available, controlling critical mud properties, and on the truly controllable drilling and completion parameters.

■ **TABLE 6–4** Factors that control total solids in weighted muds

Item	Contributes to	
	Lower Total Solids	**Higher Total Solids**
Formation		
Unconsolidated silt	Never	With all bits and PPR
Medium to very hard	PPR of 0.0625 in. or more	PPR of 0.012 in. or less
Unconsolidated sand or gravel	With all bits & PPR	Never
Drilling Hydraulics		
Mud density	Minimum practical	Above minimum
Bit jetting	Adequate	Inadequate
Annular lift	Adequate	Inadequate
Removal Techniques		
Rig shaker screen mesh	Finest that can be maintained and that will handle full flow*	Bypassed, even slightly
Full flow hydrocyclone removal of sand**	Effectiveness varies with cuttings size	Not applicable
Partial flow centrifuge removal of clays	Effectiveness varies with drill solids size	Not applicable, but reduces viscosity

* Care should be exercised running screens finer than 150 mesh, as the *median* cut of screens is far less than the size of the openings.
** If screen mesh of 150 or smaller is used, hydrocyclones for desanding may be unnecessary.

BASIC SLURRY

The equations that follow are basic to solids in drilling fluids. From them we can derive countless mathematical expressions of these relations. The various nomographs available are based on derivations from these equations. The symbols and subscripts used agree with the list recommended in API Bulletin 13C as far as that list extends.

$$\frac{100\%}{S_s} = \frac{100\%(F_c)}{S_c} = \frac{100\%(F_b)}{S_b} \tag{6.1}$$

$$\frac{100\%(D_m)}{8.345} = (\%V_s)(S_s) + (\%V_w)(1) + (\%V_p)(0.8) \tag{6.4}$$

$$W_s/bbl = \frac{(\%V_s)}{100}(S_s)(8.345)(42) \tag{6.5}$$

$$W_b/bbl = W_s/bbl \times F_b \tag{6.6}$$

$$W_c/bbl = W_s/bbl \times F_c \tag{6.7}$$

If we assume that $S_c = 2.6$ and that $S_b = 4.3$, then:

$$W_b/bbl = (8.8687)(\%V_s)(S_s - 2.6) \tag{6.8a}$$

$$\%V_s = \frac{W_b/bbl}{(8.8687)(S_s - 2.6)} \tag{6.8b}$$

$$S_s = \frac{W_b/bbl}{(8.8687)(\%V_s)} + 2.6 \tag{6.8c}$$

where:

$$D = \text{density, lb/gal}$$
$$F = \text{weight fraction of total solids}$$
$$S = \text{specific gravity}$$
$$\%V = \text{vol\% by subscript}$$
$$W = \text{lb (weight quantity)}$$

Subscripts:

$$b = \text{barites}$$
$$c = \text{low-specific-gravity solids (includes solubles)}$$
$$m = \text{slurry of mud or mud component}$$
$$p = \text{oil}$$
$$s = \text{total solids, includes solubles}$$
$$w = \text{water}$$

The variations between graphs and nomographs based on different assumed specific gravities for the low-specific-gravity solids and for the barites are not important. The margin of error in solids determination far outweighs these minor differences. Using Eqs. 6.1 to 6.7, specific gravities of any value desired can be substituted and specific equations derived to fit any situation. But beware of graphs that attempt to classify muds as "good" and "bad" on the basis of specific gravity relationships. They are an invitation to high mud bills whether or not drilling benefits result.

SHALE SHAKERS

The primary solids-removal devices in their best order of series operation are:[11] shale shaker, sand trap (or settling tank, shale tank, etc.), hydrocyclone desander, hydrocyclone desilter, and centrifuge. Regardless of drilling and mud economics improvement with the other mechanical devices, none can operate properly and continuously without the protection of these screens. The sand trap should serve to protect the downstream equipment temporarily if something happens to a screen.

The term *shale shaker* is used in drilling mud work to cover all of the devices that in another industry might be differentiated as shaking screens, vibrating screens, and oscillating screens.[12,13] All three of these types are in use, although most would probably fall in the vibrating screen classification. The old cylindrical rotating, or squirrel cage, screen is rarely seen.

The trend since 1965 to finer and finer mesh has magnified all the problems of operating shale shakers.[14,15,16] Many disappointments were suffered when solids problems often were more severe with the new fine screen shakers than with the older shakers and coarser screens. Out of the confusion and disappointments some better and long overdue information on shale shaker operation has become available.[17,18,19,20]

The particle-size separation made by a shale-shaker screen is not so simple that all particles larger than the stated screen mesh are rejected and all smaller pass through. The amplitude, or distance, the screen moves at high vibratory speed prevents many undersize particles from passing through. On the other hand, some oversize particles do pass through the mesh. We must keep in mind that the cuttings are not neat round balls and measuring and stating their size has no perfect method. Piggybacking also adds to the carryover in muds.

The median cut-size particle is that size of which half pass through and half are rejected over the top. For vibrating screens, the median particle is much smaller than mesh size, much smaller than the maximum opening dimension if the mesh is oblong, or rectangular. On the other hand, square mesh vibrating screen cloth rejects approximately 85% of the cuttings of the same size as the mesh, according to work reported by Cagle.[21] Fig. 6–11 affords a visual idea of the range of mesh sizes found on drilling-rig shaker screens.

Wire coating is a high-surface-tension liquid with an affinity for metal that tends to form on screen wires. With large open meshes the open area is so great that no liquid capacity decrease is noticed. In fine screen mesh, the open area is a problem, and the reduction of effective area through

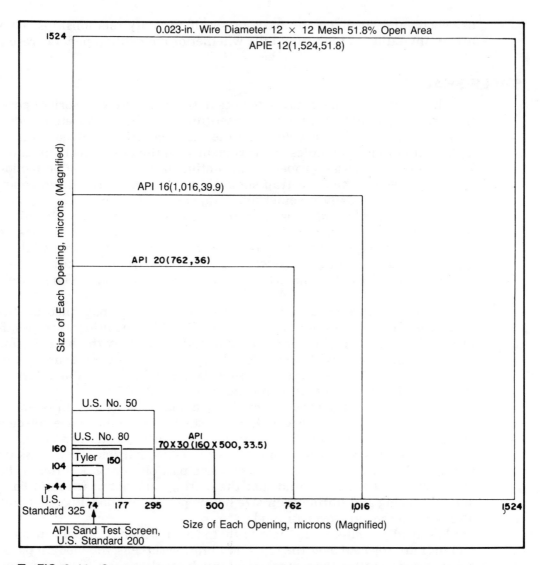

■ **FIG. 6–11** Common screen cloth openings—magnified scale comparison

which the liquid phase can pass becomes critical, due to the wire coating. Oblong mesh (Fig. 6–12) increases the percent open area without a corresponding increase in median cut size. A 40 × 80 oblong mesh screen makes *roughly* the same median cut as a 60 × 60 square mesh screen, assuming the same wire sizes as Fig. 6–12 and the same deck motion.

We tend to assume that rectangular, or oblong, opening mesh screen is always made up with the long mesh dimension transverse, or perpendicular, to the direction of solids movement over the screen. Those who examine their screens closely in the field occasionally find one with the

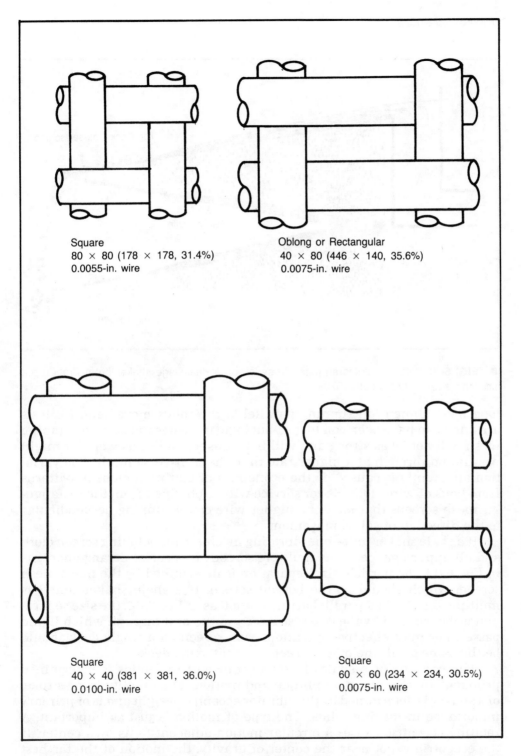

Square
80 × 80 (178 × 178, 31.4%)
0.0055-in. wire

Oblong or Rectangular
40 × 80 (446 × 140, 35.6%)
0.0075-in. wire

Square
40 × 40 (381 × 381, 36.0%)
0.0100-in. wire

Square
60 × 60 (234 × 234, 30.5%)
0.0075-in. wire

■ **FIG. 6–12** Monofilament cloth meshes, all magnified on the same scale, showing the API designation for each: mesh × mesh (microns opening × microns opening, percent open area), and also showing wire diameter in inches.

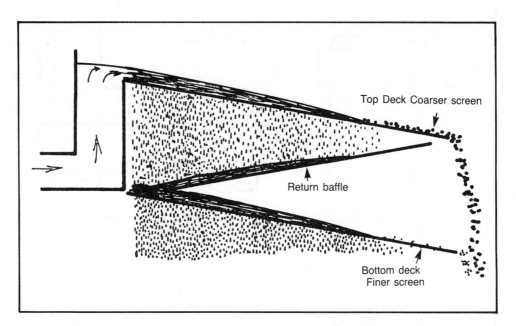

■ **FIG. 6–13** Screen cloths arranged in series on a dual-stage shaker (from Brandt) and showing a return baffle (late 1970s)

long mesh dimension running parallel to the solids movement. No tests are known to have been conducted, but logic indicates the parallel running mesh will not be as strong and will not transport solids as well but might pass liquid through at a higher rate than the transverse mesh. The variation apparently is caused by the economics of cutting screens to patterns from bolts of wire cloth. Better service will be obtained from the rectangular mesh screens that have the higher wire count running perpendicular to the direction of solids movement.

If a shale shaker unit—one vibrating mechanism including screen cloth and all appurtenances—has multiple screens in a series arrangement, as in Fig. 6–13, the particle size separation is determined by the finest mesh screen, which should be the bottom screen. If a shale shaker unit has multiple screens in a parallel arrangement, as in Fig. 6–14, the size separation is determined by the coarsest screen in the unit through which liquid passes. For most effective operation, each screen on a single deck should be the same as all the other screens on that same deck.

The capacity of a shale shaker to transport the solids is determined primarily by the screen amplitude and motion. The amplitude, or stroke, of a shaker is determined by the vibrator eccentric weight and is of primary importance in moving solids. The type of motion is just as important. A rotating eccentric causes a circular motion adjacent to its own center. If the eccentric is not near the center of gravity, the motion of the farthest parts of the machine—screen, in this case—will not be circular. (See Fig.

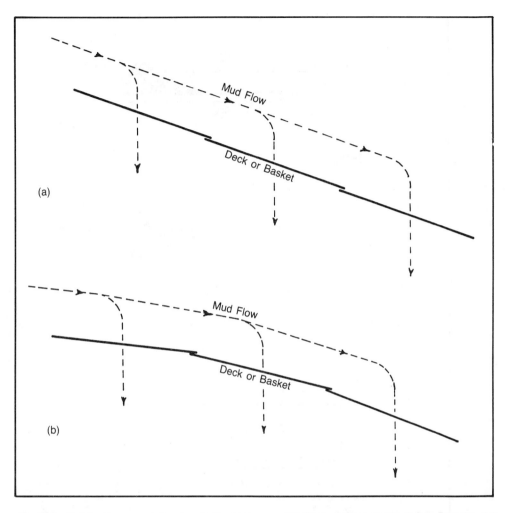

■ **FIG. 6–14** Screen cloths in single-deck, parallel-flow arrangements (after Brandt): (a) constant-angle deck, (b) differential-angle deck. All screens on one deck should be the same mesh for normal operation.

6–15a). Brandt explained the need to tilt the screens on the old-style shaker screens with this illustration.

Fig. 6–15b shows the change made possible by having the eccentric near the center of gravity (c.g.). Uniform motion still requires good design and good maintenance, especially of the resilient resonator supports.

If two eccentrics are placed adjacent to each other and rotated in opposite directions, they cancel each other's effect except for straight-line motion, back and forth, in the direction of a line passed between them, as Fig. 6–16 illustrates. The "straight-line motion" induced by this arrangement became a big advancement of the 1980s in shakers. If designed prop-

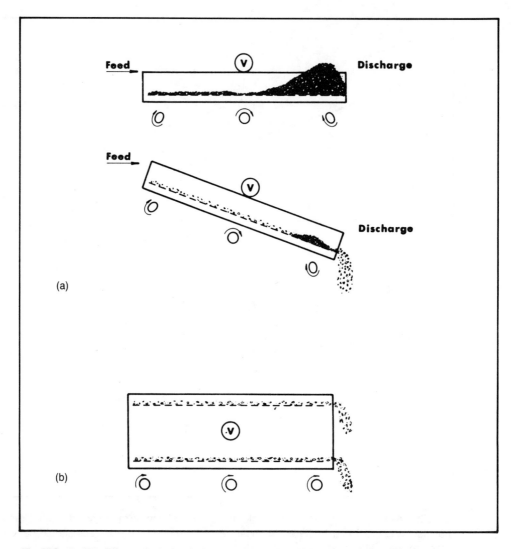

■ **FIG. 6–15** Effect of shale-shaker motion: (a) unbalanced motion creating combination circular and elliptical motion and showing the deck inclined to overcome the resulting solids pileup; (b) Balanced motion, with uniform circular motion and resultant even solids flow irrespective of deck angle.

erly, it increases solids transport and decreases screen life. This made practical the use of screen mesh finer than before.

Since amplitude and type of motion are critical to movement of solids and since some frequency of motion must be involved, it is natural that there should be a relationship between amplitude and frequency. The relation to be watched is the G forces generated, or the acceleration of gravity. The limit at which screen destruction seems unavoidable is 8 G forces.[22] The equation for this acceleration, usually called G_s, is:

$$G_s = \frac{D_a(rpm)^2}{74,000}$$ (6.9)

where:

D_a = amplitude (or diameter), in.
rpm = cycles (or revolutions)/min.

Most shakers operate near 5 Gs with the range between 4 and 6. Obviously, if the advantage of high amplitude is to be obtained, the machine cannot be high speed. Acceptance of this fact in the 1980s helped shaker operation. The figures listed below were presented by Shuman:

Speed	rpm (spm)	Amplitude
High	2,500	Low
Medium	2,500–1,500	Medium
Low	1,500	High

Liquid—or more exactly, slurry—throughput for a screen depends upon three major factors, barring disasters:

1. Percent open area, including the effect of wire coating
2. Plastic viscosity of the mud, lower PV allowing high throughput
3. Transport of solids: the fewer solids on the screen at any given time, the better the liquid throughput

Mesh has been discussed, but anyone responsible for screen purchase specification or for shaker operation should obtain a magnifying mesh counter,

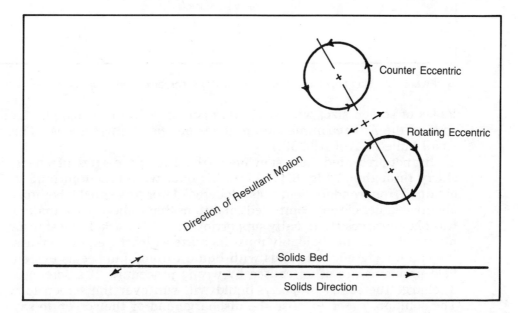

■ **FIG. 6–16** Side elevation of double counter-eccentric vibrators showing the straight-line resultant (for illustrative purposes only).

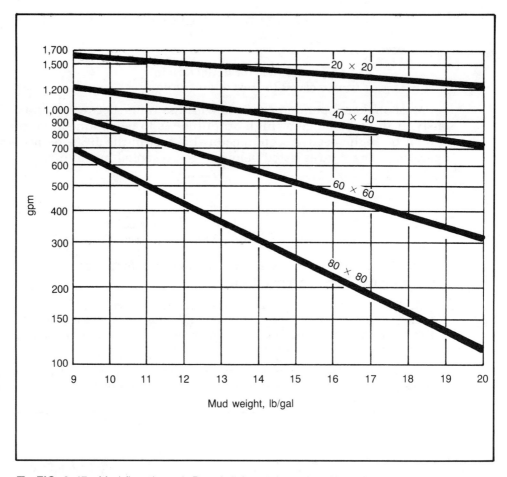

■ **FIG. 6–17**   Mud flow through Brandt 5 ft × 4 ft screens (field data averaged, 1973).

tables of screen size, wire size, and open area from his suppliers and a copy of the API recommended practice on specifying screens. (This is a small bulletin, API RP 13E.)

Brandt presented the first evidence that mud properties affected screen slurry throughput. Fig. 6–17 shows his data were based on mud weight obtained from random field tests. Hoberock later presented laboratory data obtained under closely controlled conditions that indicated the major factor was plastic viscosity, actually supporting Brandt's work. In most instances, plastic viscosity unavoidably must increase as mud weight increases.

The interference of solids with liquid throughput seems evident, but field problems are often caused by wrong procedure. If a screen cannot discharge the solids properly, liquid will run over the screen to waste. The solution is *not* to raise the discharge end of the screen to save the liquid. The problem is solids transport, and the screen discharge end should be *lowered,* assuming that is possible, to help discharge the solids. A coarser screen may be necessary, also.

Among the disasters for screens would be near-size plugging, or wedging. When this occurs, finer screens often work better than coarser, but if the liquid throughput requires more open area (when not plugged), coarser screens must be used. Sometimes with dual stage (upper and lower, series) a change in both screen sizes can avoid the problem. In any case, do not bypass the screens, even partially. This quickly makes all downstream equipment ineffective.

From casual observation of conventional shale shakers operating with screen cloths of mesh coarser than 30×30, it was assumed for years that shale shakers remove solids with relatively little loss of mud. Later preliminary investigations strongly indicate that mud loss ranges from 1 gal of mud per gallon of net dry cuttings removed under the best conditions and coarser screens to 4 gal of mud loss per gallon of net dry solids removed with severe conditions and very fine screens.[23] This range of loss was found with primary removal screens operating efficiently.

As this is written a screen cloth device to replace the vibrating screens is new on the market. First models have the ability to provide up to 600 sq ft of clean screen surface per second and to pull the drilling mud through that area under a low vacuum. This principle shows promise of making great changes in several procedures. Among the interesting features are an apparently efficient degassing capability and a lack of free liquid on the solids removed, even though the screen mesh may be as fine as 400 mesh. The future will be interesting.

Economic limits exist for all methods of separation. As conditions become more severe, the shale shaker design principles and construction details become more critical, and maintenance costs climb rapidly. Anyone responsible for shaker specifications must familiarize himself with the field operating characteristics and limits of the various machines available.

SAND TRAPS—OPERATING PRINCIPLES

If the shaker screen were always adequate—never developed a tear that passed oversize solids through, never had to be bypassed during drilling—the major justification for a sand trap (or shale trap or settling tank) would disappear. Since that is unlikely in the near future, the sand trap is extremely important. Some information has been published on the subject, but the following points are cardinal:[24,25]

1. The sand trap usually receives the liquid slurry passing through the shale shaker and should receive all liquid slurry bypassing the shale shaker and going to the active mud tanks.
2. Being a gravity settling compartment, it is not to be stirred and it must *not* be used as a suction compartment for *any* removal process.

3. A sand trap must have a discharge control easily and quickly opened and closed, so the settled solids can be dumped with minimum whole mud losses. This discharge must be visible to the valve operator.
4. The sand trap should only be dumped, not washed out. If the bottom is not sloped to the solids pile angle, the settled solids should be left to form their own sloped sides. Cleaning the bottom, other than possibly at moving time, serves no purpose but increases loss of mud and mud cost.
5. Since Stokes' law applies in a sand trap, large quantities of barites (as well as API sand) may be settled from weighted drilling fluids. Provision for bypassing the undersize screen discharge slurry from the carrying pan direct to the next processing compartment is also advisable. Since all compartments except the sand trap are stirred in a well-designed active system, this bypass prevents barite settling out. The sand trap must *not* be bypassed if there is any problem with any other solids removal apparatus.
6. The mud exit from the sand trap should be over a retaining weir to a stirred compartment, normally the degasser suction.

Rig operators who have engineered a proper sand trap into their removal systems have thereby increased the utility of the other removal equipment and often have been able to continue drilling progress under unexpectedly severe solids conditions.

DEGASSERS

Degassers sometimes are essential to the solids-removal process. Shale shakers remove a good portion of the gas from a badly gas-cut mud, especially if the yield point is lower than 10 lb/100 sq ft. It is generally accepted among experienced field drilling personnel that special degassing equipment is not necessary if the yield point of the mud is six or less.

Hydrocyclones are fed by centrifugal pumps. Slurry-handling centrifugal pumps for abrasive oilfield muds are not able to maintain efficiency when pumping gas-cut muds. Hydrocyclones do not function properly if pumping (feed) head is not constant or if there is gas or air in the feed. Therefore, provision for degassing should be in a stage between the sand trap and the first hydrocyclones.[26]

If the degasser is one of the vacuum types that requires power mud to operate the mud eductor, the power mud should be taken from the degasser discharge compartment *only*. Violation of this rule will cause bypassing of desanding or desilting equipment when the degasser is in operation. When circulating on a choke, both the degassing and the control of abrasives in the mud are important. Taking the eductor power mud from below decks on a marine rig is extremely dangerous in the event gassy mud should get by the degasser before it is detected.

HYDROCYCLONES

Hydrocyclones can perform the finest cut of any primary separation equipment operating on the full-flow circulating rate of an unweighted

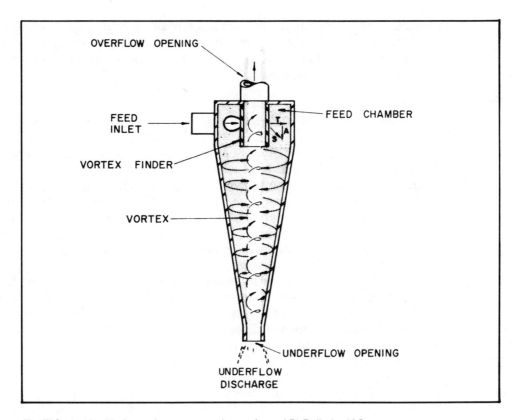

OVERFLOW OPENING

FEED INLET

VORTEX FINDER

VORTEX

FEED CHAMBER

UNDERFLOW OPENING

UNDERFLOW DISCHARGE

■ **FIG. 6–18** Hydrocyclone nomenclature from API Bulletin 13C

mud system. Some understanding of the most common designs is essential to proper operation. The hydrocyclone (Fig. 6–18) has a conical-shaped portion in which most of the settling takes place and a cylindrical feed chamber at the large end of the conical section. At the apex of the conical section is the underflow opening for the solids discharge. In operation, the underflow opening is usually at the bottom, for convenience, and *bottom* or *top* as used in this chapter concerning hydrocyclones refers to that mode.

Near the top end of the feed chamber, the inlet nozzle enters tangential to the inside circumference and on a plane perpendicular, or nearly so, to the top-to-bottom central axis of the hydrocyclone. A hollow cylinder, called the *vortex finder,* extends axially from the top into the barrel of the hydrocyclone past the inlet. The inside of the vortex finder forms the overflow outlet for the liquid discharge or effluent. The overflow opening is much larger than the underflow opening. The *size* of a hydrocyclone is determined by the largest inside diameter of the conical portion. All dimensions are critical to the operation of any specific design and size.

The hydrocyclone obtains its centrifugal field from the tangential velocity of the slurry entering the feed chamber. An axial velocity component is created by the axial thrust of the feed stream leaving the blind annular space of the feed chamber. The resultant is a downward-spiraling velocity represented by S in Fig. 6–18, in which T represents the tangential and A the axial velocity component.

Balanced Design

In a hydrocyclone the stream spirals down along the wall of the conical section toward the underflow opening, reverses axial direction, and spirals upward to exit at the overflow opening. The liquid stream is forced to the top of the hydrocyclone by the very high centrifugal force combined with the large overflow opening and small underflow opening. Properly designed, a hydrocyclone can be adjusted so that, when a clean liquid is fed, there is only a very slow drip at the underflow. This is the ideal adjustment. When separable solids particles are fed to such a hydrocyclone, solids are discharged at the underflow opening, each particle taking with it a free liquid film covering its surface area.

The solids particles in the turbulent downward-spiraling feed stream are settled outward toward the wall, following modifications of Stokes' law. When the liquid stream reverses to spiral upward, the solids particles continue spiraling downward through the underflow opening due to their greater mass and higher inertial forces (Fig. 6–19).

When a balanced design hydrocyclone is adjusted with the underflow opening too small in relation to the balance point opening, a *beach* of dry cone wall exists between the edge of the actual underflow opening and the balance point, or the point of liquid turnback. If the hydrocyclone is capable of settling any silt less than 44 microns, this adjustment or design causes sticking and plugging as these fine particles lose their liquid film attempting to traverse the beach. The trouble this undesirable adjustment causes is known as *dry plugging.*

If a balanced design hydrocyclone is opened to an underflow opening greater than the balance point diameter, a cylindrical shell of spiraling liquid will discharge unless there are enough separable solids to take up the unbalanced discharge space. This adjustment is termed *wet bottom,* and unless it is excessive, it does not result in high mud losses in normal unweighted muds and at normal drilling rates. As a matter of fact, a wet bottom adjustment runs a better drilling mud and a better hole-conditioning mud than a dry bottom adjustment or design. With an inexpensive liquid phase, the wet operation will not raise mud cost. With an expensive liquid phase—or if environmental considerations dictate—a decanting centrifuge will be necessary to salvage the liquid phase from the cyclone underflow and to put it back into the system. In an unweighted mud system, a screen must *not* be used for this. The ideal balance point adjustment is not always known.

Fig. 6–19 depicts a balanced design hydrocyclone operating in spray discharge. The air being sucked in at the underflow opening is the proof

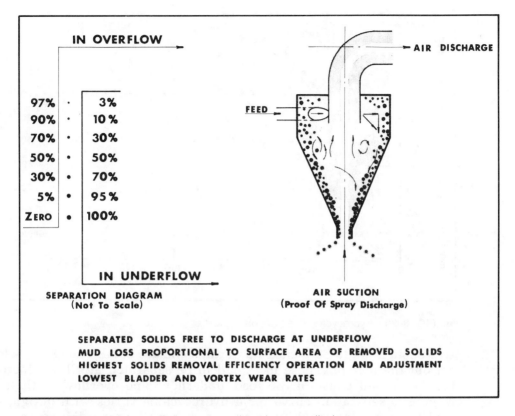

IN OVERFLOW

97%	3%
90%	10%
70%	30%
50%	50%
30%	70%
5%	95%
ZERO	100%

IN UNDERFLOW

SEPARATION DIAGRAM
(Not To Scale)

AIR DISCHARGE

FEED

AIR SUCTION
(Proof Of Spray Discharge)

SEPARATED SOLIDS FREE TO DISCHARGE AT UNDERFLOW
MUD LOSS PROPORTIONAL TO SURFACE AREA OF REMOVED SOLIDS
HIGHEST SOLIDS REMOVAL EFFICIENCY OPERATION AND ADJUSTMENT
LOWEST BLADDER AND VORTEX WEAR RATES

■ **FIG. 6–19** Balanced design hydrocyclone in spray discharge

of spray discharge, and it can be felt with the end of a finger for absolute verification. This air is simply entering the low-pressure zone created by removing the air from the vortex by the friction of the high-velocity, upward-spiraling inner stream, which takes the vortex air out with it through the overflow opening. The flow rates in mud hydrocyclones are so high that total retention time, from inlet to either exit, is less than one-third of a second.

In spray discharge the underflow opening is not a choke. It is a circular weir, or ring dam—so is the overflow opening—as in a cream separator. As long as the underflow opening acts as a circular weir, the cyclone can discharge all the solids it can settle to the wall and move to the bottom. The hydrocyclone is free to remove solids at its highest possible efficiency when it is in spray discharge.

There are two spiraling countercurrent turbulent streams, one downward near the cyclone wall and one upward near the vortex. At the adjacent boundaries of these streams, eddy currents are created somewhat as in Fig. 6–19. These currents sweep up some solids particles from near the wall and move them to the inside upward stream. Some particles are mas-

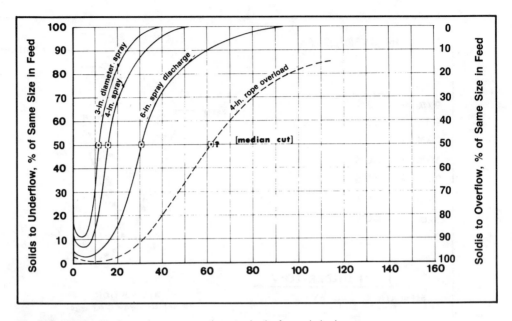

■ **FIG. 6–20** Hydrocyclone separation, typical of good design

sive enough not to be picked up at all; some are small enough to seldom get out the bottom. Those too small to be separated at all are found in the free liquid phase of the feed, underflow, and overflow in the same ratio. The separation curve for a hydrocyclone varies with design, size, feed head, solids specific gravity and shape, percent solids in the feed, feed viscosity, and adjustment. Illustrative (only) separation curves are shown in Fig. 6–20.

When the cyclone can settle and send to the underflow more solids than can pass through the underflow opening, the cyclone is *overloaded.* A balanced design underflow takes on characteristics of a rope, or sausage links, thus the terms *rope discharge* and *sausage discharge* for the overload condition. There is no air suction at the underflow opening in rope discharge, and in a balanced design this is the proof test for overload (Fig. 6–21). The underflow opening is no longer a weir; it is now a choke.

In rope discharge not all particles reaching the underflow opening can exit. Due to the inertial separation, the smaller ones must exit at the overflow immediately but not *all* of even the largest and heaviest particles can exit the underflow. The difference between rope and spray underflow discharge is predictable. Smaller particles have more surface area per pound; therefore, the rope underflow has *less* solids surface area per pound of solids than a spray discharge underflow. The liquid accompanying the solids in a balanced cyclone is a function of surface area; therefore, the rope discharge has less liquid per pound of solids and is *heavier* in slurry weight than a spray discharge underflow. It is *not* an indication of greater

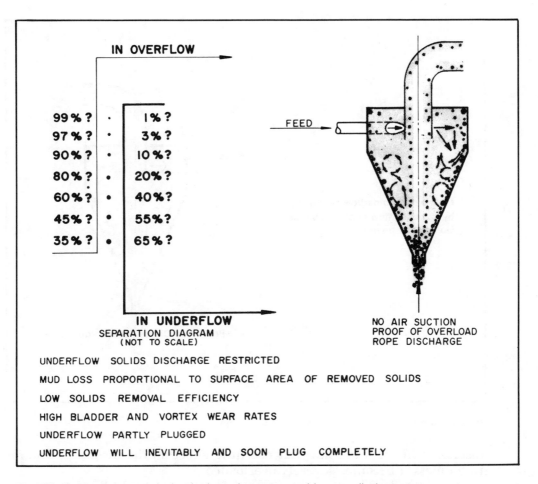

■ **FIG. 6–21** Balanced design hydrocyclone operated in rope discharge

efficiency. Unfortunately, the finer solids remaining in the system do increase mud weight just as effectively as coarser solids and the finer solids particles still take free liquid when they are finally removed, unless a decanting centrifuge is used. See Fig. 6–20 again for an example of typical separations in a 4-in. hydrocyclone operation changing from spray to rope discharge.

Choke-Bottom Design

Another design of hydrocyclones is shown in Fig. 6–22. This can be described as a *choke-bottom design* because it has no balance point and both the overflow and underflow openings act as chokes at all times under all conditions. For example, if clear liquid is fed to a choke-bottom cyclone with a 10 mm bottom opening, it will discharge a forced underflow stream 10 mm in diameter. Adding separable solids to the feed causes solids to

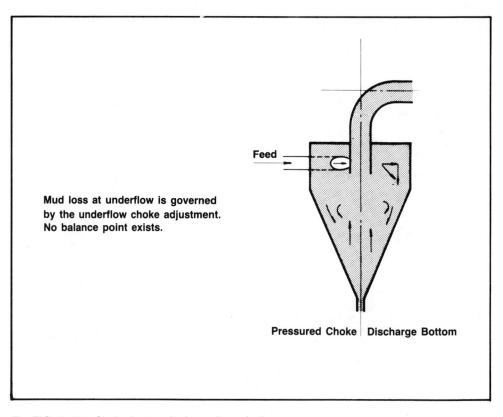

Feed

Mud loss at underflow is governed
by the underflow choke adjustment.
No balance point exists.

Pressured Choke | Discharge Bottom

■ **FIG. 6–22** Choke-bottom hydrocyclone design

discharge in the underflow but has little effect on the discharge volume
rate unless it becomes severely overloaded.

Liquid underflow discharge is not a function of solids discharge unless
an overload condition is reached so that not all reporting solids can exit
in the underflow. The best adjustment when operating a choke-bottom
design hydrocyclone is a compromise between the minimum opening required to prevent underflow discharge of all mud during the low-solids
feed period (making drillpipe connections) and the maximum opening
required to remove cuttings during the short high-solids feed period
(pumping the kelly down) in fast hole.

Hydrocyclone Capacity

As with the shale shaker and all other mechanical separation devices,
hydrocyclones have two capacities—*feed slurry* and *underflow solids discharge.* As a design problem, the two capacities are related, but the products available commercially have an extreme variance in their relative
values.

Feed capacity is the volume rate of feed slurry a specific hydrocyclone will accept at a specific feed head. Head may be expressed in feet or meters; it cannot be expressed as pressure unless feed slurry weight is also specified.

As an example, we might find some size and make of hydrocyclone rated as having a feed capacity of 100 gpm at 75 ft of head. If the feed were water, which has a hydrostatic head of 0.433 psi/ft, the feed pressure would be 32.48 psi. If the feed slurry weight were to increase to 16.68 lb/gal (specific gravity of 2.0), the hydrostatic head gradient would be doubled to 0.866 and 75 ft of feed head would amount to 64.95 psi. The volume feed rate would be approximately the same with a 16.68-lb/gal slurry at 64.95 psi as it would with a 8.34-lb/gal slurry at 32.48 psi because the head is the same.

Centrifugal pumps operating at constant speed—electric motor or closely governed engines—and feeding the same equipment automatically maintain a constant head as slurry weight changes. Further study on this subject is strongly advised for anyone operating or installing hydrocyclones.[27]

The solids-removal ability, not the feed capacity, is the real key to the value of hydrocyclones and their operation. Solids-removal ability consists of two things: first, the high separation forces and internal flow pattern to settle fine particles (most difficult ones) to the underflow opening; second, a balanced underflow opening large enough for the cyclone to discharge the settled solids at a high rate before reaching solids overload. If a cyclone does reach the overload condition, the underflow will rope and removal ability will be ruined as shown in Fig. 6–21. Obviously, a balanced design can be operated economically with a much larger opening than can a flood-bottom design, considering mud cost.

A rare series of hydrocyclone performance tests were run in 1964 under the auspices and using the facilities of the National Science Foundation Mohole project. The actual designer or head engineer was responsible for field performance, adjusting his own hydrocyclone in each test. Since that test series at least two of these hydrocyclones have been discontinued for drilling mud use, four have been modified, and these same manufacturers have introduced a total of at least four new models. Most of a rash of new hydrocyclones that entered the market in the late 1970s and early 1980s are poor performers in the field. The differences and dilemmas pointed out by the Mohole tests have not disappeared.

Table 6–5 presents unaltered summaries of performance results, in order of decreasing solids removal.[28] Rating A gives top rating for removing the highest percentage of solids entering; rating B penalizes for liquid in the underflow. These are contrary ratings, and the two multiplied together give a total performance rating, which surprisingly was in almost exactly the same order as Rating A, or percent removal only. This arrangement is in the approximate order of practical assistance to field personnel desiring to maintain minimum solids (low mud weight) as economically as

■ **TABLE 6-5** Mohole hydrocyclone evaluation
(From timed and measured data taken during tests)

Original order of removal efficiency. Also, approximate increasing order of equipment cost for equal underflow solids removal rates.

Cone			Rating A	Rating B	A × B
			UF Solids as	UF Solids	
			Vol Rate %	wt% of	
Make	Diameter, in.	Feed, gpm	of Feed Solids	Total UF	Total Ratings
W	3	16.76	11.72	56.25	670
X	2	22.03	11.69	49.30	575
W	4	57.12	8.94	56.83	507
X	6	108.39	5.95	47.26	281
Y	4	62.65	4.55	47.96	218
W*	6*	88.73*	3.28	59.31	194
Z	4	44.90	1.55	65.25	101
X	12	407.00	1.80	54.46	98
Z	8	167.70	0.746	66.70	49.8

* Company W requested, prior to any evaluation work, that their 6-in. cone test be deleted because it was not taking the proper feed volume due to an undetermined malfunction. Request denied.

possible in unweighted mud systems. The cyclones in the higher ratings literally remove solids at a cheaper rate per pound than those in lower ratings given the same feed materials. Practical limits exist in operational mechanics.

In Table 6–6 the data have been rearranged in the approximate order of increasing purchase cost per volume rate of feed. In specifying new

■ **TABLE 6-6** Mohole hydrocyclone evaluation
(Rearranged in order of increasing cost per volume rate of feed)

Cone		Feed, gpm	UF Solids as Vol Rate % of Feed Solids	Total Ratings
Make	Diameter, in.			
X	12	407.00	1.80	98
Z	8	167.70	0.746	49.8
W*	6*	88.73*	3.28	194
X	6	108.39	5.95	281
Y	4	62.65	4.55	218
W	4	57.12	8.94	507
Z	4	44.90	1.55	101
X	2	22.03	11.69	575
W	3	16.76	11.72	670

* Malfunctioning

■ **TABLE 6–7** Mohole hydrocyclone evaluation
(In original order but arranged to show effect of design)

Cone Make and Size				Feed Head, ft	$A \times B$ Total Ratings
W	X	Y	Z		
3				76.1	670
	2			105.7	575
4				76.1	507
	6			76.8	281
		4		76.8	218
*6				76.1	*194
			4	67.1	101
	12			67.1	98
			8	80.7	49.8

* Malfunctioning

equipment, this initial cost often determines the choice because the specifier is seldom field experienced and is usually pressured to hold down the initial equipment cost for totally processing (feed rate) some specific maximum rig pump capacity. It is immediately apparent that the cost per unit of feed rate capacity is almost diametrically opposed to the cost per unit of solids discharge ability.

The uninitiated often are unable to accept the big differences in performance between hydrocyclones of the same size, as shown in Table 6–5. In Table 6–7 the cyclones are in the same order of rating as in Table 6–5, but they are columned by manufacturer, or by design philosophy. It is apparent from Table 6–7 that smaller cyclones of the same design have greater separation ability, but also different design philosophies can cause even more difference than size. There is no substitute for observing comparative equipment performance in the field and issuing specifications based on performance.

Hydrocyclone Plugging Effects

The results of underflow plugging in hydrocyclones is as disastrous as plugging of shale shakers, but the effects are not as obvious. Fig. 6–23 illustrates the change in balanced hydrocyclone operation from normal to plugging at the underflow. To look at it, a cyclone plugged in this manner offers no clue to the damage it is inflicting. Actually, it is undoing the cleaning work of all other cyclones on the unit by mixing unprocessed mud directly into the cleaned overflow mud from the other cyclones on the same header. At the same time the upper parts of the cyclone are rapidly being cut to pieces by the particles that should have been dis-

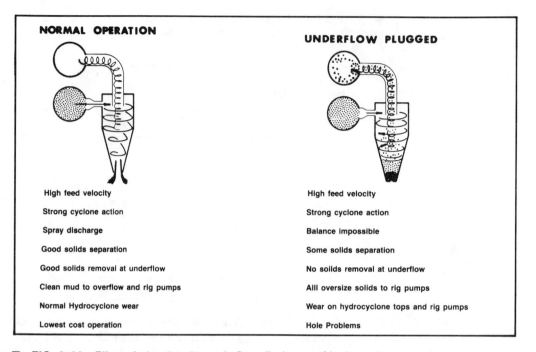

NORMAL OPERATION	UNDERFLOW PLUGGED
High feed velocity	High feed velocity
Strong cyclone action	Strong cyclone action
Spray discharge	Balance impossible
Good solids separation	Some solids separation
Good solids removal at underflow	No solids removal at underflow
Clean mud to overflow and rig pumps	Alll oversize solids to rig pumps
Normal Hydrocyclone wear	Wear on hydrocyclone tops and rig pumps
Lowest cost operation	Hole Problems

■ **FIG. 6–23** Effect of plugging the underflow discharge of hydrocyclones

charged out the bottom. A plugged cyclone must be unplugged or removed from the manifold immediately.

Underflow plugging is usually caused either by a dry beach or by solids overload. The dry beach effect can be from improper design (dry bottom), from too small an adjustment at the underflow, or from excessive feed head. Solids overload can be overcome by adjusting to a larger underflow opening if possible, installing more cyclones in the stage, removing more solids in a stage ahead of the roping hydrocyclones, or purchasing cyclones of a better solids discharge capability.

Fig. 6–24 shows the effects of feed plugging. With partial plugging, mud is feeding but has a low velocity in the feed chamber. The cyclone body acts as a swirling funnel, and the mud lost at the underflow is the same as pit mud and may be discharging at a volume rate from 10 to 200 times the volume rate of a normal balanced underflow discharge. The underflow may even contain some cleaned mud from the overflow headers. The cost of this mud loss and the aggravation of the end-of-location cleanup problem cannot be tolerated. A cyclone flooding feed and overflow mud at the bottom must be unplugged, repaired, or removed immediately.

If the inlet is plugged completely, the result will depend upon the hydrocyclone overflow header design. If the overflow header is full size (has a maximum velocity not more than 8 ft/sec) and has each hydrocyclone overflow entering the header at a 45° angle into the direction of flow,

FEED PARTLY PLUGGED

Low feed velocity

No cyclone action

Flood (pressured) bottom

No separation

Whole mud loss at underflow, high rate

No hydrocyclone wear

Very high drilling costs

FEED COMPLETELY PLUGGED

No feed velocity

No cyclone action

Flood (pressured) bottom

No separation

Whole mud (cleaned by other cones) loss

at underflow, high rate

No hydrocyclone wear

Very high drilling costs

■ **FIG. 6–24**   Effect of plugging the feed inlet of hydrocyclones

there is usually no loss at the underflow of that plugged hydrocyclone. Usually it will suck air at the bottom.

If the overflow header is undersize or if the hydrocyclone overflows enter the overflow header at a 90° angle and directly opposite each other, the underflow loss will be in a range of 20 to 80 times that of an underflow operating normally. This must be corrected immediately.

Feed plugging is caused by poor housekeeping on the mud system. Bypassed shale shakers or torn screen cloths are the most common offenders. Special large open-area suction screens installed on the end of the centrifugal pump suction inside the suction pit can reduce plugging 90%, but even this is not enough to cope with a bypassed shale shaker, especially if there is not a good sand trap.

DECANTING CENTRIFUGES

The decanting centrifuge is the only liquid-solids separation device used on drilling fluids that can remove (decant) all free liquid from the separated solids particles, leaving only adsorbed moisture on the surface area. This adsorbed moisture does not contain solubles, such as chlorides, or colloidal suspended solids, such as bentonite. The dissolved, suspended solids are associated with the continuous free liquid phase from which the decanting centrifuge separates the inert solids and they remain with that liquid. The adsorbed moisture can be removed from the separated solids only by evaporation, which has been neither desirable nor practical so far in drilling mud work.

Centrifuges of other types, notably the nozzle solids discharge type with vertical axis, have been tested periodically in this country since the early thirties for use on weighted water-base muds. The decanting centrifuge was first tested and proved adaptable as a practical field tool in 1953.[29] It moved rapidly into the rental field and is now recognized universally as a useful tool, though perhaps not as well understood as it might be.

A more complete description of present decanters would be continuously conveying decanting centrifuges. A schematic cutaway of the actual separating chamber is shown in Fig. 6–25. The size of the machines is usually given as maximum bowl diameter and length. All else equal, the larger diameter machines have higher feed capacities at equal separation

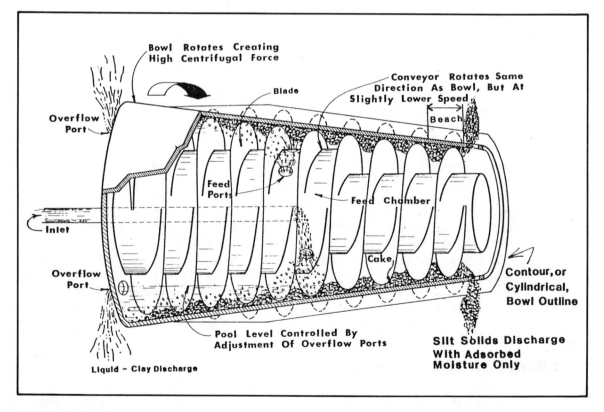

■ **FIG. 6–25** Decanting centrifuge, typical sectional operating view. Contour, or cylindrical, bowl is ghosted in. (*courtesy Stonewall Associates Inc.*)

ability and finer separation ability at the same loads. Gear box choices are critical to high solids capacities.

Operating Principles

Referring to Fig. 6–25, separation occurs inside the bowl, which is rotated at speed ranges that vary—according to the bowl size, conveyor design, application, and preventive maintenance program—from 1,500 to 3,500 rpm. Inside the bowl is usually a double-lead conveyor connected to the bowl through an 80:1 gear box so the conveyor loses one revolution for every 80 revolutions of the bowl. Thus, there is a relative rotation of 22.5 rpm when the bowl speed is 1,800 rpm. This relative rotation is the conveying speed, and it can be varied by using other gear ratios.

The cost of renewing the conveyor wear surfaces, traditionally the most expensive routine problem, increases approximately with $(rpm)^3$. This is why most centrifuge rental companies encourage proportionately slower operation of their machines and go to great lengths to justify it. No valid justification exists for operating at less than optimum separating speed according to bowl diameter, size and shape of pool, and rate of feed.

The mud slurry to be treated is metered to the centrifuge through a feed tube centered in the hollow axle to the feed chamber. From there the feed passes through feed ports into the leads, or channels, in the separation chamber. The blades—one in the single lead model—form the walls of the channel(s). As the slurry is slung outward into the annular ring of mud called the pool, it also accelerates to the approximate rotational speed of the conveyor and bowl. The depth of the pool is determined by the adjustment of the eight to twelve overflow openings or overflow ports.

The slurry flows through the channels toward the overflow ports under high centrifugal force and in laminar flow if the operation is effective. Following Stokes' law, the solids settle outward to the outer walls, the most massive particles settling first and the lesser ones further along the leads. Those not having committed to the solids cake against the wall by the time the slurry stream is increasing in velocity to exit through the overflow ports do not settle. They pass out the overflow with the liquid. In the normally recommended operating ranges, retention time in the machines varies from 10 to 80 sec.

As the solids are settled outward against the wall, the conveyor continuously plows or scrapes them toward the small end of the bowl, out of the pool, and across that portion of the wall between the pool and the solids underflow discharge openings called the *beach.* As the solids cake crosses the beach, all free liquid is removed by both *centrifugal drainage* and *centrifugal squeezing* of the cake. The free liquid drains back to the pool, taking with it the colloidal and dissolved solids. As the separated solids discharge out the underflow, they contain only adsorbed moisture.

Dilution of the feed to a decanter has only one purpose—to reduce the viscosity of the feed to aid in maintaining separation ability. The most convenient rule of dilution is to add only enough dilution water to reduce the effluent to 37 Marsh funnel seconds but not lower. More dilution decreases settling time out of proportion to the increase in settling rate.

Dilution is applicable only to weighted water-based muds. Continuous oil-phase muds respond better to heat, and stove-oil dilution for centrifuging is not advised. Heating above 90°F (32°C) for centrifuging usually is neither economical nor necessary, but separation does benefit at a decreasing rate up to at least 140°F (60°C). Hydrocyclone underflow should be diluted only if necessary for feeding and then sparingly, and with mud.

Capacities

The solids discharge capacity of a decanting centrifuge is limited by the volume rate of solids cake that can be moved by the conveyor blades through the minimum annular cross section. The liquid discharge capacity of a decanter is limited by the capacity of the overflow ports to discharge liquid at whatever pool depth adjustment they may be set. The feed-rate capacity of the decanting centrifuge is determined by the volume of separable solids in the feed or by the free liquid and colloid content of the feed, according to which discharge limit is being approached—liquid or solids.

A feed mud low in solids has the feed rate limited by liquid discharge capacity; a feed very high in separable solids—a heavily weighted mud—has the feed limit determined by the solids discharge capacity. If a very fine separation is desired, it is necessary to hold down feed volume rate to stay within laminar flow in the settling chamber and to maintain retention time for fine separation ability. "Cheating" by feeding centrifuges at liquid volume rates so high there is no time for separation became widespread in the 1980s and is causing considerable confusion and harm (increased drilling cost).

Note the outline of a bowl with a cylindrical portion ghosted over the conical bowl in Fig. 6–25. This ghosted configuration is called a *cylindrical,* or a *contour* bowl, and is the trend in drilling fluid centrifuges. It originated in the sewage industry. It permits approximately doubling the liquid feed rate while making the same median cut and/or doubling the solids discharge rate—all with the same maximum diameter and length bowl and with other operating parameters the same. The extra capacities have not been needed in weighted water-based mud (except in Iran, Iraq, and Trinidad), but the extra capacity has good application in secondary separation on hydrocyclone underflow in all other muds.

Separation Limits

An effective decanting centrifuge for weighted water-based mud work should be capable of making a separation approximately at the 2-micron barite particle size. Due to the settling behavior according to Stokes' law, light solids would be separated down to 3 microns ($1\frac{1}{2} \times 2$ microns) in the same machine under the same conditions. In a continuous oil-phase mud, the effective viscosity of the oil is greater than that of water in a water-based mud, therefore the separation in the centrifuge is not as fine.

An 18-in. (diameter) × 28-in. (length) conical bowl, double-lead centrifuge operating at 1,800–2,000 rpm on water-based mud, with a feed rate not exceeding 20 gpm, with median pool level, and with the overflow viscosity approximately 37 Marsh funnel seconds will make the 2-micron median cut on barite. Obviously, with a contour bowl of the same diameter and length and other parameters equal, a higher feed rate can be used and the same separation obtained.

This legitimate increased capacity of the cylindrical over the conical bowl has led to a ridiculous feed capacity race, to the detriment of drilling mud quality. In the 1980s mud is pumped through machines at rates that do not permit separation of anything near reasonable advertised separation sizes. Indeed, if effective separation were possible at the high rates, the conveyor would overload and shut the machine down. So much "dilution" mud is added to many hydrocyclone underflows before feeding to the decanters that most of the solids separated by the hydrocyclones are sent back to the system in the "liquid-clay" overflow of the centrifuges. Likewise, in weighted muds a large percent of weight material that should be saved is sent to discard. Five microns and even six (instead of two) is frequently mentioned and, worse, *accepted!*

In field operation the decanting centrifuge is fitted with a housing over the bowl, liquid and solids *hoppers* to take the liquid and solids discharges of each, respectively, a slurry feed pump, dilution water connection, and remix water connection if desired (Fig. 6–26). The feed is metered. The bowl speed usually can be varied, but variation is best left to the centrifuge technician because it is seldom either necessary or desirable. To reduce damage to expensive gears, shear pins, trips, limit switches, etc., shut down the equipment automatically in case of a solids overload.

Dilution water connections should be equipped with a flow gauge or meter and a flow adjusting valve. Remix connections are usually provided. *Remix water* should be added in the housing to aid in softening the underflow solids for more rapid mixing into the system or even for reducing the underflow solids slurry density to the range of system density, as desired.

PERFORATED ROTOR MUD SEPARATORS

The perforated rotor centrifugal separator was invented and developed in the middle 1960s by Mobil Oil Research personnel to do the work of a decanting centrifuge, but with portability.[30] The decanters need to be directly over a stirred section of the active mud system to return weight material to weighted water-based mud. This positioning can be expensive, impossible, or inconvenient to achieve. The perforated rotor machine over-

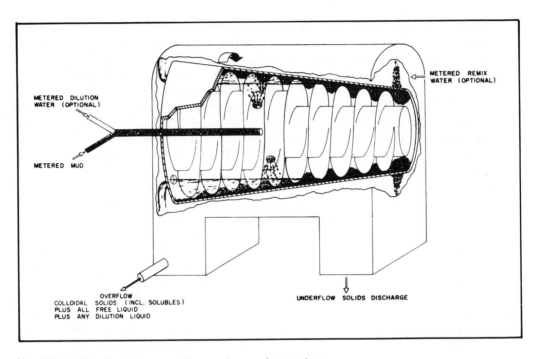

METERED REMIX
WATER (OPTIONAL)

METERED DILUTION
WATER (OPTIONAL)

METERED MUD

OVERFLOW
COLLOIDAL SOLIDS (INCL. SOLUBLES)
PLUS ALL FREE LIQUID
PLUS ANY DILUTION LIQUID

UNDERFLOW SOLIDS DISCHARGE

■ **FIG. 6–26** Decanting centrifuge, scheme of operations

comes this problem by having both discharge streams (overflow and under-flow) in flowable, liquid-slurry form. The machine is known as the RMS, for rotor mud separator.

Operation

The perforated rotor design consists of a 6-in. outside-diameter cylindri-cal shell rotor 42 in. long, with 474 ½-in. evenly spaced holes drilled in the shell, spinning in a horizontal position inside an 8-in. inside-diameter stationary pipe housing fitted with fluid seals at each end for the axles of the rotor to protrude (Fig. 6–27). A slurry to be treated is fed by a positive displacement metering pump and is mixed with dilution water—assuming water-based mud—from a second metering feed pump. A part of the dilu-tion water is used to flush the rotating seals. The diluted feed enters the annular separation chamber through an opening into one end of the annu-lar space.

Under centrifugal force, larger and heavier solids are concentrated against the outer annular wall by Stokes' law. At the opposite downstream end of the annular space and on the periphery is an underflow solids dis-charge opening. All feed material, both liquid and solids, would discharge out this opening were it not for an adjustable choke arranged to hold *back pressure* on this underflow opening and to permit the under-flow solids slurry to discharge at a controlled rate.

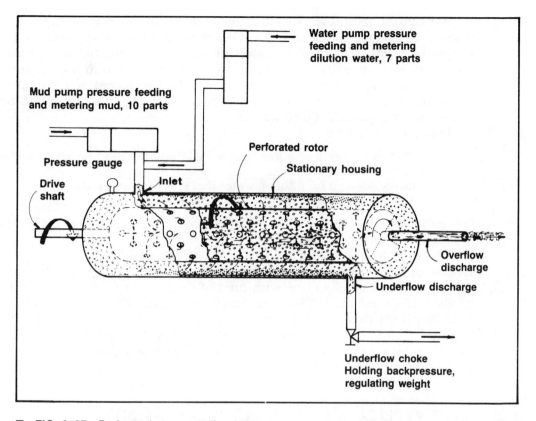

FIG. 6–27 Perforated rotor centrifugal separator

Separation Parameters

According to the adjustment of the underflow choke, a volume fraction of the feed is forced out the overflow. That volume portion not permitted to leave at the underflow discharge can only exit at the overflow. When rotating near design speed of 2,750 rpm, the particle size of the median cut varies. On water-based mud, normal operational variations within recommended feed rates result in a median cut point varying from 4.5–6.7 microns, 2.6 specific gravity—equivalent to 3–4.5 microns median cut on barites.

The principal potential benefit of the perforated rotor separator is portability. For this, it is necessary to discharge the underflow as a very fluid stream in order to be able to flow this solids discharge, under its own discharge head, from some spot on the drilling location easily accessible to a trailer over to the active mud system, without excessive line plugging problems. Operating literature recommends an underflow solids slurry weight higher than the undiluted feed (active system mud) weight by an increment that varies with the feed weight.[31]

For environmental work such as secondary separation to salvage liquid from hydrocyclone underflow, the RMS obviously has no application. For weight-material salvage while rejecting clays in a weighted water-based mud to reduce viscosity, the RMS is an excellent second to the decanter. Most offshore rigs are not planned to allow the movement of a decanting centrifuge onto or off of (for maintenance) the mud tanks. The RMS becomes the only practical alternative.

Disadvantages are:

- A slightly higher barite loss than a well-designed and properly operated decanter
- A heavy demand for dilution water that is drill-water quality
- A high rate of discard (disposal problem in some areas)

Nevertheless, this machine is a valuable tool in many areas of the world, especially on offshore operations.

The RMS has also been used to "split" weighted oil mud into two fractions of different weights, neither the same as the original mud. For example, a job has been finished with an 18-lb/gal oil mud, and the oerator needs a 15-lb/gal mud. The machine is run with the overflow adjusted to 15 lb/gal (by manipulation of the underflow choke). The underflow discharge will, of course, weigh more than 18 lb/gal. The heavier fraction usually is also saved for future use diluted with a lighter oil mud. The feed is not diluted on oil mud.

EQUIPMENT APPLICATIONS—PRIMARY SEPARATION

Primary separation, as used here, refers to the use of liquid-solids separators to *remove* either a liquid-undersize solids fraction or an oversize solids fraction from the *active mud system.* Any separator taking its feed directly from an active system and returning either the liquid (less solids than the feed stream) or solids (more solids than the feed stream) back to the active system has in effect removed a selective fraction from the system and is performing primary separation. Fig. 6–28 is a schematic diagram of typical primary separation equipment related to a system.

Since 1952, improvement in primary removal equipment has had considerable impact on general mud-weighting practices, as shown in Fig. 6–29. The lowest drilling mud weights economically possible, the materials required to maintain a given weight mud—assuming the same type of chemical system—the lowest weight at which barites are used, and the highest weight at which barites are not used have all moved to the left and downward on this diagram. At the same time, the high limit of drilling mud weight that can be maintained, as a matter of practicality, has moved to the right. It is not possible to insert actual cost figures because of innumerable variables.

This reduction in cost at any given weight is because of the greater

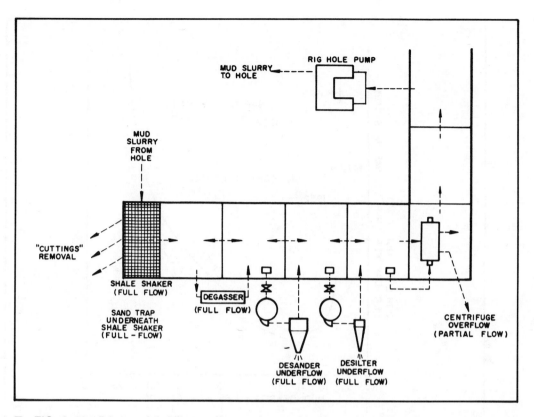

■ **FIG. 6–28** Primary separation equipment in proper schematic relationship to a drilling fluid system.

efficiency of mechanical removal devices over discard and dilution, or *watering-back,* which was once the mainstay of solids control after mud had passed the conventional shale shaker. The technique of dilution is to increase mud-pit level with liquid addition, discard a fraction of whole mud, add liquid, and discard mud. Obviously, this procedure effects no change in particle-size distribution—the good goes with the bad.

Mechanical devices selectively remove solids, permitting us to alter the particle-size distribution. They permit the removal of more of the unwanted solids with less loss of beneficial (for the purpose at a specific time) solids than is possible by dilution, and at less cost.

FULL FLOW PROCESSING—UNWEIGHTED MUD

Full flow processing of all drilling fluid from the borehole is for the removal of specific solids, selectively most of the silts. In this application the best 4-in. hydrocyclones, or desilters, can perform the ultimate removal to the economic limit. For the desilting to be performed economically and

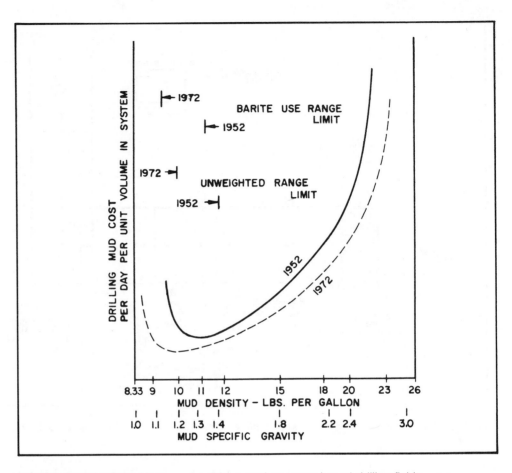

■ **FIG. 6–29** Effect of improved solids control on water-based drilling fluids

efficiently, balanced design hydrocyclones must be used with spray-type discharge. Solids discharge overload, or rope discharge, must be prevented in unweighted muds.

Whole Mud vs Selective Solids Discharge

Table 6–8 demonstrates one of the advantages of effective use of solids removal equipment. More solids always are discarded for each barrel of total discard than with discard and dilution without selective machinery. The illustration is for discarding hydrocyclone underflow of varying weights compared to discard of the mud feeding to the cyclones. Note that, in addition to more solids per barrel of underflow to be discarded, there is less water per barrel of discard.

The unseen advantage of selective removal—and all solids removal equipment is selective—is that the most damaging and least beneficial solids are selectively removed. If the desilters are effectively operated, they

■ **TABLE 6–8** Whole mud vs selective solids discharge; feed mud discard vs hydrocyclone underflow

Mud wt, lb/gal		Solids, lb/bbl		Water, vol %	
Feed Mud	Cyclone Underflow	Feed Mud	Cyclone Underflow	Feed Mud	Cyclone Underflow
9.1	11.0	55	189	93.8	78.5
9.1	10.0	55	119	93.8	86.6
9.1	9.5	55	83	93.8	90.7
9.1	9.4	55	77	93.8	91.4
9.1	9.3	55	69	93.8	92.2
9.1	9.2	55	61	93.8	93.0

reduce abrasive solids to the point of leaving only the polishing size range. They leave in the system most of those solids that can be beneficial to the hole: the clays and low-size silts.

Adjusting hydrocyclone underflows to maximum weight is a mistake. They must be adjusted for efficient removal, to the balance point if that is possible. The underflow weight varies with the size of solids being removed, mud viscosity, etc.

The knowledgeable drilling supervisor with an adquate removal system at his command will not permit whole mud discard from the active system under any normal drilling condition. In emergencies he will attempt to save as much as possible in steel storage equipped with adequate stirrers and will salvage back into the active system as quickly as practical when the emergency is over. Hydrocyclones, centrifuges, and combinations can be used for this.

Oil as a Weight Reducer

If oil were added to the system at any point, it would temporarily decrease mud weight due to its low density. Within a very short time the system weight would return to original projection and then increase above it due to: (1) the increased viscosity caused by the oil, (2) the oil particles acting as solids particles and interfering with separation, and (3) the increased underflow waste problem due to the oil. *For minimum drilling mud weights with water-based muds, oil should not be added.*

Oversize and Overload Protection

At slow drilling rates, diamond drilling for example, cuttings are small and their volume cannot possibly begin to overload a full-flow installation of well-designed desilters. If fine screens are used and never bypassed and no formation cavings can enter the feed, no other full-flow separation device is necessary for this situation. If any major part of the hole is drilled at a rapid penetration rate, protection against both oversize and solids volume overload is necessary for the desilters. Coarse screens usually are used for maximum drilling rates.

The result of various full-flow solids-removal procedures is generalized in Fig. 6–30. Optimum solids removal in unweighted muds is obtained by continuous spray discharge underflow from adequate balanced desilters handling the full circulation rate as a minimum. This presupposes a shale shaker upstream adequate to protect against oversize solids particles that could plug the desilter inlets, and whatever other solids-removal equipment (such as desanders) necessary to prevent solids overload of the desilters.

Delayed and Intermittent Desilting

The upper line in Fig. 6–30 represents the coarse-screen shaker plus dilution technique necessary before efficient hydrocyclones were available. Thinking they are saving money, operators often wait till mud weight is fairly high to begin desilting. The result of this delayed desilting is also shown. The additional solids in the system, representing the increased mud weight over a continuously desilted mud, decrease in size due to mechanical degradation, and most no longer can be removed by the desilters. Periodic desilting is a false money saver. It never results in minimum mud weight or best mechanical characteristics. Resulting weight fluctuations, especially weight decreases, can be extremely detrimental to the

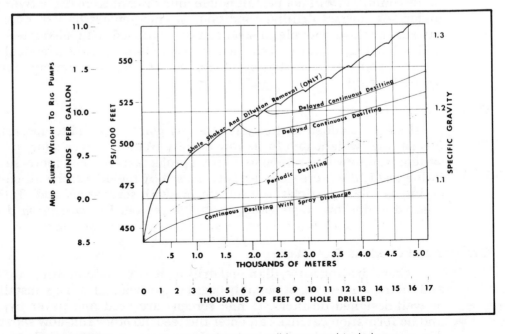

■ **FIG. 6–30** Effect on mud weight of various solids-remvoal techniques

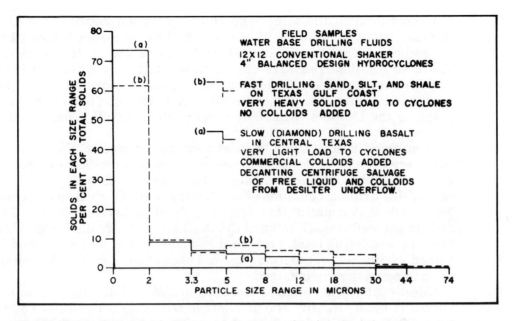

■ **FIG. 6–31** Comparing solids removal—screened and desilted slurry

stability of an open hole. With balanced cyclones the cost per pound of solids removal is minimized when desilting is continuous.

Delayed desilting has another penalty besides increased mud weight. The specific surface area of solids particles, on a new weight or volume basis, increases with decreased size particles (Table 6–2). The volume of liquid film associated with the surface area of all solids particles removed from a mud by hydrocyclones and by screens must vary directly with the surface area per pound of solids removed, i.e., more liquid phase is lost proportionately with finer solids. Delay in removing the solids always results in greater liquid loss and more expense for those that can still be removed.

Particle Sizes in Desilted Muds

Fig. 6–31 shows the particle size distribution of two desilted (low silt or low silt-size) drilling fluids from different situations. Curve (a) is the distribution of solids in a drilling fluid at the rig pump suction that has been processed first through a 12 × 12 mesh shake screen, and then through well-designed balanced desilters at a very low underflow solids load, always in spray discharge. This is probably as clean as a desilted mud will be. Note the small volume of particles that are larger than 8 microns. There were no desanders on the job, and had there been, it would not have changed noticeably the solids in the mud because the light load to the desilters permitted them to operate at maximum efficiency.

Curve (b) is the distribution of solids in a drilling fluid that has been processed through the same type of well-designed desilters but at very high underflow solids loads. Most of the drilling time—until shortly before the sample was taken—the underflow discharge was roping periodically. There were no desanders. The conventional shale shaker had 12 × 12 mesh cloth and separated very little material because the formation was poorly consolidated sand, silt, and shale. The shale shaker prevented any feed plugging the cyclones, which were operated with full opening (slightly wet bottom).

Had there been a stage of good desanders ahead of the desilters or had there been more desilters to provide greater underflow discharge capacity, the solids distribution might have been as fine as that of curve (a). In this drilling condition—fast drilling a sticky shale with a water-base mud—it is doubtful that very fine shale shaker screens could have functioned well enough to keep the desilters in spray discharge without blinding and losing mud. Very good 6-in. desanders certainly could have helped. Nevertheless, the mud was well enough desilted that low-silt mud drilling benefits were obtained.[32]

Most of the 73% colloidal material in mud (a) of Fig. 6–31 was purchased to build viscosity and gels and to reduce filtration rate. There are no good clays in the formation that would add any desirable properties to the mud. In mud (b) there are no commercial clays. The formations in that coastal area contain enough bentonite clay to provide filtration control, viscosity, and hole stabilization, but only when the clays are constantly saved in the system by good desilting practice. In any formations where beneficial clays are not present, some must be purchased and added if mud properties are needed.

Particle Sizes in Desanded Muds

Solids distributions in two desanded muds are shown in Fig. 6–32. Both these muds have passed through conventional 12 × 12 mesh shale shakers and then through full-flow 8-in. hydrocyclones. Curve (b) is from a mud being used to drill a hard formation in central southern Oklahoma using a flat-toothed bit. Drilling was slow, and the desanders never approached solids overload. There is a small amount of API sand-size particles that these particular hydrocyclones could not remove, even at the low solids load. Note the higher percentage of solids above 8 microns compared to either of the desilted muds in Fig. 6–31.

Curve (a) of Fig. 6–32 is from an unusual drilling situation found in the first few thousand feet of drilling in the delta area of Nigeria. Here the formation is coarse sand and fine gravel, with almost no formation particles less than 74 microns. Curve (a) shows the desanders have left some of the API sand-size particles. All of the colloidal material in the mud is commercial bentonite, added at the rate of one 100-lb sack per foot of hole primarily to increase gels and reduce whole mud loss through the coarse formation pores.

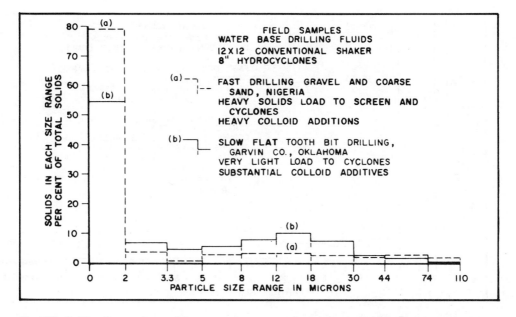

■ **FIG. 6–32** Comparing solids removal—screened and desanded slurry

If the Nigerian mud in curve (a) were more thoroughly desanded, it would have the characteristics of a desilted mud, simply because there is such a small amount of formation particles between 8 and 74 microns. Either finer screens ahead of the desanders to prevent hydrocyclones underflow solids overload or better desanders with more underflow solids capacity would have removed more of the sand-size solids. On the other hand, in curve (b) the solids load had nothing to do with the problem. Adequate desilters would have produced a low silt mud with nothing else needed except a shale shaker to protect against oversize particles. The slow drilling rate should not overload desilters.

Since there is no diagram of solids distribution in a mud treated with fine-mesh screens, an interesting deduction can be made. A 150-mesh Tyler Standard screen has openings 104 × 104 microns. Cagle's work showed that a shaker screen removes approximately 85% of the cuttings equivalent to opening size. Some material would be left larger than 104 microns. Note in Fig. 6–32 that the desanded particle size distribution had to be very close to that of a 150-mesh screen.

Particle Size in Coarse Screened Mud

Fig. 6–33 affords a study of two extreme solids distributions in unweighted muds. Curve (a) is a mud from the Niger delta area that has been through the 12 × 12 mesh shale shaker screen only. While not from the same drilling rig, it is from the same area and is very similar to the mud of Curve (a) in Fig. 6–32 *before* it had been desanded, i.e., feeding

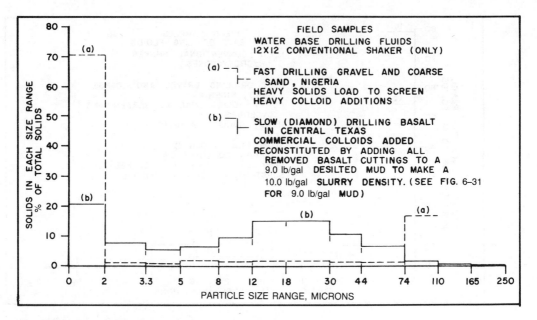

■ **FIG. 6–33** Comparing solids removal—screened only

to the hydrocyclone desanders. The small amount of solids particles between 2 and 74 microns below the shaker screen is striking and also obviously is due to there being hardly any in the unconsolidated, clean, gravel and sand formation. The commercial clays contrast with the API sand-size solids not removed by the screen.

Curve (b) of Fig. 6–33 is a good example of the cuttings size distribution from a diamond (core) bit. A considerable amount of commercial viscosifiers and filtration rate reducers were in the mud, and obviously most of the colloidal particles and at least half the particles between 2–3.3 microns are not diamond bit cuttings. Just as obviously, almost all of the particles above 5 microns *are* diamond bit cuttings. None were removed by the 12 × 12 mesh shale shaker cloth. The finest of practical full-flow shaker screens would remove only a fraction of these diamond bit cuttings.

The distribution in Fig. 6–33 illustrates the variations that can and do exist in the solids-removal problems in unweighted muds. Figs. 6–32 and 6–31 picture the change that occurs in solids distributions as the full-flow separation is improved to desilting.

WEIGHTED MUDS—FULL-FLOW PROCESSING

In muds weighted with barites, the problem of solids removal is more complicated. Study again Figs. 6–7a and 6–8a. Compare the API barite distributions, which are consistently similar, with the light solids distributions in Figs. 6–8a and 6–33. It is evident that removing light solids coarser than barites with fine screen cloths without removing any barites should be fairly effective in some cases and impossible in others, with varying results between extremes.

Problem of Barites Loss

The reduction in screen cloth liquid flow capacity, due to weighted mud effect, can be a problem. For this reason, for several years a few operators have used screen cloths coarse enough to take total flow without problems and then have processed the weighted mud through desilting hydrocyclones to remove the coarsest drill solids with acceptable losses of coarse barites. Recalling the 1½:1 ratio of diameters effect of Stokes' law at equal settling rates of barite and drilled solids, study Figs. 6–8 and 6–9 to understand how difficult it would be not to lose some of the coarser end of the barites. At the same time many of these operators report their losses of barite at the underflow to be negligible under most conditions.

Several facts are pertinent here:

1. API barite standards limit API sand-size barites, and a loss of barites (as other solids) in this size range no doubt is beneficial in reducing equipment erosion and filter-cake problems. The same is true of that portion of barites above actual 44 microns, or 66 microns equivalent to light solids.
2. Desilters are not as efficient in separation ability in heavy weighted muds, so their removal is not as severe as would otherwise be expected, especially if there is a slight restrictive adjustment at the underflow (this must not be overdone).
3. If there is a moderate load of coarse drill solids in the API sand range to remove, it is easy to adjust to lose virtually no barites below 44 microns actual size.
4. It is not difficult to test the hydrocyclones underflow to determine the cost of replacing whatever solids are being removed.

In the 1970s and '80s hydrocyclone operation on weighted muds deteriorated to the point that they could not be operated without screens on the underflow to return silt particles (includes the barites) back into the system. It then deteriorated further so it became a problem to justify screen cost. The trouble was that the feed head was not maintained to the hydrocyclones as mud weight increased. The head was decreased as the mud weight increased, keeping feed pressure in the same range. This is due partly to ignorance and partly to inferior products that cannot withstand the pressure of a cyclone handling weighted mud with proper head.

When the head is reduced on a hydrocyclone, the cyclone effect is lost; the cone becomes a swirling funnel; from 40 to 60 gpm of mud can discharge at the underflow instead of less than 4; and the solids in that underflow can amount to 600 lb/min instead of the normal maximum of 20 lb/min when the feed head is proper. This discussion concerns 4-in. hydrocyclones with a maximum ⅝-in. diameter underflow openings.

Benefits from Size Refinement

If the larger particle size fraction—some fraction of the 44–74 microns and all larger than 74—can be removed continuously from a weighted mud, any one or a combination of several possible benefits occur. Abrasion in the equipment such as rig pumps, manifolds, etc., is reduced; mud filter cake becomes smoother in character and may be slightly thinned. Differential pressure sticking is usually reduced if it has been a problem. The buildup of low-grade colloidal material, both light solids and barites, is reduced in rate, easing—but not eliminating—viscosity control problems; total solids content may be reduced in some conditions, and overall drilling progress usually speeds up. Although which of the above factors receives credit may depend upon the observer's previous benefits from desilting unweighted mud, it is no mere coincidence.[33] Such benefits in unweighted muds have been the result of *solids particle size distribution* changes, not solids average specific gravity changes.

Barites vs Light Solids Replacement

Improving the full-flow primary separation removal in weighted muds using hydrocyclone desilters downstream from the shale shakers does involve an added cost. Many persons are wary of the cost of barite losses. Most are not at all aware that every 100 lb of light solids removed must be replaced by 80 lb of barites and 20 lb of water, assuming barite specific gravity of 4.2 and light solids specific gravity of 2.6. In this process the larger end of the particle size distribution is surely removed, and the solids average specific gravity might be increased. Under no circumstances is the solids specific gravity decreased. A decrease in the coarsest material certainly improves the drilling operation. Our object is to remove the coarser end of the solids particle size distribution. An example of the two extreme removal situations possible, with one between the two, helps put the economics in perspective. Assume in each case that 1,500 lb of coarse solids are removed in the hydrocyclone underflow each hour. Assume water cost is negligible—though this is not always true—and that barites cost $8/100 lb, U.S.

Case 1: All removal is light, drilled solids. Cost to replace the weight and volume effect of 1,500 lb of light solids:

$$1,500 \times \frac{\$8}{100} \times 80\% = \$96/\text{hr}$$

Case 2: All removal is barite particles. Cost to replace the weight and volume effect of 1,500 lb of barites:

$$1,500 \times \frac{\$8}{100} \times 100\% = \$120/\text{hr}$$

Case 3: Removal is one-half barites by weight. Cost to replace the light solids:

$$\frac{1,500}{2} \times \frac{\$8}{100} \times 80\% = \$48$$

Cost to replace the barites:

$$\frac{1,500}{2} \times \frac{\$8}{100} \times 100\% = \$60$$

Total replacement cost:

$$\$48 + \$60 = \$108/\text{hr}$$

Case 1 is as unlikely as case 2, with case 3 more closely averaging real situations. The most important thing to realize is that if our objective is to remove most of the particles larger than 44 microns in a weighted mud, replacement cost is not much more for barites than for light solids and the benefits are about the same. Also, it must be realized that first barite loss on a new system is reduced as the larger barites are removed. Further barite loss is almost completely restricted to makeup barites. This should be performed on the full-flow rate, never on partial flow.

PRIMARY SEPARATION—PARTIAL FLOW

Partial-flow processing in primary removal is, or should be, only to reduce the clay, or sometimes soluble, solids content of the mud. Obviously, to remove the clay phase from the total circulating rate of a weighted drilling mud is courting immediate disaster whether the liquid phase is replaced without clay or an attempt is made to replace a liquid phase with needed properties at the full rig circulating rate.

Decanting Centrifuge

The decanting centrifuge has been such a popular clay removal machine simply because it assures, by its very nature, a complete removal of clays from the small fraction of the circulating rate it was intentionally selected to process.

Although the machine is most often referred to as a *barite salvage* machine, that is a misnomer. All primary separation machines have a removal function. Barites can be "saved" merely by not pumping mud from the pit at all. The primary problem in a water-based weighted mud is viscosity increase. The primary cause of this problem is the natural, constant increase in clays, usually not of high quality. Their presence and quantity prevent addition of better filtration control clays and may severely retard drilling progress.

Dilution and discard to remove these excess low-quality clays is effective but also expensive due to the high cost of barites to replace the total weighting material, including inert-size low-specific-gravity particles, in the discard. Unpublished data from several parts of the world indicate the *minimum* dilutions, or equivalent mechanical treatment, amounts to an average of 5% of the total system each day a weighted mud is in use, and ranges up past 20% per day. A limit apparently exists representing the maximum rate at which a colloidal phase can be rebuilt and controlled successfully. It is in the range of 10% per circulation. The larger decanting centrifuges, operated for a cut point of 2 microns of barite particles, have a feed capacity ranging from extremes of about 3–12% of the circulating rate, depending on circulation rate and mud weight.

Perforated Rotor Centrifuges

The perforated rotor centrifuge in the recommended mode of operation, as already discussed, partially separates the clay fraction. To remove the same volume of clay material it must process more of the same mud than a decanting centrifuge. The dilution, per unit of raw mud feed, averages two or three times that required for the best designs of decanting centrifuges. The overflow volume needing to be disposed of is 2 to 5 times that of a decanter to accomplish the same clay removal, or viscosity reduction. If dilution water compatible with the mud is available in quantity and the overflow volume can be utilized or disposed of easily, or if a good decanting centrifuge is not available or cannot be set up properly on the system, the perforated rotor machine can be of great assistance in controlling viscosity in a weighted mud system, with worthwhile money savings.

Hydrocyclones for Viscosity Reduction

If the average weighted mud were fed to a balanced hydrocyclone capable of separating most of the inert solids, the underflow opening could not handle the solids load and would rope. Since a very high separation ability is necessary to separate the major part of an API barite grind and since efficient separation requires spray discharge, weighted muds must be heavily diluted to reduce their total solids content before they enter special barite hydrocyclone units. Much less dilution would be required if the only problem was to reduce the viscosity enough for high separation efficiency.

This dilution must be sufficient to reduce solids content to about 4 vol% or to about 9 lb/gal feed slurry density. Such a dilution involves large quantities of water, much more than a perforated rotor centrifuge for example. At high dilution, proper head, and spray discharge conditions the best barite hydrocyclones for field work can make a median cut between 7 and 9 microns. The API standard now permits a maximum of 30% by weight of the barites to be less than 5 microns in size. With this median cut, the free liquid film on the barite surface area keeps underflow density down to the range of 14–17 lb/gal.

In heavier muds low underflow weight at maximum separation creates a problem. Care must be taken to mix new barites while the cyclones are operating to keep up mud weight, or the hydrocyclones must be operated with a rope underflow to bring underflow weight up to or above pit mud weight, with a resulting heavy loss of medium silt barites in the cyclone overflow. Special coarse barite grinds can reduce this problem but introduce other problems that are worse.

Hydrocyclones for partial flow removal of colloidals from weighted muds may be cheaper to buy and operate than decanting and perforated rotor centrifuges, but they cannot be justified over decanting centrifuges nor over perforated rotor machines, on the total mud cost basis. If good decanting centrifuges are not available or cannot be serviced, if dilution water is available in large quantities and at low cost, and if the overflow disposal is not a problem, the RMS should be considered for barite salvage. Cyclones are a last resort.

EQUIPMENT APPLICATION—SECONDARY SEPARATION, UNWEIGHTED MUDS

In some situations full-flow primary solids removal, which is always selective of the larger particle sizes, creates intolerable problems, in their turn requiring solutions.

In desilting unweighted muds, the free liquid film associated with the separated silt may create a waste disposal problem or may involve a very expensive liquid-phase loss, or both. An example of the disposal problem is desilting any drilling fluid in highly populated areas. Another is desilting an unweighted oil mud in any area. An example of both scarcity and disposal problems is desilting in the Arctic, where water is extremely expensive and liquid disposal even more so.

In these cases one solution to the problem would be to confine the removal to coarse material that does not have a high specific surface area and which can be removed by full-flow shale shaker screens. With the best designed screens of sufficient capacity, the free liquid draining from the large cuttings should not present a major problem. However, this leaves a higher solids content retained in the mud due to the particles that would not be present if the mud were being desilted. This means higher mud weight—hydrostatic head on the formation—thicker and coarser filter cake with increased tendency for differential sticking in permeable formations, slower drilling progress, shorter bit footage life, increased pump fluid end abrasion, and less hole stability if there is any natural weakness.

Hydrocyclone–Centrifuge Combination

A better solution is to desilt the mud with full-flow hydrocyclone desilters (primary separation) and to use a decanting centrifuge in secondary

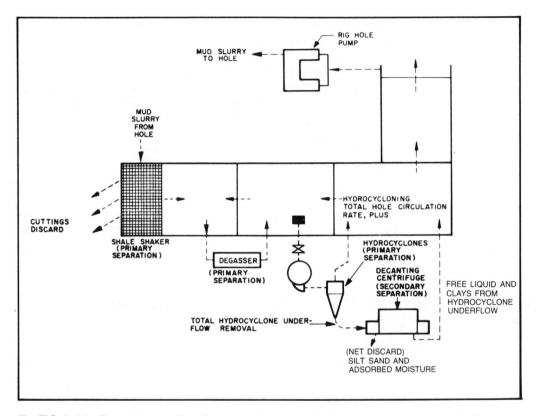

■ **FIG. 6–34** Decanting centrifuge for secondary separation from primary separation desilters

separation to separate the liquid-colloidal phase and return it to the active system (Fig. 6–34). The underflow of the decanting centrifuge, with no free liquid present, contains silt and sand removed by the hydrocyclone desilters. This decanter underflow can be handled by a dirt scoop or blade but cannot be sucked into a vacuum truck due to the lack of free liquid. Disposing of it is a simple problem compared to liquid whole mud or hydrocyclone underflow.

In expensive liquid-phase muds—water in the Arctic and the desert, oil anywhere—liquid cost alone may be economic justification for the decanting centrifuge in secondary separation. In some cities, wilderness areas, and marine environments, reducing disposal cost alone justifies the decanter. The benefits of primary separation desilting are not lost through secondary separation with an effective decanting centrifuge.

In unweighted water-based muds an increase in mud weight of 0.2 lb/gal occurs. The return of a salvaged clay phase instead of water obviously must involve some greater density in (less solids removed from) the system. Many salvaged clays in water-based muds are substituting for those that would be purchased and added to the fresh liquid if not for the salvage.

Impressive mud-cost reductions in many areas indicate the validity of this reasoning.

Perforated Rotor Machine

The fluid underflow requirement of the perforated rotor centrifuge precludes its use as a secondary separation unit on the underflow of hydrocyclones. Inherent characteristics of the separation mechanism prevent its being modified to provide an underflow devoid of free liquid.

Hydrocyclone-Screen Combinations

Special fine shale-shaker screens for secondary recovery from primary separation hydrocyclones, mud cleaners, should never be used for unweighted muds. Returning to the system any silt removed by the primary separation desilters decreases the low-silt-size solids drilling benefits and increases active system weight—compared to desilting without secondary separation of the underflow—by approximately 0.5 lb/gal (0.06 specific gravity) or more.

If the underflow liquid loss from primary separation desilting cannot be tolerated, the secondary decanting centrifuge is the only practical alternative to stopping desilting.

Hydrocyclone desilters were first run on weighted muds, up to 18 lb/gal, in 1965. Toolpushers on operator-owned rigs were the first to try it. They had been instructed on determination of equivalent weight-material loss, but their trials were not supervised by any technicians. The unanimous tone of all reports were that mud cost, specifically barite cost, was not noticeably affected, that drilling rate increased, and that differential sticking was reduced.

Those early runs were never written up, much less published. Unfortunately, most operators find these statements difficult to believe because of the low head almost universally used on hydrocyclones in weighted mud today. Also, some of the publicity introducing the mud cleaners contained unfortunate statements based on a failure to understand the limitations of hydrocyclone separation.

Hydrocyclone desilters, well designed, can be used on weighted muds without excessive loss of barites if proper feed head is maintained. Nevertheless, very few current field supervisors have tried it.

Hydrocyclones—Decanting Centrifuge

With the number of various machines available for solids control, it is small wonder that several of Murphy's famous laws regarding mishaps are the rule. The decanting centrifuge has more than its share of misapplications. Table 6–9 may help.

The decanter has legitimate application on hydrocyclone underflow in oil muds and *unweighted* water-based muds. It should never be used on hydrocyclone underflow of weighted water-based muds. (It does not make

■ TABLE 6–9 Primary and secondary separations, full and partial flow processing (normal pertinent effects on the mud system)

| | | | | | Effect on | | | | |
| | | | | | Unweighted Mud | | | Weighted Mud | |
Item	Separation	Type of Flow	Type of Equipment*	Character of the Discard	Total Solids	Mud Weight	Plastic Viscosity	Total Solids	Plastic Viscosity
1	Primary	Full	Shale-shaker screens	Wet	Down	Down	Negligible	Down	Slightly down
2	Primary	Full	Desanding** hydrocyclones	Wet	Down	Down	Negligible	Down*	Slightly down
3	Primary	Full	Desilting hydrocyclones	Very wet	Down	Down	—	Down	Slightly down
4	Primary	Partial**	Decanting centrifuges	Low-volume liquid	MA***	MA	MA	Slightly down	Down
5	Primary	Partial**	Perforated rotor centrifuge	Medium-volume liquid	MA	MA	MA	Slightly down	Down
6	Secondary	Desilter underflow	Decanting centrifuge	Damp	Slightly up	Slightly up	—	Slightly up—oil	Slightly up—oil
7	Secondary	Desilter underflow	Special screens	Wet	MA	MA	MA	—	—

* All equipment of each type assumed to be the best design available.
** A misapplication if used on oil muds; use on water base only.
*** MA—Misapplication. Full-flow, plus secondary separation if justified, is the only economic approach for unweighted muds.

sense to spend money to salvage and return to a system the liquid-clay phase that another machine is removing to keep viscosity under control.)

Hydrocyclones—Fine Screens (Mud Cleaners)

The special screens mentioned under unweighted muds (see Fig. 6–35) have an application in weighted-mud situations, for which they were developed. The process is straightforward, and the most important thing is to know or determine the drilling benefits and costs when desilting without secondary separation by the screens. Then, when the screens reduce mud cost, the savings can be weighed against the drilling benefits, which may be reduced by screening. If drilling benefits are not reduced, the savings are completely valid.

The second thing to keep in mind with secondary separation screening is the viscosity problem. Again, if the viscosity increases, the first remedy is to stop returning colloidal material to the system with the screen. Let it go with the coarses in the desilter underflow. If the viscosity still is a problem, a decanting centrifuge is recommended for primary separation of clays.

Decanting Centrifuge—Two-Stage Centrifuging

Occasionally, we hear of a double-stage centrifuging job on weighted mud. This process began on the Gulf Coast of Texas and Louisiana in

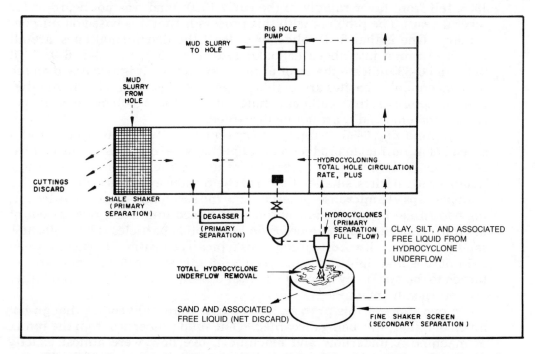

■ **FIG. 6–35** Special shale shakers for secondary separation from primary separation desilters[4]

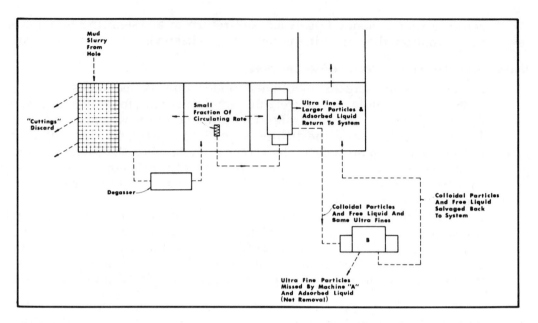

Mud
Slurry
From
Hole

Small
Fraction Of
Circulating Rate

Ultra Fine &
Larger Particles &
Adsorbed Liquid
Return To System

A

"Cuttings"
Discard

Degasser

Colloidal Particles
And Free Liquid And
Same Ultra Fines

B

Colloidal Particles
And Free Liquid
Salvaged Back
To System

Ultra Fine Particles
Missed By Machine "A"
And Adsorbed Liquid
(Net Removal)

■ **FIG. 6–36** High-speed decanting centrifuge for secondary separation from medium-speed decanter

1955, fell from favor rapidly in the early 1960s, and was not heard of for several years. The problem with this process is that it is based on misconceptions. One is that drill solids, or at least the detrimental ones, are all in a size range just above colloidal size. Figs. 6–5, 6–7, 6–8, 6–9, 6–31, 6–32, and 6–33 indicate the error of this assumption. It is also based on the misconception no barites are in this range (see Figs. 6–7 and 6–8). Another misconception is that colloidal materials, including chemicals, do not cause problems and must not be discarded.

Examine Fig. 6–36. Primary-separation centrifuge A is operated at a speed below normal to produce a cut at perhaps 4–6 microns (barite equivalent). The underflow solids particles return to the system, and overflow liquid—with barites smaller than maybe 5 microns, light solids smaller than perhaps 7½ microns—proceeding to the secondary-separation decanting centrifuge B, operated at normal full speed to make a regular cut of microns on barites and 3 microns on light solids particles. The underflow solids of machine B are discarded. This typically consists of barites between 2–5 microns and light solids between 3–7.5 microns. All colloidals are returned to the system by machine B; most ultrafines and all fines and larger are returned by machine A.

In one study of this process, more barite than light solids, though not much of either, was being discarded.[31] The mud properties from the remix of machine A underflow and machine B overflow were almost exactly identical to the feed to A from the pit. In a water-base mud, if feed dilution were used, that water alone would produce a change. The same water could be added to the system without the machines. The dual stage, as described here and as pictured in Fig. 6–36, cannot be recommended in water- or oil-based mud for maintenance.

For stripping an oil mud of all possible solids (a batch process), it is legitimate to use a heavy-duty centrifuge to pull out the major load of solids to permit a high-speed machine (that cannot handle a large solids load) to make a fine cut on the overflow of the first machine. This is *not* dual-stage centrifuging, as the underflows of *both* machines are discarded.

THE "TWILIGHT ZONE" IN MUDS

The solids problem and the problem of controlling solids in weighted muds are different from unweighted muds. This difference has long posed a dilemma to operators who have reached a depth where they do not need a weighted mud, the formation pore pressure requires more hydrostatic head than the desilted and unweighted mud in the hole, and they do not wish the drilling disadvantages of stopping the desilters. This problem has been classic in the range from a low of 9.0–9.5 lb/gal at the low weight up to 10.5 lb/gal, reflected in Fig. 6–29. It has been approached three basic ways in the past:

1. The most common method has been to stop desilting first and do basically nothing while mud weight increases from drilled fines and mediums. As higher weight is needed, barites are added and either a centrifuge is used or discard and makeup are resorted to for viscosity control. This procedure is more adaptable to slower drilling rates.
2. In faster drilled holes, with rapidly increasing transition zones, a similar but quicker method has been used often. Intensive desilting is carried on until the transition zone is evident. Then, simultaneously, desilting is stopped, decanting centrifuging for colloidal removal is started, and rapid barite weighting is begun. Fast drilling is continued. Several operators have reported increasing weight 3 lb/gal in one circulation. They also report this can be done only with a thoroughly desilted mud.
3. A third method is popular with those who have tried it, but the majority are reluctant to test it for fear of barite cost. Intensive desilting is continued, but barites are added as more weight is needed. This is usually continued to a mud weight of 11 lb/gal or more. In this range desilting usually has been stopped, or the underflow restricted more or resorted to periodically to thin the filter cake, reduce abrasion, alleviate hole problems, etc. The decanting centrifuge is used as needed for viscosity reduction, or the discard-and-makeup method is used. A decanter in this situation must be operated for a 2-micron barite cut.

Special secondary-separation screens may be of benefit for the "Twilight Zone" of neither weighted nor unweighted muds. As shown in Table 6–9, returning the liquid phase from the screen to the unweighted system causes some weight increase. If a more rapid increase is needed, barites can be added as necessary. As the transition to a weighted mud is completed, additional measures can be taken if needed. This should be tried against method 3 above to determine comparative drilling advantages.

INSTALLATION AND MAINTENANCE

The subjects of proper mud system design, equipment installation engineering, and maintenance programs are too lengthy and involved to take up here. However, some pointers are in order.

On any drilling operation in the world, poorly maintained solids-removal equipment reflects the knowledgeability and attitude of the management of that operation. No equipment is kept in operating condition if management does not make it clear that failure to maintain it in top shape and to operate it properly will not be tolerated. If an installation is not as it should be, the chances are it will be changed soon if proper maintenance is insisted upon.

Equipment inadequate for the job cannot operate properly. On rigs with the most and best solids-removal equipment, operators would not part with any of it; on rigs most poorly equipped, they would gladly give away what they have because it can't work properly. Rigs capable of drilling the fastest need the highest solids-removal capacities. Potential drilling benefits justify the best equipment available. Initial cost is of least importance.

Time and effort spent by a drilling supervisor to understand drilling mud solids, and the machinery used to control them, will return handsome rewards. Besides study, this understanding requires familiarity with the equipment, gained on the drilling rig.

■   **REFERENCES**

1. R.A. Bobo and R.S. Hoch, "The Mechanical Treatment of Weighted Drilling Muds," AIME paper no. 290-G.
2. R.L. O'Shields and Don E. Wuth, "New Mud Desander Cuts Drilling Cost," AAODC annual meeting, Houston (September 1955), *Drilling Contractor* (October 1955).
3. G.S. Ormsby, "Desilting Drilling Muds with Hydrocyclones," AAODC Rotary Drilling Conference, *Drilling Contractor,* (March–April 1965).
4. T.C. Mondshine, "New Fast Drilling Muds also Provide Hole Stability," *Oil & Gas Journal* (March 21, 1966).
5. Unpublished data sources, including Pioneer Centrifuging Co. files.
6. Official minutes, API Task Group on Determination of Liquids and Solids, D.B. Anderson, chairman (set up in 1968 by API Committee on Standardization of Drilling Fluid Materials).
7. API RP 13-C, first edition (September 1973), API Committee on Standardization of Drilling Fluid Materials, Task Group on Solids Removal Devices, Luther Bartlett, task group head.

8. *Principles of Drilling Fluid Control,* twelfth edition (Austin: Petroleum Extension Service, University of Texas, 1969).
9. R.L. Ritchey, "Effect of Particle Size on $\overline{YP}$ and Viscosity of Drilling Muds," Salt Water Control Co. (SWACO) publication (February 1959).
10. H.G.N. Fitzpatrick, private communication.
11. J.A. Gill, "Drilling Mud Solids Control—New Look at Techniques," *World Oil* (October 1966).
12. API RP 13-C, 1973.
13. John H. Perry, *Chemical Engineers Handbook,* third edition, (New York: McGraw-Hill).
14. API RP 13-C, 1973.
15. L.L. Hoberock, "Shale Shaker Selection and Operation," *OGJ* (23 November, 7 December, 21 December 1981).
16. Ron Morrison, "New Shaker Employs Straightline Motion," *Drilling Contractor* (January 1983).
17. API RP 13-C, 1973.
18. Louis K. Brandt, "Remarks on Fine Screen Shakers," *Proceedings,* Rotary Drilling Conference of IADC, Houston (1 March 1973).
19. Hoberock, *OGJ,* 1981.
20. Morrison, *Drilling Contractor,* 1983.
21. W.S. Cagle, "Solids Removal Efficiencies of Shale Shaker Screens," *Transactions,* Drilling Technology Conference of IADC/CAODC, Calgary (9–12 March 1981).
22. George Shuman, response to a question, IADC Shale Shaker Conference, Los Angeles, CA (20–22 January 1976).
23. W.F. Roper, leader of screen section of Solids Removal Task Group, verbal statements in task group meetings leading to API RP 13-C.
24. G.S. Ormsby, *Drilling Contractor,* 1965.
25. G.S. Ormsby, "Why Your Removal System Won't Remove," *Proceedings,* Rotary Drilling Conference of IADC, Houston (1 March 1973). Published also as "Proper Rigging Boosts Efficiency of Solids-Removing Equipment" and "Proper Pumps, Lines Help Cut Costs," *Oil & Gas Journal* (12 March and 19 March 1973, respectively).
26. Ormsby, *OGJ,* 1973.
27. Ormsby, *OGJ,* 1973.
28. Direct copies of original Mohole-calculated data in Pioneer Centrifuging Co. library, from Mohole test work conducted near Uvalde, TX (1964).
29. Bobo and Hoch.
30. Ralph F. Burdyn, D.E. Hawk, and F.D. Patchen, "New Device for Field Recovery of Barite: II Scale-up and Design," *SPE Journal* (June 1965).
31. M.D. Nelson, "Removal of Fine Solids from Weighted Muds," API paper no. 926-15-H, presented before the Southern District Div. of Production, Houston (4–6 March 1970).
32. Ormsby, *Drilling Contractor,* 1965.
33. Ormsby, *Drilling Contractor,* 1965.
34. L.H. Robinson and J.K. Heilhecker, "Solids Control in Weighted Drilling Fluids," SPE paper no. 4644, annual fall meeting, Las Vegas (1973).
35. George S. Ormsby, "An Analysis of a Dual Centrifuging Process on a Weighted Inverted Oil Emulsion Mud Containing Asphalt Additives," unpublished, in Pioneer Centrifuging Co. library (March 1966).

■ **ADDITIONAL REFERENCES**

"Characteristics of Particles and Particle Dispersoids," *Stanford Research Institute Journal,* Third Quarter 1961 (Department 300, Stanford Research Institute, Menlo Park, CA).

Longwell, Knopf, and Flint. *Outlines of Physical Geology,* second edition. New York: John Wiley & Sons.

U.S. Department of Agriculture, SCS. "A Guide for Interpreting Engineering Uses of Soils." Washington, DC: Director of Documents, 1971. Also, International Standard Soil Classification System, adopted by International Society of Soil Scientists in 1934.

CHAPTER 7

FLUID CIRCULATION

This chapter includes information on circulating pressure losses, a determination of the lifting capacity of drilling muds, and surge and swab pressure calculation methods. There are no exact solutions for non-Newtonian fluids such as drilling muds. For this reason flow behavior must be defined, and any calculations must be based on the equivalent thickness of the drilling mud at a specific flow rate.

THEORETICAL CONSIDERATIONS

True or Newtonian fluid behavior is defined by Eq. 7.1:

$$\tau = \mu_p \left(-\frac{\Delta v}{\Delta r} \right) \qquad (7.1)$$

where:

τ = shear stress, dynes/sq cm
μ_p = viscosity, poise
$\dfrac{\Delta v}{\Delta r}$ = shear rate, sec^{-1}

Shear stress is defined for pipe flow as force per unit of shear area. This definition is in terms of pressure, as shown in Eq. 7.2:

$$\tau = \frac{\Delta P_d r}{2l_f} \qquad (7.2)$$

where:

ΔP_d = pressure, dynes/sq cm
r = pipe radius, ft
l_f = length of flow path, ft

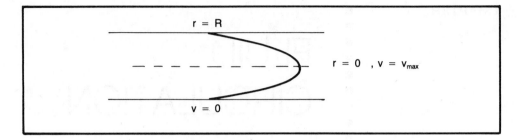

■ **FIG. 7–1** Shear rate curve for pipe flow

Eq. 7.1 defines the fluid viscosity, μ, as a proportionality constant between shear stress and shear rate. For non-Newtonian fluids, mud thicknesses changes with flow rate. Thus the conventional definition of viscosity shown in Eq. 7.1 does not apply.

Because of convention, mud thickness is defined in terms of viscosity. However, the flow rate must be determined first. Then thickness is expressed as the equivalent viscosity at a specific flow rate.

The negative sign in Eq. 7.1 is based on the fact that, as velocity is increased, the pipe radius decreases in the definition of shear rate. This is shown in Fig. 7–1. For this reason the negative sign must be used to obtain a positive shear stress.

All of the equations used in this chapter are based on the assumption that the flow behavior of the drilling fluid in laminar flow follows a so-called power-law behavior as shown in Eq. 7.3:

$$\tau = K \left(-\frac{\Delta v}{\Delta r} \right)^{n_d} \tag{7.3}$$

where:

K = proportionality constant
n_d = deviation of the fluid from Newtonian behavior

$$\log \tau = \log K + n \log \left(-\frac{\Delta v}{\Delta r} \right) \text{ (log form of Eq. 7.3)} \tag{7.4}$$

Comparing Eq. 7.3 to Eq. 7.1, K represents the friction component as μ does in Eq. 7.1. If $n = 1.0$ in Eq. 7.3, $K = \mu$ and the fluid behavior as described by Eq. 7.3 is equal to Newtonian behavior. Flow data indicate that n in Eq. 7.3 is less than 1.0 for drilling muds. Eq. 7.4 predicts that a plot of shear stress versus shear rate on log paper will be a straight line.

Other frequently used assumptions describe the flow behavior of drilling muds in laminar flow. Probably the most common is the Bingham plastic flow behavior demonstrated in Eq. 7.5:

$$\tau = Y + PV \left(-\frac{\Delta v}{\Delta r} \right) \tag{7.5}$$

where:

Y = yield point, dynes/sq cm
PV = plastic viscosity, poise

Eq. 7.5 indicates that the behavior of muds follows a straight line on regular coordinate paper, and this is untrue. Yield point and plastic viscosity still are used in operations; however, the terms have no quantitative significance. All of the input data to equations for laminar flow come from rotating viscometer data.

Rotating viscometer data are based on the assumption that the flow pattern is laminar. Readings are taken using two-speed, six-speed, and variable-speed viscometers. The preferred viscometer is the variable speed in which a digital readout shows the viscometer speed; the speed can be varied to match the annular fluid velocity. Shear rate units are sec^{-1}, and the viscometer speed is measured in revolutions per minute (rpm). The relationship between viscometer speed in sec^{-1} and rpm is shown in Eq. 7.6:

$$\text{Shear rate in } sec^{-1} = (1.703)(\text{Shear rate in rpm}) \tag{7.6}$$

Obviously, the most accurate mud properties measurement is made with a variable-speed viscometer. When using a two-speed viscometer, the 3-rpm speed (or gel strength) and the 300-rpm speed commonly are used to construct a straight line for calculation purposes, rather than plotting the straight line between the 300- and 600-rpm speeds. The shear rate for a rotating viscometer that matches the annular fluid velocity is as follows:

$$\gamma_v = \frac{2.4v}{D_h - D_{op}}\left(\frac{2n+1}{3n}\right) \tag{7.7}$$

where:

v = annular velocity, ft/min
D_h = diameter of the hole, in.
D_{op} = outside diameter of the pipe, in.

The shear rate (in rpm) for a rotating viscometer that matches the annular fluid velocity is given by Eq. 7.8:

$$\overline{\gamma} = \frac{1.41v}{D_h - D_{op}}\left(\frac{2n+1}{3n}\right) \tag{7.8}$$

where:

$\overline{\gamma}$ = shear rate for a rotating viscometer, rpm

Eqs. 7.7 and 7.8 often are shortened by assuming $n = 0.7$. The maximum error is less than 10%. This modification of Eq. 7.8 is shown in Eq. 7.9:

$$\overline{\gamma} = \frac{1.61v}{D_h - D_{op}} \tag{7.9}$$

Eq. 7.9 is very simple and easy to use. Also, Eq. 7.9 is very handy when using a variable-speed viscometer, where data are not plotted on log paper.

Turbulent flow patterns are more difficult to define because they do not depend on the mud's measurable properties. For this reason, all of the calculations for turbulent flow are based on empirical measurements and equations. A common empirical method for defining turbulent flow is to use an empirical friction factor and a dimensionless combination of variables called the *Reynolds number.*

The empirical friction factor used in this book is the classical Fanning friction factor defined by Eq. 7.10:

$$f = \frac{9.3(10^4)\Delta P_p D_p}{v^2 l_f} \tag{7.10}$$

where:

f = Fanning friction factor, dimensionless
ΔP_p = pressure loss over a given length of pipe, psi
D_p = pipe diameter, in.
v = fluid velocity, ft/min
l_f = length of flow chamber, ft

The Reynolds number is defined by Eq. 7.11:

$$Re = \frac{15.47\rho v D_p}{\mu} \tag{7.11}$$

where:

ρ = mud weight, lb/gal
μ = viscosity, cp

Note: For annular flow, $D_p = D_h - D_{op}$.

Experimental flow data are illustrated on the log-paper plot of the Fanning friction factor versus the Reynolds number in Fig. 7–2. These data show laminar flow to a Reynolds number of 2,000 and turbulent flow at Reynolds numbers of 4,000 and above. Between 2,000 and 4,000, the flow pattern is in transition, which means it can be laminar or turbulent, or both.

Pressure losses through the bit nozzles are based on changes in kinetic energy, defined in Eq. 7.12:

$$\Delta KE = \frac{M\Delta v_1^2}{2g_c} \tag{7.12}$$

where:

ΔKE = changes in kinetic energy, ft-lb$_f$
M = lb$_m$
Δv_1 = change in velocity, fps
g_c = change in lb$_m$ to lb$_f$ = $\dfrac{32.2 \text{ lb}_m\text{-ft}}{\text{lb}_f\text{-sec}^2}$

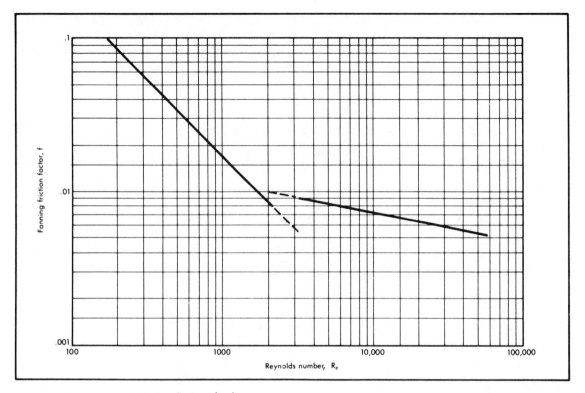

■ **FIG. 7–2** Fanning friction factor chart

In addition, pressure losses through bit nozzles generally are calculated using a nozzle coefficient of 0.95.

Pressure losses through fixed-bladed bits are based on the total flow area across the face of the bit when it is sitting on bottom of the hole. Flow technology through fixed-bladed bits is changing because of the need to cool all the cutting surfaces as well as remove cuttings below the bit. In past years, these requirements were determined by trial-and-error procedures in field drilling operations. In the future, more laboratory development work will be done. This should open the door to calculation procedures for predicting requirements.

DETERMINATION OF FLOW PATTERNS

When the Reynolds number is less than 2,000, the flow pattern is assumed to be laminar. When the Reynolds number is greater than 4,000, the flow pattern is assumed to be turbulent. The flow between Reynolds numbers of 2,000 and 4,000 may be laminar or turbulent, or both.

In general, when figuring pressure losses when the Reynolds number is between 2,000 and 4,000, the pressure loss is calculated assuming both laminar and turbulent flow, and the higher number is used. If the operator is concerned with keeping the flow pattern laminar to minimize hole erosion, then the Reynolds number simply is kept below 2,000. When considering lifting capacity, calculate slip velocity by assuming laminar and turbulent flow around a cutting and use the lower slip velocity.

If the flow pattern is assumed to change from laminar to turbulent flow at a Reynolds number of 2,000, Eq. 7.13 may be used to determine the specific fluid velocity at the transition point.

$$v_c = \left[\frac{2.585(10^4)K}{\rho} \right]^{1/(2-n)} \left[\frac{2.4}{D_h - D_{op}} \left(\frac{2n+1}{3n} \right) \right]^{n/(2-n)} \qquad (7.13)$$

where:

v_c = fluid velocity when the flow pattern changes from laminar to turbulent, fpm
ρ = mud weight, lb/gal
D_h = hole diameter, in.
D_{op} = outside pipe diameter, in.

Values for n and K are defined below.

$$n = 3.32 \log \frac{\theta_{2\gamma}}{\theta_\gamma} \qquad (7.14)$$

where:

n = slope of the viscometer data on log paper
θ = viscometer reading, lb_f/100 sq ft

$$K_i = \frac{\theta_{\gamma s}}{\gamma_s^n} \qquad (7.15)$$

where:

K_i = intercept of viscometer data on log paper
γ_s = shear rate, sec^{-1}

Eq. 7.13 can be used for other Reynolds numbers by dividing 2.585 × 10^4 by 2,000 and multiplying by the desired Reynolds number. Eq 7.14 is simply an easy way to determine the slope of a plot of viscometer readings versus shear rate on log paper. Another way to determine the slope simply is to measure it with a ruler.

Eq. 7.15 is also an easy way for determining the intercept of a plot of the viscometer readings versus shear rate on log paper. Another way to determine K is simply to read it off the log paper.

Remember that many times the data are plotted on log paper using the viscometer speed (in rpm). If this is done, n can still be calculated using Eq. 7.14, but K cannot be read directly from the plotted data, and the shear rate will have to be changed from rpm to sec^{-1} to use Eq. 7.15.

Fig. 7–3 is a diagram of the drilling fluid circulating system. The summation of pressure losses in the entire circulating system is shown on

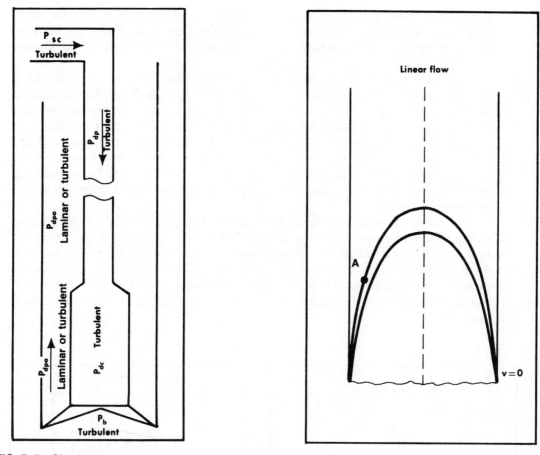

■ **FIG. 7–3** Circulation system and normal flow patterns   ■ **FIG. 7–4** Linear flow

the surface pressure gauge, normally located on the standpipe in a corner of the rig mast. This summation of pressures is given in Eq. 7.16:

$$P_t = P_{sc} + P_{dp} + P_{dc} + P_b + P_{dca} + P_{dpa} \qquad (7.16)$$

where:

P_t = standpipe pressure, psi
P_{sc} = pressure loss in surface connections, psi
P_{dp} = pressure loss through drillpipe, psi
P_{dc} = pressure loss through drill collars, psi
P_b = pressure loss through bit nozzles, psi
P_{dca} = pressure loss through hole-drill collar annulus, psi
P_{dpa} = pressure loss through hole-drillpipe annulus, psi

Note: Any other variations in pipe or hole size can be separated and added according to the lengths involved (in feet).

The total pressure gives no indication whether the flow pattern in any part of the system is laminar or turbulent. As noted in Fig. 7–3, the flow patterns inside the drillstring are generally turbulent. Flow patterns in the annulus may be either laminar or turbulent, and methods will be shown to help distinguish the flow pattern.

Laminar flow is distinguished by a smooth flow pattern, as illustrated in Fig. 7–4. The velocity of each layer of fluid increases toward the middle of the stream until some maximum velocity is reached. It is not uncommon to encounter special cases of laminar flow in which the center of the flow pattern is flat, as shown in Fig. 7–5. In this flat portion of the stream there is no shear of fluid layers; this type of flow is called *plug flow*. In hole cleaning it is often desirable to flatten the velocity profile by increasing the mud thickness; however, this practice generally would increase annular pressure losses.

Turbulent flow can be described as *random flow*. There is no viscous shear of fluid layers as shown in Fig. 7–4. A typical pattern for turbulent flow is illustrated in Fig. 7–6. Fluid velocity at the hole wall is zero; however, fluid velocity in the core of the stream is essentially flat. There are different degrees of turbulence. Thus, a simple statement that the flow pattern is turbulent is not very definitive. As turbulence increases, the pressure losses also increase.

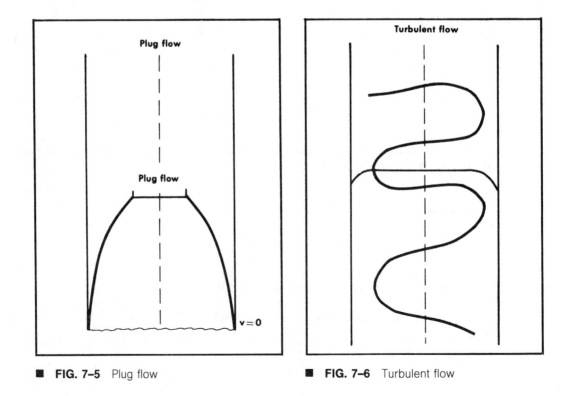

■ **FIG. 7–5** Plug flow   ■ **FIG. 7–6** Turbulent flow

An adverse effect of fluids in turbulent flow is hole erosion. Significantly, the viscous flow properties of the mud have little effect on pressure losses in turbulent flow. This fact is used to determine changes in annulus pressure losses during routine drilling operations.

Equations for Calculating Pressure Losses

Laminar Flow   Pressure losses in laminar flow for the annulus may be calculated using either Eq. 7.17 or Eq. 7.18.

$$P_{la} = \left(\frac{2.4v}{D_h - D_{op}} \frac{2n+1}{3n} \right)^n \frac{Kl_f}{300(D_h - D_{op})} \tag{7.17}$$

$$P_{la} = \frac{l_f \theta_1}{300(D_h - D_{op})} \tag{7.18}$$

where:

l_f = length of annulus under consideration, ft
θ_1 = viscometer reading at a specific shear rate that equals the annular fluid velocity

The annular velocity, v, can be calculated using Eq. 7.19

$$v = \frac{24.5Q}{D_h^2 - D_{op}^2} \tag{7.19}$$

where:

v = annular velocity, fpm
Q = volume flow rate, gpm

Pressure losses in turbulent flow are determined from empirical correlations. In Fig. 7–20 an equation for the straight-line portion of the curve in turbulent flow is shown as follows:

$$f = \frac{0.046}{Re^{0.2}} \tag{7.20}$$

where:

Re = Reynolds number, dimensionless

Substituting Eq. 7.20 into Eqs. 7.10 and 7.11 gives the following equation for turbulent flow pressure losses inside the pipe:

$$P_t = \frac{7.7(10^{-5})\rho^{0.8}Q^{1.8}PV^{0.2}l_f}{D_i^{4.8}} \tag{7.21}$$

where:

PV = plastic viscosity, cp
D_i = inside pipe diameter, in.

The plastic viscosity enters Eq. 7.21 through the Reynolds number. There are no actual viscous effects in turbulent flow; however, an increase in solids content increases pressure losses in turbulent flow. The plastic viscosity is used to determine a change in the solids content of the drilling mud. An approximation of this effect is shown in Eq. 7.22:

$$\mu_T = \frac{PV}{3.2}$$ (7.22)

where:

μ_T = solids content effect on fluid properties in turbulent flow
$PV = \theta_{600} - \theta_{300}$

To calculate annular pressure losses in turbulent flow, Eq. 7.23 should be used.

$$P_{ta} = \frac{7.7(10^{-5})\rho^{0.8}Q^{1.8}PV^{0.2}l_f}{(D_h - D_{op})^3 \ (D_h + D_{op})^{1.8}}$$ (7.23)

Turbulent flow pressure losses are seldom calculated. Instead, most operators utilize either charts or hydraulic slide rules. The charts generally are prepared using the following assumptions:

- Mud weight, ρ, = 10.0 lb/gal
- μ_T = 3.0 cp or PV = 9.6 cp

To correct for a different mud weight, often the pressure loss for a 10-lb/gal mud is multiplied by a ratio of the mud weight increase. If that is the procedure used, multiplying the pressure loss for a 10-lb/gal mud by a ratio of the mud weights to the 0.8 power would be more accurate.

Commonly included with these charts for turbulent flow pressure losses is a correction for plastic viscosity. This correction is based on the assumption that the plastic viscosity is 9.6 cp. The effect of plastic viscosity is shown in Eqs. 7.21 and 7.23.

Bit Nozzle Pressure Losses

Pressure losses through the bit nozzles may be calculated using Eq. 7.24.

$$P_b = \frac{\rho v_n^2}{1,120}$$ (7.24)

where:

P_b = pressure loss through bit nozzles, psi
v_n = nozzle velocity, fps

The nozzle velocity may be calculated using Eq. 7.25.

$$v_n = \frac{0.32Q}{A_n}$$ (7.25)

where:

A_n = area of all nozzles in use, sq in.

Applications

Example 7.1 shows the general procedure for calculating annular pressure losses in laminar flow using Eq. 7.17 for power-law behavior and Eq. 7.18 where shear stress is determined directly.

☐   **EXAMPLE 7.1**

Well depth = 10,000 ft
Hole size = 8½ in.
Drillpipe size = 4½-in. OD
Rate of circulation = 300 gpm
Mud weight = 10 lb/gal
Viscometer data:

$\theta_{600} = 127$
$\theta_{300} = 94$
$\theta_{200} = 78$
$\theta_{100} = 54$
$\theta_{6} = 8$
$\theta_{3} = 5$

SOLUTION:

The viscometer readings versus the viscometer speed (in rpm) for Example 7.1 are plotted in Fig. 7–7. Note that the readings at 200, 300, and 600 form a straight line in Fig. 7–7, and the readings at 3, 6, and 100 form a straight line. If only a two-speed viscometer were available, the calculated shear stress for any annular shear rate less than a 100 rpm or 170 sec^{-1} would be higher than that predicted by the actual measurements.

The annular pressure losses will be calculated by Eqs. 7.17 and 7.18. From Fig. 7–7, n and K are determined as follows:

$$n = 3.32 \log \frac{\theta_{100}}{\theta_{50}} = 3.32 \log \frac{54}{34} = 0.666$$

$$K = \frac{\theta_{100}}{170^{0.666}} = \frac{54}{170^{0.666}} = 1.77$$

$$v = \frac{(24.5)(300)}{(8.5)^2 - (4.5)^2} = 141 \text{ fpm}$$

$$P_{la} = \left[\frac{(2.4)(141)(1.166)}{4} \right]^{0.666} \frac{(1.77)(10,009)}{(300)(4)} = 314 \text{ psi}$$

Using Eq. 7.18 the shear rate (in rpm) must be determined.

$$\gamma_{rpm} = \frac{(1.41)(141)(1.166)}{4} = 58 \text{ rpm}$$

$$\theta_{58} \text{ (Fig. 7–7)} = 37 \text{ lb}_f/100 \text{ sq ft}$$

$$P_{la} = \frac{l_f \theta_{58}}{300(D_h - D_{op})} = \frac{(10,000)(37)}{(300)(4)} = 308 \text{ psi}$$

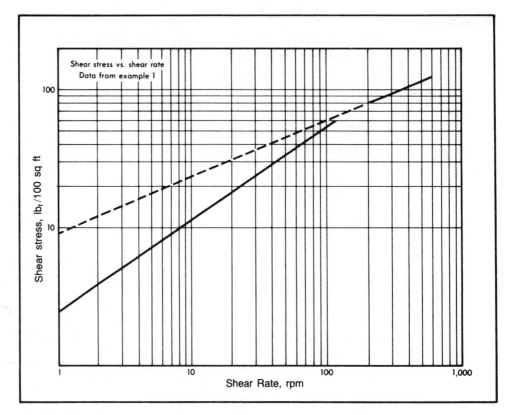

■ **FIG. 7–7** Shear stress vs shear rate (data from Example 7.1)

The difference between 308 and 314 psi is simply the reading accuracy from plotted data. The answers should be exactly the same.

Another procedure for determining annular pressure losses involves rearranging Eq. 7.9 as shown in Eq. 7.26:

$$P_{at} = P_t - P_{sc} - P_{dp} - P_{dc} - P_b \qquad (7.26)$$

where:

P_{at} = total pressure loss in annulus, psi

The annular pressure loss, P_{at}, may be determined by reading the standpipe pressure directly and subtracting the calculated pressure losses inside the drillstring and bit.

The advantages of this procedure are: (1) The inside measurements of the drillstring are more precise than hole-size estimates; (2) Pressure losses through the bit are considered accurate; (3) Flow patterns inside the drillstring and through the bit are generally turbulent, and viscous flow properties have a minor effect on the total pressure loss in this part of the string; (4) Flow patterns in the annulus are generally laminar, and viscous flow properties of the mud affect pressure losses substantially. Thus, downhole changes in annular mud properties result in higher pressure readings on the standpipe pressure gauge.

These same changes in mud properties have only a minor effect on pressure losses inside the drillstring and through the bit. Therefore, in most cases if the standpipe pressure at a given flow rate increases above normal levels, most of the increase in pressure loss is in the annulus. This, of couse, assumes no mechanical problems or bit plugging.

The disadvantages of using Eq. 7.26 are: (1) Most of the pressure loss in the circulating system occurs inside the drillstring and through the bit. Thus, a 5% error in these calculations results in a potential 30–40% error in the determination of annular pressure losses; (2) The procedure is limited to determining total annular pressure losses, and the point of interest is often just below the last casing seat to the surface, not from total depth to the surface.

Considering the advantages and disadvantages of Eq. 7.26, we believe the greatest advantage in using Eq. 7.26 is to determine changes in annular pressure and to check for substantial variations between calculated annular pressure losses and those obtained using Eq. 7.26.

The accuracy of annular pressure loss equations is limited because the flow properties of the mud must be measured at a given surface temperature and these measurements are used in the equations. Because the flow properties of the mud are sensitive to temperature changes, this practice of using surface flow properties introduces a potential error. However, a comparison with downhole measurements of actual annular pressure losses has shown calculated values surprisingly accurate at times.

Measurements that have been inaccurate are those where the downhole mud is thick due to flocculation. Also, calculated values immediately following trips are generally inaccurate because downhole properties may differ substantially from those measured at the surface.

Example 7.2 shows a comparison between calculating annular pressure losses and using Eq. 7.26.

□   **EXAMPLE 7.2**

Assumed Conditions

Well depth = 12,000 ft
Casing program:
13⅜-in. surface casing at 3,500 ft
9⅝-in. protective casing at 9,000 ft, ID = 8.92 in.
8½-in. hole to 12,000 ft

Drillstring

Drill collars: 90 ft of 6¼-in. OD, 2¾-in. ID
Heavyweight pipe: 1,500 ft, 4½-in. OD, 3-in. ID
Drillpipe: 10,410 ft, 4½-in. OD, 3.82-in. ID
Bit: 3½-in. nozzles

Mud properties at 120°F
Weight = 15 lb/gal
Filtration rate: 20cc @ 300°F
Viscometer readings: $\theta_{600} = 100$
$\theta_{300} = 60$
$\theta_{200} = 44$
$\theta_{100} = 22$
$\theta_6 = 4.4$
$\theta_3 = 3.0$

Hole Conditions

Essentially no hole enlargement
Fracture gradient just below casing seat at 9,000 ft = 0.82 psi/ft
Drilling in sand and shale

Circulating Conditions

Q = 350 gpm
Measured pressures versus circulation rate just before pulling pipe
are shown in Table 7–1.

■ **TABLE 7–1** Measured
standpipe pressures vs flow rates

Q, gpm	P_t psi
350	2,150
300	1,625
250	1,160
200	770

Determine:
    a. Total annular pressure loss at 350 gpm using Eq 7.17 and 7.18.
    b. Total annular pressure loss using Eq. 7.26.
    c. Annular pressure losses from 9,000 ft to the surface.
    d. The maximum permissible standpipe pressure while drilling
    at 12,000 ft with 15 lb/gal mud.

SOLUTION

Viscometer data are plotted in Fig. 7–8.
P vs Q data are plotted in Fig. 7–9.

(a) Total annular pressure loss at 350 gpm using Eq. 7.18.

$$n = 3.32 \log \frac{\theta_{100}}{\theta_{50}} = 3.32 \log \frac{22}{14.5} = 0.60$$

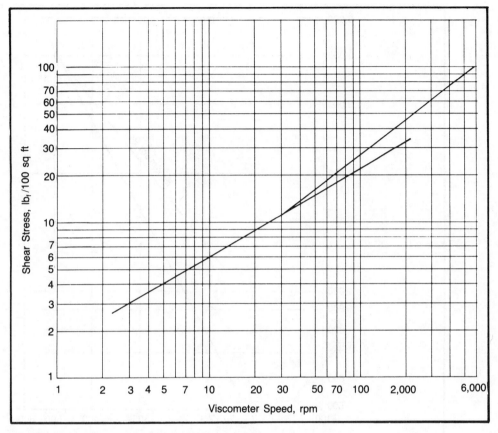

■ FIG. 7–8 Viscometer data for Example 7.2

$$K = \frac{22}{170^{0.6}} = 1.33$$

$$\frac{2n+1}{3n} = \frac{2.2}{1.8} = 1.33$$

8½-in. hole, drill collar annulus
Determine the flow pattern using Eq. 7.13.

$$v_c = \left[\frac{(2.585)(10^4)(1.33)}{15} \right]^{1/1.4} \left[\frac{(2.4)(2.2)}{(2.24)(1.8)} \right]^{0.6/1.4}$$

$$v_c = (250.7)(1.12) = 281 \text{ fpm}$$

The fluid velocity necessary to convert the flow pattern from laminar to turbulent is 281 fpm. The actual annular velocity is shown as follows:

$$v = \frac{(24.5)(350)}{72-39} = \frac{(24.5)(350)}{33} = 260 \text{ fpm}$$

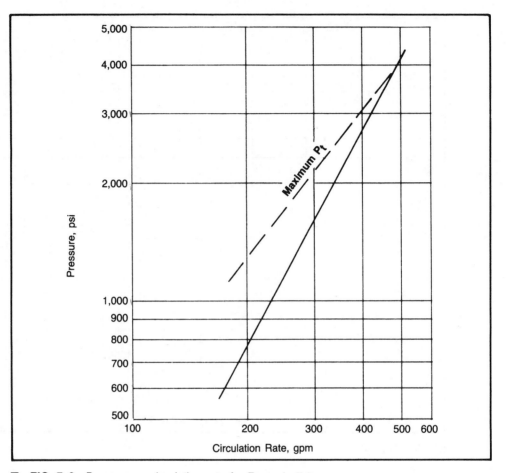

■ **FIG. 7–9** Pressure vs circulation rate for Example 7.2

This shows the flow pattern around the drill collars to be laminar. If it is laminar around the drill collars, then the flow pattern in other parts of the annulus is assumed to be laminar.

$$\text{Shear rate} = \frac{(1.41)(260)(1.22)}{2.25} = 199 \text{ rpm}$$

Shear stress from Fig. 7–3 = 44 lb$_f$/100 sq ft

$$P_{dc} = \frac{(90)(44)}{(300)(2.25)} = 6 \text{ psi}$$

8½-in. hole, drillpipe annulus

$$v = \frac{(24.5)(350)}{72.3 - 20.3} = \frac{(24.5)(350)}{52} = 165 \text{ fpm}$$

$$\text{Shear rate} = \frac{(1.41)(165)(1.22)}{4} = 71 \text{ rpm}$$

Shear stress from Fig. 7–3 = 18 $lb_f/100$ sq ft

$$P_{dp} = \frac{(2{,}910)(18)}{(300)(4)} = 44 \text{ psi}$$

Casing, drillpipe annulus

$$v = \frac{(24.5)(350)}{79.6 - 20.3} = \frac{(24.5)(350)}{59.3} = 145 \text{ fpm}$$

$$\text{Shear rate} = \frac{(1.41)(145)(1.22)}{4.42} = 56 \text{ rpm}$$

Shear stress from Fig. 7–3 = 15.5 $lb_f/100$ sq ft

$$P_{dp} = \frac{(9{,}000)(15.5)}{(300)(4.42)} = 105 \text{ psi}$$

$$P_{at} = 6 + 44 + 105 = 155 \text{ psi}$$

(b) Total annular pressure loss determined by subtracting circulating system losses from the standpipe pressure are shown in Table 7–2.

■ **TABLE 7–2**   Annular pressure losses vs flow rate

Q, gpm	P_t, psi	P_{sc}	P_{dp}	P_{hw}	P_{dc}	P_b	P_{at}
350	2,150	44	937	435	35	490	209
300	1,625	33	708	330	30	360	164
250	1,160	24	510	238	22	250	116
200	770	16	338	159	14	160	83

(c) Annular pressure losses from 9,000 ft to the surface. From part a, P_{dp} at 9,000 ft = 105 psi.

The 105 psi represents 68% of the total annular pressure drop of 155 psi. Using this percentage in part b, P_{at} from 9,000 ft to the surface is shown in Table 7–3.

■ **TABLE 7–3**   Estimated annular pressure losses from 9,000 ft to surface

Q	P_{at} at 9,000 ft
350	142
200	112
250	79
200	56

(d) The maximum permissible standpipe pressure from 9,000 ft to surface while drilling at 15,000 ft with 15 lb/gal mud.

$$\begin{aligned} \text{Fracture gradient} &= 0.82 \text{ psi/ft} \\ \text{Hydrostatic pressure} &= \underline{0.78 \text{ psi/ft}} \\ \text{Difference} &= 0.04 \text{ psi/ft} \end{aligned}$$

Permissible annular pressure loss, 9,000 ft to surface $= \dfrac{(9,000)(0.04)}{0.68} = 529$ psi

Using the annular pressure data from part c, the maximum standpipe pressure is shown in Table 7–4.

■ **TABLE 7–4** Maximum permissible standpipe pressure

Q, gmp	P_t	P_{at}	Permissible P_{at}	Additional P_{at}	Maximum Permissible P_t
350	2,150	142	529	387	2,537
300	1,625	112	529	417	2,042
250	1,160	79	529	450	1,610
200	770	56	529	473	1,243

Note that the maximum permissible surface pressure is based on the annular pressure losses determined using Eq. 7.26 rather than the annular pressure losses using either Eq. 7.17 or 7.18. The only reason for selecting the higher annular pressures is safety. An operator may decide Eq. 7.17 or 7.18 provides a more accurate representation of annular pressure losses.

Using Eq. 7.21 for turbulent flow pressure losses inside drillpipe is shown in Example 7.3.

□   **EXAMPLE 7.3**

Given:

$$\begin{aligned} \text{Drillpipe size} &= 4\tfrac{1}{2}\text{-in. OD and 3.78-in. ID} \\ \text{Drillpipe length} &= 1,000 \text{ ft} \\ \text{Mud weight} &= 10 \text{ lb/gal} \\ Q &= 300 \text{ gpm} \\ PV &= 25 \text{ cp} \end{aligned}$$

Determine the pressure loss in psi.

SOLUTION:

$$P = \frac{7.7(10^{-5})(10)^{0.8}300^{1.8}25^{0.2}(1,000)}{3.78^{4.8}}$$

$$P = \frac{7.7(10^{-5})(6.31)(2.892)(10^4)(1.902)(10^3)}{592} = 45.1 \text{ psi}$$

A hydraulic slide rule manufactured and distributed by a service company shows a pressure loss of 43.5 psi for Example 7.3.

These equations for pipe flow are considered accurate enough for general use. Charts and hydraulic slide rules available from major service companies are generally reliable for determining turbulent-flow pressure losses inside pipe. There are small variations in calculated pressures from these sources because of variations in the friction factor relationship used by each. Because pressure losses in turbulent flow depend on the degree of turbulence, which is affected by fluid type and pipe roughness, a general solution for pressure losses in turbulent flow is difficult to develop.

Many polymers suppress the degree of turbulence and thus reduce pressure losses in turbulent flow. To determine the degree to which these polymers are effective requires experimental testing and a wide range of results from this testing. Many companies selling polymers claim to have done this type of testing and have published results. These results should be checked in field operations when possible for accuracy.

Annular flow Annular pressure losses in turbulent flow are difficult to determine accurately. It is also difficult to determine if flow in the annulus is laminar or turbulent. In laminar-flow pressure-loss calculations, fluid velocity affects the pressure loss less than it does in turbulent flow. Because annular velocity is affected by hole size, determining annular velocities in open-hole sections with any degree of accuracy may be difficult.

Caliper surveys of open-hole sections are helpful, but unless a four-point caliper is used, the accuracy of caliper surveys is questionable. These questions are introduced not to discourage the determination of flow patterns and turbulent-flow pressure losses in the annulus but to emphasize the importance of careful analysis when these pressures become critical to further drilling.

If the flow pattern in the annulus is determined to be turbulent, then Eq. 7.23 should be used to determine annular pressure losses. Again, solutions to equations similar to Eq. 7.23 are available, using charts and hydraulic slide rules with the limitations already mentioned.

Bit Nozzle Pressure Losses

Pressure losses through the bit nozzles may be calculated using Eq. 7.24. Example 7.4 illustrates the procedure for calculating bit nozzle pressure losses.

□ **EXAMPLE 7.4**

Assume:

Flow rate, Q, = 350 gpm
Bit nozzle size = $^{12}/_{32}$ in.
Total nozzles in use = 3
Mud weight = 10.0 lb/gal

Determine the pressure loss through the bit.

SOLUTION:

Use Eq. 7.25 to calculate the nozzle velocity (in fps).

$$v_n = \frac{(0.32)(350)}{0.331} = 338 \text{ fps}$$

$$P_b = \frac{(10)(338^2)}{1,120} = 1,020 \text{ psi}$$

A hydraulic slide rule commonly used in the oil industry shows the same bit nozzle pressure losses. In most cases charts or slide rules are considered accurate for determining bit nozzle pressure losses because both follow the same basic equations. Also, the bit nozzle pressure losses are directly proportionate to mud weight, and the pressure losses inside the pipe are not, as shown by Eq. 7.23.

Practical Application

All of the calculations for pressure losses may seem to be unnecessary in most field drilling operations. While this may be true, there are times when calculations of pressure losses may reduce trial-and-error situations that are expensive, at best, and that may result in losing the well, at the worst. This type of situation is shown in Example 7.5, which illustrates with minor modifications an actual field condition.

□   **EXAMPLE 7.5**

Directional well
Projected well depth = 16,000 ft
Drilling depth = 14,750 ft
Hole size = 8½ in.
Drillpipe size = 4½ in.
Mud weight = 10.7 lb/gal
10 lb/bbl of lost circulation material
Formation type at 14,750 ft = shale
Partial loss of circulation has happened in a formation at 12,500 ft.
Shale sloughing has occurred with numerous bridges and fill on bottom blocking drilling progress.

Attempts to orient and run a downhole motor have been unsuccessful; there has been no drilling progress for two months. Mud viscosity is too thick to measure through a funnel.

Circulation rate = 300 gpm
Annular velocity = 141 fpm

SOLUTION:

Thin the mud, use viscous sweeps, and raise the mud weight to 11.0 lb/gal. Use a variable-speed viscometer to determine circulating pressure losses, and use a simple approach, assuming $n = 0.7$.

Utilize Eq. 7.9 to determine the viscometer speed that equals the annular velocity.

$$\gamma = \frac{(1.61)(141)}{8.5 - 4.5} = 50 \text{ rpm}$$

Viscometer reading at 50 rpm:

$$\theta_{50} = 42 \text{ cp}$$

The circulating pressure loss in the annulus from 12,500 ft to the surface is determined by Eq. 7.18.

$$P_{la} = \frac{(12,500)(42)}{(300)(4)} = 438 \text{ psi}$$

The mud was thinned until the viscometer reading at 50 rpm was 13.0 cp, and viscous sweeps of mud were used to clean the hole. Also, the mud weight was increased to 11.0 lb/gal to help control shale sloughing.

The circulating pressure loss in the annulus with the thinner mud is as follows:

$$P_{la} = \frac{(12,500)(13)}{(300)(4)} = 135 \text{ psi}$$

The increase in hydrostatic pressure due to the increase of 0.3 lb/gal in mud weight is shown below:

$$\frac{(12,500)(0.3)}{19.25} = 194 \text{ psi}$$

The circulating pressure loss in the annulus was reduced as follows:

$$438 - 135 = 303 \text{ psi}$$

The reduction in total pressure against the formation at 12,500 ft while circulating is given by:

$$303 - 194 = 109 \text{ psi}$$

Note: The viscous sweeps were as thick as the centrifugal precharge pump would pick up. The viscous sweeps were pumped 10 bbl at a time and were spaced to allow one sweep for each complete cycle of the mud. In gauge hole the feet of annulus covered by the viscous sweep is as follows:

$$\frac{10 \text{ bbl/ft}}{0.05 \text{ bbl}} = 200 \text{ ft}$$

The lost-circulation material was screened out of the mud system. After four complete cycles of the mud, the downhole motor was run and drilling continued. Total time required to correct the two-month-old problem was two days. The quick solution was possible because trial-and-error attempts were eliminated, and a specific solution was used.

Summary of Pressure Losses in the Circulating System

This discussion has emphasized accurately determining annular pressure losses. It is difficult to determine annular pressure losses accurately because of (1) hole size variations, (2) changes in fluid behavior, (3) changes in mud properties due to temperature changes, and (4) a general lack of concern with the problem. Reason 4 is changing because methods have been developed to determine pore pressures and fracture gradients more accurately, which places a premium on the determination of the total annulus pressure imposed at any point.

Many times specific calculations are made to estimate changes rather than quantitative magnitudes. This was shown in the field example in which a two-month-old problem was solved in two days.

LIFTING CAPACITY OF DRILLING FLUIDS

Initially, the primary purpose of the drilling fluid was to remove formation solids continuously. No time was spent on a scientific evaluation of the carrying capacity of the fluid, and little effort was made to control the fluid properties. Even with the advancement of science in mud treating, operators ignored the lifting capacity of muds.

If an operator felt that the quantity of solids being removed was less than that being generated, he increased the fluid velocity or made the mud thicker. Thus, the days of high annular velocities and thick muds were introduced with an almost painless effort.

The introduction of jet bits in 1948 reversed the trend toward high circulation rates. Jet bits brought the need for increased pressure losses through the bit nozzles for adequate bottom-hole cleaning. One way of meeting this need was to reduce pressure losses in other parts of the drillstring by reducing circulation rates. Pressure losses were then increased through the bit by using small nozzle sizes.

This philosophy did not occur suddenly, but the needed direction was recognized. In 1950 Williams and Bruce published a paper on the lifting capacity of fluids, which has become a classic of technical literature.[1] Based on many months of controlled research on the lifting capacity of fluids, Williams and Bruce conclude the following:

1. Turbulent flow in the well annulus is the most favorable flow pattern for removing formation cuttings.
2. Low-viscosity or thin fluids are generally more desirable than thick fluids for good hole cleaning.
3. Pipe rotation aids in removing formation cuttings.
4. Annular fluid velocities of 100–125 fpm using water are adequate for removing formation solids.

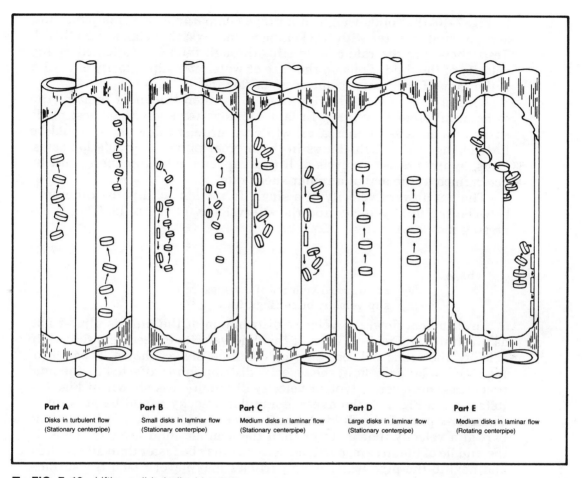

Part A	Part B	Part C	Part D	Part E
Disks in turbulent flow (Stationary centerpipe)	Small disks in laminar flow (Stationary centerpipe)	Medium disks in laminar flow (Stationary centerpipe)	Large disks in laminar flow (Stationary centerpipe)	Medium disks in laminar flow (Rotating centerpipe)

■ **FIG. 7–10** Lifting solids in liquid solutions

These conclusions are based on tests performed in an actual well and in plastic pipe in the laboratory. Fig. 7–10 shows results of laboratory tests where the actual behavior of the transported solids could be photographed. Parts A, B, and C show the reason for conclusion 1—the solids are being removed continuously while the fluid is in turbulent flow, as shown in part A. In parts B and C, the solids next to the centerpipe and the wall of the hole are falling back. Part E shows the effect of pipe rotation and the advantages for lifting solids next to the rotating pipe. Part D simply illustrates the lifting of large disks, which are too thick to be turned on their side next to the wall of the pipe and slip back as shown in parts B and C.

Formation cuttings weigh about 21.0 lb/gal, which means the solids tend to slip downward through any fluid that is lighter than 21.0 ppm.

In Fig. 7–10, the solids weigh in a range comparable to formation solids, and the fluid is water with thickening agents, which weighed 8.33 lb/gal. Tests show that the rate a solid slips through fluid is unaffected by the velocity of the fluid. Thus, in the case of water, the slip velocity of a solid is the same if the water is quiescent or flowing.

This is also true for drilling mud if the mud thickness remains the same at different flow rates. In laminar flow, we know that most muds shear thin as velocity increases, while in turbulent flow there would be little change in mud thickness in the normal range of circulation rates. Thus, in any precise analysis of lifting capacity, mud thickness must be determined at the actual annular shear rate.

The net upward velocity of a single formation solid is the difference between the fluid velocity and the slip velocity of the solid. This is expressed mathematically in Eq. 7.27:

$$v_p = v - v_{sc} \tag{7.27}$$

where:

v_p = net upward velocity of cuttings, fpm
v_{sc} = slip velocity of cuttings, fpm

In turbulent flow, the velocity distribution of the fluid is almost flat, as illustrated in Fig. 7–6. Thus, if the fluid velocity exceeds the solids slip velocity, the solids will be removed continuously, as shown in part A of Fig. 7–10. In laminar flow, the velocity distribution is affected by the mud properties; however, a typical velocity distribution is shown in Fig. 7–4. Referring to Fig. 7–4, the average annular velocity would be at point A. Therefore, part of the fluid is traveling at a velocity that is higher and part at a velocity that is lower than the average fluid velocity. Thus, in the middle of the stream, particles recovery may be faster than anticipated, and next to the pipe walls some particles may never reach the surface.

The practice of injecting dye or oats in the mud to determine travel time from bottom gives erroneous results if the fluid is in laminar flow. Some of the dye or oats will be in the middle of the stream and the time from bottom will be substantially less than the normal time based on the average annular velocity.

The importance of the velocity distribution was recognized by Williams and Bruce, and some quantitative effects were later emphasized by Walker in 1963.[2] Walker considered the velocity distribution as a function of the mud properties. In general, an increase in the ratio of yield point to plastic viscosity or a decrease of the line slope, n, for the power law behavior of fluids results in a flattening of the velocity profile as shown in Fig. 7–5.

The flat part of the velocity profile shown in Fig. 7–5 is referred to as plug flow. In the plug flow region there is no shearing of fluid layers. Thus, as the width of the plug flow region increases, low-velocity parts of the stream decrease, and the cleaning action of the fluid is increased. Equations may be developed to determine velocity profiles.

However, an estimate of the shape of an annulus profile is impossible because pipe rotation disturbs the profile, the drillstring is not concentric in the hole, and the shape of the hole is irregular. It is enough to know that increases in the yield point or a decrease in the n factor will flatten the velocity profile and assist in hole cleaning.

The actual lifting capacity of a fluid is related directly to the rate a solid slips through the fluid. The slip velocity of a particle may be estimated using Eq. 7.28:

$$V_{sc} = 113.4 \left[\frac{D'_p(\rho_p - \rho)}{C_d\rho} \right]^{1/2} \tag{7.28}$$

where:

D'_p = equivalent diameter of cuttings, in.
ρ_p = weight of cuttings, lb/gal
C_d = drag coefficient, dimensionless
ρ = mud weight, lb/gal

The particle diameter may be estimated from a visual inspection, or if more precision is needed, the equivalent diameter may be determined by a screen analysis. The particle density is generally considered constant at 21.0 lb/gal for Eq. 7.28. The drag coefficient is the frictional drag between the fluid and the particle.

No methods are available to determine this frictional drag precisely; however, Fig. 7–11 shows the drag coefficient curve vs the particle Reynolds number. Fig. 7–11 has been prepared using limestone and shale cuttings from field drilling operations. Eq. 7.28 is used to calculate the drag coefficient after the slip velocity has been determined experimentally. The particle Reynolds number is calculated using Eq. 7.29:

$$R_p = \frac{15.47\rho v_{sc}D'_p}{\mu} \tag{7.29}$$

where:

R_p = particle Reynolds number, dimensionless
μ = equivalent thickness of drilling mud, cp

Note: The equivalent particle diameter (in inches) may be determined using screens. The equivalent thickness is determined as shown in chapter 5.

Water and glycerin mixed with water are used as the fluid base for determining Fig. 7–11. Above a particle Reynolds number of approximately 2,000 the drag coefficient remains constant at about 1.50. Thus, when the flow pattern around the particle is turbulent, a drag coefficient of 1.50 can be used in Eq. 7.28 and the slip velocity calculated directly. When the fluid flow pattern is laminar around the particle, the drag coefficient varies with the particle Reynolds number. For a particle Reynolds number of 1.0 or less, the drag coefficient can be determined using Eq. 7.30:

$$C_d = \frac{40}{R_p} \tag{7.30}$$

Substituting this value of the drag coefficient in Eq. 7.28 gives Eq. 7.31 for the slip velocity:

$$v_{sc} = \frac{4,980 D_p^2 (\rho_p - \rho)}{\mu} \tag{7.31}$$

Eq. 7.31 has limited application because in most cases the particle Reynolds number exceeds 1.0 when the flow around the particle is laminar. Drawing the best approximate straight line between the particle Reynolds numbers 10 and 100 gives Eq. 7.32 for the drag coefficient:

$$C_d = \frac{22}{R_p^{0.5}} \tag{7.32}$$

Using this value of drag coefficient in Eq. 7.28 gives Eq. 7.33 for the slip velocity:

$$v_{sc} = \frac{175 D_p (\rho_p - \rho)^{0.667}}{\rho^{0.333} \mu^{0.333}} \tag{7.33}$$

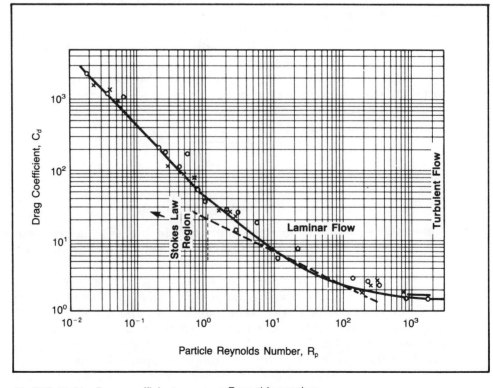

■ **FIG. 7–11** Drag coefficient curve vs Reynolds number

Eq. 7.33 should be used in routine solutions. After determining slip velocity, the particle Reynolds number may be calculated. If the particle Reynolds number is more than 2,000, Eq. 7.28 should be used with a drag coefficient of 1.50. If the operator is in doubt about whether the flow around the particle is turbulent or laminar, the equation giving the lowest slip velocity for the specific problem should be used. This procedure always provides the most nearly correct solution.

The use of these equations is illustrated by Examples 7.6–7.8. Example 7.6 shows the slip velocity when the fluid around the particle is in turbulent flow. Example 7.7 shows the particle slip velocity in a thick low-weight mud, and Example 7.8 shows the particle slip velocity in a weighted mud.

☐   **EXAMPLE 7.6:**

Well depth = 10,000 ft
Hole diameter = 7⅞ in.
Drillpipe diameter = 4½ in.
Fluid = water weight = 8.33 lb/gal
Shale cuttings
Weight of formation solids = 21 lb/gal
Average particle size = 0.3 in.

SOLUTION (using Eq. 7.28):

$$v_{sc} = 113.4 \left[\frac{0.3(32 - 8.33)}{(1.5)(8.33)} \right]^{1/2} = 63 \text{ fpm}$$

☐   **EXAMPLE 7.7:**

Well depth = 10,000 ft
Hole diameter = 7⅞ in.
Enlarged sections of hole = 20 in.
Drillpipe diameter = 4½ in.
Fluid, weight = 9.5 lb/gal
Viscometer = $\theta_{600} = 100$
$\theta_{300} = 60$
Weight of formation solids = 21 lb/gal
Average particle size = 0.3 in.
Circulation rate = 300 gpm

SOLUTION (using Eqs. 7.28, 7.31, and 7.33):

Eq. 7.28

$$v_{sc} = 113.4 \left[\frac{(0.3)(21 - 9.5)}{(1.5)(9.5)} \right]^{1/2} = 56 \text{ fpm}$$

Eq. 7.31

$$v_{sc} = \frac{(4,980)(0.09)(21 - 95)}{89} = \frac{(4,980)(0.09)(11.5)}{89} = 57 \text{ fpm}$$

$$\mu = \left[\left(\frac{2.4\overline{v}}{D_h - D_p}\right)\left(\frac{2n + 1}{3n}\right)\right]^n \frac{200K(D_h - D_p)}{v}$$

$\overline{v} = 175$ fpm for 7⅞-in. hole

$$n = 3.32 \log \frac{100}{60} = (3.32)(0.223) = 0.74$$

$$K = \frac{60}{511^{0.74}} = \frac{60}{101} = 0.595$$

$$\mu = \left[\frac{(2.4)(175)(1.12)}{3.375}\right]^{0.74} \frac{(200)(0.595)(3.375)}{175}$$

$$\mu = (39)(2.29) = 89 \text{ cp}$$

Eq. 7.33

$$V_{sc} = \frac{(175)(0.3)(21 - 9.5)^{0.667}}{(9.5)^{0.333}(89)^{0.333}} = \frac{268}{9.4} = 28.5 \text{ fpm}$$

Eq. 7.33 represents the most nearly correct solution. Note that the slip velocity is about one-half that calculated by Eqs. 7.31 and 7.28. This shows the fluid to be in laminar flow around the particle. These solutions would be changed because of the change in equivalent viscosity in the enlarged hole section. Considering the 20-in. hole, the slip velocity is determined as follows using Eq. 7.33:

$$\mu = \left[\frac{(2.4)(19)(1.12)}{15.5}\right]^{0.74} \frac{(200)(0.595)(15.5)}{19}$$

$$\mu = (2.42)(97.1) = 235 \text{ cp}$$

$$V_{sc} = \frac{(175)(0.3)(11.5)^{0.667}}{(9.5)^{0.333}(235)^{0.333}} = \frac{(175)(0.3)(5.1)}{(2.12)(6.16)} = 20.5 \text{ fpm}$$

The mud is substantially thicker, 235 cp compared with 89 cp, in the larger hole because of the reduction in shear rate. However, in the gauge hole section the average annular velocity of 175 fpm is substantially above the slip velocity of 28.5 fpm, so there should be no problem cleaning the hole. In the enlarged hole section, the average annular velocity of 19 fpm is just below the slip velocity of 20.5 fpm; thus, the mud would have to be thickened to remove particles that are 0.3 in. in diameter or larger. The average annular velocity is the velocity at only one point in the flow stream; part of the fluid is moving at a higher velocity, and part of the fluid is moving at a lower velocity.

This introduces the question, what is the required average annular velocity for adequate hole cleaning? To give a complete answer, the velocity profile would have to be determined, the size of the cuttings to be removed would have to be known, and downhole mud thickness would have to be predicted. In most practical applications there is no way to make these precise determinations; thus, the required lifting capacity is generally based on observations of hole cleaning while the drilling operation is underway.

Again, consider the case of the 20-in. hole. In this case the slip velocity was slightly greater than the annular velocity. This condition could result in bridges in the hole and plugging of the annulus; regardless of the calculations made, the operator would be aware of the problem. If corrections are not made quickly, he may suffer the potential consequences of stuck pipe or other associated problems. The useful part of the calculation is that the operator is able to define the precise problem and is aware of the required treatment.

We assume hole problems are being experienced in Example 7.7. The operator knows he must either increase the annular velocity or thicken the mud. Considering the problem to be associated directly with the enlarged hole section, we can also calculate the slip velocity using Eq. 7.31:

$$V_{sc} = \frac{(4,980)(0.09)(21 - 9.5)}{235} = 21.9 \text{ fpm}$$

The particle Reynolds number is determined using Eq. 7.29:

$$R_p = \frac{(15.47)(9.5)(21.9)(0.3)}{235} = 4.1$$

If the mud is thickened further, the slip velocity should be calculated using Eq. 7.31. We can see that the slip velocity in Eq. 7.31 is inversely proportional to viscosity, while in Eq. 7.33 the slip velocity is inversely proportional to viscosity to the 0.333 power. The required viscosity, assuming the slip velocity must be at least 25% less than the average annular velocity, is calculated as follows:

$$v_{sc} = (19)(0.75) = 14 \text{ fpm}$$

Using Eq. 7.31:

$$\mu = \frac{(4,980)(0.09)(11.5)}{14} = 368 \text{ cp}$$

This shows that for a 32% reduction in slip velocity, the mud thickness must be increased 57%. These calculations have been based on particle sizes of 0.3 in. in diameter. If the formations are sloughing, the cuttings may be substantially larger and the mud may have to be thickened even more. In any event such calculations provide a basis for solving the problem.

Lifting capacity with weighted muds is examined in Example 7.8.

☐ **EXAMPLE 7.8**

Well depth = 15,000 ft
Hole diameter = 7⅞ in.
Drillpipe diameter = 4½ in.
Maximum hole enlargement = 10 in.

Fluid weight = 16 lb/gal
Weight of formation solids = 21.0 lb/gal
Average particle size = 0.3 in.
Circulation rate = 250 gpm
Viscometer = $\theta_{600} = 90$
$\theta_{300} = 50$

SOLUTION (for the gauge hole section):

$$v = 143 \text{ fpm}$$

$$n = 3.32 \log \frac{90}{50} = (3.32)(0.255) = 0.85$$

$$K = \frac{50}{(511)^{0.85}} = \frac{50}{200} = 0.25$$

$$\mu = \left[\frac{(2.4)(143)(1.06)}{3.375}\right]^{0.85} \frac{(200)(0.25)(3.375)}{143}$$

$$\mu = (53)(1.179) = 62.4 \text{ cp}$$

Using Eq. 7.33

$$V_{sc} = \frac{(175)(0.3)(21-16)^{0.667}}{(16)^{0.333}(62.4)^{0.333}}$$

$$V_{sc} = \frac{(175)(0.3)(2.93)}{(2.5)(3.96)} = 15.5 \text{ fpm}$$

For the enlarged hole section

$$\overline{v} = 75 \text{ ft/min}$$

$$\mu = \left[\frac{(2.4)(75)(1.06)}{5.5}\right]^{0.85} \frac{(200)(0.25)(5.5)}{75}$$

$$\mu = (20.38)(3.67) = 74.7 \text{ cp}$$

$$V_{sc} = \frac{(175)(0.3)(21-16)^{0.667}}{(16)^{0.333}(75)^{0.33}}$$

$$V_{sc} = \frac{(175)(0.3)(2.93)}{(2.5)(4.2)} = 14.65 \text{ fpm}$$

The slip velocity of the cuttings changed very little from the gauged to the enlarged hole section. Average annular velocity in the enlarged hole section is 75 fpm, which is still five times higher than the calculated slip velocity. From this analysis we see that mud can be thinned substantially and still be thick enough to remove cuttings. In fact, the required viscosity is only that required to support barite. This is the normal case for weighted muds of 15 lb/gal or heavier because (1) the density of the mud provides lifting capacity and (2) hole enlargement in deeper hole sections where weighted muds are normally used is less than in the normal-pressure upper hole sections.

Downhole Mud Properties

Surface measurements of mud properties are not always indicative of downhole mud properties. Muds may thin with increased temperature, or they may thicken by flocculation. Unweighted muds are more likely to thin as the temperature increases, so what appears to be an adequate thickness at the surface may be insufficient in enlarged downhole sections. An example was noted while drilling in 600 ft of water offshore. The water temperature cooled the mud to about 60°F, and the measured thickness of the mud was considered adequate to remove cuttings. When the mud thickness was corrected for downhole temperature, calculations showed the mud was too thin to clean the hole. This corresponded with the actual problem at the well, where problems of hole cleaning were occurring daily. Fortunately, before an expensive, unnecessary corrective procedure was initiated, the real problem was discovered. After thickening the mud, at a small additional cost, problems of hole sloughing and hole cleaning were corrected.

Examples 7.9 and 7.10 show how temperature and flow rate affect mud thickness.

☐ **EXAMPLE 7.9**

Assume: Mud weight = 9.5 lb/gal

■ Viscometer readings at the following temperature

Viscometer Speed, rpm	Viscometer Readings, °F			
	80°	**120°**	**150°**	**200°**
600	100	54	39	(25)
300	60	32	23	(15)
200	45	24	16.5	(11)
100	27	14	10.5	(7)
6	3.5	1.9	1.40	(0.9)
3	2.1	1.1	0.77	(0.5)

Note: Data in parentheses were obtained from Fig. 7–12.
The data are plotted in Fig. 7–12.

Example 7.10 shows an estimate of mud thickness at 10,000 ft where the temperature is estimated to be 200°F.

☐ **EXAMPLE 7.10**

Assume: 100-rpm viscometer speed is equal to annular flow velocity, 3-rpm viscometer speed is equal to mud pit properties. Hole problems are occurring at 10,000 ft where the temperature of the mud is estimated to be 200°F.

Determine: The difference in mud thickness in the annulus while the mud is flowing at 200°F and the mud in the surface pits at 80°F

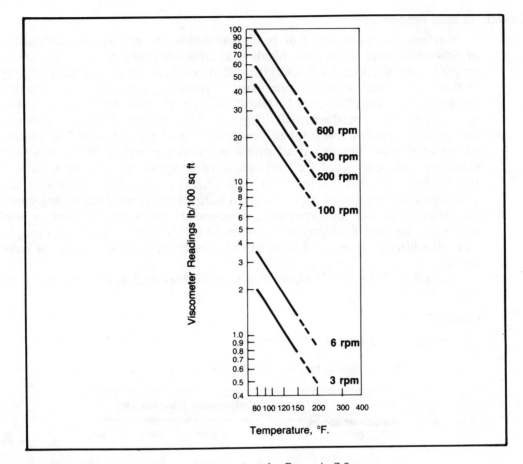

■ **FIG. 7–12**   Mud thickness vs temperature for Example 7.9

Solution: Using the data in Fig. 7–12, the mud thickness equals 23.5 cp in the annulus and equals 235 cp in the surface pits. The difference is so substantial that it can be readily seen why operators are often misled by the physical appearance of the mud in the surface mud tanks.

A better understanding of hole cleaning and mud properties will help operators evaluate whether some of the routine, time-consuming field practices are necessary.

One practice that seems to continue is the short trip before logging. Experience shows that logs are likely to go to bottom just as often when the pipe is pulled without making the short trip. If problems are encountered with getting logs to bottom, filter cake thickness should be controlled, and a batch of thick mud should be mixed and circulated for one cycle while drilling just before pulling pipe. Note that short trips may be necessary to determine the degree of potential swabbing in exploratory areas. This latter practice is a judgment decision, and the short trip may or may not be necessary.

Poor hole cleaning is responsible for 50% of hole problems. For some reason convincing operators to thicken the mud enough to clean the hole is almost impossible. Also, the problem may be more complicated. Thickening all of the mud may increase the annular circulating pressures enough to cause a loss of circulation. If all of the mud is thickened, it is possible that enlarged sections of the hole may be bypassed when circulating and give up cuttings to form bridges when circulation is stopped and the drillstring is pulled. This potential phenomenon is shown in Fig. 7–13.

When all of the muds get very thick, the thick mud will bypass the enlarged hole sections. When the pipe is pulled, the cuttings fall out of the enlarged hole into the gauge hole sections and form bridges. Better hole cleaning can be achieved by using viscous sweeps of mud. Viscous sweeps are mixed as follows:

- Mix up a batch of 50–100 bbl of mud to the level of thickness needed to remove cuttings. The upper limit of mud thickness is the point at which the centrifugal precharge pump will no longer pick up the mud.

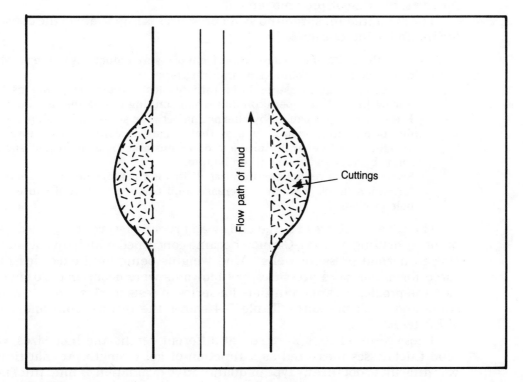

■ **FIG. 7–13** Cuttings removal from enlarged holes

- Pump enough of the viscous sweep to cover 200–300 ft of the annulus.
- The frequency of the viscous sweeps should be as needed to clean the hole; however, we suggest that only one batch of viscous mud be in the annulus at a time.

It is very difficult to understand why those in charge of wells are reluctant to use viscous sweeps when better hole cleaning is needed. One reason may be the frequently quoted concept of shocking the hole with a sudden batch of viscous mud. There is absolutely no foundation for this concept. In fact, experience in all cases shows the viscous sweeps to be essential to improving hole cleaning.

SURGE AND SWAB PRESSURES

Surge pressures are associated with fluid flow, caused by running equipment into a liquid-filled borehole. Swab pressures are associated with fluid flow, caused by pulling equipment out of a liquid-filled borehole. Procedures used for estimating the magnitude of these pressures are similar to those for estimating the pressure losses in conventional fluid circulation. To reduce calculation problems, swab pressure is estimated by calculating surge pressure and assuming that this is equal to the swab pressure for the same rate of pipe movement.

The magnitude of surge and swab pressures is important to the operator for the following reasons:

- More than 25% of blowouts result from pressure reductions in the borehole due directly to swabbing when pulling pipe.
- Excessive surge pressures have initiated lost circulation problems both during the drilling operation and when running casing into the hole.
- Pressure changes caused by alternating between surge and swab pressures due to pipe movements, such as those made on connections, cause hole sloughing and generally promote other unstable hole conditions such as solids bridges and solids fill on bottom.
- Swab pressure reductions may result in mud contamination by entry of formation fluids. This may increase mud treating costs and cause other hole problems.

The adverse effects of surge and swab pressures were recognized early in rotary drilling. In 1934, Cannon became concerned with blowouts occurring in normal pressure wells.[3] Mud weights being used exceeded measured formation pore pressures, yet blowouts were occurring. To investigate the problem, Cannon initiated a series of tests to measure the actual surge and swab pressures. Table 7–4 shows the results from one series of his tests.

These series of tests were run at different depths and hole sizes, with mud thicknesses measured as a function of gel strength. At that time it was difficult to obtain any type of quantitative evaluation of mud thickness. The magnitude of the pressure surges was surprising. At 7,000 ft, with a

■ **TABLE 7–4**   Surge and swab pressures

Annulus Size	Depth, ft	Gel Strength	Pressure Surge, psi
10¾-in. casing,			
4½-in. drillpipe	7,000	36	275
	7,000	12	125
	3,000	36	125
	3,000	12	62
7-in. casing,			
3½-in. drillpipe	7,000	60	487
	7,000	36	462
	7,000	6	362
	3,000	60	212
	3,000	36	200
	3,000	6	160

gel strength of 36, the surge pressure in the 7-in. hole using 3½-in. drillpipe was 462 psi. This is comparable to an increase in mud weight of about 1.3 lb/gal and, if considered when pulling pipe, easily shows the potential hazard of a blowout.

Under these conditions a normal weight mud of 10.0 lb/gal would be insufficient to control normal formation pore pressures of 9.0 lb/gal, which are common in the coastal areas. These tests also show that the surge pressures are directly proportional to depth. For example, note that the surge pressure in the 7-in. cased hole was 200 psi at 3,000 ft, compared with 462 psi at 7,000 ft for a gel strength of 36. This would be expected because the surge pressure is simply the pressure required to overcome friction at the displacement rate of the fluid.

In 1951, Goins et al. related the problems of lost circulation in coastal area wells directly to surge pressures associated with pipe movement.[4] They also noted that the surge pressures often exceeded equivalent mud weights of 1.0 lb/gal. They noticed that surge pressures could be reduced significantly by reducing the pipe running speed.

In 1953, Cardwell presented information on surge pressures and pulling suction and provided a chart for estimating the magnitude of these pressure changes.[5] This chart was related to hole and pipe geometry and pipe running speed. He assumed an equivalent viscosity of 300 cp for the mud. The chart is not included because subsequent work has shown that fluid properties may be even more important than running speed.

In 1956, Clark further emphasized the effect of surge pressures on problems of lost circulation and presented a good review on factors that contributed to the surge pressures.[6] He also presented a comprehensive mathematical analysis of the problem.

In 1960, Burkhardt presented an excellent résumé of how to determine

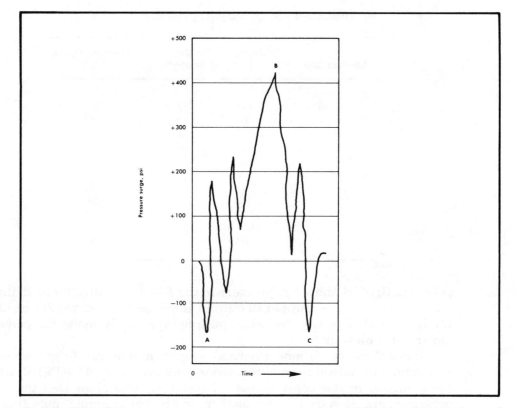

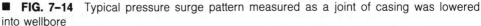

■ **FIG. 7–14** Typical pressure surge pattern measured as a joint of casing was lowered into wellbore

surge pressures.[7] His work included test results and a mathematical prediction method that compared favorably with actual pressure surge measurements. Fig. 7–14 shows a chart of the measured pressure changes in the borehole while running one joint of pipe at about 1,850 ft. Fig. 7–15 shows the velocity and acceleration of this joint of pipe as it is lowered into the hole. Lettering indicates the same time period on both figures.

At point A the pipe is lifted from the slips. The swab pressure reduction is almost 200 psi as noted in Fig. 7–14, and the pipe velocity is negative as shown in Fig. 7–15. At point B the surge pressure is 400 psi; note that this is the point of maximum pipe velocity. At point C swabbing again occurs; this point corresponds to the maximum deceleration of the pipe.

Of interest is the fact that the swab pressure reduction at point C is almost 200 psi. This occurred when running pipe into the hole. The deceleration pressure indicates a well can be swabbed when running pipe into the hole. Other changes in pressure are probably due to the changes in pipe speed.

Figs. 7–14 and 7–15 show that (1) pressure can change drastically when

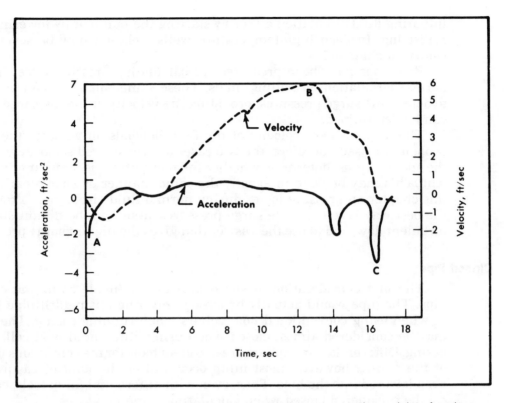

■ **FIG. 7–15** Typical pipe velocities and accelerations measured as a joint of casing was lowered into wellbore

running pipe, (2) running speed changes substantially when running pipe, and (3) pressure surges can be very high at shallow depths.

Surge pressures or swab pressure reductions may be caused by breaking the gel strength, viscous shear, fluid displacement in turbulence, or pipe acceleration and deceleration. These effects will be considered separately.

Surge Gel Factors

The surge pressure required to break the gel stength may be estimated using the definition of shear stress as a function of pressure. This relationship is shown in Eq. 7.34:

$$P_{su} = \frac{D_w \theta}{300(D_h - D_{op})} \qquad (7.34)$$

The gel strength in Eq. 7.34 is gel strength of the mud in $lb_f/100$ sq ft as determined from the rotating viscometer. The surge pressure recorded when breaking the gel strength is generally less than that experienced at the maximum pipe running speed, as indicated in Fig. 7–14. However, this may not always be the case, and the operator should be careful when initiating fluid circulation either by starting the pump or by lowering the drillstring. In deep high-temperature wells, gelation may be severe, as shown in chapter 5.

Equations have been presented for calculating pressure losses in the normal circulation of drilling muds. These same equations can be used to determine surge pressures, providing the velocity of the displaced fluid can be determined.

The displacement velocity of the fluid depends on whether the pipe is open or closed; therefore, the two cases are considered separately. Also, the operator must determine whether the flow pattern is laminar or turbulent. This may be done in exactly the same manner as shown earlier in this chapter on pressure losses during normal fluid circulation. Probably the best way is to calculate surge pressures, assuming both laminar and turbulent flow, and to use the answer that gives the highest surge pressure.

Closed Pipe

The first consideration of surge pressures is made assuming closed pipe. The pipe would actually be closed only when using drillpipe floats or when using conventional float collars when running casing. The pipe may be considered almost closed when using differential float collars in casing. Differential float collars are excellent tools for the continuous filling of the casing; however, most fillup occurs after the joint of casing has been lowered into the hole. Thus, even with differential fillup tools, casing can be considered closed when calculating surge pressures.

Fluid displaced by pipe volume is easily determined. However, friction losses are determined by relative movement of fluid to the pipe wall, so pipe movement into the hole must also be considered. The rate of fluid movement by the pipe is increased by the rate the pipe is lowered into the hole. The assumed velocity profile is shown by Fig. 7–16.

Burkhardt introduced the fact that the mud velocity when running pipe was due to pipe displacement and to mud clinging to the wall of the pipe. The mud clinging to the pipe wall displaced other mud up the annulus when pipe was run. Fig. 7–17 shows the measured mud-clinging constants as functions of the ratio of pipe diameter to hole diameter. As shown, the mud-clinging constants vary from 0.39 to 0.48 for laminar flow and from 0.44 to 0.50 for turbulent flow. Because the accuracy involved is somewhat questionable, assume that the mud-clinging constant is 0.45. From this assumption the fluid velocity when the pipe, equipped with float equipment, is being run may be calculated using Eq. 7.35.

$$v = \left(0.45 + \frac{D_p^2}{D_h^2 - D_{op}^2}\right)\overline{v}'_p \qquad (7.35)$$

where:

$\overline{v}'_{pi} = $ running velocity of the pipe, fpm

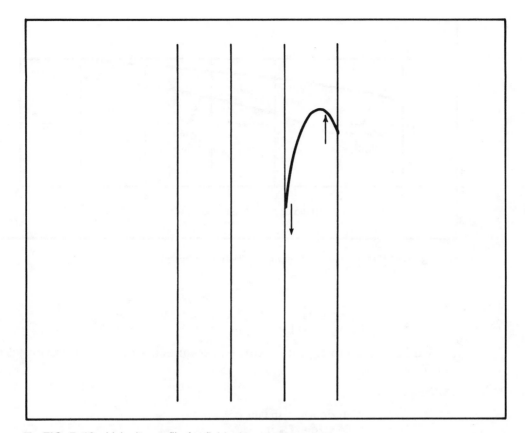

■ **FIG. 7–16**   Velocity profile for fluid when running pipe

The pipe running velocity used in Eq. 7.35 is the average running speed of the pipe. An accepted maximum running velocity is shown in Eq. 7.36.

$$v'_{pi} = 1.5\overline{v}'_{pi} \tag{7.36}$$

where:

$$v'_{pi} = \text{maximum pipe velocity, fpm}$$

Open Pipe

Pressure surges with open pipe are difficult to define. The relative flow in the annulus and inside the drillstring depends on the relative pressure forces and cross-sectional areas. Pressure forces in the annulus depend many times on mud properties, while those through the bit and inside the drillstring are generally independent of mud properties because the flow is turbulent. As a first step, assume that the pipe is completely open; for this purpose Eq. 7.37 can be used to calculate the displacement velocity:

$$v_d = \left[0.45 + \frac{D_{op}^2 - D_i^2}{D_h^2 - D_{op}^2 + D_i^2} \right] v_{pi} \tag{7.37}$$

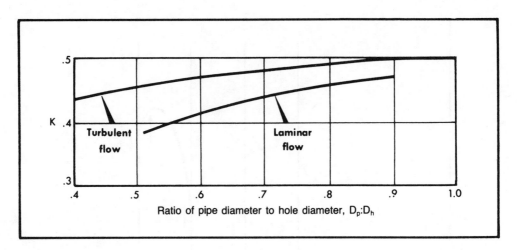

■ **FIG. 7–17** Mud "clinging" constant (from Burkhardt)

where:

v_d = displacement velocity, fpm
v_{pi} = pipe velocity, fpm

Methods used to calculate surge pressures are shown in Examples 7.11 and 7.12.

☐ **EXAMPLE 7.11**

Well depth = 15,000 ft
Hole size = 7⅞ in.
Drillpipe size = 4½ in.
Drill-collar length = 700 ft
Drill-collar size = 6¼-in. OD
          2¾-in. ID
Mud properties: Mud weight = 15 lb/gal
          Viscometer readings, $\theta_{600} = 140$
          $\theta_{300} = 80$
Average pipe running speed, $\nabla'_{pi} = 270$ fpm
Bit nozzle size = 3¹¹⁄₃₂ in.

Determine the surge pressure at 15,000 ft, assuming plugged pipe.

SOLUTION:

$$n = 3.32 \log \frac{140}{80} = (3.32)(0.243) = 0.805$$

$$K = \frac{80}{(511)^{0.805}} = \frac{80}{152} = 0.527$$

$$v'_{pi} = (1.5)(270) = 405 \text{ fpm}$$

Maximum fluid velocity is as follows:

$$v = \left(0.45 + \frac{20}{42}\right)(405) = 375 \text{ fpm}$$

Around drillpipe (14,200 ft)

Assume laminar flow:

$$P_{su} = \left[\frac{(2.4)(375)(1.08)}{(3.375)}\right]^{0.805} \frac{(0.527)(14.3)(10^3)}{0.3(10^3)(3.375)} = (95)(7.44) = 707 \text{ psi}$$

Assume turbulent flow:

$$P_{su} = \frac{(7.7)(10^{-5})(15^{0.8})(642^{1.8})(60^{0.2})(14.3)(10^{1.8})}{(3.375)^3(12.375)^{1.8}} = 690 \text{ psi}$$

Because P_{su} is about the same, the flow pattern may be either laminar or turbulent.

Around drill collars (700 ft)

$$v = \left(0.45 + \frac{39}{23}\right) 405 = 870 \text{ fpm}$$

At the higher velocity, the flow around the drill collars can be assumed to be turbulent.

$$P_{su} = \frac{7.7(10^{-5})(15^{0.8})(816^{1.8})(60^{0.2})(700)}{(1.625)^3(14.125)^{1.8}} = 375 \text{ psi}$$

The total annular surge pressure at 15,000 ft is shown as follows:

$$690 + 375 = 1,065 \text{ psi}$$

$$\text{Equivalent increase in mud weight} = \frac{(1,065)(19.25)}{15,000} = 1.37 \text{ lb/gal}$$

Next consider the same data as shown in Example 7.11 except leave the floats out of the drillstring. This set of conditions is shown in Example 7.12.

□ **EXAMPLE 7.12**

Assume:

The fluid velocity around the drillpipe equals the fluid velocity inside the pipe.

$$v'_{pi} = (1.5)(270) = 405 \text{ fpm}$$

$$v_{pi} = \left(0.45 + \frac{5.9}{56}\right) 405 = 225 \text{ fpm}$$

Assume laminar flow around the drillpipe because at a velocity of 375 fpm the flow pattern was at the transition point.

$$P_{su} = \left[\frac{(2.4)(225)(1.08)}{3.375}\right]^{0.805} \frac{(0.527)(14,300)}{(300)(3.375)} = 476 \text{ psi}$$

Fluid velocity around the drill collars:

$$v = \left(0.45 + \frac{31.5}{30.5}\right) 405 = 600 \text{ fpm}$$

Assume turbulent flow.

$$P_{su} = \frac{7.7(10^{0.5})(15^{0.8})(562^{1.8})(60^{0.2})(700)}{(1.625^3)(14.125^{1.8})} = 188 \text{ psi}$$

Total annular surge = 476 + 188 = 664 psi.

Now consider the pressure required to create the same fluid velocities inside the pipe.

The flow inside the pipe is laminar; for laminar flow inside pipe use Eq. 7.38 to determine pressure losses.

$$P_{su} = \left[\frac{1.6v}{D_i} \frac{3n+1}{4n}\right] \frac{KD_w}{(300)D_i} \tag{7.38}$$

$$P_{su} = \left[\frac{(1.6)(225)(1.06)}{3.82}\right]^{0.805} \frac{(0.527)(14,300)}{(300)(3.82)} = 268 \text{ psi}$$

Assume the flow pattern is turbulent through the drill collars.

$$P_{su} = \frac{7.7(10^{-5})(15^{0.8})(134^{0.8})(60^{0.2})700}{2.75^{4.8}} = 56 \text{ psi}$$

Determine the pressure drop through the bit.

$$v_n = \frac{(0.32)(134)}{0.278} = 154 \text{ fps}$$

$$P_b = \frac{(15)(154^2)}{1,120} = 317 \text{ psi}$$

Total pressure inside pipe = 317 + 56 + 268 = 641 psi

These calculations show the annular surge will be between the 641 psi, considering the pipe completely open, and 664 psi, which represents the pressure required to flow fluid inside the pipe at the indicated rate. A reasonable estimate of annular surge in Example 7.11, with the pipe open, would be about 650 psi. This is 61% of the surge with the pipe plugged and indicates a substantial reduction in annular surge pressure by omitting float collars even with small jet nozzles.

The calculation assuming open pipe is a rough estimate, and the relative effects of having open pipe will change with mud properties and drillstring geometry. In any event these calculations do show a reduction in annular surge pressures by omitting the use of drillstring floats.

In addition to the surge or swab pressures associated directly with pipe velocity, note from Fig. 7–14 that acceleration and deceleration effects may need to be considered. Of particular interest is the swab pressure reduction noted at point C in Fig. 7–14, when the pipe is stopped suddenly.

This simply shows that a well could be swabbed in while running pipe and emphasizes the need for the operator to be conscious of this effect. Eq. 7.39 may be used to estimate acceleration or deceleration pressures associated with pipe movement if the pipe is plugged, and Eq. 7.40 may be used for the same purpose if the pipe is open.

Plugged pipe:

$$P_{su} = \frac{0.00162\rho l D_p{}^2 a_p}{D_h{}^2 - D_{op}{}^2} \qquad (7.39)$$

Open pipe:

$$P_s = \frac{0.00162\rho(D_p{}^2 - D_i{}^2)a_p D_w}{D_h{}^2 - D_{op}{}^2 + D_i{}^2} \qquad (7.40)$$

where:

a_p = pipe acceleration, fps²

Example 7.13 illustrates the use of these equations.

☐ **EXAMPLE 7.13:**

Same conditions as given in Example 7.11. Assume a_p = 4.5 fps², which is comparable to stopping the pipe in one second while running the pipe into the hole at a velocity of 4.5 fps. Determine the surge pressure assuming plugged and open pipe.

Around drillpipe:

$$P_{su} = \frac{1.62(10^{-3})(15)(14.3)(10^3)(20)(-4.5)}{42} = -745 \text{ psi}$$

Around drill collars:

$$P_{su} = \frac{1.62(10^{-3})(15)(0.7)(10^3)(39)(-4.5)}{23} = -130 \text{ psi}$$

Total negative surge = −875 psi

Around open pipe:

$$P_{su} = \frac{1.62(10^{-3})(15)(14.3)(10^3)(6)(-4.5)}{56} = -168 \text{ psi}$$

Around drill collars:

$$P_{su} = \frac{1.62(10^{-3})(15)(0.7)(10^3)(31.4)(-4.5)}{(30.6)} = -79 \text{ psi}$$

Total negative surge = −168 + (−79) = −247 psi

The negative surge probably would be higher than −247 psi, and to be precise, the actual quantities of fluid entering the pipe would have to be determined.

The deceleration losses noted in Example 7.13 are rough estimates and at best simply show why the negative surge pressures occur and provide a warning signal on their potential order of magnitude. This warning becomes even more important as mud weights are reduced to levels close to pore pressure.

SPECIAL CONSIDERATIONS

In deep high-temperature wells, the mud may gel considerably on the bottom of the hole. The surge pressures may be trapped in the mud and have an additive effect to circulating pressure losses when circulation is commenced. Therefore, it is always important to pick up pipe slowly when pumping is commenced. Also, surface pressures should be watched carefully when breaking circulation to avoid excess pressure on the formation.

If problems should be experienced with lost circulation, remember that the crucial point in slowing down pipe running speed is when the bit first reaches the weak zone. In coastal areas, this generally will be just below the protective casing seat because at this point the drill collars are above the weak zone and this restriction in annulus size as compared with drillpipe increases fluid velocity over the collar interval.

We can see that pressure fluctuations when handling pipe can be very high. This places a premium on careful pipe handling practices in areas where hole stability is a problem.

■   **REFERENCES**

1. C.E. Williams, Jr. and G.H. Bruce, "Carrying Capacity of Drilling Muds," *Petroleum Transactions* Reprint Series, Drilling No. 6.
2. R.E. Walker, "Practical Oil-Field Rheology," Southern District, API Div. of Production, San Antonio, March 1964.
3. George E. Cannon, "Changes in Hydrostatic Pressure Due to Withdrawing Drill Pipe from the Hole," Drilling and Production Practice, 42 (1934).
4. W.C. Goins, Jr., J.P. Weichert, J.L. Burba, Jr., D.D. Dawson, Jr., and A.J. Teplitz, "Down-the-Hole Pressure Surges and their Effect on Loss of Circulation," Southwestern District, Div. of Production, Beaumont, Texas, 1951.
5. W.T. Cardwell, Jr., "Pressure, Changes in Drilling Wells Caused by Pipe Movement," Drilling and Production Practices API (1953).
6. E.H. Clark, Jr., "A Graphic View of Pressure Surges and Lost Circulation," Drilling and Production Practices, API (1956).
7. J.A. Burkhardt, "Wellbore Pressure Surges Produced by Pipe Movement," SPE of AIME paper No. 1546-G, Denver, Colorado, 1960.

CHAPTER 8

HYDRAULIC HORSEPOWER IN ROTARY DRILLING

Horsepower is a defined rate of doing work: one horsepower does 33,000 foot-pounds of work in one minute. This definition is universal and, other than changes in units, applies all over the world. In rotary drilling, the engines that supply power are rated on output horsepower, sometimes called brake horsepower. Fluid pumps that receive power are rated on the basis of input horsepower. For this reason, a 1,600-hp pump classification means the horsepower fed into the pump should not exceed 1,600. Output horsepower from pumps used in rotary drilling is determined from charts of maximum permissible surface pressure and maximum circulation rate. These maximums are dictated by the pump manufacturer using Eq. 8.1.

$$Hp_s = \frac{P_s Q_c}{1,714} \qquad (8.1)$$

where:

Hp_s = Surface hydraulic horsepower
P_s = pressure at surface, psi
Q_c = circulation rate, gpm

In actual drilling operations there may be some confusion about maximum pump pressure and maximum circulation rates. Pump manufacturers publish what are called a maximum liner rating and a maximum circulation rate for a specific pump. These maximums are seldom, if ever, used in drilling operations.

MAXIMUM PUMP PRESSURE

The manufacturer's published maximum pressure is based on the maximum permissible force on the power-end bearings. This force, F_m, in lb force, is determined by Eq. 8.2.

$$F_m = \Delta PA \tag{8.2}$$

where:

ΔP = differential pressure across the fluid-end piston, psi
A = area of the piston or internal area of the liner, sq in.

Considering F_m a constant maximum, then, as A is reduced, ΔP may be increased.

In field drilling operations, the maximum pump pressure is rarely achieved. Many arbitrary standards are used. One common standard utilizes a fixed percentage of the maximum liner rating pressure. Most operators would not exceed 90% of the liner rating. Some rig operators ignore percentages completely and simply place a maximum on surface pump pressure, which is well below the liner rating.

Pump maintenance costs go up as pump pressures are increased. Showing a direct mathematical relationship between pump pressures and maintenance costs is difficult because so many other variables, such as mud properties, also have a direct bearing on pump maintenance expense. However, pump maintenance costs rise much faster than the increase in pump pressures. Field observations show pump maintenance costs often more than double by increases in pump pressures from 2,500 to 3,000 psi. Precise numbers for specific rigs or operations must be determined at the drilling rig.

MAXIMUM CIRCULATION RATE

The maximum circulation rate, Q_m, in gallons per minute for a triplex single-acting pump can be determined by Eq. 8.3.

$$Q_m = \frac{(D_l^2 \; l) \; (spm)}{98} \tag{8.3}$$

where:

D_l = internal diameter of the liner, in.
l = stroke length, in.
spm = strokes per minute

Given a triplex pump with a 6-in. liner, a 12-in. stroke length, and a pumping speed of 150 spm, the pump output would be 660 gpm. These conditions assume 100% fluid efficiency. In actual practice, the pumping speed may be reduced substantially below the maximum shown on pump charts. The displacement fluid efficiency is usually between 95 and 100%

if a centrifugal pump is used to precharge the triplex pump. Maximum speeds for triplex pumps sometimes exceed 150 spm. In field operations the normal maximum pumping speed is 100–110 spm. Because the maximum surface pressures are almost always below the manufacturers stated maximums and the maximum circulation rate is seldom used, the output hydraulic horsepower from a mud pump is generally less than one-half the rated input horsepower.

An operator's primary concern with hydraulic horsepower is to achieve adequate cleaning below the bit, important for the following reasons:

- Drilling rates depend on cleaning below the bit.
- Some bits, particularly fixed bladed bits, may overheat if cuttings accumulate below them.
- Poor cleaning below the bit may make it impossible to detect lithology and pore pressure changes from drilling rate.

Removal of cuttings from below a bit is based on the following design criteria:

- Jet velocity
- Bit hydraulic horsepower
- Hydraulic impact force

There is no guideline for choosing velocity. The minimum jet velocities suggested are based on minimum cooling requirements for cutting surfaces. For many years rock bit manufacturers suggested minimum jet velocities of at least 225 feet per second (fps) for adequate cooling. Cutting structures and requirements for cooling change. For these reasons there are no mandatory minimum nozzle velocities for any bit type. Maximum jet velocities are based on the material for nozzles. Tungsten carbide is normally used in jet rock bits. Tests show considerable nozzle erosion when the nozzle velocity exceeds 450 fps. Other materials used for nozzles include ceramics. Thus, the maximum nozzle velocity is always subject to change.

Nozzle velocity with fixed bladed bits is unknown. With any fixed bladed bit the key to success is the removal of cuttings and cooling below the bit. Cooling depends on fluid volume as well as fluid velocity, so fixed bladed bits are in a state of flux.

Bit hydraulic horsepower was introduced as a design criterion in 1949 or 1950; it is a measure of the work required to push mud through the bit nozzles. This work measurement is related to the removal of cuttings from below the bit. Bit hydraulic horsepower is the most common design procedure, probably because it was used first.

Hydraulic impact force as a design criterion was introduced in 1956 (see chapter 11). Hydraulic impact force is a measure of the force exerted by the fluid at the exits of the bit nozzles. This fluid force cleans by direct erosion on the hole bottom and by cross flow beneath the bit. McClean studied bottom-hole cleaning and showed that fluid cross flow beneath the bit was directly proportional to the hydraulic impact force.

Hydraulic impact force below the bit seems more logical when considering design procedures for bottom-hole cleaning. In recent years, jet nozzles for rock bits have been extended, putting them closer to the bottom of the hole. Both laboratory and field tests show better bottom-hole cleaning with extended bit nozzles. Extending the nozzles did not change the bit hydraulic horsepower for given conditions. The hydraulic impact force on the bottom of the hole is increased with these nozzles. This force relates directly to the erosional force of the fluid.

Arguments for using bit hydraulic horsepower for design are based on the fact that lower circulation rates are often required. Lower circulation rates reduce the chances for turbulent fluid flow patterns in the annulus. This in turn minimizes hole enlargement due to fluid erosion. There is some merit in keeping fluid flow patterns in the annulus laminar; however, this should be a limiting consideration in designing a hydraulics program, not a design criterion.

The argument about which design criterion to utilize may be moot because either may be used to optimize bottom-hole cleaning requirements. Optimum cleaning requirements are based on drilling tests in actual drilling operations. Thus, if the bottom-hole cleaning requirements are determined using bit hydraulic horsepower, then bit hydraulic horsepower should be the design base. The same holds true for hydraulic impact force.

There are two basic steps in a hydraulics program design. First, the bottom-hole cleaning requirements should be determined by field tests. Second, bottom-hole cleaning should be maximized based on the hydraulic horsepower available.

BOTTOM-HOLE CLEANING REQUIREMENTS

In very soft formations, it may be difficult to determine the bottom-hole cleaning required for maximum penetration rates. It is possible in soft formations, such as those found in some coastal areas, to generate hole by the jetting action of the bits. In these areas, maximum penetration rates are achieved with maximum bottom-hole jetting action. Thus, the problem is one of using the maximum jetting action that is economically feasible. Economic feasibility depends on the maximum penetration rates possible, based on hole conditions and other activities that have an equalizing effect, such as connection time and the maintenance of support equipment.

In formations of normal hardness, when there is no specific breaking point, the amount of bottom-hole cleaning necessary may be determined directly in field operations.

A method of determining the amount of bottom cleaning needed is indicated by the field tests shown in Fig. 8–1. In this series of tests, the nozzle velocity was increased while holding circulation rate constant. By

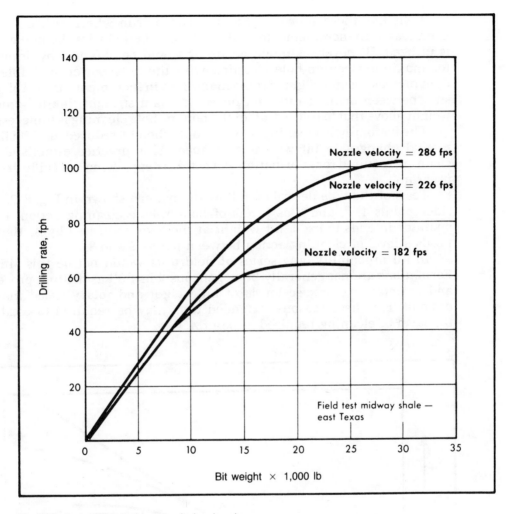

■ **FIG. 8–1** Effect of bottom-hole cleaning

this method both bit hydraulic horsepower and hydraulic impact increased. However, the surface horsepower also increased.

Fig. 8–1 shows that drilling rate at 20,000 lb of bit weight increased 25% when nozzle velocity increased by 24.2%. A further 26.5% increase in nozzle velocity resulted in a 12% increase in drilling rate. This test shows that bottom-hole cleaning was inadequate at the highest nozzle velocity. Note that the top curve is a straight line only to a bit weight of 7,500 lb. This indicates that more bottom-hole cleaning was needed.

The drilling rate should be directly proportional to bit weight if bottom-hole cleaning is adequate. Exceptions are those cases when low bit weights are not high enough to fracture the rock. This field test suggests one of several methods for determining bottom-hole cleaning requirements.

Tests like those shown in Fig. 8–1 can be run where nozzle velocity is increased in increments until all of the available hydraulic horsepower is utilized. However, as much accuracy would be obtained by designing for maximum bottom-hole cleaning with the horsepower available and then increasing bit weight to some maximum in increments. If the relationship between drilling rate and bit weight is a straight line to some bit weight above that to be used, all of the bottom-hole cleaning is unnecessary.

The bottom-hole cleaning power should then be reduced and the incremental changes in bit weight continued. This practice establishes the cleaning required for a formation at a certain depth using a specific drilling program.

Tests of the type described follow the pattern shown in Fig. 8–2. Consider points 1, 2, and 3 as levels of bottom-hole cleaning action. If the operator intends to use a bit weight at the level indicated by A, then he needs a level of cleaning action between points 2 and 3.

In all probability, any series of field tests would not be this simple. Formations are not generally homogeneous, bits dull as drilling proceeds, and maintaining precise levels of bit weight and rotary speed may be difficult. For these reasons, extended tests may be required to establish the level of cleaning required at point A.

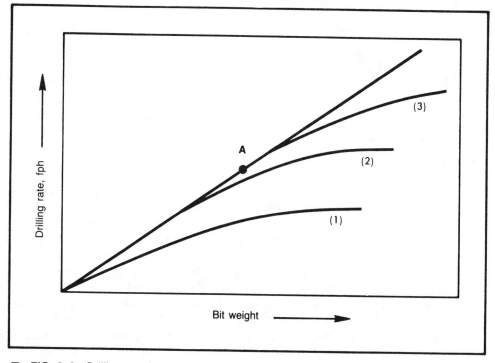

■ **FIG. 8–2** Drilling rate increases with bit weight

A suggested procedure is to determine drilling rates over 5-ft intervals at different bit weights on an alternating basis during one 24-hr period. The specific problem may be simplified by using insert bits with friction bearings because the maximum loading on these bits is generally fixed, rather than selected on an economic basis.

In coastal operations using mill-tooth bits, a standard number is 5.2 bit hhp/sq in. of hole. Using diamond bits, a 3.1 bit hhp/sq in. of hole is frequently used. In west Texas, where formations are hard, less bottom-hole cleaning is required than in the soft formations of coastal areas.

Operators are cautioned to consider rules of thumb as starting points only. There are far too many variables and drilling costs are too high to permit continuous guessing.

DESIGN OF THE HYDRAULICS PROGRAM

Available surface hydraulic horsepower is based on the maximum permissible discharge pressure and the maximum flow rate. Example 8.1 illustrates the calculation of maximum surface hydraulic horsepower.

□   **EXAMPLE 8.1**

Maximum permissible surface pressure = 3,000 psi
Maximum flow rate = 500 gpm

Available surface horsepower: $Hp_s = \dfrac{(3,000)\,(500)}{1,714} = 875$

Remember that the size of drilling rig pumps is based on input horsepower to the pump. Example 8.2 illustrates a typical field situation.

□   **EXAMPLE 8.2**

Rig pump size = 1,500 hp
Using 6½-in. liner, maximum surface pressure = 4,118 psi
Recommended maximum pumping rate = 500 gpm
Available surface horsepower = 1,200

In actual practice, the maximum surface pressure is usually limited to some value less than the maximum. In this case, if the pressure is limited to 3,000 psi, the surface horsepower available is 875. Thus, under these conditions, the 1,500-hp pump delivers only 875 hp. Many operators, or the design of the hydraulics program, sometimes reduce the circulation rate below the maximum. Assume the maximum circulation rate is reduced to 350 gpm at a reduced pressure of 3,000 psi. The available surface horsepower is now only 613.

Example 8.2 shows that the pump selection is very important in the design of a hydraulics program. Of equal importance are the maximum pressure and circulating rates that can or will be used. In addition, remember that the prime movers available must be able to supply the required input horsepower. The operating practices of the drilling rig operator are as important as the type of equipment included in the rig inventory.

When the available surface horsepower has been determined, the next step in designing the hydraulics program is to select the design method. Two methods for designing hydraulics programs are considered: *hydraulic impact* and *bit hydraulic horsepower*. In each method, there are three design criteria: (1) the situation of no limit on surface pressure; (2) the case of limited surface pressure when reductions in surface horsepower are necessary to maximize bottom-hole cleaning; and (3) the intermediate case between the two maximums stated in (1) and (2).

Regardless of the hydraulics design method used, commonly the fluid flow pattern is assumed to be turbulent in all parts of the circulating system. In many cases the fluid flow pattern in the annulus is laminar, and the turbulent flow assumption introduces a small error in the predicted pressure losses for the circulating system. Because the annular pressure losses are generally less than 10% of the total circulating pressure losses, this error has virtually no effect on the hydraulics program.

Assuming turbulent flow, pressure losses inside the drillstring and in the annulus can be determined using equations given in chapter 7. Pressure losses through the bit can be calculated using equations from the same chapter. In normal field practice, charts or hydraulic slide rules are misused because the engineer fails to recognize that the pressure losses are only for turbulent flow. If there is no change in mud weight or the solids content of the mud, drillstring geometry, and hole size, the circulating pressure (excluding the bit) may be shown as a function of circulating rate only by Eq. 8.4.

$$P_c = KQ_c^{1.8} \text{ or } \log P_c = \log K + 1.8 \log Q_c \qquad (8.4)$$

where:

P_c = circulating pressure, psi
K = proportionality constant
Q_c = circulation rate, gpm

This equation relates the pressure losses in the circulating system, excluding the bit, to changes in circulation rate. It predicts that the plot of P_c vs Q_c on log paper will be a straight line with a slope of 1.8. In field operations a common practice is to run the pump at three or more speeds, read the pump pressure, and plot P_c vs Q_c on log paper. This type of plot is generally made for pressure control purposes. The pressure loss in the circulating system, excluding the bit, can also be obtained from Eq. 8.5.

$$P_c = P_s - P_b \qquad (8.5)$$

where:

P_s = pressure at surface, psi
P_b = pressure drop through bit, psi

The pressure loss through the bit, P_b, can be determined using equations from chapter 7, from charts, or from hydraulic slide rules. The bit nozzle pressure loss relates directly to the change in fluid velocity and to a well-defined nozzle coefficient. As a result, the pressure losses through the bit are calculated more accurately than pressure losses in other parts of the fluid circulating system.

After determining P_c, a plot of it vs Q_c on log paper is usually a straight line. The slope of the line can be measured with a ruler or can be calculated from Eq. 8.6.

$$n_c = 3.32 \log \frac{P_q}{P_{0.5q}} \tag{8.6}$$

where:

n_c = slope
P_q = pressure at the circulation rate, psi
$P_{0.5q}$ = pressure at one-half the circulation rate, psi

In Eq. 8.6, the pressure, P_q, is read from the straight line of P_c vs Q_c. If P_c vs Q_c is not a straight line on log paper, check the low flow rate pressure. The flow pattern inside the drillpipe can be laminar at low flow rates;* this results in considerable error. The best procedure is to use the pressure losses at the higher rates of flow and drop the pressure loss at the low flow rate. This practice assumes that the low flow rate is below the actual rate used in the hydraulics program.

The slope of P_c vs Q_c determines the required pressure distribution in the circulating system for maximum bottom-hole cleaning for the specific design procedure.

Because hydraulic impact and bit hydraulic horsepower design procedures are frequently used, both procedures are discussed.

Hydraulic Impact

Hydraulic impact force is based on Newton's second law of motion, $F = ma$. In field units, impact force at the exit of the bit nozzles is defined by Eq. 8.7.

$$F_i = \frac{\rho Q_c V_n}{1,932} \tag{8.7}$$

where:

ρ = mud weight, lb/gal
Q_c = circulation rate, gpm
V_n = nozzle velocity, fps

Eq. 8.7 also determines the maximum pump-off force** on the bottom of the hole.

---

* Low flow rate may fall below the straight line on the P_c vs Q_c graph. Should this occur, do not use this rate.
** This force is the lifting force of the fluid on the drillstring.

If F_i vs Q_c is plotted as in Fig. 8–3, a circulation rate in both the unlimited and limited pressure cases results in the maximum impact force. If an equation of F_i vs Q_c (see Eq. 8.8) is differentiated, and that differential is set equal to zero, the maximum impact force can be determined. This holds true for both the limited and unlimited cases (see Fig. 8–3). The maximum impact force for the unlimited-surface-pressure case shown in part *a* of Fig. 8–3 is calculated from Eq. 8.8:

$$P_c = \frac{P_s}{n+2} \text{ or } Hp_c = \frac{Hp_s}{n+2} \tag{8.8}$$

where:

P_c = pressure loss in circulating system (excluding bit), psi
P_s = pressure at surface, psi
n = slope of F_i vs Q_c
Hp_c = horsepower required in circulating system (excluding bit)
Hp_s = surface hydraulic horsepower

The maximum impact force in the reducing-surface-horsepower, limited-surface-pressure case shown in part *b* of Fig. 8–3 is calculated from Eq. 8.9.

$$P_c = \frac{2P_s}{n+2} \tag{8.9}$$

The maximum impact force for the intermediate case falls between the boundary conditions of Eq. 8.8 and 8.9.

The boundary conditions given by these equations and the intermediate condition are shown in Example 8.3, with assumed parameters.

☐ **EXAMPLE 8.3**

Maximum available surface horsepower = 600
Maximum permissible surface pressure = 2,000 psi

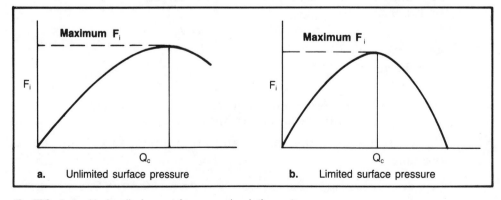

a. Unlimited surface pressure   **b.** Limited surface pressure

■ **FIG. 8–3** Hydraulic impact force vs circulation rate

Slope of the P_c vs Q_c line on log paper $= 1.8$
Required minimum annular velocity $= 120$ fpm
Well depth $= 11,000$ ft

Thus, for the unlimited surface pressure case,

$$\text{Hp}_c = \left(\frac{1}{1.8+2}\right)\text{Hp}_s = 0.26\ \text{Hp}_s.$$

For the limited surface pressure case,

$$P_c = \left(\frac{2}{1.8+2}\right)P_s = 0.53\ P_s$$

The assumed conditions in Example 8.3 are illustrated in Fig. 8-4.

For the data plotted in Fig. 8-4, it is impossible to reach the unlimited surface pressure case because the maximum pump circulation rate is 514 gpm. At a depth of 2,000 ft to reach the point where the pressure in the circulating system, excluding the bit, is $0.26P_s$ would require a circulation rate greater than 514 gpm. Thus, the unlimited-pressure case is simply an unreachable boundary condition. The intermediate case is applicable for depths of 3,000 and 5,000 ft. In the intermediate case, surface pressure and circulation rate remain constant. At 7,000 and 9,000 ft, the limited-surface-pressure case applies. Surface pressure remains constant, and circulation rate is lowered, reducing the surface horsepower utilized. Note that $P_c = 0.53P_s$, or 1,060 psi, as would be calculated using Eq. 8.9. In Fig. 8-4, a circulation rate of 250 gpm is assumed as necessary to maintain a minimum annular velocity. For this reason, the impact force at 11,000 ft cannot be maximized because circulation rate cannot be reduced to 232 gpm as required.

The circulation rate that would be selected at each depth is included in Table 8-1.

Bit Hydraulic Horsepower

Bit hydraulic horsepower is defined mathematically in Eq. 8.10:

$$\text{Hp}_b = \frac{P_b Q_c}{1,714} \tag{8.10}$$

This equation shows the horsepower required to push the fluid through the bit nozzles at a given circulation rate. Bit hydraulic horsepower can also be calculated by the following equation:

$$\text{Hp}_b = \frac{P_s Q_c}{1,714} - \frac{P_c Q_c}{1,714} \tag{8.11}$$

In Eq. 8.11, the hydraulic horsepower utilized in the circulating system (excluding the bit), Hp_c, is subtracted from the available surface hydraulic horsepower, Hp_s, to give the bit hydraulic horsepower, Hp_b.

In the unlimited surface pressure case, the maximum horsepower oc-

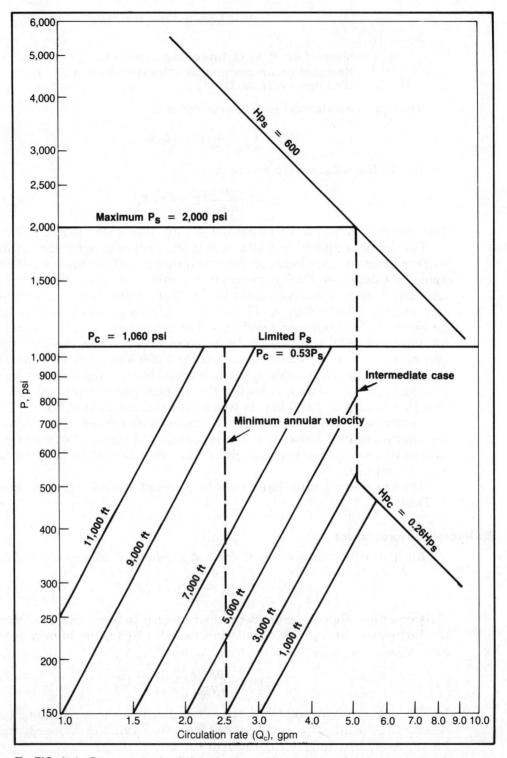

■ FIG. 8–4 Pressure vs circulation rate for maximum impact force

■ **TABLE 8–1** Selected circulation rates (hydraulic impact force design method)

Depth, ft	Circulation Rate, gpm	Case
1,000	514	Unlimited P_s
3,000	514	Intermediate
5,000	514	Intermediate
7,000	450	Limited P_s
9,000	295	Limited P_s
11,000	250	Minimum Annular Velocity

curs when Hp_c approaches zero. Hp_c approaches zero as the circulation rate, Q_c, approaches zero. So the maximum Hp_b is obtained in field practice for the unlimited-surface-pressure case by first selecting the minimum possible circulation rate, calculating the pressure losses in the circulating system (excluding the bit), and selecting nozzle sizes that will use the remaining available pressure at the bit.

For the limited-surface-pressure case, if Hp_b is plotted against circulation rate, the curve will look like Fig. 8–5. The circulation rate, Q_c, at which bit hydraulic horsepower, Hp_b, reaches a maximum can be determined by Eq. 8.12:

$$P_c = \frac{P_s}{n+1} \tag{8.12}$$

After the pressure loss in the circulating system (excluding the bit), P_c, is calculated using Eq. 8.12, circulation rate can be established for a specific system.

The intermediate case lies between the two boundary conditions of unlimited and limited surface pressure.

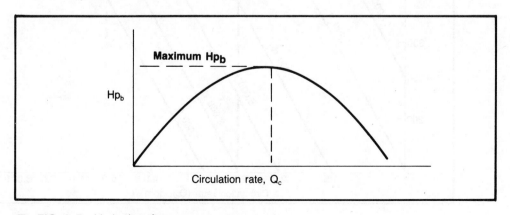

■ **FIG. 8–5** Limited-surface-pressure case

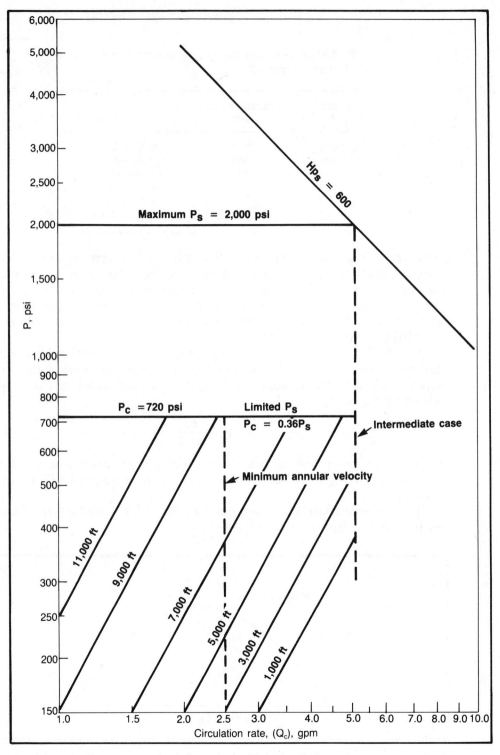

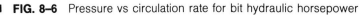

■ **FIG. 8–6** Pressure vs circulation rate for bit hydraulic horsepower

Example 8.4 shows the boundary conditions of the unlimited-surface-pressure condition and Eq. 8.12, using assumed parameters.

☐ **EXAMPLE 8.4**

Maximum available surface horsepower = 600
Maximum permissible surface pressure = 2,000 psi
Maximum circulation rate = 514 gpm
Slope of the P_c vs Q_c line on log paper = 1.8
For the unlimited-surface-pressure case,

$$Q_c = 250 \text{ gpm}$$

For the limited-surface-pressure case,

$$P_c = \left(\frac{1}{1.8 + 1}\right) P_s = 0.36 P_s$$

The assumed conditions in this example are illustrated in Fig. 8–6.

In the unlimited P_s case, the minimum annular velocity occurs when the circulation rate is 250 gpm (see Fig. 8–6). If the hydraulics program is designed initially for 250 gpm, only 292 surface horsepower will be used, rather than the available surface horsepower of 600. The reduction in surface horsepower would reduce bit hydraulic horsepower. For the 1,000- and 3,000-ft depths, the intermediate case should be used. For the 5,000- and 7,000-ft depths, the limited-pressure case applies. To obtain the maximum bit hydraulic horsepower at the 5,000- and 7,000-ft depths, reduce the circulation rate to limit P_c to 720 psi. At 9,000 and 11,000 ft, the maximum bit hydraulic horsepower calculation utilizes circulation rates of 240 and 180 gpm; however, the minimum permissible circulation rate is 250 gpm.

The circulation rate selected at each depth is listed in Table 8–2.

Note that the selected circulation rates are the same in Tables 8–1 and 8–2 at the depths of 1,000 and 3,000 ft. As shown, the selected circulation rates using the bit hydraulic horsepower design method are equal to or less than those selected using the hydraulic impact force design method.

■ **TABLE 8–2** Selected circulation rates (bit hydraulic horsepower design method)

Depth, ft	Circulation rate, gpm	Case
1,000	514	Intermediate
3,000	514	Intermediate
5,000	475	Limited P_s
7,000	362	Limited P_s
9,000	250	Minimum Annular Velocity
11,000	250	Minimum Annular Velocity

This is always true because the circulating pressure losses allowed in the circulating system for the hydraulic impact force design method are equal to or greater than the circulating pressure losses allowed in the circulating system for the bit hydraulic horsepower design method.

PRACTICAL APPLICATIONS

Examples 8.5 and 8.6 illustrate the design of a hydraulics program. The first, Example 8.5, assumes the line slope of P_c vs Q_c on log paper is 1.8. Example 8.6 uses field date to determine the line slope of P_c vs Q_c and to design the hydraulics program. In addition, the following special cases are illustrated.

- The effect of surface pressure on bottom-hole cleaning
- The selection of surface pressure and circulation rates after bottom-hole cleaning requirements are known
- The limitations imposed if circulation rates in the hydraulics program are too high or too low

Example 8.5 shows the classic design of a hydraulics program based on the available surface horsepower and the assumption that P_c vs Q_c on log paper is 1.8.

☐ **EXAMPLE 8.5**

Well depth: 15,000 ft
Surface casing: 13⅜-in. O.D., 12½-in. I.D., set at 4,000 ft
Intermediate casing: 9⅝-in. O.D., 8⁷⁄₁₀-in. I.D., set at 12,000 ft
Hole size: 12,000–15,000 ft, 8½-in.
Drillpipe: 4½-in. O.D., 3⁴¹⁄₅₀-in. I.D.
Drill collars: 7 in. O.D., 3 in. I.D., 900 ft
Pump: Triplex 6¼-in. × 8 in.
     Liner size = 6 in.
     Maximum surface pressure = 2,000 psi
     Maximum pump speed = 120 spm
     Assume pump efficiency = 100%
Mud properties, lb force/100 sq ft: $\theta_{600} = 85$
                                $\theta_{300} = 50$
                                $\theta_{200} = 36$
                                $\theta_{100} = 23$
                                $\theta_6 \;\;= \;\;7.9$
                                $\theta_3 \;\;= \;\;6.0$
Mud weight = 10 lb/gal from 4,000 to 12,000 ft
                    16 lb/gal from 12,000 to 15,000 ft

SOLUTION:

$$Q_{max} = 352 \text{ gpm}$$

$$Hp_s = \frac{(2,000 \times 352)}{1,714} = 411$$

where:

$$Q_{max} = \text{maximum pumping rate, gpm}$$

Fig. 8–7 shows the plotted data for Example 8.3.

The maximum available surface horsepower is 411, and the maximum surface pressure is 2,000 psi. The limited pressure line for hydraulic impact is determined using Eq. 8.9:

$$P_c = \frac{2}{(1.8 + 2)} P_s = (0.53)\,(2,000) = 1,060 \text{ psi}$$

The limited pressure line for bit hydraulic horsepower is calculated by Eq. 8.12:

$$P_c = \frac{1}{(1.8 + 1)} P_s = (0.36)\,(2,000) = 720 \text{ psi}$$

Pressure losses in the circulating system (excluding the bit) are shown by the depth lines in Fig. 8–7. Table 8–3 shows the hydraulics program based on hydraulic impact force.

The circulation rate, Q_c, in Table 8–3 was taken from Fig. 8–7. The maximum circulation rate of 352 gpm was used at depths of 4,000, 8,000, and 12,000 ft. The limited surface pressure case applies at 13,000 and 15,000 ft. The circulation rate was reduced to prevent P_c from exceeding 1,060 psi. At each depth the pressure loss in the circulating system was subtracted from the surface pressure of 2,000 psi to determine the pressure remaining for use through the bit. For example, consider the depth of 4,000 ft.

$$P_c = 405 \text{ psi}$$
$$P_b = 2,000 - 405 = 1,595 \text{ psi}$$

Nozzle sizes can be determined using equations in chapter 7; however, in this case they were read directly from a hydraulic slide rule. The actual pressure drop through the bit results from the best possible selection of nozzle sizes. The actual bit nozzle pressure losses are kept below those

■ **TABLE 8–3** Hydraulics program, based on hydraulic impact force

Depth, ft	Q_c, gpm	P_b, psi	Nozzle Sizes, 1/32nd in.	Actual P_b, psi	Nozzle Velocity, fps	Impact Force, lb
4,000	352	1,595	3–11	1,460	406	740
8,000	352	1,343	2–11, 1–12	1,290	382	696
12,000	352	1,091	3–12	1,030	340	619
13,000	275	940	2–12, 1–13	900	252	574
15,000	260	940	3–12	905	252	543

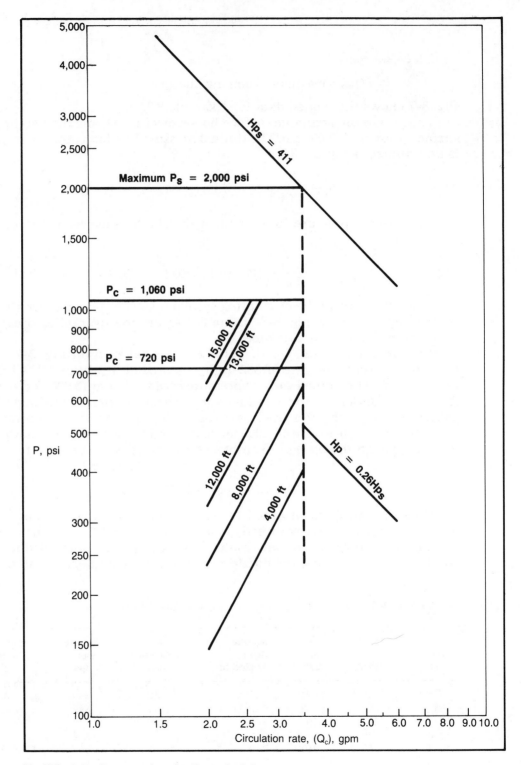

■ **FIG. 8–7** Pressure data for Example 8.5

allowed because the drillstring pressure losses increase as more hole is made.

The nozzle velocity also can be calculated using chapter 7's equations, but, in this case, the nozzle velocity, too, was read from a hydraulic slide rule. Impact force is calculated by Eq. 8.7. For example, consider the depth of 4,000 ft.

$$F_i = \frac{(10 \times 352 \times 406)}{1,932} = 740 \text{ lb}$$

Table 8–4 shows the hydraulics program based on bit hydraulic horsepower.

The circulation rate, Q_c, in Table 8–4 was taken from Fig. 8–7. The maximum circulation rate of 352 gpm was used at depths of 4,000 and 8,000 ft. The limited surface pressure case applies at 12,000, 13,000, and 15,000 ft, and the circulation rate was reduced to prevent P_c from exceeding 720 psi. At each depth the pressure loss in the circulating system was subtracted from the 2,000-psi surface pressure to determine the pressure left for use through the bit. For example, consider the depth of 4,000 ft.

$$P_c = 405 \text{ psi}$$
$$P_b = 2,000 - 405 = 1,595 \text{ psi}$$

Note that the circulation rate, pressure loss through the bit, and bit nozzle sizes are the same for both the hydraulic impact force and bit hydraulic horsepower design procedures at depths of 4,000 and 8,000 ft. This will always be so, as long as both design procedures are in the intermediate range. The primary difference in the two procedures occurs in the limited surface pressure case. The bit hydraulic horsepower in Table 8–4 was calculated using Eq. 8.10. At the depth of 4,000 ft:

$$Hp_b = \frac{(1,460 \times 352)}{1,714} = 300 \text{ hp}$$

■ **TABLE 8–4** Hydraulics program, based on bit hydraulic horsepower

Depth, ft	Q_c, gpm	P_b, psi	Nozzle Sizes, 1/32nd in.	Actual P_b, psi	Bit Hydraulic Horsepower
4,000	352	1,595	3–11	1,460	300
8,000	352	1,343	2–11, 1–12	1,290	265
12,000	310	1,280	2–11, 1–10	1,270	230
13,000	225	1,280	2–10, 1–11	1,220	160
15,000	210	1,280	3–10	1,220	149

EFFECT OF PRESSURE ON BOTTOM-HOLE CLEANING

Consider the bit hydraulic horsepower of Example 8.5 that is reported in Table 8–4. If the required amount of bottom-hole cleaning is 5.2 bit hhp/sq in. of hole, then it is below that required for all depths below 4,000 ft. In the 12¼-in. hole for 4,000–12,000 ft, the required bit hydraulic horsepower is 614. In the 8½-in. hole from 12,000 to 15,000 ft, the required bit hydraulic horsepower is 295. As shown in Table 8–4, the only way bit hydraulic horsepower from 12,000 to 15,000 ft could be increased would be to increase the maximum permissible surface pressure. The circulation rate has already been reduced, so increasing pump output without raising the allowable surface pressure is of no value at 12,000 ft and deeper. If the maximum surface pressure is increased to 3,000 psi, the hydraulics program is changed as illustrated in Table 8–5. The limit on circulating pressure (excluding the bit) is raised to 1,080 psi, and the maximum output horsepower is increased from 411 to 616.

■ **TABLE 8–5**   Hydraulics program based on bit hydraulic horsepower

Depth, ft	Q_c, gpm	P_b, psi	Nozzle Sizes, 1/32nd in.	Actual P_b, psi	Bit Hydraulic Horsepower
			Maximum P_s = 3,000 psi		
4,000	352	2,595	1–9, 2–10	2,450	503
8,000	352	2,343	3–10	2,150	442
12,000	352	2,090	2–11, 1–12	2,090	429
13,000	280	1,920	2–10, 1–11	1,900	310
15,000	265	1,920	2–10, 1–11	1,700	297

The increase in surface pressure to 3,000 psi provided adequate bottom-hole cleaning from 12,000 ft to total depth. The only way bottom-hole cleaning would be adequate in the 12¼-in. hole would be to increase the circulation rate. This rate increase could be attained by increasing the pump speed or paralleling pumps. An increase in circulation rate also requires a surface pressure increase. If the minimum annular velocity with 10-lb/gal mud must be 100 fpm, the minimum circulation rate would need to be 530 gpm. Otherwise, the annular velocity at the maximum circulation rate of 352 fpm will be only 66 fpm. Thus, pump pressure and circulation rate must be increased above 12,000 ft to give enough cleaning below the bit and adequate annular velocity for removing cuttings.

For more accuracy, use a field hydraulics design method. The primary difference between the field and the classical procedures is the effect of circulation rate on circulating pressure losses. Example 8.5 assumed that pressure losses in the circulating system (excluding the bit) were as shown in Eq. 8.4, where the slope of the P_c vs Q_c line was 1.8. This relationship of P_c vs Q_c may change in the field and can be determined by measurement just before pulling the bit. The effect of changing 1.8 to another number may be small and can be quickly checked by Eqs. 8.8, 8.9, or 8.12, depending on the hydraulics design method used.

The field method of designing a hydraulics program is summarized as follows:

1. Measure and record the standpipe pressure at three or more rates of flow.

2. Determine the pressure loss through the bit nozzles at the flow rates used in step 1.

3. Subtract the bit nozzle pressure losses from the recorded standpipe pressures. This difference represents the pressure losses in the circulating system (excluding the bit).

4. Plot the pressure losses in the circulating system (excluding the bit) on log paper against circulation rate. Measure the slope of this plot.

5. Use the slope of step 4 in Eq. 8.8, 8.9, or 8.12 to calculate the percentage of the standpipe pressure that should be used in the circulating system (excluding the bit).

6. After the pressure losses in the circulating system (excluding the bit) are determined, refer to step 4 to find the circulating rate.

7. If the circulating rate selected in step 6 is above the maximum obtainable circulating rate, then simply use the maximum. This is the intermediate case. If the circulation rate selected in step 6 is below the maximum obtainable circulating rate, then use that rate. This is the limited surface pressure case.

8. From the circulation rate chosen in step 7, record the pressure losses in the circulating system (excluding the bit).

9. Subtract the pressure losses in the circulating system (excluding the bit) from the maximum permissible standpipe pressure. This difference is the pressure loss through the bit nozzles.

10. Using the circulation rate found in step 7 and the bit nozzle pressure losses, select the nozzle sizes.

Note that in most cases nozzle sizes do not match those required for the bit nozzle pressure losses from step 9. The best selection procedure is to choose the next larger size because more hole will be drilled with the next bit run, and that will increase standpipe pressures some. Also notice that the program of design calls for total nozzle areas, so the nozzle areas can be matched using two, three, or more nozzles, depending on the operator's preference.

The field design procedure is illustrated in Example 8.6 and Table 8–6.

□ **EXAMPLE 8.6**

Well depth: 14,000 ft
Hole size: 8½ in.

Contract depth: 16,000 ft

Protective casing: 9⅝-in. O.D., 8⁹⁄₁₀-in. I.D. at 11,000 ft

Drillstring: I.D. drillpipe = 4½-in. O.D., 3⁴¹⁄₅₀-in. I.D.

Hevi-wate = 4½-in. O.D., 3.0-in. I.D., at 1,500 ft

Drill collars = 7-in. O.D., 3-in. I.D., at 90 ft

Bit nozzles = three ⁹⁄₁₆ in.

Mud: Weight = 15.0 lb/gal

$\qquad$ Viscometer = θ_{600} = 90 lb force/100 sq ft

$\qquad\qquad$ θ_{300} = 50 lb force/100 sq ft

Pump data: Triplex 7¼ × 12 in.

$\qquad$ Input horsepower = 1,600

$\qquad$ Liner size = 6 in.

$\qquad$ Pump speed = 100 spm

$\qquad$ Pump fluid efficiency = 95%

$\qquad$ Pump output = (440 × 0.95) = 418 gpm

$\qquad$ Maximum permissible surface pressure = 3,500 psi

■ **TABLE 8–6** Pressure vs circulation rate for Example 8.6

Q_c, gpm	P_s, psi	P_b, psi	P_c, psi	Nozzle Velocity (V_n), fps
418	2,700	435	2,265	180
350	2,000	305	1,695	155
300	1,540	225	1,315	130
250	1,140	155	985	108

Column 1 of Table 8–6 represents arbitrarily chosen pumping rates. Column 2 includes the standpipe pressures at the desired flow rates. Column 3 lists the calculated pressure losses through the three ¹⁸⁄₃₂-in. bit nozzles at the chosen flow rates. Column 4 was obtained by subtracting the bit nozzle pressure losses from the standpipe pressures. Column 5 includes the calculated nozzle velocities.

SOLUTION:

1. The bit hydraulic horsepower and hydraulic impact force available at the observed conditions are determined as follows:

$$Hp_s = \frac{P_sQ_c}{1,714} = \frac{(2,700 \times 418)}{1,714} = 658$$

Bit hydraulic horsepower in use, using data from Table 8–6:

$$Hp_b = \frac{(435 \times 418)}{1,714} = 106$$

$$Hp_b/\text{sq in. of hole} = \frac{106}{56.7} = 1.87$$

Hydraulic impact force available, using data from Table 8–6:

$$F_i = \frac{Q_c V_n}{1,932} = \frac{(15 \times 418 \times 180)}{1,932} = 584 \text{ lb force}$$

$$F_i/\text{sq in. of hole} = \frac{584}{56.7} = 10.3 \text{ lb force/sq in.}$$

2. The maximum bit hydraulic horsepower and hydraulic impact force that can be obtained by adjusting the circulation rate and changing bit nozzle size with no change in the 2,700-psi surface pressure are as follows:

P_s and P_c vs circulation rate are plotted in Fig. 8–8. The slope of the P_c vs Q_c line is 1.6.

Maximum bit hydraulic horsepower using a surface pressure of 2,700 psi:

$$\text{From Eq. 8.12, } P_c = \frac{1}{(1 + 1.6)} \times 2,700 = 1,026 \text{ psi}$$

$P_b = 2,700 - 1,026 = 1,674$ psi
From Fig. 8–8, $Q_c = 260$ gpm
Nozzle sizes = two 10/32, one 11/32 in.
Actual $P_b = 1,550$ psi

$$\text{Bit horsepower} = \frac{(1,550 \times 260)}{1,714} = 235$$

This is an increase of 122% over the 106 bit hhp available before the adjustment in circulation rate and nozzle size. This increase in bit hydraulic horsepower was obtained while reducing the surface horsepower output from 658 to 409.

Maximum hydraulic impact force available using a surface pressure of 2,700 psi:

$$\text{Using Eq. 8.7, } P_c = \frac{2}{(1.6 + 2)\,2,700} = 1,512 \text{ psi}$$

$P_b = 2,700 - 1,512 = 1,188$ psi
From Fig. 8–8, $Q_c = 330$ gpm
Nozzle sizes = one $12/32$, two $13/32$ in.
Actual $P_b = 1,105$ psi
$V_n = 288$ fps

$$F_i = \frac{\rho Q_c V_n}{1,932} = \frac{(15 \times 330 \times 288)}{1,932} = 738 \text{ lb force}$$

This is an increase of 26.4% over the hydraulic impact force of 584 lb force before the adjustments in circulation rate and nozzle size. This increase in hydraulic impact force was obtained while reducing the surface horsepower output from 658 to 520.

3. The maximum bit hydraulic horsepower and hydraulic impact force

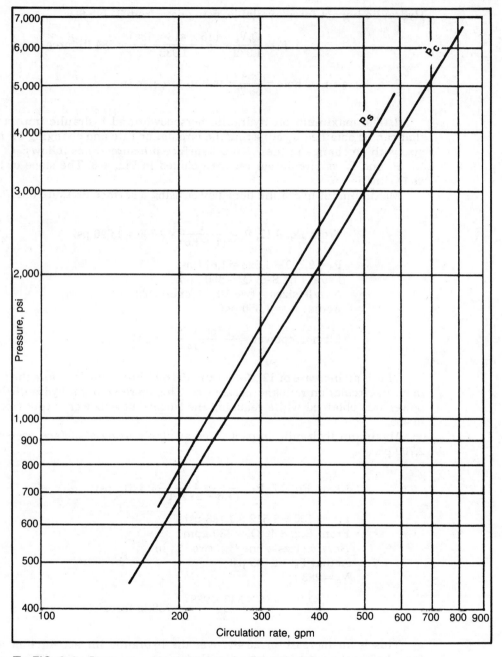

■ **FIG. 8–8** Pressure vs circulation rate for Example 8.6

that can be obtained if the surface pressure is increased to 3,500 psi are shown below:

Maximum bit hydraulic horsepower using a surface pressure of 3,500 psi:

$$\text{Using Eq. 8.10, } P_c = \frac{1}{(1 + 1.6)\,3,500} = 1,330 \text{ psi}$$

$P_b = 3,500 - 1,330 = 2,170$ psi
From Fig. 8–8, $Q_c = 300$ gpm
Nozzle sizes = two $^{10}/_{32}$, one $^{11}/_{32}$ in.
Actual $P_b = 2,050$ psi

$$Hp_b = \frac{(2,050 \times 300)}{1,714} = 359$$

$$Hp_s = \frac{3,500 \times 300}{1,714} = 612$$

The bit hydraulic horsepower is now increased to 359, and the surface horsepower in use is still below the 658 utilized before making adjustments.

Maximum hydraulic impact force using a surface pressure of 3,500 psi:

$$P_c = \frac{2}{(1.6 + 2)\,3,500} = 1,960$$

$P_b = 3,500 - 1,960 = 1,540$ psi
From Fig. 8–8, $Q_c = 385$ gpm
Nozzles sizes = two $^{13}/_{32}$, one $^{12}/_{32}$ in.
Actual $P_b = 1,505$ psi
$V_n = 337$ fps

$$F_i = \frac{(15 \times 385 \times 337)}{1,932} = 1,007$$

$$Hp_s = \frac{(3,500 \times 385)}{1,714} = 786$$

4. The required combination of surface pressure and circulation rate to obtain a bit hydraulic horsepower of 5.2/sq in. of hole is determined as follows:

$$\text{Using 3,500 psi, } Hp_b/\text{sq in.} = \frac{359}{56.7} = 6.33$$

Assume $P_s = 3,000$ psi
$P_c = (0.38)\,(3,000) = 1,140$ psi
$P_b = 3,000 - 1,140 = 1,860$ psi
From Fig. 8–8, $Q_c = 275$ gpm
Nozzle sizes = two $^{10}/_{32}$, one $^{11}/_{32}$ in.
Actual $P_b = 1,730$ psi

$$Hp_b = \frac{(1,730 \times 275)}{1,714} = 278$$

$$\mathrm{Hp_b/sq\ in.} = \frac{278}{56.7} = 4.9$$

Assume $P_s = 3,100$ psi
$P_c = (0.38)\ (3,100) = 1,178$ psi
$P_b = 3,100 - 1,178 - 1,922$ psi
From Fig. 8–8, $Q_c = 280$ gpm
Nozzle sizes = two $^{10}\!/_{32}$, one $^{11}\!/_{32}$ in.
Actual $P_b = 1,795$ psi

$$\mathrm{Hp_b} = \frac{(1,795) \times 280}{1,714} = 293$$

$$\mathrm{Hp_b/sq\ in.} = \frac{293}{56.7} = 5.17$$

5. The required combination of surface pressure and circulation rate to obtain an impact force of 14.0/sq in. of hole is determined as follows:

$$\text{Using 3,500 psi, } F_i/\text{sq in.} = \frac{1,007}{56.7} = 17.8$$

Assume $P_s = 3,000$ psi
$P_c = (0.56)\ (3,000) = 1,680$ psi
$P_b = 3,000 - 1,680 = 1,320$ psi
From Fig. 8–8, $Q_c = 350$ gpm
Nozzle sizes = two $^{13}\!/_{32}$, one $^{12}\!/_{32}$ in.
Actual $P_b = 1,240$ psi
$V_n = 305$ fps

$$F_i = \frac{(15 \times 350 \times 305)}{1,932} = 829$$

$$F_i/\text{sq in.} = \frac{829}{56.7} = 14.6$$

In Example 8.6, the minimum circulation for any of the suggested programs is 260 gpm. The annular velocity for 260 gpm is about 120 fpm, and the mud weight is 15.0 lb/gal. The 120 fpm is far more than necessary to remove cuttings. The maximum circulation rate for any of the programs is 350 gpm. For Example 8.6, any circulation rate over 295 gpm results in a turbulent flow pattern in the annulus. If hole erosion is or has been a problem, some operators may choose to reduce the flow rate to maintain laminar flow.

OTHER SPECIAL CONSIDERATIONS IN HYDRAULICS

Many questions arise in designing hydraulics programs, and some are listed below:

1. What are the effects of blanking-off one nozzle and using only two nozzles?
2. Are extended nozzle bits beneficial to bottom-hole cleaning?

3. Do small nozzles in bits increase surge pressure when running pipe?
4. Do jet bits contribute to hole enlargement?
5. Can hole be made by the jetting action of the bit?
6. What are the practical limitations to hydraulics programs in field operations?

Blanking-Off One Nozzle

Using one blank nozzle is common in our industry. The practice was started to prevent the need for very small jets in some hydraulic programs. For example, two ⅜-in. jets are about equal to three 5⁄16-in. jets and in many cases are preferable because of the smaller jets' potential for plugging. The question arises, will bottom-hole cleaning be affected by using two instead of three jets?

A study by Sutko and Myers indicated bottom-hole cleaning is improved at constant power levels by reducing the number of nozzles.[1] Unpublished field work shows improvements in drilling rates, at constant power levels, when using two instead of three bit nozzles. There have been reports of overheating in bearings using only one jet and some isolated incidences of the same problem using two jets.

In general, there seems to be no disadvantage to using two instead of three nozzles in three-cone bits. The advantage of two nozzles appears to be a reduction in the danger of plugging when compared to the same area divided among three nozzles.

Extended Nozzles

The concept of extending nozzles on three-cone bits is an old one. Laboratory results by Sutko and Myers[2] and unpublished work show improvements in bottom-hole cleaning from extended nozzles. Field tests, particularly in Prudhoe Bay and other parts of Alaska, confirm the laboratory tests. The initial problem of breaking the extended nozzles has been overcome.

Currently most bit manufacturers charge more for extended nozzle bits, so the extra cost must be justified in field operations.

Extended nozzle bits should see increased usage. Used properly, these bits permit a reduction in pump pressure when bottom-hole cleaning is adequate and an increase in bottom-hole cleaning when pumps are already used at their capacities.

Small Nozzle Effects on Pressure Surges

Pressure surges are increased by using jet nozzles, but much less than is generally believed. An example of this effect is shown in chapter 7.

Jet Bit Effect on Hole Enlargement

No reported laboratory or field tests show the effect of jet bits on hole enlargement. In most jet bits, the fluid contacts the bottom of the hole toward the outer edge as shown in Fig. 8–9.

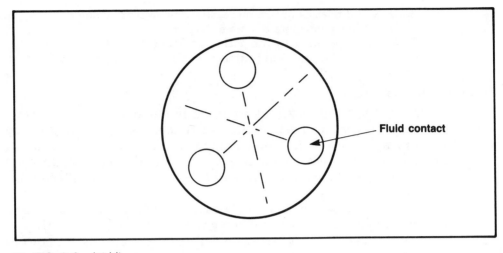

■ **FIG. 8–9** Jet bit

Possibly some hole erosion takes place, particularly at high jet velocities. However, the jetting action of the fluid is not a major reason for hole enlargement. Caliper logs indicate that most hole enlargement is progressive and does not occur when the hole is first opened.

Hole-making Action of Jet Bits

The jetting action of jet bits can make hole. This is confirmed in soft formations in coastal areas where hole is made in many cases with almost no bit rotation. Also, new erosion drilling techniques with very high surface pressures (up to 20,000 psi) resulted in drilling rates of three times normal rock-bit rates in very hard formations. Economic feasibility of erosion drilling is being studied.

PRACTICAL LIMITATIONS TO HYDRAULICS PROGRAMS IN FIELD OPERATIONS

The first limitation was discussed. Many operators limit pump pressure arbitrarily. The potential effects of this on bottom-hole cleaning are shown in Examples 8.5 and 8.6. Also, the specified circulation rates may be difficult to obtain. Power-driven duplex pumps cannot be slowed very much, so reduced circulation rates would require liner changes in the pump. Triplex pumps have more flexibility. However, with direct power drives, smaller liners may still be necessary. The diesel-electric type of drive mechanism offers more flexibility. These can be slowed to reduce circulation rates.

■   **REFERENCES**

1. Sutko, A.A., and Myers, G.M., "The Effect of Nozzle Size, Number, and Extension on the Pressure Distribution under a Tricone Bit," SPE paper no. 3109, fall meeting of SPE of AIME, Houston, Texas, October 1970.
2. Ibid.

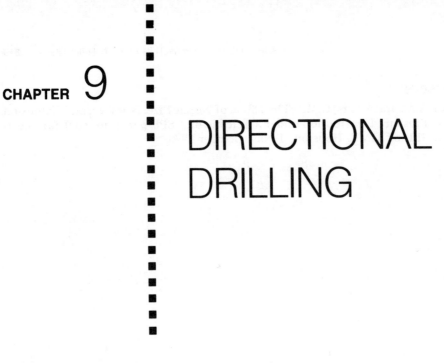

CHAPTER 9

DIRECTIONAL DRILLING

The common procedure of directional drilling must be part of the complete drilling program. Primary emphasis should not be placed on directional control.

CONTROLLING WELL DIRECTION

There are no straight holes. All wells deviate from vertical and experience direction changes. The first step in drilling any well is to determine the desired bottom-hole location. Next, establish the limits that will be placed on hole inclination and deviation during the drilling. Then plan the types of drilling assemblies that will be needed to accomplish the desired objectives. Make alternate plans in case the original plan must be changed.

Operators prefer to control direction with the proper type of bottom-hole drilling assembly. This is not always possible; they may be forced to use tools designed to change both the inclination and direction of the hole. Changes in either hole inclination or direction may require some patience. Gradual changes, if objectives permit, are preferable to abrupt changes.

One secret in controlling hole inclination and direction is to make measurements often. Because time is required, most operators prefer to make the minimum number of measurements. Generally some type of compromise results.

Bottom-hole measuring devices include both magnetic and gyroscopic devices. Also, measurements while drilling, called MWD, commonly are taken. The MWD are recorded from mud pulse readings that are translated at the surface. Additionally, magnetic instruments locate casing in wells blowing out from nearby directional wells. In general, the directional well location needs to be within 50 ft of the casing in the well blowing out before reliable readings can be obtained.

Decisions on drilling practices are based on a series of compromises. Always try to reach the desired objectives, in a trouble-free environment, in the shortest period of time possible.

This chapter dwells on the drilling of directional wells; however, many problems associated with direction control are common to the so-called straight hole. Well direction control is a part of the total drilling program, not a separate consideration.

This discussion emphasizes:

- The common types of directional wells
- Terminology used
- A general survey of directional drilling equipment
- Methods for controlling inclination and hole direction
- Dogleg severity and estimated limits

TYPES OF DIRECTIONAL WELLS

The more common types of directional wells are as follows:

- *Slant-hole.* A special case for directionally drilling shallow wells. The hole is drilled using a slant hole rig with the traveling block on a track attached to the mast.
- *Kickoff and hold a constant angle.* At the selected kickoff point, the well is deviated to the desired angle, and that angle is maintained constant to total depth. The angle buildup section is controlled to minimize any dogleg effects.
- *S-shaped hole.* The well is deviated to the desired angle and then returned to vertical before entering the potential pay zones. This procedure may be used when more than one pay zone in the same vertical plane is to be penetrated.

Fig. 9–1 illustrates the type of directional wells.

DIRECTIONAL DRILLING TERMINOLOGY

Measured depth (MD). Actual drillstring measurement.

Course length. Measured length between survey points.

Drift or inclination angle. Deviation from vertical in degrees.

True vertical depth (TVD). True vertical depth as calculated from the directional survey.

Departure or course deviation. Horizontal distance well is deviated from vertical.

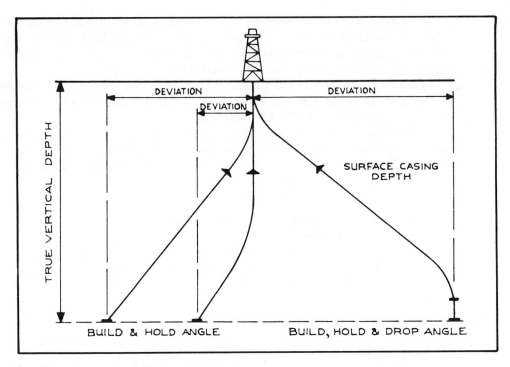

■ **FIG. 9–1** Types of directional wells

Drift direction. The well direction measured in degrees, i.e., N60°E.

Buildup angle. The buildup angle usually is given in degrees per unit of length.

Kickoff point. The well depth at which deflection of the hole is initiated.

Monel collar. A nonmagnetic drill collar used to locate the device for measuring hole inclination and direction.

Closure. Horizontal distance and direction to any specified point in the hole, i.e., 3,000 ft N60°E.

Lead. The practice of anticipating the normal direction and angle at which the bit will drill.

Dogleg. Total change in hole angle due to both the deviation from vertical and a change in hole direction.

Bottom-hole orientation (BHO). Method used to orient directional tool in the desired direction.

Declination. Difference between magnetic and true north lines.

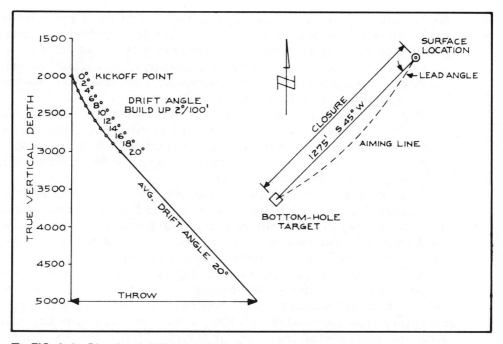

■ **FIG. 9–2** Directional drilling terminology

Bottom-hole location (BHL). The true vertical depth and closure, i.e., TVD 10,000 ft, 3,000 ft N60°E.

Bottom-hole assembly (BHA). The combination of stabilizers and reamers used to control hole deviation and direction.

Fig. 9–2 illustrates the more common terms used in directional drilling.

SURVEY EQUIPMENT

Surveying equipment includes the following:

- Wireline surveys run to determine only inclination
- Single-shot magnetic surveys
- Multishot magnetic surveys
- Gyroscope surveying instruments
- Measurements while drilling, MWD

Wireline surveys are taken routinely in most wells. Most drilling rigs are equipped with wireline units. The frequency with which wireline surveys are run depends on the operator, state regulations, and operating area. Also, many of the inclination measuring instruments are dropped into the hole and pulled with the bit. Most drilling personnel do not like to stop and run a wireline tool to measure hole inclination. During the down time they are afraid of sticking the drillstring. Thus, most of the hole inclination readings are taken when the bit is pulled. However, this practice of running the inclination surveys only on bit trips should be watched closely. One or two misruns on bit trips may result in very long intervals of hole that have not been surveyed. The result may be excessive hole angles that are undetected.

Single-shot magnetic surveys are the foundation instruments in directional drilling. They measure inclination and direction. They may be run on a wireline or dropped into the hole and pulled with the bit. The survey instrument measures magnetic north and must be landed in a nonmagnetic drill collar located just above the bit.

Depending on the area and hole angle, several nonmagnetic collars may have to be run. Both higher hole angles and shorter distances to magnetic north pole increase the required number of nonmagnetic collars. All direction measurements are corrected to the true north pole, because the magnetic north pole changes slightly with time.

All other directional measurements generally are compared with the single-shot measurements. To improve accuracy, the instrument should have maximum resolution for the hole being drilled. This may cause trouble if there has been a sudden large increase in hole angle. To maximize accuracy, increase the frequency of surveys. Most operators compromise on accuracy and run the surveys only when specified during well planning. Some items that must be watched closely when running surveys are listed below:

- The instrument must be set and centralized in the open hole or in a nonmagnetic drill collar.
- The precise depth of measurement must be known.
- Instruments must be calibrated and checked frequently.

The magnetic multishot surveys are run primarily to check single-shot readings. Some difference may be noted in survey results if the multishot surveys are not taken at the same depth as the single-shot surveys. Any difference should be small; if there is a large difference, additional checks are necessary. Under normal circumstances a multishot survey is run routinely before setting casing in directional wells.

There are single-shot and multishot gyroscopic measuring tools. The gyroscope tools must be oriented to true north; these measure direction from the orientation point. The measuring mechanism is actuated by a high-speed rotating mass.

The key to accuracy is the orientation process. The gyroscope may be oriented at the surface or when in place down the hole. Surface orientation is approximate, and downhole orientation is generally preferred. The gyroscopic tools are not affected in any way by magnetic fields. Thus, these tools may be used inside casing or in the open hole.

A new generation of measuring tools allows the determination of hole inclination and direction while drilling, commonly known as MWD. Actually, drilling is stopped and measurements are taken through mud pulse readings at the surface. The measuring time required is about 90 seconds. The forerunner of this technology was the teledrift instrument, which measured only inclination by the same procedure.

The obvious advantage of MWD tools is the savings in rig time and the alacrity with which the survey can be run. In directional drilling particularly there is substantial savings in rig time. Probably the most important savings is a reduction in hole trouble. Sticking pipe when running wireline surveys is not uncommon because of the long shutdown times required. When long-running bits were introduced, more wireline surveys were necessary because surveys had to be taken more frequently than each bit run.

There was some question about the accuracy of MWD, so these were compared with single-shot and multishot magnetic surveys. The accuracy of established MWD companies is now accepted and cross checks are not required.

Initially, the primary disadvantage of MWD tools was the daily cost. They are still much more costly on a per-run basis than regular magnetic surveys. However, the cost of MWD tools is being reduced by competition, as several companies market them.

In addition to measuring inclination and direction, some MWD tools also provide a continuous lithology log of the hole. Also, there are claims that it will be possible to measure actual downhole weight on the bit, downhole rotary speed, circulating annular pressure losses, and mud temperature in the vicinity of the bit. All of these measurements will be invaluable to the development of technology and the early recognition of any downhole problems. With such measuring systems symptoms of impending problems can be diagnosed early, treatments and solutions can be prescribed quickly, and most serious drilling problems can be prevented.

PROCEDURES USED TO DEVIATE A WELL

Procedures used to deviate a well from vertical include the following:

- Jetting techniques
- Bent sub and bottom-hole motor
- Whipstock
- Bit guide tools

Jetting

The jetting technique is shown in Fig. 9–3. The procedure includes the use of one large nozzle and two small ones. The large jet is oriented in the desired direction, pumping is initiated, and a high bit weight is applied. The fluid through the large jet erodes the hole while the string is reciprocated. After making 5–7 ft, the hole is conventionally drilled about

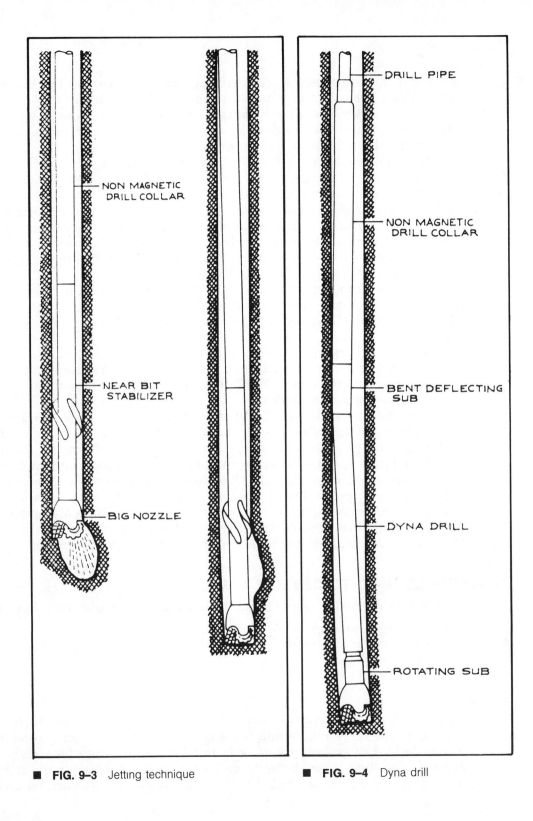

NON MAGNETIC DRILL COLLAR

NEAR BIT STABILIZER

BIG NOZZLE

DRILL PIPE

NON MAGNETIC DRILL COLLAR

BENT DEFLECTING SUB

DYNA DRILL

ROTATING SUB

■ **FIG. 9–3** Jetting technique

■ **FIG. 9–4** Dyna drill

20 ft. The procedure is repeated until the desired angle and direction are obtained.

As with any deflection technique, control is maintained by frequent surveys. The jetting procedure applies primarily to soft formations where hole erosion is possible. A rule-of-thumb has been that jetting can be applied in shale zones where the drilling rate exceeds 50 ft/hr.

Bent Sub and Downhole Motor

This technique is shown in Fig. 9–4. As shown, a sub is bent, generally no more than about 2 degrees, and is run just above the rotating turbine. The orientation procedure is the same as with the jetting technique. Surveys are generally made every 20–30 ft, and any changes in orientation are made as needed. Drilling is continued until the desired deflection is achieved.

Whipstock

The conventional whipstock application is shown in Fig. 9–5. The whipstock is wedge-shaped with a concave channel for guiding the bit. A small bit is run and attached to the whipstock by a shear pin. The whipstock and bit are then lowered to the bottom of the hole and oriented in the desired direction. Weight is applied to force the chisel point of the whipstock into the formation. The shear pin is sheared by applying additional weight to the bit, and the bit is rotated slowly to drill off the set whipstock. After drilling about 20 ft of hole, the bit and whipstock are pulled and the hole is enlarged to full size. After drilling about 30–50 ft, the hole is surveyed and the procedure repeated or drilling continued, depending on the results.

METHODS USED TO CONTROL DIRECTION AND DEVIATION

After obtaining the desired angle and direction, surveys must be continued to ensure that the inclination angle and direction are maintained. In general, maintaining a given inclination angle is not difficult; however, the direction may be another story. A packed-hole assembly is the best method of controlling inclination and direction.

Most bits tend to walk to the right. Sometimes this can be anticipated, and experience will show the initial direction that will be required to hit the target area. Selective stabilization may change the direction of walk. For example, if the contact force guiding the bit is on the low side of the hole, the bit is very likely to walk to the right. If the contact force guiding the bit is on the high side of the hole, the bit is likely to walk to the left.

The rate of walk and direction can often be changed by variations in rotary speed. A slow rotary with other conditions normal generally results in maximum right-hand walk. A fast rotary may reduce the rate of right-

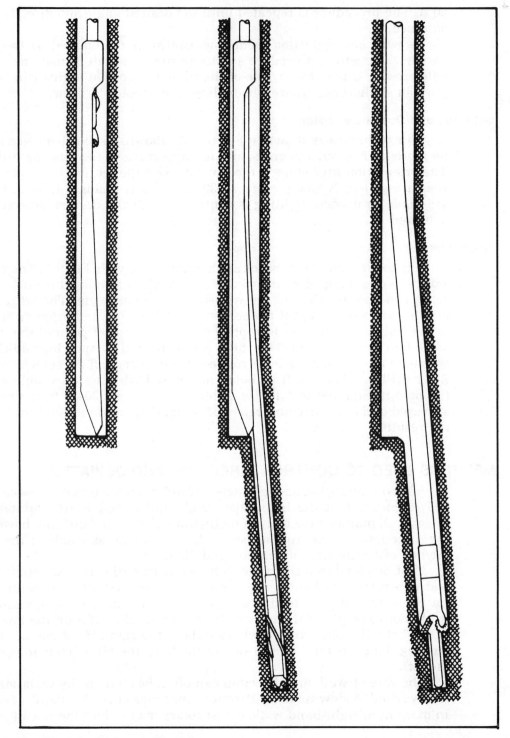

■ **FIG. 9–5** Setting a whipstock

hand walk or, in some cases, results in change of the direction. In all probability, this variation in rotary speed, when successful in changing the direction of walk, succeeds because the forces guiding the bit have been changed.

Sometimes changes in bit weight may be beneficial. Other remedies have been tried, such as using two-cone rather than three-cone bits. Supposedly, right-hand walk can be reduced with a two-cone bit because the bit is better balanced. No specific results in literature or field operations substantiate that claim.

One logical procedure to change the direction of walk is using a drillstring with left-hand threads. One company had left-hand tool joints made along with other equipment including bits to permit turning the drillstring to the left. After several field tests, the project was discontinued because two times out of five right-hand walk occurred while turning the drillstring to the left. Even this result might not have caused the project to be discontinued; however, problems with crews, rig equipment and general attitudes were going to make the use of left-hand makeup pipe uneconomical even if it worked 100% of the time.

A packed-hole assembly is what the name implies. Every effort is made to pack the hole above the bit to keep the bit from changing direction or angle. Some typical packed-hole assemblies are illustrated in Fig. 9–6. Large spiral drill collars may be used as a part of the assembly. Some operators are reluctant to used oversized drill collars because these would be hard to recover if a twistoff occurs. This argument is outdated.

Rebel Tool

The Rebel tool is shown in Fig. 9–7. The primary purpose of the tool is to cause the bit to walk to the left or right as desired. For the tool to work effectively, the hole needs to be at or near gauge with enough support to cause the desired walk. The adjustments of the top and bottom paddles (see Fig. 9–7) are necessary to control the direction of walk.

PROCEDURE TO DETERMINE BOTTOM-HOLE LOCATION

There have been many mathematical procedures introduced to determine the bottom-hole location. Three methods in use are the average tangential, the balanced tangential, and the radius of curvature. The average tangential and balanced tangential are common field procedures, and the radius of curvature generally is used to record the more precise bottom-hole location.

Field personnel sometimes use the balanced tangential procedure instead of angle averaging. The method simply involves taking half the distance between survey points using the first angle and the remaining half of the distance at the final angle. This procedure gives about the same

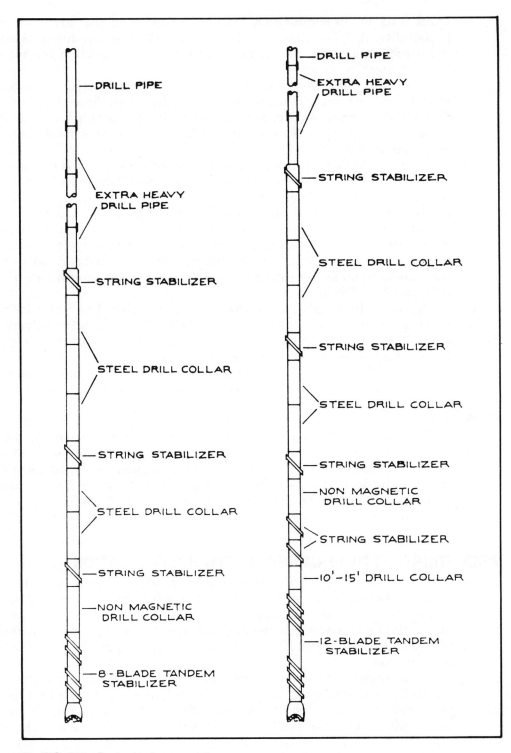

■ **FIG. 9–6** Packed-hole assemblies

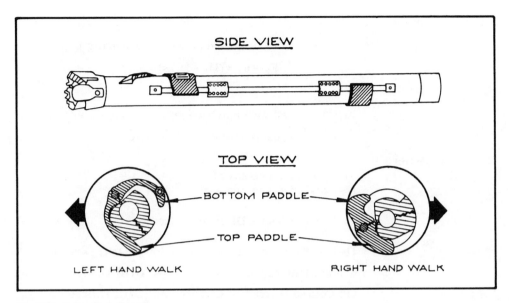

■ **FIG. 9–7** Rebel tool

result as the average tangential procedure and makes the trigonometric tables easier to use.

There are unjustified claims that the angle buildup is never constant, that the hole follows a helical pattern, and that more accurate calculation procedures are needed. If more accuracy is a requirement, more frequent surveys will have to be taken or some type of continuous recording device used. Otherwise, radius of curvature provides an adequate degree of accuracy.

The basic equations used for each procedure are listed as follows:

Average Tangential

The top angle, θ_I, and the bottom angle, θ_{II}, between survey points are averaged, $(\theta_I + \theta_{II})/2 = \overline{\theta}$:

$$V_2 - V_1 = (M_2 - M_1) \cos \overline{\theta} \tag{9.1}$$

where:

$\quad V_2 - V_1 =$ vertical distance between measuring points, ft
$\quad M_2 - M_1 =$ measured distance between measuring points, ft
$\quad \theta_I =$ inclination angle at the first measuring point, degrees
$\quad \theta_{II} =$ inclination angle at the second measuring point, degrees
$\quad \overline{\theta} =$ average direction angle, degrees

$$D_2 - D_1 = (M_2 - M_1) \sin \overline{\theta} \tag{9.2}$$

where:

$\quad D_2 - D_1 =$ horizontal distance between measuring points, ft

$$N_2 - N_1 = (D_2 - D_1) \cos \overline{\theta} \tag{9.3}$$

where:

$$N_2 - N_1 = \text{north distance between measuring points, ft}$$

$$E_2 - E_1 = (D_2 - D_1) \sin \bar{\theta} \qquad (9.4)$$

where:

$$E_2 - E_1 = \text{east distance between measuring points, ft}$$

$$\text{Closure Distance} = \sqrt{E^2 + N^2} \qquad (9.5)$$

where:

$$E = \text{total distance east, ft}$$
$$N = \text{total distance north, ft}$$

$$\text{Closure Direction} = \arctan \frac{E}{N} \qquad (9.6)$$

$$\text{Drift Difference} = \text{Target Direction} - \text{Drift Direction} \qquad (9.7)$$

$$\text{Section Difference} = (D_2 - D_1) \cos (\text{Drift Difference}) \qquad (9.8)$$

$$\text{Off-Course District} = (\text{Closure}) \sin (\text{Drift Difference}) \qquad (9.9)$$

$$\text{Buildup Angle} = D_2 - D_1 = (\text{Buildup Distance})$$

$$\sin \frac{\theta_I + \theta_{II}}{2} + (\text{Constant Angle Distance}) \sin \theta_{II} \qquad (9.10)$$

Radius of Curvature

$$V_2 - V_1 = \frac{180}{\pi} \frac{M_2 - M_1)}{(\theta_{II} - \theta_1)} \left[\sin \theta_{II} - \sin \theta_I \right] \qquad (9.11)$$

$$D_2 - D_1 = \frac{180}{\pi} \frac{M_2 - M_1)}{(\theta_{II} - \theta_1)} \left[\cos \theta_{II} - \cos \theta_{II} \right] \qquad (9.12)$$

$$N_2 - N_1 = \frac{180}{\pi} \frac{(D_2 - D_1)}{(\theta_{II} - \theta_1)} \left[\sin \theta_{II} - \sin \theta_I \right] \qquad (9.13)$$

$$E_2 - E_1 = \frac{180}{\pi} \frac{(D_2 - D_1)}{(\theta_{II} - \theta_1)} \left[\cos \theta_I - \cos \theta_{II} \right] \qquad (9.14)$$

$$\text{Buildup angle} = D_2 - D_1 = \frac{180}{\pi \theta_{II}} (\text{Buildup Distance})$$

$$\times (1 - \cos \theta_{II}) + (\text{Constant-Angle Distance}) \sin \theta_{II}$$

The use of these equations is shown in Example 9.1.

☐ **EXAMPLE 9.1**

Total depth of well = 10,000 ft
Surface casing to be set at 3,000 ft
Target radius = 200 ft
Center of bottom-hole location: N53°E, departure 2,550 ft

$$\text{Kickoff point} = 1,800 \text{ ft}$$
$$\text{Build angle for } 1,000 \text{ ft}$$

Determine: A. Buildup angle

B. Bottom-hole location

A. *Average Tangential*

Buildup angle from 1,800 to 2,800 feet

$$2,550 = 100 \sin \frac{0 + 20}{2} + 7,200 \sin 20$$

$$2,550 = 174 + 2,463 = 2,637$$

Reduce buildup angle to 19°

$$2,550 = 100 \sin \frac{0 + 19}{2} + 7,200 \sin 19$$

$$2,500 - 165 + 2,344 = 2,509$$

B. *Radius of Curvature*

$$2,550 = \frac{180}{\pi} \frac{(1,000)}{(19)} (1 - \cos 19) + 7,200 \sin 19$$

$$2,500 = 164 + 2,344 = 2,508$$

An angle buildup of 19° of 1.9° per 100 ft was used. Table 9–1 shows the solution for the average tangential procedure. Table 9–2 shows the solution for the radius of curvature procedure.

■ **TABLE 9–1** Average tangential procedure

Measured Depth, ft	Course Length, ft	Incli-nation	Vertical Section, ft	Total Vertical Section, ft	Course Departure, ft	Drift Angle, θ	Coordinate Differences, N	Coordinate Differences, E	Rectangular Coordinates, N	Rectangular Coordinates, E
1,800	1,800	0	1,800	1,800		N				
2,300	500	9	498.45	2,298.45	39.23	N10°E	39.08	3.42	39.08	3.42
2,800	500	19	485.15	2,783.60	120.96	N20°E	116.84	31.31	155.92	34.73
3,800	1,000	19	945.52	3,729.12	325.57	N30°E	295.07	137.59	450.99	172.32
4,800	1,000	19	945.52	4,674.64	325.57	N40°E	266.69	186.74	717.68	359.06
5,800	1,000	19	945.52	5,620.16	325.57	N50°E	230.21	230.21	947.89	589.27
6,800	1,000	19	945.52	6,565.68	325.57	N60°E	186.74	266.69	1,134.63	855.96
7,800	1,000	19	945.52	7,511.20	325.57	N70°E	137.59	295.07	1,272.22	1,151.03
8,800	1,000	19	945.52	8,456.72	325.57	N80°E	84.26	314.48	1,356.48	1,465.51
9,800	1,000	19	945.52	9,402.24	325.57	S88°E	22.71	324.78	1,379.19	1,790.29
10,435	635	19	600.40	10,002.64	206.73	S75°E	−30.56	204.46	1,348.63	1,994.75

Closure = $(1,348.63^2 + 1,994.75^2)^{1/2}$ = 2,407.87 ft

Closure Direction = arctan $\frac{E}{N}$ = arctan $\frac{1,994.75}{1,348.63}$ = N55.94°E

Drift Difference = Target Direction − Drift Direction = N53°E − N55.94°E = N(−2.94°)E

Off-Course Distance − (Closure) sin (Drift Distance) = 2,407.87 sin 2.94 = 123.50 ft

■ **TABLE 9–2** Radius of curvature

Measured Depth, ft	Course Length, ft	Incli-nation	Vertical Section, ft	Total Vertical Section, ft	Course Departure, ft	Drift Angle	Coordinate Differences, N	E	Rectangular Coordinates, N	E
1,800	1,800	0	1,800	1,800	0	N				
2,300	500	9	497.95	2,297.95	39.19	N10°E	39.03	3.41	39.03	3.41
2,800	500	19	484.53	2,782.53	120.80	N20°E	117.06	31.23	156.09	34.64
3,800	1,000	19	945.52	3,728.00	325.57	N30°E	294.69	137.42	450.78	172.06
4,800	1,000	19	945.52	4,673.52	325.57	N40°E	266.35	186.50	717.13	358.56
5,800	1,000	19	945.52	5,619.04	325.57	N50°E	229.92	229.92	947.05	588.48
6,800	1,000	19	945.52	6,564.56	325.57	N60°E	186.50	266.35	1,133.55	854.83
7,800	1,000	19	945.52	7,510.08	325.57	N70°E	137.42	294.69	1,270.97	1,149.52
8,800	1,000	19	945.52	8,455.60	325.57	N80°E	84.16	314.08	1,355.13	1,463.60
9,800	1,000	19	945.52	9,401.12	325.57	S88°E	22.66	324.18	1,377.79	1,787.78
10,435	635	19	600.40	10,001.52	206.73	S75°E	−31.05	204.02	1,346.74	1,991.80

Closure $= (1,346.74^2 + 1,991.80^2)^{1/2} = 2,404.37$

Closure Direction $= \arctan \dfrac{1,991.80}{1,346.74} = 55.94$

Drift Difference $=$ N53°E $-$ N55.94°E $=$ N($-2.94°$)E

Off-Course Difference $= 2,404.37 \sin 2.94 = 123.32$ ft

PROBLEMS ASSOCIATED WITH DIRECTIONAL DRILLING

The primary problems associated with directional drilling are:

- Specific drilling assemblies that can be used to build, drop, or hold angles
- The effect of drilling practices on hole angles
- The effect of hole deviation on hole cleaning
- The determination of dogleg severity
- Critical doglegs for various grades of pipe

Build Hole Angle

Bottom-hole assemblies used to build hole angle are shown in their order of suggested use in Fig. 9–8.

The near-bit stabilizer is the primary angle-building tool. The normal weight of the drillstring tends to help drop angle because any assembly has a tendency to hang in a vertical plane. When hanging free in part B of Fig. 9–8, stabilizer number 2 would simply reduce the rate of building angle. For example, if the rate of building angle is just above the desired rate, then, in addition to the near-bit stabilizer, another stabilizer could be run 90 ft above the bit. To reduce the rate of building angle, the second stabilizer can be moved closer to the bit. Stabilizer number 3 is optional and may or may not be used. Stabilizer 3 gives more stability to the string and sometimes improves angle control.

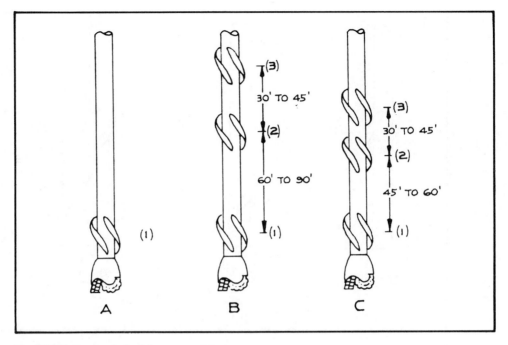

■ **FIG. 9–8** Angle-building assemblies

Reduce Hole Angle

Bottom-hole assemblies used to reduce hole angle are shown in their order of suggested use in Fig. 9–9.

The secret of dropping angle is to space stabilizer 1 as far above the bit as possible and still have no wall contact by the pipe between the stabilizer and the bit. A normal distance is 60 ft as shown in part A; however, in some cases with low bit weights in low-angle holes, more angle may be dropped by using stabilizer 1 more than 60 ft above the bit.

The hole angle drop would be more moderate with part B because the length of the pendulum has been reduced. However, if high bit weights are utilized, it may be necessary to move the stabilizer from 60 ft to 45 ft to reduce hole angle.

Part C is a compromise assembly and may include more stabilizers than shown. Part C is used when some hole angle reduction is desired but not essential with the next bit run. Also, an assembly such as part B with a second stabilizer spaced about 30 ft above stabilizer 1, which is placed 45 ft above the bit, may replace part C if that assembly fails to drop angle an acceptable amount.

Angle Holding Assemblies

Bottom-hole assemblies used to hold angle are shown in their order of suggested use in Fig. 9–10.

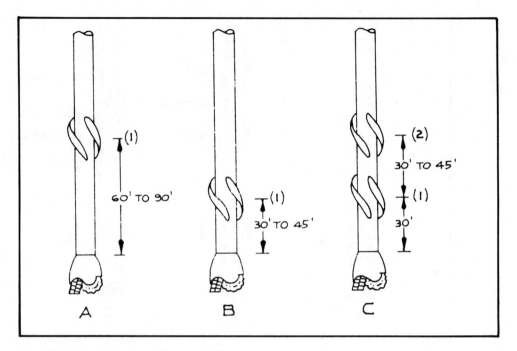

■ **FIG. 9–9** Angle-dropping assemblies

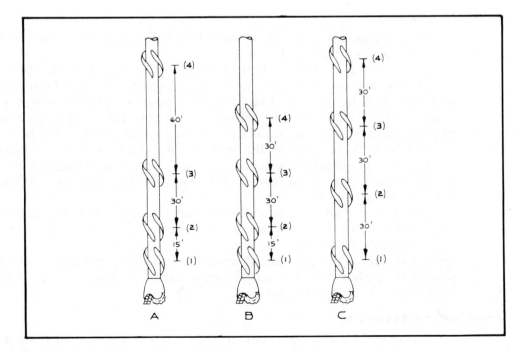

■ **FIG. 9–10** Angle-holding assemblies

The stiffer the bottom-hole assembly near the bit, the more difficult for the bit to move off course. The near-bit stabilizer is generally spaced about 3–5 feet above the bit. Part A in Fig. 9–10 represents a commonly used assembly. Sometimes stabilizer 3 is moved to 15–20 ft above stabilizer 2. In some cases, part B may be better than part A. Also, square drill collars may be used in place of stabilizers. More contact area means greater control.

One other word of warning is necessary in hole stabilization. If stabilizers wear or if the hole enlarges above gauge size, the results may be changed substantially from those expected.

DRILLING PRACTICES

Regardless of the drilling practices, formation bed dips and formation hardness substantially contribute to changes in hole direction and deviation. The general tendency for a bit is to drill up-dip up to a bed dip of about 45 degrees. Above 45 degrees, the bit may tend to drill down-dip. This does not mean that at 44 degrees the bit tends to drill up-dip each time, and at 46 degrees the bit tends to drill down-dip. However, it does mean that in this critical range, the operator has to be more careful.

Changes in formation hardness also affect results. For example, if the bit is drilling in hard formations and a sudden reduction in formation hardness is encountered, the general tendency is to lose angle. The reverse may be true if the formations change from soft to hard. For these reasons, a sudden change in drilling rate with no changes in drilling practices should be carefully checked when hole inclination and direction are critical.

To some extent, drilling practices affect hole direction. Drilling practices that may affect, or have been claimed to affect, hole direction are bit type, bit weight, and rotary speed.

Bit Type

There is no conclusive evidence that the bit type plays an important role in hole deviation. Claims have been made that two-cone or four-cone bits are better than three-cone bits in controlling hole deviation. There are no specific drilling results that justify the claims. Evidence does exist that an increase in offset in a specific bit increases the tendency of the bit to walk to the right and may also contribute to an increase in hole inclination. Generally, an increase in offset in a bit is used as a means to increase drilling rate. Thus, the general tendency is always to work toward an increase in bit offset as new bits are introduced.

There is a trend toward diamond compact bits. These fixed-blade bits do not walk to the right like the roller-cone bits. Right-hand walk with roller-cone bits is basically predictable and is used in planning directional wells.

Predicting the direction of walk for a fixed-blade bit is difficult. As the fixed-blade bits become a standard item, probably the direction of walk for a given drilling assembly will be predictable.

Bit Weight

General beliefs are that an increase in bit weight increases hole angle and that a decrease in bit weight decreases hole angle. Practice shows that an increase in bit weight with a holding assembly may have no effect on hole direction, particularly if a very stiff bottom-hole assembly is being used.

On the other hand, if a pendulum assembly is implemented to drop angle, an increase in bit weight reduces the effectiveness of the pendulum and would be expected to reduce the rate hole angle is dropped. If hole angle is being built, an increase in bit weight may have no effect, or it may increase or decrease the rate of building hole angle. Again, if a very stiff bottom-hole assembly is being used, bit weight may not affect hole angle.

Rotary Speed

Another common belief in straight-hole drilling is that an increase in rotary speed and/or a decrease in bit weight help keep the hole straight. There is no conclusive evidence that a change in rotary speed affects hole direction or walk predictably. However, changes in rotary speed are used often to obtain a change in hole inclination or walk.

Actually, any change in drilling practices may result in a change in hole inclination or walk by changing the points of wall contact with the drilling assembly. For example, if the primary point of wall contact is on the low side of the hole, as is normal, the bit will walk to the right if the pipe is turned to the right. If the primary point of contact between the drilling assembly and the wall can be changed to the high side of the hole, the bit will walk to the left when the pipe is rotated to the right.

EFFECT OF HOLE DEVIATION ON HOLE CLEANING

Hole cleaning will be affected by hole angle because cuttings removal depends on the vertical component of fluid velocity rather than normal calculated annular velocity. The vertical component of fluid velocity can be calculated using Eq. 9.15:

$$\overline{v}_v = \overline{v} \cos \theta \qquad (9.15)$$

where:

$$\overline{v} = \text{vertical component of velocity, ft/min}$$

The use of this equation is shown in Example 9.2.

☐ **EXAMPLE 9.2**

Assume:

$$\bar{v} = 150 \text{ ft/min}$$
$$\theta = 60°$$
$$\bar{v}_v = 150 \cos 60° = 75 \text{ ft/min}$$

Example 9.2 shows the effective annular fluid velocity to be only 75 ft/min. Actually, that may be a problem in hole cleaning; however, the biggest problem will probably be cuttings that are bypassed on the low side of the hole. These cuttings may be removed by viscous sweeps. Also, it will be helpful to use heavyweight drillpipe.

DOGLEG SEVERITY

A change in deviation or direction creates a dogleg. Dogleg severity may be calculated using Eq. 9.16. In addition, the effect of doglegs on drillstring failures and graphs of critical dogleg severity have been published by Lubinski and Nicholson.[1,2,3]

$$\text{Dogleg severity} = \frac{100a}{M_2 - M_1} \qquad (9.16)$$

where:

$$a = \cos^{-1}\{\cos\phi_1 \cos\phi_2 + (\sin\phi_1 \sin\phi_2)[\cos(\theta_2 - \theta_1)]\}$$

Dogleg severity is generally given in degrees/100 ft. Eq. 9.16 permits the expression of dogleg severity in any terms. Example 9.3 shows the use of Eq. 9.16.

☐ **EXAMPLE 9.3**

First survey:

Inclination angle $\phi_1 = 10°$
Direction angle $\theta_1 = N30°E$
Depth = 3,500 ft

Second survey:

Inclination angle $\phi_2 = 13°$
Direction angle $\theta_2 = N40°E$
Depth 3,600 ft
$$a = \cos^{-1}[\cos 10° \cos 13° + (\sin 10° \sin 13°)(\cos 10°)] = 3.59°$$
$$\text{Dogleg severity} = \frac{(100)(3.59)}{100} = 3.59°/100 \text{ ft}$$

API Bulletin D8 shows dogleg severity to be 3.59°/100 ft. Eq. 9.16 is in common use and is generally used as the official procedure in determining dogleg severity.

Doglegs in the hole affect drillstring wear, the running of wireline tools, the running of casing, and, if severe, could affect production operations. Drillstring wear has been considered the most serious problem; generally, the control mechanism is used to limit dogleg severity.

Lubinski initiated work in the areas of dogleg severity and has been instrumental in developing field technology in controlling hole deviation.[4] Nicholson in many cases has expanded and added to work done by Lubinski.[5]

Two considerations are necessary when considering fatigue of the drillstring. First is the potential failure of tool joints. Fig. 9–11, taken from Nicholson, shows a determination of permissible dogleg severity on tool joints. The dogleg severity numbers given in Figs. 9–12 and 9–13 are considered conservative. However, most pipe is exposed to corrosive environments, and it is not uncommon to have some damage to pipe in the normal handling process. Corrosion and pipe damage reduce the fatigue resistance, and most operators prefer to be conservative.

Sometimes the maximum permissible doglegs are determined prior to drilling, particularly in the buildup portion of a directional well. In fact, dogleg limitations may determine the kickoff point and the length of the buildup portion of the hole.

The use of Figs. 9–11, 9–12, and 9–13 is shown in Example 9.3.

☐ **EXAMPLE 9.3**

> Mud weight = 16 lb/gal
> Dogleg at 3,000 ft
> Well depth = 12,000 ft
> Drillpipe = 4½-in., 16.6 lb/ft, grade E
> Length of drill collars = 1,000 ft
> Collar weight = 60,000 lb in 16-lb/gal mud
> Weight on bit = 60,000 lb

Determine:
    The maximum permissible dogleg severity based on tool joint fatigue

The answer is shown in Fig. 9–11 to be 3.5°/100 ft for Range 2 drillpipe and less than 2.5°/100 ft for Range 3 drillpipe. If the lateral force on the tool joints is allowed to go to 3,000 lb, the permissible dogleg is 5.2°/100 ft for the Range 2 pipe and 3.4°/100 ft for the Range 3 pipe.

Using the same data shown in Example 9.3, the maximum acceptable dogleg for the drillpipe from Fig. 9–12 is 4.5°/100 ft. Whether the operator chooses to use 3.5° or 4.5°/100 ft will probably depend on his past experience and the problems involved in correcting any dogleg.

In straight-hole or directional drilling, remember that keeping the bottom-hole assembly simple and being patient are two of the best ways to reach the desired results. Direction and inclination changes in the hole are affected by bed dips and lithology changes. If the bottom-hole assembly is changed frequently, this can result in hole problems and raise the drilling cost substantially. If changes in bottom-hole assembly are made to either build or drop angle, the results may not be immediate. Thus, before

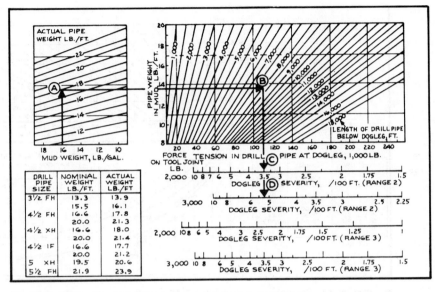

■ **FIG. 9–11** Permissible dogleg severity on tool joints

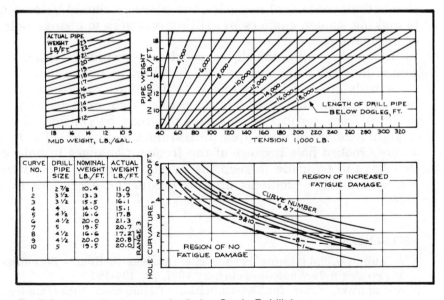

■ **FIG. 9–12** Dogleg severity limits, Grade E drillpipe

making further corrections, exercise patience, particularly if the change
has a history of being successful.

The inertia to change is often a function of the density of the medium
in which it is operating. Change in direction occurs quickly in air. An
oceanliner changes direction slowly. The bit changes direction much more

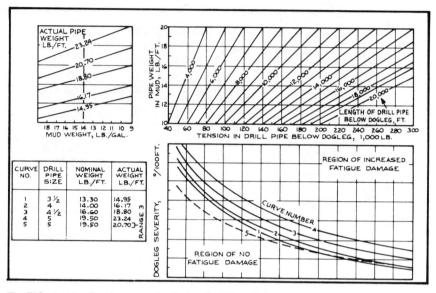

■ **FIG. 9–13** Dogleg severity limits, S-135 drillpipe

slowly than the oceanliner. Also, the operator should be careful relative to making sudden changes in drilling practices to change hole direction or inclination.

Remember, one of the biggest problems encountered in a hole is doglegs caused by sudden changes in hole direction or inclination.

A truly crooked hole is one in which there are frequent changes in hole inclination and direction. Even in straight-hole drilling, a hole that deviates from vertical as much as 20° may be considered a relatively straight hole. In the same area, a maximum inclination of 6° may produce a very crooked hole because of the frequent changes in hole inclination that have to be made to keep the maximum to 6°. The latter case will also create more problems in later production operations than the 20° hole.

■ **REFERENCES**

1. A. Lubinski, "Maximum Permissible Dog-legs in Rotary Boreholes." *Journal of Petroleum Technology,* February 1961, pp. 194–195.
2. A. Lubinski, "Chart for Determination of Hole Curvature (Dog-leg Severity)." *Petroleum Engineer,* February 1957.
3. Robert W. Nicholson, "Acceptable Dogleg Severity Limits," *Oil & Gas Journal,* 15 April 1974, pp. 73–82.
4. Lubinski, *JPT,* 1961.
5. Nicholson, op. cit.

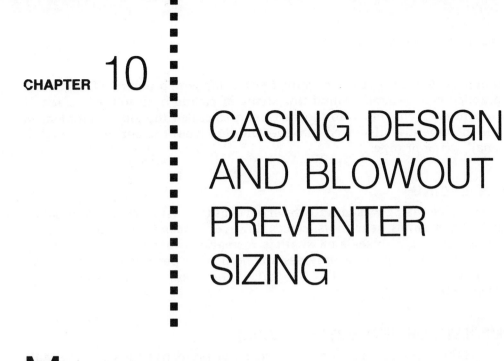

CHAPTER 10

CASING DESIGN AND BLOWOUT PREVENTER SIZING

Many questions arise when sizing the equipment for a job. Often safety factors must be altered so available equipment can be used. Because there are no established rules for these, safety factors are varied frequently to meet specific requirements. This chapter's discussion includes design criteria for casing and blowout preventers and instructions on how to select casing seats based on pore pressure requirements. Casing design concepts relate to using the casing string that meets requirements at the lowest possible cost.

Before considering economics, you need to know the setting depths of the casing strings, whether plans must be made to control corrosion, and whether special considerations may be necessary to account for casing wear, which is related to rotating time. The setting depths of casing strings may be based on governmental regulations, hole requirements, or routine practices. Casing strings are generally identified by their purposes in the hole. The types of casing string are listed below.

STRUCTURAL CASING OR DRIVE PIPE

This string of structural casing or drive pipe provides initial hole support close to the surface and usually varies in length from about 30 to 300 ft. The casing may be driven in or a hole drilled. If it is necessary to drill a hole, the casing may be cemented in place.

CONDUCTOR CASING

Conductor casing in some areas may be the first casing string set and cemented. In others, particularly offshore, this is the first casing string that provides any protection against formation pore pressures. The general practice is to cement around this string of casing back to the surface. If cement does not return to the surface, a so-called top-job is common, in which cement is displaced into the annulus from the surface through a small string of pipe.

SURFACE CASING

Surface casing was initially referred to as the casing string set to isolate freshwater sands near the surface. However, depending on the location, surface casing also seals off shallow formations that may cause drilling problems. In addition, it provides a conduit that permits the use of mud weights in the range of 12–14 lb/gal in the pressure transition zone of deep wells.

INTERMEDIATE OR PROTECTIVE CASING

As the name implies, the intermediate or protective casing string seals off open formations so the well can be drilled deeper. In some instances, such as when production liners are used, this casing may also be part of the production casing. More than one string of intermediate casing may be required to reach the total depth.

PRODUCTION CASING

The string of production casing is set and cemented to isolate the productive formations. The amount of cement depends on company preference. This amount is based on the volume needed to bring the cement back to a predetermined height. The cement volume may be determined from caliper log calculations or may be a given amount of excess based on experience. Remember that cement travels more readily up the large side of the hole. As a result, more excess than calculations indicate may have to be used to cover the casing adequately in the small part of the hole if the casing is not centralized in the hole.

LINERS

Liners are strings of pipe that do not extend back to the surface. A primary advantage of using liners is the savings in steel costs. Also, liners may be run more quickly than full casing strings, which improves the chances of getting the liner to the desired depth.

Both protective and production liners are set; their purpose is the same as that of a full string of casing. One problem with liners is getting a good cement seal around them. Liners are often neither moved nor centralized when cementing; as a result, getting cement on all sides of the liner is difficult. Liner hangers that permit liner movement during cementing are available, and to ensure a good cement job liners should be moved. Other problems associated with liners include normal clearances, which are sometimes small, particularly with deep production liners.

Casing string designs are based on tension, collapse pressure, and burst pressure. Compression may be considered when landing casing. Most casing design criteria use the minimum weight and grade of casing that meets the hole requirements. Specific designs often depend on the casing available either from company stock or from a company's normal supplier. Before designing the casing the following information is necessary:

- Setting depths
- Formation pore pressures and mud weights
- Formation behavior, such as salt sections, that may result in collapsed casing during later production operations
- Corrosive characteristics of the fluids to be produced
- Rotating time in the casing
- Auxiliary operations that may be necessary, like squeeze cementing

Casing setting depths depend on the following:

- Governmental requirements
- Routine practices
- Hole problems
- Pore pressures

Governmental regulations generally meet minimum requirements. The rules may be specific, based on past problems, or flexible, depending on what happens while drilling. Governmental rules may be altered with special permission or changed if hole information justifies this.

Routine practices follow casing programs that have been used in the field. Sometimes routine depths for surface casing are based more on practice than on requirements. Protective casing is sometimes set by the amount of open hole rather than by a drilling requirement. With the emphasis on environmental protection, efforts always are made to be completely safe.

One safety theory includes the concept that each casing string must provide complete protection from a potential blowout from open formations below the casing. Carrying this idea a step further, assume that the hole is emptied when the well kicks and that the only hydrostatic pressure is the weight of gas. Also assume that all formations will fracture if the pressure against them reaches the rock overburden gradient of 1.0 psi/ft.

Using a gas gradient of 0.1 psi/ft and a normal pore pressure of 0.465 psi/ft, the setting depth of each subsequent string of casing is determined by dividing the last casing depth by 0.365, or (0.465 − 0.10). If the last casing string was set at 4,000 ft, the next string of casing would be set, regardless of other conditions, at 10,959 ft. This would be a ridiculous requirement because, when drilling into a pressure transition zone, there would be no way to set enough casing strings.

For example, assume the casing is set at 4,000 ft, and the pore pressure at 10,000 ft is comparable to a gradient of 0.73 psi/ft. On the same basis, the casing should have been set at 6,349 ft, or (4,000 ÷ 0.63). However, the pressure at 6,349 ft or, for that matter, at 9,000 ft was normal, and no casing string was actually required. For this reason, the setting depths of casing for pressure control should be based on the fracture gradients, the pore pressures, the differential pressures across exposed formations, and the volumetric capabilities of the exposed formations.

Take precautions so that, if mistakes are made, the well can be contained without losing control either at the surface or underground. If by accident control is lost, the well can be brought under control without excessive pollution underground or at the surface.

One procedure for selecting the casing setting depth from the pressure control requirements is to determine the maximum requirements at the bottom of the hole and work toward these requirements from the surface casing setting depth. This procedure is shown in Example 10.1.

☐ **EXAMPLE 10.1**

Planned well depth = 15,000 ft
Bottom-hole pressure anticipated = 13,800 psi
Surface casing to be set at 4,000 ft in normal pore pressure zone
Production casing or liner at 15,000 ft = 4½ in.
Anticipated top of the pressure transition zone = 9,800 ft

Assuming the necessary mud weight is at least 0.5 lb/gal more than that required to control the bottom-hole pressure, the total hydrostatic pressure exerted by the mud is as follows:

$$(0.05 \text{ lb/gal})(0.052)(15,000) + 13,800 = 14,190 \text{ psi}$$

or,

$$\text{the mud weight required} = \frac{14,190}{(15,000)(0.052)} = 18.2 \text{ lb/gal}$$

To reach this level, all open sections of the hole must withstand a mud weight equal to 18.2 lb/gal. This is a pressure gradient of 0.946 psi/ft. To be precise, the pore pressures and fracture gradients at all depths must be known before the exact setting depths of the casing can be determined.

In this example, Eaton's procedure for calculating fracture gradients in the Gulf Coast is used.[1] The procedure used to select a casing seat is valid in hard-rock areas if fracture gradients are determined for hard-rock areas. Just below the surface casing seat in normal pressure formations, the fracture gradient is estimated as follows:

Poisson's ratio = 0.39
Overburden gradient = 0.895 psi/ft

Normal pore pressure gradient = 0.465 psi/ft

$$F_{mo} = (0.895 - 0.465)\left(\frac{0.39}{0.61}\right) + 0.465 = 0.74 \text{ psi/ft}$$

If the maximum mud weight is assumed to be 0.5 lb/gal less than the fracture gradient equivalent in pounds per gallon, the maximum mud weight is limited to 13.73 lb/gal. Also assume that the mud weight should exert a pressure equivalent to 0.5 lb/gal in excess of the pore pressure at the point where the first string of protective casing is set. This means the pore pressure gradient at the protective casing seat cannot exceed an equivalent mud weight of 13.23 lb/gal. For this problem, assume that the protective casing must be set at 11,000 ft where the pore pressure gradient is 0.688 psi/ft.

The fracture gradient at 11,000 ft is calculated using Eaton's procedures.

Poisson's ratio = 0.45
Overburden gradient = 0.96 psi/ft
Pore pressure gradient = 0.688 psi/ft

$$F_{mo} = (0.96 - 0.688)\left(\frac{0.45}{0.55}\right) + 0.688 = 0.91 \text{ psi/ft}$$

The required mud weight gradient at total depth was 0.946 psi/ft, which is considerably larger than the fracture gradient of 0.91 psi/ft at 11,000 ft. Another string of protective pipe must be set before reaching the total depth. If room is left for circulating the fluid, the mud gradient cannot exceed 0.884 psi/ft; the pore pressure of the formation where the next string of pipe is set should not exceed 0.858 psi/ft. Based on the assumed pore pressure data in Fig. 10–1, the next string of pipe should be set at 13,920 ft. The fracture gradient at 13,920 is calculated as follows:

Poisson's ratio = 0.46
Overburden gradient = 0.97 psi/ft
Pore pressure gradient = 0.858 psi/ft

$$F_{mo} = (0.97 - 0.858)\left(\frac{0.46}{0.54}\right) + 0.858 = 0.953 \text{ psi/ft}$$

Based on the required mud weight of 18.2 lb/gal or 0.946 psi/ft at 15,000 ft, reaching 15,000 ft without problems will be difficult. The fracture gradient of 0.953 psi/ft below the casing seat at 13,920 ft limits the maximum mud weight to just above 17.8 lb/gal, or 0.927 psi/ft, if the mud weight is maintained at a level of 0.5 lb/gal below the fracture gradient.

In this case, some compromise is necessary. The difference between the fracture gradient at 13,920 ft and the pore pressure gradient at 15,000 ft is 0.033 psi/ft, or (0.953 − 0.920). This simply means that the margins of safety must be reduced to about 0.32 lb/gal. At 13,920 ft the circulating

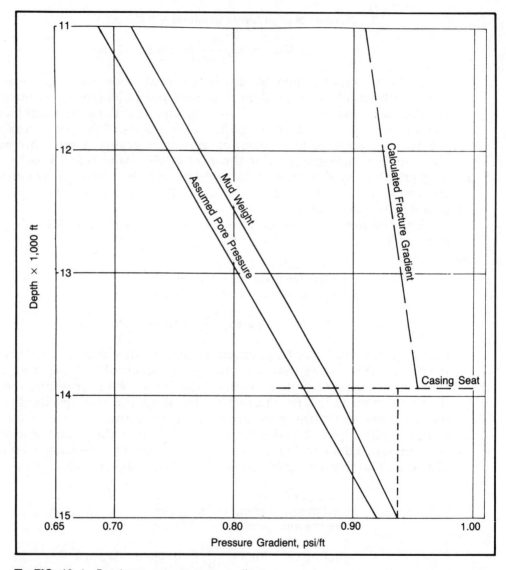

■ **FIG. 10–1** Depth vs pore pressure gradient

pressure cannot exceed 239 psi, and at 15,000 ft the swab pressure reductions must be limited to 248 psi. To accomplish this, the operator will have to keep the mud extremely thin, pull the pipe slowly, and watch all indications of trouble very carefully.

The mud weight and estimated fracture gradients are given in Fig. 10–1. Information such as that shown in Fig. 10–1 can be preplanned; however, there is no substitute for listening to the well as drilling progresses and correcting the plan when the need arises.

If the operator arbitrarily set the protective pipe at 10,000 ft in Example 10–1, reaching the total depth with only one additional string of protective pipe would be difficult.

The pore pressure gradient at 10,000 ft is 0.57 psi/ft. In soft-rock areas, the calculated fracture gradient would be 0.87 psi/ft. If the mud weight is kept at a level equivalent to 0.5 lb/gal less than 0.87 psi/ft, the maximum mud weight that can be used is 16.23 lb/gal; the maximum pore pressure gradient of the open formations will be 0.818 psi/ft. According to Fig. 10–1, this pore pressure will be reached at 13,230 ft. Thus, the 7-in. pipe must be set at 13,230 ft. The calculated fracture gradient at 13,230 ft is 0.947 psi/ft.

The mud weight at 15,000 ft must exceed the pore pressure gradient of 0.92 psi/ft at 15,000 ft. To be safe, the mud weight should be at least 0.5 lb/gal more than the pore pressure. If the fracture gradient is 0.947 psi/ft at 13,230 ft, circulating a mud that imposed a pressure gradient of 0.946 psi/ft at 15,000 feet would be impossible. Actually, the operator might manage it by reducing his mud weight a little and keeping the mud very thin. However, the chances are that he will lose the well.

Another alternative in this case is to set the surface casing at a greater depth, which would permit setting the first string of protective pipe deeper. The time to make these decisions is in the well planning stage. In this particular case, if surface casing had been set at 6,000 ft instead of 4,000 ft, total depth could have been reached with the desired safety factors.

The operator determines whether to set liners or casing after setting the 9⅝-in. protective casing string. If liners are used, the 9⅝-in. casing must be designed to withstand the pore pressures at 15,000 ft. To reach 15,000 ft and set a 4½-in. liner or casing string requires the following strings:

- Drive or structural pipe
- Conductor pipe
- Surface pipe
- First string of protective pipe
- Second string of protective pipe
- Production pipe

The required casing and hole sizes depend on the requirements of a given area; however, one combination is shown in Table 10–1.

All of the facts in these examples have been assumed. Fracture gradients have been calculated. In actual practice, particularly in coastal areas, use casing seat leakoff tests to determine fracture gradients. The proper selection of all casing seats may be the key to reaching the required depth with the desired casing size. Even in development areas, there are no guarantees that the formation and pore pressures expected will be those encountered. For this reason, plan alternative courses of action for unexpected situations.

■ **TABLE 10–1** Sample casing design

Pipe String	Hole Size, in.	Casing Size, in.	Setting Depths, ft
Drive	—	30	100
Conductor	24	20	800
Surface	17½	13⅜	4,000
Protective	12¼	9⅝	11,000
Protective	8½	7	13,920
Production	6⅛	4½	15,000

The casing seat in Example 10–1 was selected from pore pressures and fracture gradients. Other factors may have to be considered when choosing casing seats. Many times hole problems force the setting of casing before the pore pressure or fracture gradient indicates it should be set. Hole problems that may force casing setting include severe sloughing or heaving shale, open salt sections, and pipe sticking problems caused by high pressure differentials across permeable zones.

Salt sections may alter casing and mud programs. At times, casing is set to seal off open salt sections, and then freshwater muds are used. Some salt sections may be penetrated and left open if saltwater or oil-based muds are used. The final decision could depend on both economic and environmental conditions.

Differential pipe sticking can force the setting of casing. In some areas, pipe sticking problems can be related directly to the pressure differentials across open permeable zones. For example, with permeable zones open, a pressure differential of 1,500 psi in some Gulf Coast areas signals a potential pipe sticking problem. Because many other factors (discussed in chapter 3) also affect pipe sticking, the pressure differential is only one criterion. Consequently, the operator may decide that the best economic choice is to set casing.

Remember that while many factors affect the selection of casing seats, most of them leave the operator a choice. However, pore pressures and fracture gradients generally do not leave an option; reaching the total depth may depend on setting the last casing string at a depth that allows the setting of the next string of casing deep enough to reach total depth before the operator runs out of hole. As demonstrated by Example 10–1, the operator may run out of alternatives if casing is set too shallow. Many times this is the reason the total depth is not reached.

CASING DESIGN

Designing casing strings is generally related to minimizing cost. The minimum performance properties of the casing are set by the American Petroleum Institute (API) standards or, in the case of non-API grades, by the manufacturer. These performance ratings are affected by the tension imposed when setting casing. Changes in performance are considered in the design program.

Casing designs differ with various operators. Some consider buoyancy effects on the casing; others do not. Safety factors vary, but those commonly used are listed below:

Tension	1.6–1.800
Collapse	1.0–1.125
Burst	1.0–1.250

Sometimes the safety factor in tension is reduced to 1.6 and, in some cases, to 1.3. Not all operators consider buoyancy, and some use a safety factor less than 1.0 for collapse because they assume that the casing is completely empty, which is seldom, if ever, the case. Thus, rather than estimate the potential pressure imposed inside the casing, safety factors of less than 1.0 are used.

Safety factors for burst are less likely to be reduced than those for collapse. Burst becomes an important consideration in high-pressure gas wells when the bottom-hole pressure is reduced only by the pressure imposed by a column of gas. In oil wells or wells that produce high quantities of liquids, burst safety factors are often lowered.

The first step in the design of a casing string is to determine the available casing. Table 10–2 shows the API grades of casing, and Table 10–3 lists some of the more common non-API grades.

Note that, in general, the lower grades of steel are more resistant to hydrogen sulfide embrittlement than the higher grades. Special efforts must be made to minimize corrosion in hydrogen sulfide environments. As a result, it may be necessary to use lower grades of steel and higher weights. The higher casing weights restrict the internal clearances and may cause the operator to use smaller-sized bits for the remainder of the

■ **TABLE 10–2** API casing grades

	Yield strength, psi	
Grade	Minimum	Ultimate
H–40	40,000	60,000
J–55	55,000	75,000
K–55	55,000	75,000
C–75	75,000	95,000
N–80	80,000	100,000
C–95	95,000	105,000
P–110	110,000	125,000

■ **TABLE 10–3** Non-API casing grades

Grade	Yield strength, psi	
	Minimum	Ultimate
S80	55,000	95,000
S0090	90,000	105,000
SS95	95,000	100,000
S95	95,000	110,000
S105	95,000	110,000
S00125	125,000	135,000
S00140	140,000	150,000
V150	150,000	160,000
S00155	155,000	165,000

hole. Also, higher casing weights increase the tension support requirements for casing close to the surface.

The general plan for designing casing strings is to start at the bottom of the hole and work toward the surface. At the bottom of the hole, consider collapse and then tension. Burst becomes the primary consideration toward the surface. Compression is also evaluated near the surface.

When considering collapse, the collapse strength of the casing must be reduced for tension. Table 10–4 shows the effect of tension on collapse resistance. In this table, the ratio of unit tensile stress to minimum yield strength, R_t, is defined by Eq. 10.1.

$$R_t = \frac{\text{Total tensile load}}{\text{Total yield of pipe}} \qquad (10.1)$$

The total tensile load is determined by the hanging weight of the steel below the point under consideration. The total yield of the pipe, in pounds, is determined by multiplying the minimum yield stress, in pounds per square inch. Thus, R_t becomes a decimal fraction of the actual load to the minimum load that could cause the pipe to fail. The percent of full collapse as a function of the R_t value has been calculated experimentally.[2]

The question of how to treat buoyancy when designing casing strings often arises. For example, in considering collapse, the casing is assumed to be completely empty. If this were true, buoyancy would be based on the complete displacement of the pipe rather than on just metal displacement. However, though the casing might be completely empty, a substantial portion of the casing is supported by cement, and the casing is not hanging free. Eq. 10.2 calculates the buoyancy factor when only the displacement of the metal is concerned.

$$B_f = (1 - 0.015\rho) \qquad (10.2)$$

■ **TABLE 10–4** Effect of tension on collapse resistance (*data from Holmquist and Nadai, ref. 2*)

Ratio of Unit Tensile Stress to Minimum Yield Strength, R_t	Percent of Full Collapse Pressure	Ratio of Unit Tensile Stress to Minimum Yield Strength, R_t	Percent of Full Collapse Pressure	Ratio of Unit Tensile Strength to Minimum Yield Strength, R_t	Percent of Full Collapse Pressure
0.005	99.8	0.220	87.1	0.435	70.8
0.010	99.5	0.225	86.8	0.440	70.3
0.015	99.3	0.230	86.7	0.445	69.9
0.020	99.0	0.235	86.3	0.450	69.5
0.025	98.7	0.240	85.6	0.455	69.1
0.030	98.4	0.245	85.3	0.460	68.6
0.035	98.2	0.250	85.0	0.465	68.2
0.040	97.9	0.255	84.7	0.470	67.8
0.045	97.6	0.260	84.3	0.475	67.3
0.050	97.3	0.265	83.9	0.480	66.9
0.055	97.0	0.270	83.6	0.485	66.4
0.060	96.8	0.275	83.2	0.490	65.9
0.065	96.5	0.280	82.8	0.495	65.5
0.070	96.2	0.285	82.5	0.500	65.0
0.075	95.9	0.290	82.1	0.505	64.5
0.080	95.6	0.295	81.7	0.510	64.1
0.085	95.3	0.300	81.3	0.515	63.6
0.090	95.1	0.305	81.0	0.520	63.1
0.095	94.8	0.310	80.6	0.525	62.6
0.100	94.5	0.315	80.3	0.530	62.1
0.105	94.2	0.320	79.9	0.535	61.7
0.110	93.9	0.325	79.6	0.540	61.2
0.115	93.6	0.330	79.2	0.545	60.7
0.120	93.3	0.335	78.8	0.550	60.2
0.125	93.0	0.340	78.4	0.555	59.7
0.130	92.7	0.345	78.0	0.560	59.2
0.135	92.4	0.350	77.7	0.565	58.7
0.140	92.1	0.355	77.3	0.570	58.1
0.145	91.8	0.360	76.9	0.575	57.6
0.150	91.5	0.365	76.5	0.580	57.1
0.155	91.2	0.370	76.1	0.585	56.6
0.160	90.9	0.375	75.8	0.590	56.1
0.165	90.6	0.380	75.4	0.595	55.5
0.170	90.3	0.385	75.0	0.600	55.0
0.175	89.9	0.390	74.6	0.605	54.5
0.180	89.6	0.395	74.2	0.610	54.0
0.185	89.3	0.400	73.7	0.615	53.5
0.190	89.0	0.405	73.3	0.620	52.9
0.195	88.7	0.410	72.9	0.625	52.4
0.200	88.4	0.415	72.5	0.630	51.9
0.205	88.1	0.420	72.0	0.635	51.4
0.210	87.7	0.425	71.6	0.640	50.9
0.215	87.4	0.430	71.2	0.645	50.3

where:

$$B_f = \text{buoyancy factor}$$
$$\rho = \text{mud weight, lb/gal}$$

To determine buoyancy for the total displacement of empty casing, the total weight of the displaced fluid must be considered. This creates a potential situation in which the casing is floating, so tension effects are zero. These hypothetical situations were introduced simply to show the difficulty of deciding how to apply safety factors and design requirements for casing.

Some of the same problems with collapse design also apply for tension. However, in tension, the casing is supported as it is run into the hole; buoyancy effects are pronounced whether the casing is being run or pulled. Thus, buoyancy factors generally are used in tension design. At times, the casing is run only partially filled, which reduces its suspended weight.

Burst design is based on high pressure inside the casing and low pressure outside of it. The general situation is that the bottom-hole pore pressure reduced by a gas gradient might be inside the pipe with very little pressure on the outside of the pipe toward the surface. Other conditions affecting burst requirements are the pressures required to fracture formations, break down perforations, or squeeze cement around the tops of the liners.

When applying high surface pressures, pipe and a packer are often used to prevent the pressures from being applied to most of the casing. However, a problem with potential burst limitations may occur inside the casing above the formations to be fractured when squeeze cementing the liner top. While the casing may not be designed for burst in the lower sections of the hole, burst must be considered when applying pressure. Also, experience may dictate that burst be considered for pipe in the hole's lower sections because of repeated problems.

Compression of casing is generally considered when landing or setting casing. The casing is commonly hung or landed in tension. Some operators hang the entire buoyed weight of the casing in tension, while others hang only 80% of the tension. The string of casing supporting the hanging casing is placed in compression. Generally, if burst is considered, it is unnecessary to also design for compression loading. In oil wells in which burst may not be a primary design criteria, the compression limits of casing may have to be evaluated.

Example 10.2 illustrates a typical casing design program. Compression is not considered in this example because the design takes into account burst requirements.

□ **EXAMPLE 10.2**

Given:

9⅝-in. protective casing to be set at 10,000 ft
Mud weight = 14.0 lb/gal

Fracture gradient @ 10,000 ft = 0.90 psi/ft
5½-in. production casing to be set at 15,000 ft
Mud weight = 16.0 lb/gal
Safety factors
to be used: Collapse = 1.0
                   Tension = 1.8
                   Burst   = 1.0

Design:

   (a) 9⅝-in. protective casing
   (b) 5½-in. production casing

SOLUTION:

(a) Design of the 9⅝-in. protective casing
    Collapse pressure @ 10,000 ft = (14)(0.052)(10,000)
                                  = 7,280 psi
    Collapse pressure at surface = 0

Burst pressure: Determining burst pressure depends on the operator's objectives. If he wants the protective casing to withstand the anticipated bottom-hole pressure before setting the production pipe, then he must have a heavier design. If he wants the protective casing to withstand only the pressures at 10,000 ft, then he can use minimum requirements. For this problem, burst requirements are based on the pressure conditions before setting the 5½-in. casing. These conditions are shown below.

Maximum surface pressures based on the fracture gradient at 10,000 ft:

$$(10,000)(0.90) - (10,000)(0.10) = 8,000 \text{ psi}$$

Maximum surface pressures based on the bottom-hole pressure at 15,000 ft:

$$(15,000)(15.5)(0.052) - (15,000)(0.10) = 10,590 \text{ psi}$$

Because of the fracture gradient at 10,000 ft, the surface pressure cannot exceed 8,000 psi before setting the 5½-in. casing. This is assumed to be the maximum surface pressure. The maximum burst pressure at 10,000 ft would be:

$$9,000 - 7,280 = 1,720 \text{ psi}$$

Fig. 10–2 shows the maximum collapse and burst pressures versus depth and the casing to be used.

$$\text{Buoyancy factor} = 1 - (0.015)(14) = 0.79$$

$$\text{Axial neutral point in casing} = (10,000)(0.79) = 7,900 \text{ ft}$$

Step 1
Using API grades of casing:
Select first P–110, 53.5 lb/ft casing, collapse resistance = 7,930 psi.
Note: API performance properties for casing are given in Appendix.

Step 2
Next, select P–110, 47.0 lb/ft casing, collapse resistance = 5,310 psi.

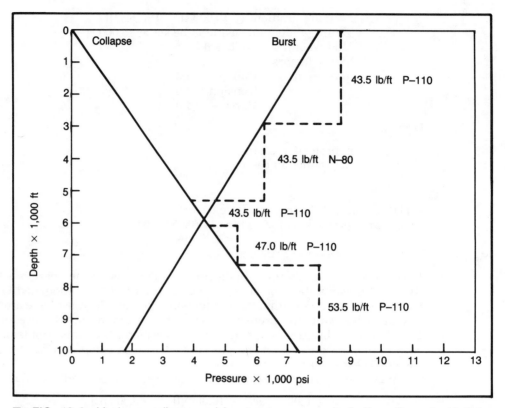

■ **FIG. 10–2** Maximum collapse and burst pressures vs depth (from Example 10.1) for 9⅝-in. casing

Setting depth (not considering tension) = 7,294 ft. This is above the neutral point at 7,900 ft and the collapse resistance must be reduced for tension. Assume:

Setting depth = 7,000 ft
Total tension = (7,900 − 7,000)(53.5 lb/ft) = 48,150 lb

$$R = \frac{48,150}{(110,000)(13.57)} = 0.032$$

Percent of full collapse pressure (from Table 10–4) = 0.984
Actual collapse pressure with tension = (5,310 psi)(0.984) = 5,225 psi

$$\text{Collapse design factor} = \frac{5,225 \text{ psi}}{(7,000)(0.052)(14)} = 1.03$$

Step 3
Next, select P–110, 43.5 lb/ft casing, collapse resistance = 4,430 psi
Setting depth (not considering tension) = 6,085 ft
This depth is above the neutral point at 7,900 ft; thus, the collapse resistance must be reduced for tension effects.

Assume:

Setting depth = 5,800 ft
Total tension = 48,150 + [(7,000 − 5,800)47] = 104,550

$$R_t = \frac{104,550}{(12.559)(110,000)} = 0.08$$

Percent of full collapse pressure (from Table 10–4) = 0.956
Actual collapse pressure with tension = (4,430)(0.956) = 4,235 psi

$$\text{Collapse design factor} = \frac{4,235}{(5,800)(0.052)(14)} \approx 1.00$$

Step 4
Next, select N–80, 43.5 lb/ft casing, collapse resistance = 3,810 psi
Assume:

Setting depth = 4,600 ft
Total tension = 104,550 + [(5,800 − 4,600)(43.5)] = 156,750 lb

$$R = \frac{156,750}{(12.559)(80,000)} = 0.16$$

Percent of full collapse (from Table 10–4) = 0.909
Actual collapse pressure with tension = (3,810)(0.909) = 3,463 psi

$$\text{Safety factor} = \frac{3,463 \text{ psi}}{14\,(0.052)(4,600)} = 1.03$$

Step 5
Consider burst next. Burst resistance of the N–80, 43.5 lb/ft pipe 6,330 psi.
This permits the use of the N–80, 43.5 lb/ft pipe to a depth of 2,700 ft.

Step 6
Next, select P–110, 43.5 lb/ft pipe, burst = 8,700 psi
The maximum burst pressure that can occur at the surface = 8000 psi. The
P–110, 43.5 lb/ft casing will be used from the surface to 2,700 ft (see Table
10–5).

■ **TABLE 10–5**   Summary of casing setting depths

Size, in.	Casing Weight, lb	Grade	Setting Depth, ft	Length, ft
9⅝	53.5	P–110	10,000	3,000
9⅝	47.0	P–110	7,000	1,200
9⅝	43.5	P–110	5,800	1,200
9⅝	43.5	N–80	4,600	1,900
9⅝	43.5	P–110	2,700	2,700
5½	23	P–110	15,000	1,683
5½	20	P–110	13,317	4,917
5½	17	P–110	8,400	1,400
5½	17	N–80	7,000	3,100
5½	17	P–110	3,900	3,900

Burst resistance increases with tension; however, the effects of tension on burst are seldom considered in casing design.

Check the casing design in Table 10–5 for tension:

Below 4,600 ft, tension = 156,750 lb
Total tension = 156,750 + (1,919)(43.5) + (2,727)(43.5) = 358,860 lb
(358,860)(1.8) = 645,947 lb
Joint strength of the 43.5 lb/ft N–80 = 825,000 lb
Joint strength of the 43.5 lb/ft P–110 = 1,106,000 lb

The tension requirements are less than the burst requirements.

(b) Design of the 5½-in. production casing
Collapse pressure at 15,000 ft = (16)(0.052)(15,000) = 12,480 psi
Burst pressure at surface = 10,590 psi
Burst pressure at 12,000 ft is as follows:
Maximum pressure inside casing − Maximum pressure outside casing

$$10,590 + (12,000)(0.10) - (16)(0.052)(12,000) = 1,806 \text{ psi}$$

Fig. 10–3 shows the maximum collapse and burst pressures versus depth and the casing to be used.

Buoyancy factor = 1 − (0.015)(16) = 0.76
Neutral point in casing = (15,000)(0.76) = 11,400 ft

Step 1
Using API grades of casing:
Select first P–110, 23.0 lb/ft casing, collapse resistance = 14,520 psi

Step 2
Next, select P–110, 20 lb/ft casing, collapse resistance = 11,080 psi

Setting depth from Fig. 10–3 = 13,317 ft

Step 3
Next, select P–110, 17.0 lb/ft casing, collapse resistance = 7460 psi
Setting depth (not considering tension) = 8,966 ft
This is above the neutral point at 11,400 ft, and the effect of tension on collapse resistance must be considered.
Assume setting depth, considering tension = 8,400 ft

Using this assumption, tension load (11,400 − 8,400)(17.0) = 51,000 lb

$$R = \frac{51,000}{(4.962)(110,000)} = 0.093$$

Percent of full collapse resistance (from Table 10–4) = 0.950
Actual collapse with tension = (7,460)(0.950) = 7,087

$$\text{Design factor} = \frac{7,087}{(8,400)(0.052)(16)} = 1.01$$

Step 4
Next, select N–80, 17.0 lb/ft casing, collapse resistance = 6,780 psi
Setting depth (not considering tension) = 8,000 ft

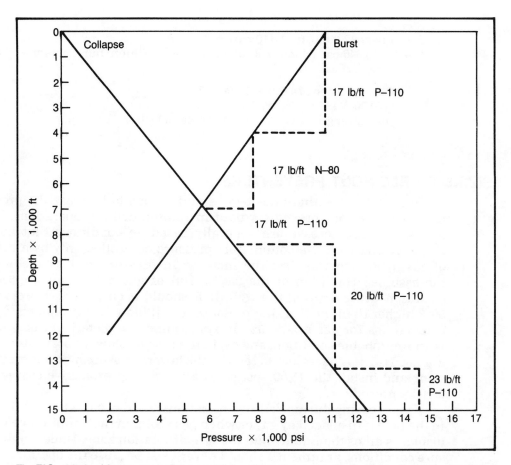

■ **FIG. 10–3** Maximum collapse and burst pressure vs depth (from Example 10.1) for 5½-in. casing

Assume:

Setting depth considering tension = 7,000 ft
Tension load = 51,000 + [(8,400 − 7,000)(17) = 74,800 lb

$$R_t = \frac{74,800}{(4.962)(80,000)} = 0.188$$

Percent of full collapse (from Table 10–4) = 0.891
Actual collapse with tension = (6,780)(0.891) = 6,040 psi

$$\text{Design factor} = \frac{6,040}{(7,000)(16)(0.052)} = 1.04$$

Step 5
Next consider burst. The burst resistance of the 17.0 lb/ft N–80 = 7,740 psi
The 17.0 lb/ft N–80 can be used to 3,900 ft for burst.

Step 6
Select next 17.00 lb/ft, P–110 casing
Burst = 10,640 psi, which will suffice for the maximum burst pressure of 10,590 psi at the surface

Check tension : 74,800 + (7,000)(17) = 170,000
(170,000)(1.8) = 306,000 lb
Joint strength of 17.0 lb/ft P–110 = 445,000 lb

SIZING OF BLOWOUT PREVENTERS

In recent years, there has been a major push to increase the pressure ratings of blowout preventers used in routine drilling operations. Arguments for this increase are generally based on conditions that include empty casing with the bottom-hole pressure controlled only by a column of gas. Arguments against the increase in pressure ratings include: (1) the assumed condition of the casing full of gas only is an improbable condition while drilling the well. If it should occur, the test pressure is 50% higher than the working pressure for all blowout preventers; (2) that calculations for gas gradients always assume completely dry gas, which is an improbable condition; and (3) in the case of blowout preventers rated at 15,000 psi, it is necessary to remove the lower-pressure blowout preventer stack and install the 15,000-psi stack after setting protective casing, also a risky procedure.

Empty casing during drilling operations could occur when all the mud in the hole is displaced by gas or when gas rising in the annulus eventually displaces all of the liquids into the lost circulation zone. Under high-pressure conditions, argument 1 should never exist because the operator would accept an underground blowout rather than keep the well open until only gas remained.

Argument 2 is a possibility and generally only occurs when a well is left unattended. One possible situation is on an offshore location when the rig must be moved off of the location because of bad weather. Under these conditions, the hole's fluid level may drop due to filtration, thereby reducing the effective head and permitting gas to flow into the wellbore. An extremely unlikely situation is if a well kicked and an underground blowout occurred at the time the rig had to be moved off of the location.

To illustrate specific conditions, consider Example 10.1, in which the second string of protective casing was set at 13,920 ft. The pore pressure at 13,920 ft was 11,943 psi. The total pressure required to fracture the formation just below the casing seat is 13,266 psi. The estimated bottom-hole pressure is 13,800 psi. The primary question to be answered in this case is whether or not a blowout preventer stack rated at 15,000 psi is needed to drill this well.

To answer this, examine all the facts, the possible maximum surface pressure, the equipment used, and the risks involved in changing blowout preventer stacks after setting 7-in. casing at 13,920 ft.

In the first place, if gas occupied the entire wellbore, there would not be an underground blowout because the pressure at 13,920 ft is less than the fracture pressure of 13,266 psi. Therefore, the maximum pressure at the surface is based on the maximum bottom-hole pressure minus the pressure exerted by a borehole full of gas. Example 10.3 illustrates the potential maximum pressure conditions at the surface.

☐ **EXAMPLE 10.3**

Assume:

Same conditions shown in Example 10.1
Bottom-hole temperature = 250°F
Surface temperature = 70°F
Bottom-hole pressure = 13,800 psi
Gas specific gravity = 0.7

SOLUTION:

$$\text{Gas gradient, psi/ft} = \frac{SP_p}{53.3zT_R}$$

where:

S = specific gravity of gas, dimensionless
P_p = pore pressure, psi
T_R = temperature, °R

Using averages:

$$T_{avg} = \frac{250 + 70}{2} = 160°F$$
$$= 160 + 460 = 620°R$$

Assume:

$$P_s = 11{,}500 \text{ psi}$$

Then,

$z_s = 1.55$
$z_b = 1.75$
$z_{avg} = 1.65$
$P_{avg} = 12{,}650 \text{ psi}$
$T_{avg} = 620°R$

$$\text{Gas gradient} = \frac{(0.7)(12{,}650)}{(53.3)(1.65)(620)} = 0.162 \text{ psi/ft}$$

Pressure exerted by the gas = (15,000)(0.162) = 2,430 psi
Maximum surface pressure = 13,800 − 2430 = 11,370 psi

This is a rough estimate of the maximum possible surface pressure, which assumes no accumulation of liquids. In all probability, liquids will accumulate, at least below the casing seat at 13,920 ft. In this case, the maximum surface pressure would probably never exceed 10,000 psi. Assuming that it did, remember that 10,000-psi working pressure blowout preventers were tested to 15,000 psi. While under normal conditions the working pressure should not exceed 10,000 psi, the preventers will hold more pressure in an emergency.

At times, the attitude of "no risks will be taken" develops. This is fine if risks are completely avoidable. Consider the conditions given in Examples 10.1 and 10.3. If the operator decides to change to blowout preventers rated at 15,000 psi after setting the 7-in. casing, he must do this after cementing and before drilling out of the 7-in. casing.

During this change-over, suppose the cement job fails around the 7-in. and the well kicks with no surface control. The argument that the cement job should be tested first is justified except that there is no guarantee that the cement job will not fail later. In fact, it is safer to assume the blowout preventers, which have been tested to 15,000 psi, will hold a pressure of 11,370 psi than it is to change preventers. The requirement that the test pressure should be 50% above the working pressure for blowout preventers is arbitrary. The requirement could be set at 25% above the working pressure, in which case the working pressure for blowout preventers tested to 15,000 psi would be 12,000 psi. The operator and those who regulate the industry should think about the alternative risks before imposing conditions that, in the long run, increase costs and problems with no risk reduction.

■ **REFERENCES**

1. Eaton, Ben A., "Graphical Method Predicts Geopressures Worldwide," *World Oil*, July 1976.
2. Holmquist and Nadai, *Ellipse of Biaxial Yield Stress*, 1940.

CHAPTER 11

ROTARY DRILLING BITS

DAVID S. ROWLEY
Terra Tek Inc.

More than 95% of oilfield footage is drilled today with *roller bits,* the most widely used form of the rotary drilling bit. Roller bits usually have three bearing-supported rolling cones (Fig. 11–1) with teeth or cutters distributed in different geometrically controlled patterns over each cone. Teeth are made of hard faced steel or are formed by pressed-in sintered tungsten carbide inserts.

A second basic form of a rotary bit is a *diamond bit* (Fig. 11–2). Diamond bits, which drill about 5% of oilfield footage, also have distributed cutters. Normally, these are natural industrial diamonds, or the cutters may be synthetic polycrystalline diamond compacts (PDC) designed into controlled geometric patterns over the drilling surfaces of the bit.

Drag bits, the third basic drill bit form, are the oldest type of rotary drilling bit and are used rarely now. Drag bits do not have distributed cutters; instead these bits have hardfaced blades (Fig. 11–3), usually two blades (the "fishtail" bit) or three.

Rotary drill bits are manufactured in 26 standard diameters (see Table 11–1), ranging from 3¾ in. up to 26 in., though many nonstandard sizes are also available.[1] Bit designs are adjusted and different materials used in each manufacturer's product line to cover as much of the spectrum of widely varying formations and drilling conditions as possible. For example, five manufacturers of roller bits (Hughes, Reed, Sandvik, Smith, Security) have added PDC bits or natural industrial diamond bits, or both, to their product lines. Other oilfield bit manufacturers—like Diamant Boart, Dowdco, and NL Hycalog—have specialized in PDC and natural industrial diamond bits. Norton Christensen manufactures natural industrial diamond bits and PDC bits and recently added a line of roller bits. Varel

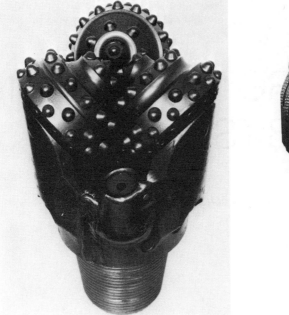

■ **FIG. 11–1** Sealed journal bearing roller bit with tungsten carbide inserts for low-compressive-strength formation (IADC code 5–1–7) (*courtesy Security Div.*)

■ **FIG. 11–2** D331 diamond drill bit set with natural industrial diamonds (*courtesy Norton Christensen Inc.*)

■ **FIG. 11–3** Two-way jet drag bit circa 1963 (*courtesy Norton Christensen Inc.*)

■ **TABLE 11–1** Standard oilfield drill bit sizes, in inches

3¾	6	7⅞	9⅞	14¾
3⅞	6⅛	8⅜	10⅝	17½
4⅛	6¼	8½	11	20
4¾	6½	8¾	12¼	24
5⅞	6¾	9½	13½	26
			13¾	

and RBI are roller bit manufacturing specialists. A product line of roller bits integrated with a diamond bit product line extends the manufacturer's range of product applications all the way from the shallow, soft formations and gumbo shales down to the medium-hard formations and on deeper to hard, abrasive, more expensive drilling conditions.

Technological and marketing competition is intense between the rotary drill bit manufacturers on a worldwide scale; this benefits the oilfield drilling industry. The competition brings in a practically uninterrupted flow of improved bits, capable of more cost effective performances, i.e., faster drilling rates, longer rotating hours, thus lower drilling costs.

Generally, the softer formations are geologically younger and are shallower. Although these can often be drilled with the less expensive steel-tooth roller bits, tungsten-carbide-insert (TCI) roller bits also are competing in the very soft formations. Medium-hard formations and the harder shales can be drilled effectively either with steel-tooth roller bits or with TCI roller bits, and some of these same formations can be drilled economically using PDC bits. The harder, deeper, more expensive, and often abrasive formations are normally penetrated using either TCI roller bits or natural industrial diamond drill bits.

CLASSIFICATION OF ROTARY DRILLING BITS

The International Association of Drilling Contractors (IADC) in 1972 adopted generalized rotary drill bit classification codes. The IADC code for roller bits has also been adopted and published by the American Petroleum Institute (API) as Recommended Practice 7G, but the IADC diamond bit code is under revision. These codes enable each bit manufacturer to preserve its proprietary style and model codes (e.g., Christensen R482, Hughes HH99), while broadly grouping bits intended for the same application under a single IADC code number.

In the IADC system for roller bits, the first of three figures is the series number. Series numbers 1, 2, and 3 are reserved for steel, milled-tooth roller bits in soft-, medium-, and hard-formation categories. Series numbers 4 through 8 are for TCI roller bits in soft, soft-to-medium, medium, hard, and extremely hard formations, respectively. The types (the second figure of the code) are numbered 1 through 4. These break down grades of formation hardness within each series. A 5–1 formation is softer than a 5–2 formation, and a 6–1 formation is harder than a 5–1 formation. The third figure of the code provides for design options.

For example, the softest formation could be drilled with a standard steel-tooth roller bit that would be designated IADC code 1–1–1, whereas the hardest formations would be drilled with a guage-protected, sealed journal bearing TCI roller bit, IADC code 8–4–7.

IADC code 5–1–7 rotary drill bits, for instance, include competitive products from different manufacturers, all intended for the same general application in series 5 applications, which is "Carbide Tooth Bits (TCI) and Soft to Medium Formation with Low Compressive Strength."

Some competitive IADC series 5 drill bits are listed below:

Manufacturer	IADC Generalized Code Assigned	Manufacturers' Product Designation
Hughes	5–1–7	J22
RBI	5–1–7	C2
Reed	5–1–7	FP51A
Sandvik	5–1–7	CFS20
Security	5–1–7	S84–F
Smith	5–1–7	F2

The series classifications are assigned as follows:

IADC Bit Classification

First Figure = Series	Roller Bit	Formation
1	Steel-tooth	Soft
2	Steel-tooth	Medium
3	Steel-tooth	Hard
4	TCI	Very soft
5	TCI	Soft
6	TCI	Medium
7	TCI	Hard
8	TCI	Extra hard

The design option classes are listed:

IADC Bit Classification

Third Figure = Design Option

1	Standard product
2	Air drilling
3	Gauge protected
4	Sealed bearing
5	Sealed bearing and guage protected
6	Friction, sealed (journal) bearing
7	Friction, sealed (journal) bearing and gauge protected
8	Directional
9	Other

A comparison chart of major roller bit manufacturers' products is included on the following page, courtesy of Security Division.

The IADC diamond drill bit classification system is being revised. In the meantime, diamond bits and PDC drill bits together drill a broad spectrum of formations similar to the roller bit spectrum. Equivalents worked out by the diamond bit manufacturers provide a rough guide for diamond bit selection compared to the IADC codes for roller bits. For example, a PDC diamond bit manufactured by Strata Bit Corp. to drill an IADC roller bit series 5 formation (insert design, soft-strength rock) would be similar to the PD–2, illustrated in Fig. 11–4, and by Diamant Boart would be similar to the LX–27, illustrated in Fig. 11–5.

ROLLER DRILL BITS—DESIGN AND NOMENCLATURE

Roller bits have three component groups: the rolling cones, their bearings, and the bit body. The body is a forged and welded structure, initially having three pieces, called legs, with a bearing pin on the lower end of every leg. Each leg also has a nozzle boss and a one-third circular arc-shaped piece at the top. After welding and turning, these three arc-shaped pieces form the API threaded pin connection to the bottom-hole collar. Design features and nomenclature for two modern, sealed bearing roller bits are shown in Fig. 11–6.[2]

The legs are forged generally from a nickel-chrome-molybdenum alloy steel like AISI 8720 and are then machined. Cones are also forged or may be cast from a nickel-molybdenum alloy steel like AISI 4815 and machined. Nozzles and TCI teeth are formed of sintered tungsten carbide, and the roller bearings and ball bearings are tool-steel-grade alloy.

For "friction" bits or journal bearing bits, welded and machined inlays with proprietary compositions or smooth cylindrical bearing race inserts, are used to decrease sliding friction and to increase the lubricity of the

ROCK BIT COMPARISON CHART

The chart relates the bit types of four manufacturers (SECURITY, HUGHES, REED, SMITH) by bit classification (Series 1–8) and Type (1–4). The manufacturers' bit categories are: Standard (1), Air (2), Gauge Protected (3), Sealed Bearing (4), Sealed Bearing & Gauge (5), Friction Sealed Bearing (6), Friction Bearing & Gauge (7), Directional (8), Other (9).

SECURITY

Series – Class – Type	Standard (1)	Air (2)	Sealed Bearing (4)	Sealed Bearing & Gauge (5)	Friction Sealed Bearing (6)	Friction Bearing & Gauge (7)	Directional (8)	Other (9)
1 STEEL TOOTH SOFT – 1	S3S		S33S		S33SF		2S3S,JD / S3S,JD	M3S3S
1 – 2	S3/S3T		S33	S33G	S33F			
1 – 3	S4/S4T		S44	S44G	S44F	S44GF	DS DSS	
1 – 4								
2 STEEL TOOTH MEDIUM – 1	M4N	M4N	M44N	M44NG	M44NF	M44NGF	DM DMM	
2 – 2	M4							
2 – 3	M4L		M44L					
3 STEEL TOOTH HARD – 1	H7	H7	H7					
3 – 2			H77	H77SG				
3 – 3								
4 INSERT VERY SOFT – 1						*S82F		
5 INSERT SOFT – 1			S94			*S84F		
5 – 2						S85F		
5 – 3			S86			*S86F		
5 – 4			S88			*S88F	DS88	GS88
6 INSERT MEDIUM – 1	M8JA		M88			*M84F		GM88 M8MJA
6 – 2						M89TF	DS88F	
6 – 3						M88F		
6 – 4						M90F		
7 INSERT HARD – 1						H87F		
7 – 2	H8JA		H88			H88F		
7 – 3						H99F		
8 INSERT EXTRA HARD – 1	H9JA							
8 – 2	H10JA		H100			H100F		

HUGHES

Series – Type	Standard (1)	Air (2)	Sealed Bearing (4)	Sealed Bearing & Gauge (5)	Friction Sealed Bearing (6)	Friction Bearing & Gauge (7)	Other (9)
1 – 1	R1		X3A		J1		MX3A
1 – 2	R2		X3	X1G	J2	JD3	MX1G
1 – 3	R3		X1G	XDG	J3	JD3	
1 – 4	R4		XV	XDV	J4	JD4	
2 – 1	DR5						
3 – 1	R7/DR7				J7		
3 – 2	R8				J8	JD8	
4 – 1				X11	J11		
5 – 1		HH33		X22		J22 J22C	
5 – 3				X33		J33 J33H	
6 – 1		HH44		X44		J44	
6 – 2						J55R J44C	
6 – 3		HH55				J55	
7 – 1		HH77		X77		J77	
7 – 2						J88	
8 – 1		HH88				J99	
8 – 2		HH99					

REED

Series – Type	Standard (1)	Air (2)	Gauge Protected (3)	Sealed Bearing (4)	Sealed Bearing & Gauge (5)	Friction Sealed Bearing (6)	Friction Bearing & Gauge (7)	Directional (8)	Other (9)
1 – 1	Y11	Y11		S11		HP11	HP11	Y11JD	Y11JD
1 – 2		Y12T	Y13G	S12	S13G	HP12	HP12		
1 – 3		Y13/Y13T		S13		HP13	HP13 HP13G		
2 – 1	Y21	Y21	Y21G	S21	S21G	HP21	HP21G		
2 – 2		Y22/Y22R							Y22RAP
2 – 3	Y23	Y23	Y23G	S23	S23G				
3 – 1	Y31	Y31	Y31G	S31	S31G	HP31G	HP31G		Y31RAP
3 – 2	Y34	Y34	Y34G	S34	S34G				
5 – 1				S52		S52	FP51A HS51		
5 – 3				S53		S53	FP53 FP53A HFSM		
6 – 1		Y62JA Y62BJA		S62		S62	FP62 HPM		
6 – 2							FP62B FP62X		
6 – 3		Y63JA		S63		S63	FP63 HPMH		
6 – 4				S64		S64			
7 – 1		Y73JA		S72		S72	FP72 HPH		Y73RAP
7 – 2				S73		S73	FP73		
7 – 3				S74		S74	FP74		
8 – 1		Y83JA		S83		S83	FP83		

SMITH

Series – Type	Standard (1)	Air (2)	Gauge Protected (3)	Sealed Bearing (4)	Sealed Bearing & Gauge (5)	Friction Sealed Bearing (6)	Friction Bearing & Gauge (7)	Directional (8)	Other (9)
1 – 1	DS			SDS		FDS	FDS	DJ	MSDT
1 – 2	D7/DT7			SDT	SDG	FDT	FDT	BHD	
1 – 3	DG/DGT			SDG	SDGH	FDG	FDG FDGH		
2 – 1	V2		V2H	SV	SVH	FV	FVH		
3 – 1	L4		L4H	SL4	SL4H		F1		
3 – 2							F15		
5 – 1					2JS		F2 F2H		
5 – 2							F27		
5 – 3					3JS		F3		
5 – 4							F37		
6 – 1		4JA			4JS		F4 F45		4GA
6 – 2		5JA			5JS		F5 F47H		5GA
6 – 3							F57		
7 – 1		7JA					F6		A1
7 – 2							F7		
8 – 1		9JA					F9		

NOTE: Bit classifications are general and are to be used only as simple guides. All bit types will drill effectively in formations other than those specified. This chart shows the relationship between the specific bit types.

■ **FIG. 11–4** 12¼-in. PD–2 drill bit designed to penetrate IADC code 5–1–7 formations (*courtesy Strata Bit Corp.*)

■ **FIG. 11–5** LX–27 PDC drill bit designed to penetrate IADC code 5–1–7 through 5–3–7 formations (*courtesy Diamant Boart USA*)

materials in the mating metal bearing surfaces. A small amount of silver, for example, may be used in the welded inlay.

Roller Bit Cone Designs Related to Cutting Action

Although the cones of a roller bit appear to roll on the hole bottom surface as the bit is turned, there is always some sliding, tearing, and gouging action by the teeth on the bottom. Softer formations require more action than the harder formations. The sliding and gouging action is designed into the drilling mechanics of roller bits and is caused by the fact that the cones each have multiple rolling centers. But a cone cannot have pure rolling motion around more than one rolling center at a time. For instance, consider the TCI cone cross section in Fig. 11–7. This typical design has three rolling centers for the one cone.[3] The imaginary cone, formed tangent to the heel-row TCI teeth, would like to roll about its own apex, far off the bit centerline. Similarly, the inner rows of TCI teeth form a second imaginary cone, also offset from the bit centerline.

However, the bit structure is mechanically constrained by the drill collar above, so it must rotate on the bit centerline. The pure conical rolling action of the heel- and inner-row cones cannot occur around both their centers at the same time, so the TCI teeth must slide as they roll, self-

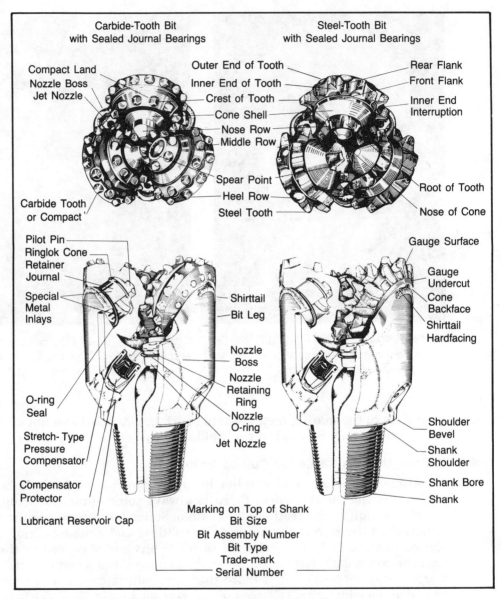

Carbide-Tooth Bit with Sealed Journal Bearings

Steel-Tooth Bit with Sealed Journal Bearings

Compact Land
Nozzle Boss
Jet Nozzle

Outer End of Tooth
Inner End of Tooth
Crest of Tooth
Cone Shell
Nose Row
Middle Row

Rear Flank
Front Flank
Inner End Interruption

Carbide Tooth or Compact

Spear Point
Heel Row
Steel Tooth

Root of Tooth
Nose of Cone

Pilot Pin
Ringlok Cone Retainer
Journal

Special Metal Inlays

Shirttail
Bit Leg

Gauge Surface

Gauge Undercut
Cone Backface

Shirttail Hardfacing

O-ring Seal

Stretch-Type Pressure Compensator

Compensator Protector

Lubricant Reservoir Cap

Nozzle Boss
Nozzle Retaining Ring
Nozzle O-ring
Jet Nozzle

Shoulder Bevel
Shank Shoulder
Shank Bore
Shank

Marking on Top of Shank
Bit Size
Bit Assembly Number
Bit Type
Trade-mark
Serial Number

■ **FIG. 11–6** Roller bit elements (*courtesy Hughes Tool Co.*)

adjusting to the rotating positions of the bearing pins. For harder formations, the overhangs of the heel-row TCI cone and the inner-row cone are smaller, as shown in Fig. 11–8. The sliding motion of the teeth on the hole bottom is greatly reduced in hard-formation bit designs to control wear. The cone actions are still not pure rolling, however.

The sliding action creates controlled tearing, gouging, and ripping mo-

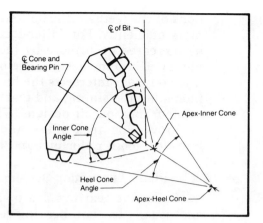

■ **FIG. 11–7** Soft-formation cone design (*courtesy Hughes Tool Co.*)

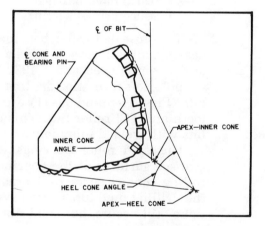

■ **FIG. 11–8** Hard-formation cone design (*courtesy Hughes Tool Co.*)

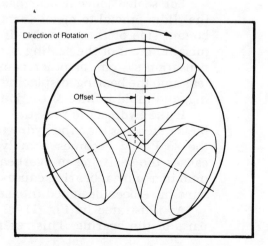

■ **FIG. 11–9** Cone bearing centerline offset soft-formation roller bit (*courtesy Hughes Tool Co.*)

tions of the teeth on the bottom, which help remove rock chips faster and more efficiently. For softer-formation roller bits, this ripping action can be increased even more by incorporating a cone offset, as illustrated in Fig. 11–9, into the design. The offset angle is such that the bearing centerline no longer intersects the bit and hole centerlines. Increasing the offset intensifies the sliding and tearing action. This means faster drilling with softer-formation bit designs. The principles of bearing centerline offset, heel-row cone apex offset, and inner-row cone apex offset are designed into steel-tooth roller bits and into TCI roller bits.

Roller Bit Bearings, Lubrication Systems, and Pressure Compensators

Roller bit bearings are manufactured in one of three configurations and usually use ball bearing retainers:

- Unsealed roller bearings
- Sealed roller bearings
- Sealed journal bearings

The unsealed roller bearing shown in Fig. 11–10 is the conventional bearing, initially grease-filled and exposed to the mud during drilling. The design, introduced about 1931, is still used on some steel-tooth bits for top hole, soft-formation drilling. Sealed and self-lubricated journal bearing bits (Figs. 11–6 and 11–11) are now the premium design both for steel-tooth and TCI roller bits. This design was introduced about 1970 for TCI bits with welded inlay journals.

Unsealed roller bearings wear due to metal fatigue of the highly stressed surfaces of the bearing races and through abrasive wear of the rolling contact surfaces. Abrasive wear results from solids like sand in the drilling mud lubrication system. For sealed roller bearings, metal fatigue still occurs, but sand abrasion is effectively eliminated as long as the seals hold.

For sealed journal bearings, wear occurs from galling or seizing of the sliding metal-to-metal surfaces that are in contact on the high-stress bottom side of the bearings. If the seal fails, due to high temperature, misalignment of the sealing surfaces, malfunctions of the pressure compensation system, or other cause, abrasives from the drilling mud will leak into the bearing, displacing the grease. This causes overheating, galling, and rapid failure of the journal bearing.

If the hydrostatic pressure on the bearing grease of a sealed bearing bit is much less than the hydrostatic mud column pressure in the borehole, the seal ring becomes heavily loaded. Excessive differential pressure causes the seal ring to overheat, to deform, and possibly to leak. Since mud in a journal bearing causes rapid failure, pressure compensating systems have been designed into sealed bearing roller bits.

As illustrated by Fig. 11–12, the grease reservoir is connected to all parts of the bearing. This reservoir is covered by a flexible diaphragm

■ **FIG. 11–11** Sealed, self-lubricated journal bearings in 7⅞-in. soft-formation F27 TCl roller bit (*courtesy Smith Tool*)

■ **FIG. 11–10** Unsealed, conventional bit roller bearing with ball bearing retainers (*courtesy Hughes Tool Co.*)

made from an elastomer that is protected from damage and held in place by a steel reservoir cap and snap-ring assembly. The diaphragm flexes inward to keep the grease under the same pressure as the drilling mud, especially on the trip into the hole. The diaphragm also compensates for the loss of some grease past the seal ring while drilling or may expand outward, if necessary, to make up for the increased grease volume when the bit temperature rises.

Journal bearing steel-tooth rock bits have proven to be cost effective for soft, low-strength formations. Applications include the U.S. Gulf coast, both offshore and onshore. The bearings in one representative design (see Fig. 11–13) use floating, silver-plated beryllium copper bushings and low-compression radial seals to permit running the higher rotary speeds necessary for a lower drilling cost. The bit has long, widely spaced steel teeth that are hardfaced with tungsten carbide. The hardfacing retards tooth wear and enables the bit to retain greater tooth height throughout its run to maximize penetration rate.

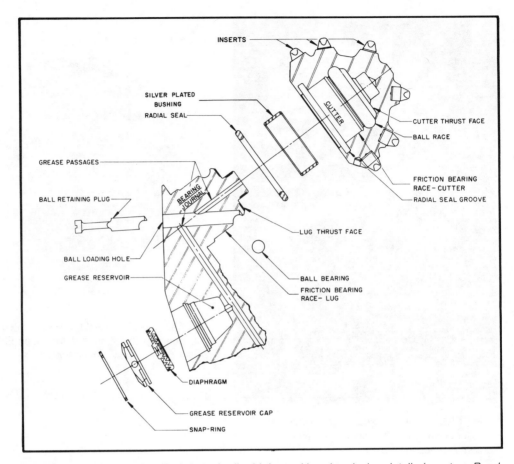

■ **FIG. 11–12** Sealed, self-lubricated roller bit journal bearing design details (*courtesy Reed Tool Co.*)

Extended Nozzles

Laboratory drilling tests and field experience demonstrate that extended nozzles, which position the three nozzle discharge openings closer to the hole bottom, increase the efficiency of bottom-hole scavenging and correspondingly increase drilling rates. Design and manufacturing concepts of an extended nozzle bit are shown in Fig. 11–14. Several manufacturers supply extended nozzle two- and three-cone roller bits. Nozzle extensions made in the form of short pipes or tubes replace the conventional jet nozzle bosses.

An intermediate approach is illustrated in Fig. 11–15, in which the nozzle boss has been extended somewhat toward the hole bottom. To increase scavenging efficiency, the nozzle's axis has also been tipped backward a few degrees. In this way the jet discharge touches the leading cutters on the cone on their way into the formation, so the cutting action

■ **FIG. 11–13** Journal bearing steel-tooth roller bit (*courtesy Reed Tool Co.*)

occurs with cutters, and the formation surface is cleaned just before contact. Field results indicate improved penetration rates and reduced drilling costs.

Operating Practices for Roller Bits

Bit weight and rotary speed practices vary, depending on whether a soft-formation (longer teeth) or hard-formation (shorter teeth) bit is being run. Higher weights and lower rotary speeds are used in the harder and more abrasive formations, with lower bit weights and higher rotary speeds in soft formations. Excessive weight on soft-formation TCI or steel-tooth bits breaks off teeth; too much bit weight may also cause hole deviation problems. This happens when the bottom collar bends into contact with the hole wall below the control stabilizer.

Soft-formation roller bits have lower recommended bit weights than hard-formation bits, not only because the drilling is easier but also to help prevent breaking the longer teeth used on soft-formation bits. Both TCI and steel-tooth soft-formation bits are run with about 4,000 lb bit weight/in. of bit diameter. This translates into 40,000 lb weight on a 9⅞-in. bit.

For hard-formation bits, permissible bit weights are substantially higher, ranging up to 8,000 lb/in. of diameter for steel-tooth bits.

■ **FIG. 11–15** Extended nozzle boss and canted nozzle axis for cone and bottom cleaning improvements (*courtesy Reed Rock Bit Co.*)

■ **FIG. 11–14** Extended jet nozzle (cutaway view) steel-tooth roller bit (*courtesy Smith Tool*)

Permissible weights are slightly lower for TCI bits than for steel-tooth bits. This prevents impact failures and cracking of the carbide inserts. A 7⅞-in. hard-formation TCI sealed bearing roller (IADC classification 7–3–7), for example, should be run with no more than 48,000 lb and, for the hardest formation TCI 7⅞-in. bit (IADC classification 8–3–7), no more than 59,000 lb.

Under normal conditions, roller bits drill faster if more weight is applied or if the bit is rotated faster, or both. Too much weight, however, can lead (1) to foundering or "balling up" of a soft-formation bit, (2) to premature failures of roller bearings due to spalling of the roller races, (3) to galling and seizure of journal bearings, or (4) breakage of teeth or tungsten carbide inserts. When both weight and speed are increased, the rotary power expended at the bit rises and produces a higher drilling rate, with corresponding acceleration of mechanical wear on the bit parts.

A bearing capability number, or WN number, has been developed as a means for rating journal bearing designs. The WN number is the product of the bit weight in thousands of pounds multiplied by the rotary speed in revolutions per minute (rpm) that the particular bearing design should be able to withstand without sustaining abusive wear or seizure. For example, the WN number for a 6-in. Hughes J44 TCI journal bearing bit is 2,650 and for a 12¼-in. J44 is 7,300. At a rotary speed of 60 rpm, bit weights that exceed 44,167 and 121,700 lb, respectively, will cause "initial seizure" of the bearings.

Table 11–3 translates the WN number ratings for one manufacturer's journal bearing bits into ranges of corresponding weights and speeds.

On hard formations, therefore, for which extra weight is needed to get the teeth to penetrate the formation, drillers use caution and run lower rotary speeds, like 40–50 rpm, to prevent premature bearing damage and accelerated wear. In fact, TCI journal bearing bits for the hardest formations are recommended to be run at even lower speeds, 35–50 rpm. Conversely, soft-formation TCI bits run at higher rotary speeds, up to 80 and 100 rpm. Recommended rotary speeds for steel-tooth bits are only moderately higher, usually 120 rpm or less. But, in the special case of the long-tooth, high-offset, jet steel-tooth roller bits for softest formations, speeds to 200 and 250 rpm are acceptable.

Field bit practices for bit weight and rotary speeds for TCI roller bits in the size range of 7⅞–8¾ in. are shown by Fig. 11–16. Comparable data on weight-speed practices for steel-tooth roller bits are illustrated by Fig. 11–17.

DIAMOND DRILL BITS AND PDC DRILL BITS

Diamond drill bits (Fig. 11–18) are cost effective and are the accepted standard rotary drill bits for many deep, hard, abrasive, and expensive drilling situations. Diamond bits are especially applicable when drilling

JOURNAL BEARING INSERT BITS
RECOMMENDED ENERGY LEVELS - WEIGHT ON BIT/ROTARY SPEEDS

IADC Code	437	517, 527	537, 547	617, 627	637, 647	737, 747	817, 837
Rotary Speed (rpm)	120-55	120-50*	90-40	70-40	60-40	55-30	50-25
Bit Size (inches)	Weight on Bit**						
4½-4¾				10-25			
5⅞-6¼		10-25	15-30	15-30	15-30		
6½-6¾		15-25	20-35	20-35	20-35	25-40	25-40
7⅝-7⅞	20-35	20-40	25-45	25-45	25-50	25-55	25-65
8⅜-9	20-40	20-45	25-50	25-50	25-55	30-60	30-65
9½-9⅞	25-45	25-50	30-55	30-60	30-65		35-65
10⅝-11		30-55	35-65	35-65			
12-12¼	35-65	35-65	40-70	40-70	40-75		45-80
13¼-15		40-75	45-85				
17½-18½		40-85	45-90	45-90	45-90		

*Note Exception: 14¾″ (110-50). 17½″ (100-50).   **Bit Weights (lbs.) × 1000

The energy level ranges listed are based on general drilling conditions. The weight – rpm factor should be controlled with lighter weights and higher rpm, or heavier weights and slower rpm. It is recommended that drill-off tests be performed on a frequent basis to establish the energy level (bit weight and rotary speed) that will result in the optimum penetration rate and drilling economics.

costs are high. Elevated hole temperatures or small hole sizes are additional applications. Diamond bits have no moving parts and so are widely and economically implemented when rotary speeds are high, i.e., above the fatigue limits of roller bit bearings. These applications include directional drilling when running on mud motors and straight-hole motor and turbine drilling.

Conventional diamond drill bits use natural industrial diamonds in sizes from about 15 stones/carat up to 1 stone/carat, cast in an erosion-resistant matrix that forms the bit shape and mud flow passages. These diamonds are hand-set individually into preplotted, geometrically controlled patterns, with one diamond per pattern plot mark. This covers the hole bottom, provides for wear and breakage with some redundancy in the pattern, and channels the mud flow to cool the cutters while removing chips.

Industrial diamonds have the same superlative hardness and exceptionally great strength of gem-quality diamonds, but are graded as industrials because of impurities, crystal defects, or dark color. An industrial diamond is not valuable as a gemstone. A 3-stones/carat-sized diamond is representative of oilfield sizes for medium-formation drilling. Such a diamond or "stone," if nearly spherical, would have a diameter of 0.13 in. Similarly, an 8-stones/carat-sized diamond, which is a smaller size suitable for harder-formation drilling, would have a diameter of 0.094 in.

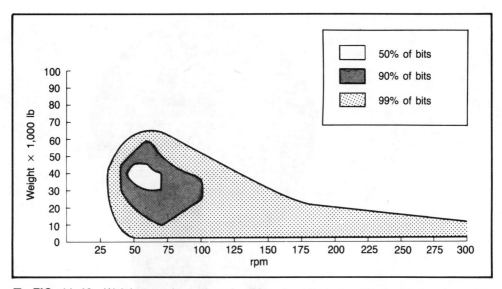

■ **FIG. 11–16** Weight-speed practices for TCI roller bits in the 7⅞- to 8¾-in. size range (*courtesy Hughes Tool Co.*)

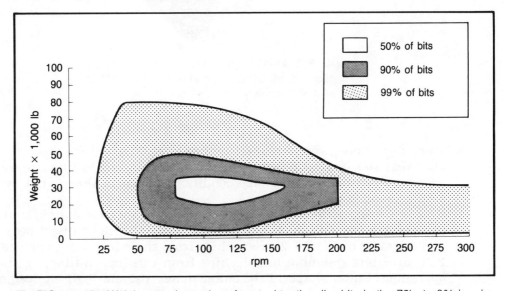

■ **FIG. 11–17** Weight-speed practices for steel-tooth roller bits in the 7⅞- to 8¾-in. size range (*courtesy Hughes Tool Co.*)

About 1978, diamond bit manufacturers introduced new, higher performance products using polycrystalline diamond compact (PDC) cutters, as shown in Fig. 11–19. A PDC cutter has a table or layer of sintered polycrystalline diamond about 0.030 in. thick mounted on an integral, sintered tungsten carbide backing. The diamond does the drilling. The carbide gives

■ **FIG. 11–18**   Medium to medium-hard formation diamond drill bit: drilling and service style 901 (*courtesy NL Hycalog*)

strength, stiffness, and toughness to the cutter. PDC cutters start out sharp and wear sharp. Most other forms of cutters and teeth that are used on drill bits start drilling in sharp condition, or nearly sharp, then wear to a dull, rounded condition.

The cutting action of the PDC bit shears the rock formation, analogous to metal cutting with ductile materials, hence the nickname "shear bits." As illustrated by Fig. 11–20, rock chips from shale formations drilled with PDC drill bits resemble metal chips from turning, milling, or shaping operations. Cutter shapes, PDC cutter materials, and PDC bit designs are being developed to maximize drilling results with PDC bits. While PDC bits are utilized as "standard" for certain field drilling situations, these are generally considered special-purpose drilling tools.

Standard applications developed for PDC drill bits include North Sea Jurassic and Cretaceous 12¼-in. bit sections below surface casing, running on turbo-drills in claystones and mudstones. Also, the Frio sand and shale series in south Texas is a standard application for PDC bits, drilling on rotary with oil-based muds. And another standard PDC drill bit application

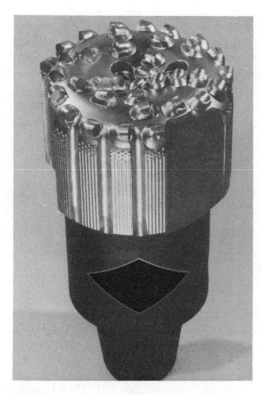

■ **FIG. 11–20** Smooth and wrinkled sides of shale chips drilled by a PDC bit at 4,400 ft, laboratory-simulated drilling conditions (*courtesy Terra Tek Inc.*)

■ **FIG. 11–19** Diamond drill bit with polycrystalline diamond compact (PDC) cutters (*courtesy Norton Christensen Inc.*)

is the Austin Chalk below 14,000 ft in the Tuscaloosa trend of Louisiana. Diamond drill bits are manufactured in several basic forms:

- Matrix body with industrial diamond cutters (Fig. 11–2)
- Matrix body with brazed-in PDC cutters (Fig. 11–19)
- Matrix body with polycrystalline, thermally stable, cast-in blocky cutters (Fig. 11–21)
- Steel body with pressed-in PDC studs, and cutters (Fig. 11–22)

Design and Nomenclature for Diamond Drill Bits

Diamond bits use distributed cutters, arranged in controlled geometric patterns to cover the hole bottom, provide for redundant cutters near the hole gauge, and still leave space for nozzles and mud flow passages. Fig. 11–23 shows the basic configuration and standard nomenclature.

Matrix diamond bit bodies are castings fired at temperatures near 2,200°F in a process called liquid-phase sintering. This process produces a strong, ductile metal part that also has the required high resistance to

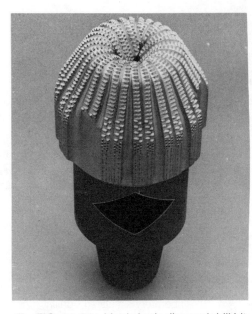

■ **FIG. 11–22**   Steel body PDC drill bit with pressed-in PDC stud-mounted cutters (*courtesy NL Hycalog*)

■ **FIG. 11–21**   Matrix body diamond drill bit with cast-in thermally stable PDC cutters (*courtesy Norton Christensen Inc.*)

fluid erosion. The diamond-supporting matrix must withstand mud stream erosion for several hundred hours and must prevent the structure holding the diamond from being washed away before the useful life of the cutters has been attained. The upper end of the casting, containing a steel core, is machined and then assembled by welding to an alloy-steel, API pin-connecting sub, completing the bit manufacture. Since there are no moving parts, diamond bits are more effective on mud motors and turbines at rotary speeds in the range from 350 to as high as 1,000 rpm. Roller bit seals and bearings cannot withstand such speeds due to metal fatigue and overheating.

As in the case of roller bits, standard diamond bit designs are manufactured for a range of different formation and drilling conditions. Smaller diamonds like 8 or 12 stones/carat size are used for harder formations like the Cotton Valley and Bromides. The larger diamonds, like the 3-stones/carat size, work for medium formations like the Frio in south Texas and in Gulf coast Miocene sandy shale formations.

The cross-sectional area of the fluid passages is designed to produce a pressure drop across the bit face sufficient to expend close to 3 hydraulic horsepower (hhp)/sq in. of hole cross sectional area. For an 8½-in. bit drilling with 10.5-lb/gal mud, this means a bit pressure drop of 833 psi at 350 gpm, using a total flow area of 0.41 sq in. for a projected bit area of 56.7 sq in.

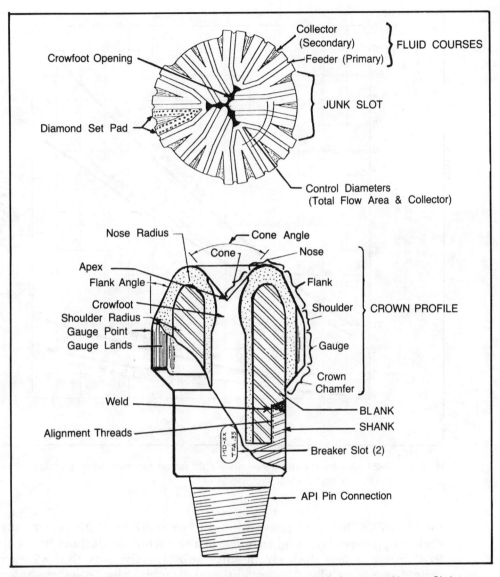

■ **FIG. 11–23** Nomenclature for conventional diamond bit (*courtesy Norton Christensen Inc.*)

Operating Practices for Diamond Bits

Cost effective diamond drilling practices utilize rig hydraulics to best advantage to gain the benefits of high mud velocities across the bit face. They also employ bit weights in the range of 2,000 to about 4,000 lb/in. of bit diameter and the highest safe and practical rotary speed available. Fig. 11–24 shows the maximum and minimum recommended bit weights

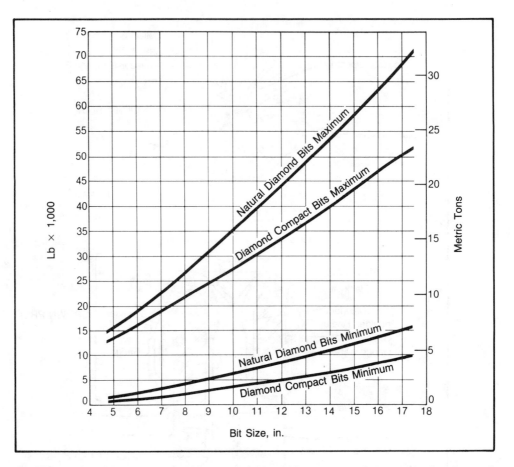

■ **FIG. 11–24** Maximum and minimum weight on bit recommended practice for diamond drill bits (*courtesy Norton Christensen Inc.*)

for both PDC bits and conventional diamond bits. With fewer cutters to divide up the weight and in view of the softer formations being drilled, PDC bits use less than (frequently substantially less than) 75% of the weight recommended for conventional diamond bits. In softer shales, PDC bits are sometimes run with a total bit weight as low as 4,000–6,000 lb.

Bit weight is added slowly in 2,000-lb increments while starting diamond bits. Drill-off tests should then be performed to find the fastest-drilling bit weight, after the maximum safe rotary speed has been established. As with roller bits, applied bit weight values must be consistent with any overriding requirements there may be for lighter loads due to drill-collar stabilization and hole-deviation correction programs.

Hydraulics for competitive conventional diamond bits are designed (1) to deliver high-velocity mud flow to all diamonds on the drilling "face" or surface, (2) to transport the cuttings away to the collar annulus, (3) to help lift cuttings off the borehole bottom, and (4) to cool the diamonds.

Cooling removes friction heat generated on the sliding cutting surfaces of the diamonds. The cooling function is emphasized because the drilling surfaces of the diamonds or PDC cutters never lose contact with the rock when the bit is drilling correctly and is running "flat" on the hole bottom. The friction heat must travel through the diamonds or PDC cutters from their wear-flat surfaces to their exposed side surfaces. There the heat is removed by conduction to the drilling mud. Preventing excessive diamond temperature is important to avoid rapid wear through oxidation and graphitization of the diamond material.

Deeper, harder, more fine-grained abrasive formations require higher drilling stresses on the diamonds in order to penetrate. Under these conditions, friction heat on the diamonds can be reduced to acceptable levels by slowing down the rotary speed while maintaining bit hydraulics at or near the guideline 3 HSI level (hhp/sq in. of projected hole area) for fluid course bit designs (see Fig. 11–2). While not a maximum—and probably not an optimum—value either, 3 HSI has been shown to yield more cost-effective diamond drilling results than lower values, especially in more shaly or dolomitic or limestone formations. HSI values up to 15 have been reported in the PDC and diamond drilling literature for bits equipped with jet nozzles, and bit hydraulics in the range of 5–9 HSI have been used on large rigs where bottom-hole scavenging and diamond cooling were thought to limit bit performance. However, many rigs cannot put up even 3 HSI under some drilling conditions. When drilling sandy formations under these conditions, HSI values as low as 1.5 can be effective.

Rotary speed for diamond bits is ordinarily not limited, except under conditions of extreme formation abrasiveness. In these cases, short bit life may result from high diamond temperatures generated by sliding friction. PDC and conventional diamond bits are routinely used in directional drilling applications on mud motors (running about 350 rpm) and in straight-hole drilling applications on positive displacement motors or turbines (with speeds as high as 600–1,000 rpm). For example, PDC bit and turbodrill combinations have been used in 8½- and 12¼-in. hole sections through Cretaceous and Jurassic formations in the North Sea developments regularly since 1979.

CUTTING ACTION OF ROTARY DRILL BITS

The borehole hydrostatic mud pressure increases with depth. Similarly, the formation pore pressure and the mechanical stresses in the rock also increase as a function of depth. The mechanical stresses increase due to geostatic loads imposed by the overburden of rock piled above. In general, as the compressive pressures or stresses applied to the rock rise, the rock's strength also increases. This phenomenon is a unique characteristic of rocks and accounts for the fact that deeper formations are harder, slower, and more difficult to drill, with any type of bit.

If the applied pressures or stresses are great enough, the rock undergoes a transition from the familiar hard, brittle, and elastic behavior we recognize at atmospheric pressure to a different condition, one more ductile that exhibits elastic-plastic behavior, while simultaneously becoming stronger. Therefore, the cutting action required at great depths in many types of rocks is very different than for the same rocks drilled at shallow depths.

For example, a typical sandstone exhibits brittle, elastic failure characteristics when drilled by the crushing and chipping actions of a roller bit's teeth at atmospheric pressure, as shown in Fig. 11–25 (right).[4,5] The same bit drilling the same Colton sandstone at a depth of 13,000 ft (full-scale laboratory simulation) penetrated at only 13.75 ft/hr, compared to 76 ft/hr at atmospheric conditions. The resulting bottom-hole pattern [Fig. 11–25 (left)] clearly showed ductile, elastic-plastic behavior of the rock.

Under these conditions at the deeper depth, the chip formation process is more difficult because the rock is stronger and because the bit action cannot easily lift the deformed rock off the bottom-hole surface. When the rock is more ductile and plastic under deep drilling conditions, the bit's teeth tend to displace the rock material laterally, i.e., to form the chip and move it parallel to the bottom-hole surface. However, the teeth do not necessarily gouge or lift the formed chip far enough to disengage it from the parent formation below and get it into the turbulent mud flow for removal. Thus, there is considerable inefficient recutting of the formation due to plasticity effects. More hydraulic energy expended at the bit

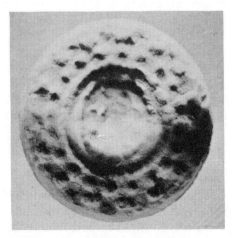 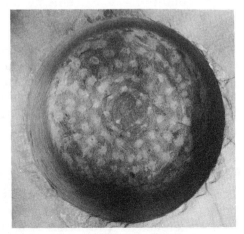

■ **FIG. 11–25** Roller bit bottom-hole patterns: plastic-elastic at depth (left) and brittle-elastic at atmospheric conditions (right)

to increase the kinetic energy and the momentum of the mud flow in the immediate vicinity of the forming rock chips helps to lift the cuttings and thus to increase the drilling rate.

These effects are illustrated in Fig. 11–26 for Mancos shale drilled with a 7⅞-in. Smith F–3 under laboratory-simulated depth conditions of 4,400 ft.[6] Similarly, Fig. 11–27 shows that lower-density drilling muds with higher energy levels produce faster penetration rates (mud B weighs 9.1 lb/gal compared to 8.33 for fresh water).

If the recutting conditions on the hole bottom, due to the rock's plasticity or to other unfavorable conditions, continue very long, the bottom-hole rock surface layers become deformed. These layers then consist of subdivided rock flour, accumulated by cutting, recutting, and regrinding the rock over and over again. Warren calls this a "cuttings bed" (see Fig. 11–28).[18]

For these reasons, bits like the PDC designs that drill with a shearing action frequently perform better in the more plastic and ductile formations because recutting and regrinding are minimized. In deep drilling, the formations are primarily much stronger, but also exhibit ductility and plasticity. Under these conditions PDC cutters may overheat and wear prematurely. Conventional diamond drill bits will usually drill such formations effectively. The individual diamonds act as indenters or penetrators under the influence of bit weight, then plow the rock into chips as the bit is rotated.

The drag bit (Fig. 11–3) is the original form of "shear bit." The modern PDC drill bit (Figs. 11–4, 11–5, and 11–19) is also classified as a shear bit, even though the shearing action takes place at the leading edge of each of the distributed cutters instead of at the leading edges of the blades, as in a drag bit.

The relationships between rock properties and bit performance for drag bits and diamond bits have been investigated by Appl[8,9,10,11] and for PDC bits by Holster,[12] Feenstra,[13] Zijsling,[14] and Glowka.[15,16,17]

The roller-bit cutting action has recently been studied by Warren,[18] Black,[19,20] and Tibbitts.[21] It was also investigated earlier and reported in well-known references by Cunningham,[22] Van Lingen,[23] and Feenstra.[24]

Brittle, elastic rock properties are associated with crushing and chipping roller-bit cutting actions in shallower wells. Shearing action is associated with PDC drill bits and drag bits, penetrating the more plastic formations like shales. Gouging, tearing and ripping actions are tied to long-tooth, high-offset roller bits drilling softer formations. The plowing type of cutting action is associated with conventional diamond drilling, when the stresses under the individual diamond indenters convert the deeper, stronger rocks into plastic-like materials in order to penetrate.

Sliding is a cutting-action component associated with all rotary drill bits. Sliding of the teeth or inserts of a roller bit occurs under each roller cone due to geometry constraints of the cone designs. Drag bit, PDC bit,

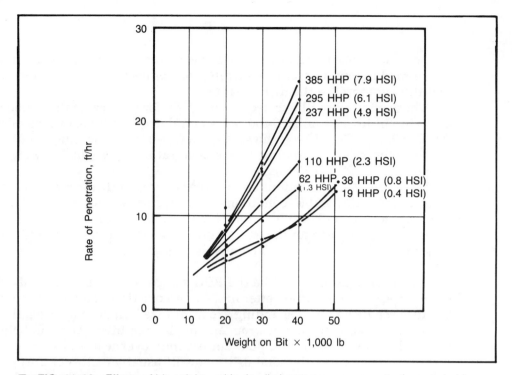

■ **FIG. 11–26** Effects of bit weight and hydraulic horsepower on penetration rate in Mancos shale with a 7⅞-in. TCI roller bit (*courtesy Terra Tek Inc., ref. 6*)

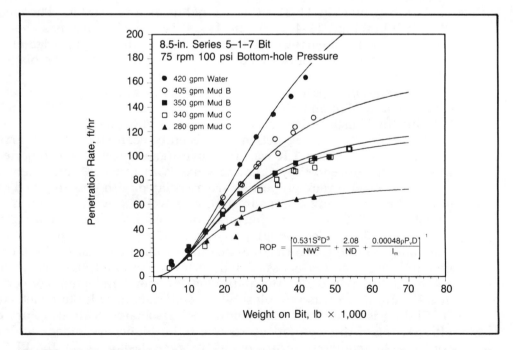

■ **FIG. 11–27** Effect of mud flow volume and type on penetration rate in limestone formation (*after Warren, ref. 18*)

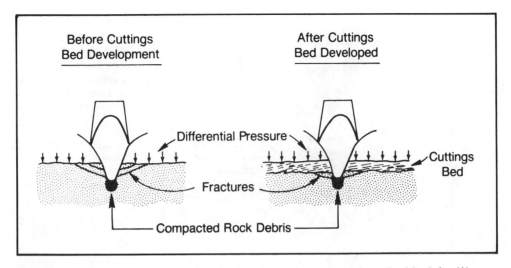

■ FIG. 11–28 Schematic of cuttings bed and cratering process for roller bits (*after Warren, ref. 18*)

and diamond bit cutters all also slide in contact with the rock during drilling. Fig. 11–29 is a representation of the chip-forming processes for the basic bit types.

DULL BIT EVALUATIONS

An analysis of the dull bits helps in selecting the best following bits for the current drilling operation and assists in selections for later offset wells by providing reference data on bit performance and wear. Bit manufacturers provide dull-bit grading manuals. The IADC has also developed a dull-bit grading code to help quantify the appearance and wear condition of dull bits.

Typical wear patterns and the manufacturers' interpretations of dull-bit conditions are given in the following sections for a few roller bits and diamond bits. These cases illustrate useful information that affects drilling costs or that can be derived from dull-bit analysis.

The steel-tooth roller bit in Fig. 11–30 is dull but has the desired appearance, indicating correct bit selection and good operating practices. The tooth hardfacing has held its cutting edge on the flanks of the teeth and provided self-sharpening. The bit was probably pulled because it had run the programmed hours or perhaps because the penetration rate was off slightly. On the other hand, the dull steel-tooth bit in Fig. 11–31 exhibits excessive tooth breakage. The probable causes include improper break-in procedures (too much weight too early), junk in the hole, teeth too long for this formation, or excessive bit weight.

A similar abusive case is the dull bit in Fig. 11–32, which shows "brad-

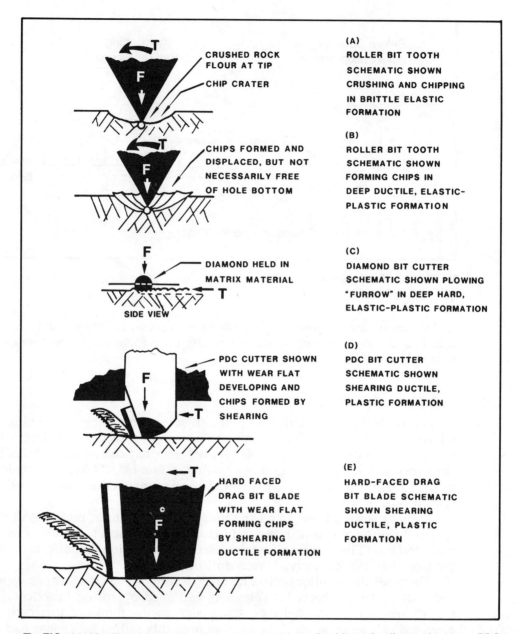

■ FIG. 11–29 Illustrations of cuttings generated by roller bit teeth, diamond cutters, PDC cutters, and shear bit blade

ded" teeth, a condition in which the more ductile tooth core deforms over the harder case surfaces of the teeth. Possible causes include using too much bit weight, running the bit too long, and increasing the weight toward the end of the run to make the dull bit drill. Using a harder-formation bit is a likely solution.

■ **FIG. 11–30** Steel-tooth roller bit showing the desired dull bit wear pattern. Teeth are partly worn down, not broken, not bradded. Bit has not been skidding or drilled off center (*courtesy Smith Tool*)

For TCI bits, evaluation of the dull products also provides data to help improve results with the next bits run in the hole or in offset wells. A well-worn TCI bit demonstrating the correct bit selection for the formation drilled plus good operating practices—the right energy level—is shown in Fig. 11–33. The cones are worn and the inserts dulled, but there are no broken inserts. Conversely, the dull TCI bit in Fig. 11–34 has been used under abusive conditions, meaning that the energy level was too high and, possibly, the formation too hard for this bit. Running a harder-formation bit (more inserts and less insert exposure) is indicated, along with a reduction in rotary speed.

The TCI bit in Fig. 11–35 exhibits different failure modes. The bearings are worn out, allowing the cones to interfere and locking a cone so it skidded and flattened some of the inserts. This is, however, a normal failure condition at the end of a long TCI bit run. The driller can detect this problem by a penetration rate reduction accompanied by an increase in torque.

Evaluating worn and dull diamond bits yields similar benefits, helping to reduce drilling costs and improve selection of following bits and of bits for offset wells. As with TCI roller bits, running on junk may damage the cutting structure of a diamond bit. Poor stabilization at the bit or in the bottom section of the collar string likewise leads to premature failures and short runs for conventional or PDC diamond bits.

Erosion of the bit matrix holding the diamonds is another possible source of diamond bit wear, usually occurring after 100 or more hours in the hole. Erosion increases as the bit mud velocity and the sand content of the mud under the bit rise. The conventional diamond bit in Fig. 11–36 clearly exhibits erosion of the bit matrix after a long run, and this process has started to overexpose the diamonds. The bit shown was run

■ **FIG. 11–31** Dull bit evaluation: excessive tooth breakage (*courtesy Smith Tool*)

■ **FIG. 11–32** Dull bit evaluation: bradded teeth (*courtesy Smith Tool*)

■ **FIG. 11–33** Example of dull TCI roller bit with low wear of inserts, indicating bit selection correct for formation drilled. No broken inserts. (*courtesy Security Div.*)

■ **FIG. 11–34** Dull bit evaluation: a dull TCI roller bit exhibiting broken inserts due to excessive energy level or formation too hard (incorrect bit selection), or both (*courtesy Security Div.*)

■ **FIG. 11–35** Dull bit evaluation: dull TCI roller bit with evidence of failed bearing, which locked cone and skidded. Normal failure at end of a long bit run (*courtesy Security Div.*)

■ **FIG. 11–36**   Dull conventional diamond drill bit exhibiting erosion of diamond-supporting matrix material (*courtesy Norton Christensen Inc.*)

in oil-based 17.0 lb/gal mud in Louisiana and produced a highly cost effective run of 3,520 ft in 236 hr drilling below 10,000 ft in sandy shale. It was pulled when the penetration rate reduced.

PDC bits have still different wear patterns. The most prevalent wear mode is abrasion, which produces wear flats on each cutter (see Fig. 11–37). The bit shown drilled to total depths (TD) of 10,000 to 11,000 ft in four different wells with oil-based muds—also favorable for PDC bit performance—and totalled 13,725 ft in 643 hr.[25]

DRILLING COSTS RELATED TO BIT SELECTION

The drill bit's performance determines much of the drilling costs, and performance, in turn, depends on the bits selected and how they are run. For example, a medium-formation bit run in a hard formation may lose

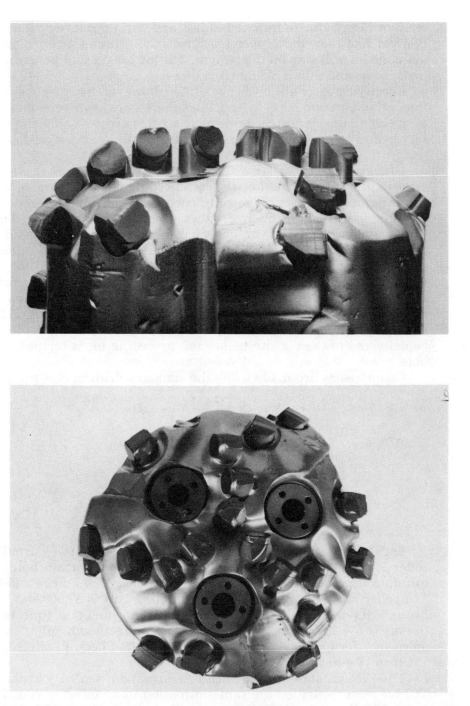

■ **FIG. 11–37** PDC drill bit exhibiting wear flats on PDC cutters due to abrasion (*courtesy Smith Tool*)

teeth or inserts due to impact failures and will therefore drill at a higher cost per foot than the proper bit. Similarly, running a roller bit when a diamond bit will stay on the bottom, drilling longer and penetrating well, costs more, and thus drilling costs increase.

Specifying the right bit is not simple, however. Because the formation interval in question is yet to be drilled, there's always an element of uncertainty about the choice. Several selection aids are available, including offset bit records, previous bit runs in a certain hole, and cost per foot calculations.

Often data are available for bit performances in offset wells in the same or in similar formations. Experienced rig personnel and bit suppliers can interpret the offset bit records, correcting for mud differences, depth changes, variations in bit hydraulics practices, and the like. Then they can reasonably forecast the expected performance of the candidate bit selections. The predicted performance and net cost of each candidate bit should then be used to calculate its expected average drilling cost per foot. The candidate drill bit with the lowest drilling cost per foot under normal circumstances is the bit selected to run. These comparisons of bit records and drilling cost calculations are carried out beforehand, so rig personnel can ensure that the chosen drill bit is available and can be delivered to the rig site before the preceding bit is tripped out of the hole.

Comparisons are made using the standard drilling cost equation:

$$C_d = \frac{O}{R} + \frac{1}{F_b}(B + T_{rt}O) \qquad (11.1)$$

where:

C_d = Drilling cost, $/ft
O = Rig operating cost, $/hr
R = Rate of penetration, fph
F_b = Bit footage, ft
B = Bit cost (net), $
T_{rt} = Round-trip time, hr

Let's take a sample case and compare a prospective TCI journal bearing roller bit run to steel-tooth bits that were run in a certain hole. For this example, assume the steel-tooth roller bits cost $750 apiece and that the TCI roller bit costs $3,500. The rig operating cost is $7,200/day, or $300/hr, and, for the depth being considered, the round-trip time to change the bit is 6½ hr. Assume that the previous steel-tooth roller bit slowed down in a formation that was sandier than had been predicted, drilling 121 ft in 14.4 hr at 8.4 fph.

The records show that a medium-formation, sealed journal bearing TCI bit recently drilled the same formation in an offset well at 7.3 ft/hr, using 15.8 lb/gal mud. However, in this hole, we have 13.5 lb/gal mud, so we assume the TCI bit will drill ½ ft/hr faster, or 7.8 ft/hr.

The bit records available also show that we should expect a run of 149 hr with the sealed bearing TCI bit. Thus, the bit would drill $(7.8 \times 149) = 1,162$ ft. The TCI drill bit would also be a good choice to overcome premature dulling of the teeth, as the tungsten carbide inserts are highly resistant to abrasion.

With these figures, the drilling costs for the two cases can be compared to help us decide whether or not the TCI bit will be an economically reasonable bit selection for the probable drilling conditions.

For the *steel-tooth bit:*

$$O = \$300/hr$$
$$R = 8.4 \text{ ft/hr}$$
$$F_b = 121 \text{ ft}$$
$$B = \$750$$
$$T_{rt} = 6.5 \text{ hr}$$

Substituting these values into the drilling cost equation gives:

$$C_d = \frac{300}{8.4} + \frac{1}{121} (750 + 6.5 \times 300)$$
$$= 35.71 + 6.20 + 16.12$$
$$= \$58.03/ft$$

And, for the selected *TCI roller bit:*

$$O = \$300/hr$$
$$R = 7.8 \text{ ft/hr}$$
$$F_b = 1,162 \text{ ft}$$
$$B = \$3,500$$
$$T_{rt} = 6.5 \text{ hr}$$

Substituting these values into the drilling cost equation gives:

$$C_d = \frac{300}{7.8} + \frac{1}{1,162} (3,500 + 6.5 \times 300)$$
$$= 38.46 + 3.01 + 1.68$$
$$= \$43.15/ft$$

Therefore, over the next 1,162 ft, if the assumptions and figures prove to be right, considerable money will be saved, running the somewhat slower and more expensive (initially) TCI roller bit, compared to steel-tooth roller bits for the same interval. The savings per foot are $\$58.03 - \$43.15 = \$14.88$. For the 1,162 ft, savings will total \$17,290.

Prospective performances for PDC drill bits and conventional diamond bits can be estimated in a like manner, working from offset bit records for the area in question or based upon performances in different geologic basins with comparable conditions (depth, mud weight, formation type, hole size).

Using this simple cost equation for steel-tooth roller bits, TCI roller bits, PDC drill bits, and conventional diamond drill bits, the cost performances of selected bits can be calculated and compared to help rig personnel find ways to reduce drilling costs and increase rig productivity.

■ **REFERENCES**

1. *Rock Bit Technology Manual,* published by Security Div., Dresser Industries, Dallas, 1982, 61 pp.
2. *Tricone Bit Handbook,* published by Hughes Tool Div., Hughes Tool Co., Houston, 1983, 45 pp.
3. Newman, E.F., "Design and Application of Softer Formation Tungsten Carbide Rock Bits," IADC/SPE 11386, presented at 1983 IADC/SPE Drilling Conference, New Orleans, 20–23 February 1983, 8 pp.
4. Black, A.D., Green, S.J., and Williams, C.R., "Drillability of a Sandstone and Dolomite at Simulated Depths," presented at ASME Petroleum Div. Conference, Houston, 18–22 September 1977, 8 pp.
5. Black, A.D. and Green, S.J., "Laboratory Simulation of Deep Well Drilling," *Petroleum Engineers International,* March 1978, pp. 40, 42, 46, 48.
6. Tibbitts, G.A., Sandstrom, J.S., Black, A.D., and Green, S.J., "The Effects of Bit Hydraulics on Full-Scale Laboratory Drilled Shale," SPE 8439, presented at 54th Annual Fall Technical Conference, Las Vegas, 23–26 September 1979, 12 pp.
7. Appl, F.C., *et al,* "Drilling Stresses on Drag Bit Cutting Edges," ed. Fairhurst, C., *Rock Mechanics: Proc. 5th Symposium on Rock Mechanics, May 1962,* Pergamon Press, 1963, pp. 119–136.
8. Ibid.
9. Appl, F.C., *et al,* "Analysis of the Cutting Action of a Single Diamond," *JPT,* September 1968, pp. 269–280.
10. Appl, F.C., *et al,* "Theoretical Analysis of Cutting and Wear of Surface Set Diamond Cutting Tools," published by Christensen Diamond Products Co., Salt Lake City, Utah, 1967, 216 pp.
11. Appl, F.C., *et al,* "Analysis of Surface Set Diamond Bit Performance," *JPT,* September 1969, pp. 301–310.
12. Holster, J.L. and Kipp, R.J., "Effect of Bit Hydraulic Horsepower on the Drilling Rate of a Polycrystalline Diamond Compact Bit," *JPT,* December 1984, pp. 2110–2118.
13. Feenstra, R. and Zijsling, D.H., "The Effect of Bit Hydraulics on Bit Performance in Relation to the Rock Destruction Mechanism at Depth," SPE 13205, presented at SPE- AIME 59th Annual Technical Conference and Exhibition, Houston, 16–19 September, 1984, 11 pp.
14. Zijsling, D.H., "Analysis of Temperature Distribution and Performance of Polycrystalline Diamond Compact Bits Under Field Drilling Conditions," SPE 13260, presented at 59th Annual Technical Conference and Exhibition, Houston, 16–19 September 1984, 16 pp.
15. Glowka, D.A., "Thermal Limitations on the Use of PDC Bits in Geothermal Drilling," *Geothermal Resources Council Transactions,* vol. 8, August 1984, pp. 261–266.
16. Glowka, D.A. and Stone, C.M., "Effects of Thermal and Mechanical Loading on PDC Bit Life," SPE 13257, presented at 59th Annual Technical Conference and Exhibition, 16–19 September 1984, 16 pp.

17. Glowka, D.A. and Stone, C.M., "The Thermal Response of Polycrystalline Diamond Compact Cutters Under Simulated Downhole Conditions," SPE 11947, presented at SPE Technical Conference and Exhibition, San Francisco, 8 October 1983, 16 pp.

18. Warren, T.M., "Penetration Rate Performance of Roller Cone Bits," SPE 13259, presented at 59th Annual Technical Conference and Exhibition, Houston, 16–19 September 1984, 12 pp.

19. Black, Green, and Williams, op. cit.

20. Black and Green, op. cit.

21. Tibbitts, Sandstrom, Black, and Green, op. cit.

22. Cunningham, R.A. and Eenink, J.G., "Laboratory Study of Effects of Overburden, Formation, and Mud Column Pressures on Drilling Rate of Permeable Formations," *JPT,* January 1959, pp. 9–17.

23. Van Lingen, N.H., "Bottom Scavenging—A Major Factor Governing Penetration Rates at Depth," *Journal of Petroleum Technology (JPT)*, February 1962, pp. 187–196.

24. Feenstra, R. and van Leeuwen, J., "Full-Scale Experiments on Jets in Impermeable Rock Drilling," *JPT,* March 1964, pp. 329–336.

25. Forrest, S. and Kuhn, K., "Near-perfect Combination Generates Record Bit Run," *Oil and Gas Journal, (OGJ)*, 19 March 1984, pp. 152–53.

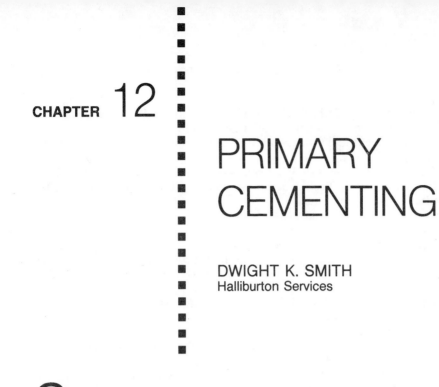

CHAPTER 12

PRIMARY CEMENTING

DWIGHT K. SMITH
Halliburton Services

Oilwell cementing is the process of mixing and displacing a cement slurry down the casing and up the annular space behind the pipe where it is allowed to set, thus bonding the pipe to the formation. No other operation in drilling or completion is as important in the producing life of the well as a successful primary cementing job (Fig. 12–1).

The first verified use of portland cement in an oil well, to shut off water that could not be held with a casing shoe, was in 1903.[1] After placing the cement, the operator normally waited 28 days before drilling the cement and testing. Improvements in cements, understanding WOC* times, and the use of admixes have reduced WOC time to a few hours under present-day practices.

Cementing procedures are classified into primary and secondary phases. *Primary cementing* is performed immediately after the casing is run into the hole. Its objective is to obtain effective zonal separation and to help protect the pipe itself. Cementing also helps to:

- Bond the pipe to the formation
- Protect producing strata
- Minimize the danger of blowouts from high-pressure zones
- Seal off lost-circulation zones or other troublesome formations as a prelude to deeper drilling

Secondary cementing, or *squeeze cementing,* is the process of forcing a cement slurry into holes in the casing and cavities behind the casing. These operations are usually performed for repairing or altering a completed well at some later date, or they may be used during the initial drilling

---

* Waiting on cement.

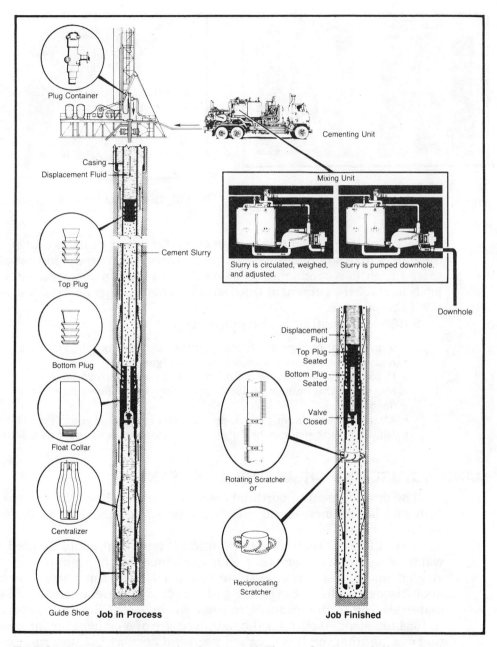

Plug Container

Cementing Unit

Casing
Displacement Fluid

Mixing Unit

Slurry is circulated, weighed, and adjusted.

Slurry is pumped downhole.

Cement Slurry

Top Plug

Downhole

Displacement
Fluid
Top Plug
Seated
Bottom Plug
Seated

Bottom Plug

Valve
Closed

Float Collar

Rotating Scratcher
or

Centralizer

Reciprocating
Scratcher

Guide Shoe **Job in Process**

Job Finished

■ **FIG. 12–1** Primary cementing (*courtesy Halliburton Services, ref. 9*)

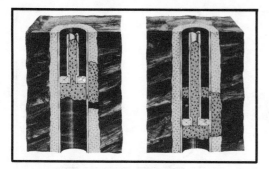

■ **FIG. 12–2** Channel behind the casing after primary cementing job

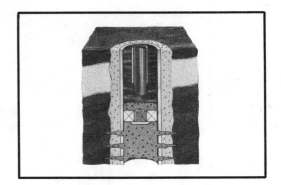

■ **FIG. 12–3** Abandoning a depleted oil or gas zone

process. Squeeze cementing is necessary for many reasons, but probably the most important one is to segregate hydrocarbon producing zones from those formations producing other fluids. The goal on a squeeze cementing job is to place the cement at the desired point or points necessary to accomplish the purpose.

Squeeze cementing is also employed to:[2]

- Supplement or repair a faulty primary cementing job (Fig. 12–2)
- Reduce the gas:oil, water:oil, or water:gas ratio
- Repair defective casing or improperly placed perforations
- Minimize the danger of lost circulation in an open hole while drilling deeper
- Abandon permanently a nonproductive or depleted zone (Fig. 12–3)
- Isolate a zone prior to perforating for production or prior to fracturing

MANUFACTURE AND CHEMISTRY OF CEMENT

The development of portland cements started when man calcined gypsum and later limestone to produce a bonding material for stone[3] (Fig. 12–4).

A result of efforts to find hydraulic cements that could be used under water was the discovery that limes produced from impure limestones yielded mortars superior to those produced from the purer limestones. Such discoveries led to the burning of blends of calcareous and argillaceous materials and to the granting of a patent by Great Britain's government in 1824 to Joseph Aspdin for the manufacture of a cement for the construction of a lighthouse.[4] It was called *portland cement* because concrete produced from it resembled stone quarried on the Isle of Portland off of the coast of England.

To understand the nature of the cement-hydration process, a brief explanation of the chemical reactions that occur in the cement kiln is in order (Fig. 12–5).[5]

EGYPTIANS —	PLASTER OF PARIS ($CaSO_4$ + HEAT)
GREEKS —	LIME ($CaCO_3$ + HEAT)
ROMANS —	POZZOLAN-LIME REACTIONS
MIDDLE AGES —	STONE CUTTING
ENGLISH —	NATURAL CEMENT-(1765 JOHN SMEATON) PORTLAND CEMENT-(1824 JOSEPH ASPDIN)
UNITED STATES —	PORTLAND CEMENT (FIRST MANUFACTURED 1872)

■ **FIG. 12–4** Development of cement

The kiln feed is a blend of finely ground calcareous and argillaceous materials. These are primarily limestone or other materials high in calcium carbonate content, clay or shale, and some iron and calcium chlorides, if there is not a sufficient quantity present in the clay or shale.[6,7] These dry materials are finely ground and mixed thoroughly in the correct proportions either in the dry condition (dry process) or mixed with water (wet process) (see Fig. 12–6).

This raw mixture is then fed at a uniform rate into the upper end of a sloping, rotary kiln and slowly travels to the lower end. The operating temperature at the fired end of the kiln is between 2,600 and 2,800°F. During its travel through the kiln, the material is converted to cement clinker, which varies in size from dust to about 2 in. in diameter. This clinker generally passes from the kiln to either a grate or a rotary cooler where it is quenched with air.

In the hot zone of the kiln, about 25% of the clinker is in a liquid state. Some of this fails to crystallize during the quenching process and

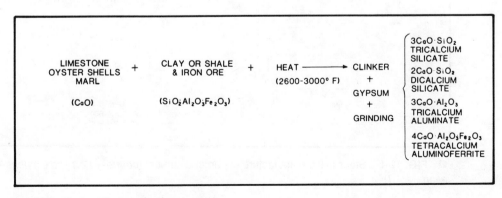

■ **FIG. 12–5** Formula for cement

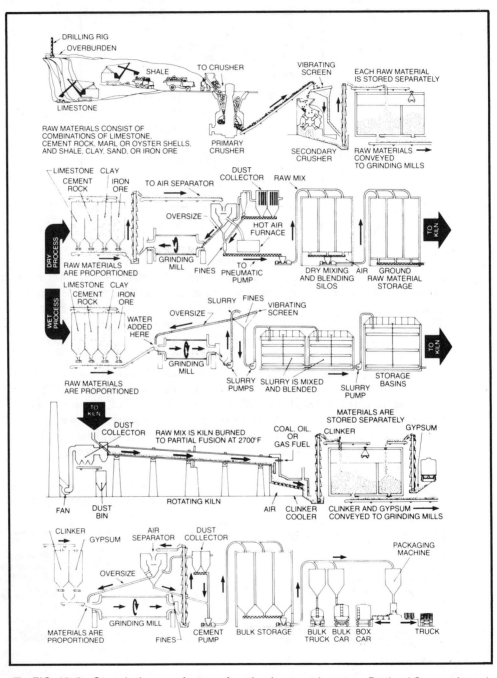

■ **FIG. 12–6**   Steps in the manufacture of portland cement (*courtesy Portland Cement Association, ref. 3*)

is present in the cooled clinker in the form of supercooled liquid or glass. Microscopic methods determine that well-quenched clinkers are as much as 10–15% glass.

The clinker contains four compounds that are the principal cementing materials, i.e., those that hydrate to form or aid in the formation of a rigid structure (Figs. 12–5 and 12–7). These compounds serve the following functions in the cement:

- Tricalcium aluminate ($3CaO \cdot Al_2O_3$) promotes rapid hydration and controls the initial set and the thickening time of the cement. It is also responsible for the cement's susceptibility to sulfate attack. To be classified as a high sulfate-resistant cement, the cement must have 3% or less tricalcium aluminate.
- Tetracalcium aluminoferrite ($4CaO \cdot Al_2O_3 \cdot Fe_2O_3$) is a low-heat-of-hydration compound in the cement. The addition of excess iron oxide increases the amount of tetracalcium aluminoferrite and decreases the amount of tricalcium aluminate.
- Dicalcium silicate ($2CaO \cdot SiO_2$) is the slow-hydration compound and accounts for the gradual gain in strength that occurs over an extended period.
- Tricalcium silicate ($3CaO \cdot SiO_2$) is the most prevalent compound in cement and the principal strength-producing material. This compound is responsi-

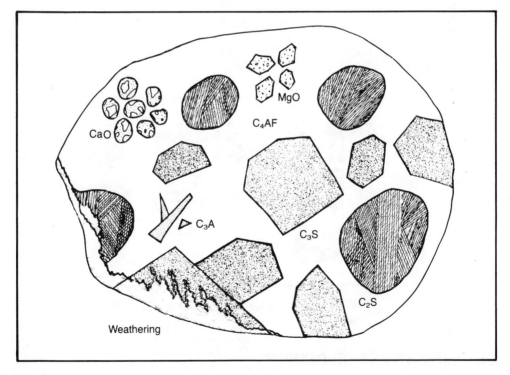

■ FIG. 12–7 Crystalline structure of cement compounds (*courtesy Halliburton Services, ref. 8*)

ble for the early strength ranging from 1 to 28 days. High-early-strength cements have a higher percentage of this compound than do portland or retarded cements.

■ **TABLE 12–1** Typical composition of portland cement[9]

API Class	Compounds, %				Wagner Fineness, sq cm/g
	C_3S	C_2S	C_3A	C_4AF	
A	53	24	8+	8	1,600–1,800
B	47	32	5–	12	1,600–1,800
C	58	16	8	8	1,800–2,200
D & E	26	54	2	12	1,200–1,500
G & H	50	30	5	12	1,600–1,800

Property	How Achieved
High early strength	Increase the C_3S content by finer grinding
Better retardation	Control C_3S, C_3A contents and grind coarser
Low heat of hydration	Limit the C_3S and C_3A contents
Resistance to sulfate attack	Limit the C_3A content

After a period of storage, the clinker is ground with gypsum to form the final cement. Gypsum controls the rate of setting and hardening of the cement paste. It is used in amounts of about 1.5–3.0% by weight of the cement (Table 12–1 and Fig. 12–8).

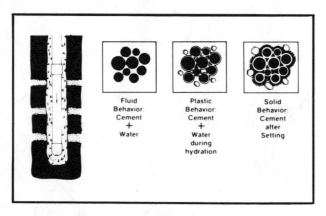

■ **FIG. 12–8** Hardening process of cement slurry (*courtesy Halliburton Services, ref. 9*)

CLASSIFICATIONS OF CEMENT

The portland cements for oilwell cementing carry the API classifications shown in Tables 12–2 and 12–3.[10]

The useable depth of API cements is a function of temperature and

■ **TABLE 12–2** API cement classification

API Class	Mixing Water, gal/sack	Slurry Weight, lb/gal	Well Depth,[a] ft	Static Temperature, °F
A (Portland)	5.2	15.6	0–6,000	80–170
B (Portland)	5.2	15.6	0–6,000	80–170
C (High Early)	6.3	14.8	0–6,000	80–170
D (Retarded)	4.3	16.4	6,000–10,000	170–230
E (Retarded)	4.3	16.4	6,000–14,000	170–290
F (Retarded)	4.3	16.4	10,000–16,000	230–320
G (Basic, California)	5.0	15.8	0–8,000	80–200
H (Basic, Gulf Coast)	4.3	16.4	0–8,000	80–200

[a] Depth based on API casing well simulation schedule.

■ **TABLE 12–3** API classifications for oilwell cements

Class A: Intended for use from surface to 6,000 ft depth,* when special properties are not required. Available only in ordinary type (similar to ASTM C 150, Type I).†

Class B: Intended for use from surface to 6,000 ft depth, when conditions require moderate to high sulfate resistance. Available in both moderate (similar to ASTM C 150, Type II) and high sulfate-resistant types.

Class C: Intended for use from surface to 6,000 ft depth, when conditions require high early strength. Available in ordinary and moderate (similar to ASTM C 150, Type III) and high sulfate-resistant types.

Class D: Intended for use from 6,000 to 10,000 ft depth, under conditions of moderately high temperatures and pressures. Available in both moderate and high sulfate-resistant types.

Class E: Intended for use from 10,000 to 14,000 ft depth, under conditions of high temperatures and pressures. Available in both moderate and high sulfate-resistant types.

Class F: Intended for use from 10,000 to 16,000 ft depth, under conditions of extremely high temperatures and pressures. Available in both moderate and high sulfate-resistant types.

Class G: Intended for use as a basic well cement from surface to 8,000 ft depth as manufactured or can be used with accelerators and retarders to cover a wide range of well depths and temperatures. No additions other than calcium sulfate or water, or both, shall be interground or blended with the clinker during manufacture of Class G well cement. Available in moderate and high sulfate-resistant types.

Class H: Intended for use as a basic well cement from surface to 8,000 ft depth as manufactured, and can be used with accelerators and retarders to cover a wide range of well depths and temperatures. No additions other than calcium sulfate or water, or both, shall be interground or blended with the clinker during manufacture of Class H well cement. Available in moderate and high sulfate-resistant types.

---

* Depth limits are based on the conditions imposed by the casing-cement specification tests and should be considered as approximate values.
† ASTM C 150: Standard Specification for Portland Cement.

■ **TABLE 12–4**   Basis for API well-simulation test schedules

Well Depth, ft	Static Bottom-hole Temperature, °F	Circulating Bottom-hole Temperature, °F		
		Casing	Squeeze	Liner
2,000	110	91(*9)	98(4)	91(4)
6,000	170	113(20)	136(10)	113(10)
8,000	200	125(28)	159(15)	125(15)
12,000	260	172(44)	213(24)	172(24)
16,000	320	248(60)	271(34)	248(34)
20,000	380	340(75)	—	—

* Minutes to reach bottom-hole circulating temperature.

pressure. In areas of subnormal temperatures, API cements can be used at greater depths. In areas of abnormally high temperatures, they may be limited to shallower depths. Normal API temperature gradient is 1.5°F/ 100 ft of depth[11] (Table 12–4).

FIELD HANDLING AND STORAGE OF CEMENT

A large percentage of the world's cementing jobs utilize bulk systems rather than manual handling in sacks. Bulk systems enable the preparation and supply of tailor-made compositions to suit any well-condition requirement. The cementing service companies offering bulk materials obtain the basic cement from manufacturers and transport it in hopper-bottom railway cars, in trucks, or by barges to their field bulk-storage stations.

At the bulk blending stations, cement is unloaded by pneumatic systems operated under 30–40 psi air pressure into weather-tight bins or tanks of various designs. Land or marine transports handle cement at the wellsite at controllable feed rates.

Water-borne service vessels utilizing the pneumatic pressure system are usually equipped with their own weighing and blending plants or may obtain weighed and blended materials from a support vessel or shore stations within their operating areas (Figs. 12–9 through 12–14).

Bulk materials supplied under these conditions have contributed much to progress in well cementing, both technically and economically. Bulk cement blending stations provide many cement compositions that otherwise would be impractical. Jobs requiring a very large number of sacks would be difficult to complete because of the excessive time required for handling sacked cement.

For convenience and storage space, especially for high-volume jobs, several types of field storage bins are used. The bins may be located at the wellsite, either on land or water, filled in advance of the cement opera-

■ **FIG. 12–9** Bulk rail transports

■ **FIG. 12–10** Land-based bulk storage plant

tion, and used as needed. This limits the transport equipment on the jobsite and conserves storage at the time of cementing.

CEMENT ADDITIVES AND THEIR EFFECTS

Bulk handling and the manufacture of basic cements has led to the tailoring of cementing compositions for specific well conditions. (See Table

■ **FIG. 12–11** Marine-based bulk storage and blending plant

■ **FIG. 12–12** Bulk transport for land operations

12–5.) This is accomplished by the use of additives with API classes A, B, G, or H cements. Additives are used to

- Reduce slurry density and increase slurry volume
- Increase thickening time and retard setting
- Reduce waiting-on-cement time and increase early strength
- Reduce water loss, help protect sensitive formations, and help prevent premature dehydration
- Increase slurry density to restrain pressure
- Restore circulation with increased fillup in annulus and reduce cost

Accelerators

The average well in the U.S. is less than 5,000 ft deep. Conductor and surface casing cements have lower temperatures, requiring an accelerator

■ **FIG. 12–13** Bulk cement and/or mud system

■ **FIG. 12–14** In remote arctic areas cement is transported by aircraft

to promote the setting of the cement and to reduce excessive waiting times for cement to set in cold weather.

Early work by Farris demonstrated that a compressive strength of 100 psi or tensile strength of 8 psi will support casing and prevent communication of formation fluids.[12] Bearden and Lane suggested that, under ideal conditions, a 4-psi tensile strength is sufficient.[13]

Most state and regulatory bodies accept a compressive strength of 500 psi as adequate for resuming drilling operations, which is a considerable safety factor, based on the other investigations.[14] Accelerators may be added to the cement or mixing water to shorten the waiting-on-cement

■ **TABLE 12–5** Summary of oilwell cementing additives

Type of Additive	Use	Chemical Composition	Benefit	Type of Cement
Accelerators	Reducing WOC time Setting surface pipe Setting cement plugs Combatting lost circulation	Calcium chloride Sodium chloride Gypsum Sodium silicate Dispersants Sea water	Accelerated setting High early strength	All API Classes Pozzolans Diacel systems
Retarders	Increasing thickening time for placement Reducing slurry viscosity	Lignosulfonates Organic acids CMHEC Modified lignosulfonates	Increased pumping time Better flow properties	API Classes D, E, G, and H Pozzolans Diacel systems
Weight-reducing additives	Reducing weight Combatting lost circulation	Bentonite-attapulgite Gilsonite Diatomaceous earth Perlite Pozzolans High Strength Microspheres Nitrogen Foam	Lighter weight Economy Better fillup Lower density	All API Classes Pozzolans Diacel systems
Heavy-weight additives	Combatting high pressure Increasing slurry weight	Hematite Ilmenite Barite Sand Dispersants	Higher density	API Classes D, E, G, and H
Additives for controlling lost circulation	Bridging Increasing fillup Combatting lost circulation	Gilsonite Walnut hulls Cellophane flakes Gypsum cement Bentonite-diesel oil Nylon fibers Brine-contact gel	Bridged fractures Lighter fluid columns Squeezed fractured zones Minimized lost circulation	All API Classes Pozzolans Diacel systems
Filtration-control additives	Squeeze cementing Setting long liners Cementing in water- sensitive formations	Polymers Dispersants CMHEC Latex	Reduced dehydration Lower volume of cement Better fillup	All API Classes Pozzolans Diacel systems
Dispersants	Reducing hydraulic horsepower Densifying cement slurries for plugging Improving flow properties	Organic acids Polymers Sodium chloride Lignosulfonates	Thinner slurries Decreased fluid loss Better mud removal Better placement	All API Classes Pozzolans Diacel systems
Special cements or additives				
Salt	Primary cementing	Sodium chloride	Better bonding to salt, shales, sands	All API Classes
Silica flour	High-temperature cementing	Silicon dioxide	Stabilized strength Lower permeability	All API Classes
Mud Kil	Neutralizing mud-treating chemicals	Paraformaldehyde	Better bonding Greater strength	API Classes A, B, C, G, and H
Radioactive tracers	Tracing flow patterns Locating leaks	Sc 46	—	All API Classes
Pozzolan lime	High-temperature cementing	Silica-lime reactions	Lighter weight Economy	—
Silica lime	High-temperature cementing	Silica-lime reactions	Lighter weight	—
Gypsum cement	Dealing with special conditions	Calcium sulfate Hemihydrate	Higher strength Faster setting	—
Hydromite	Dealing with special conditions	Gypsum with resin	Higher strength Faster setting	—
Latex cement	Dealing with special conditions	Liquid or powdered latex	Better bonding Controlled filtration	API Classes A, B, G, and H
Thixotropic	Treating lost circulation Preventing gas migration	Organic gel Inorganic gel	Fast set Stop lost circulation Stop gas migration	All API Classes
Spacers and Flushes	Stop contamination of cement Displace drilling fluid Separate incompatible fluids Water-wet formations		Uniform cement jobs Better hydraulic seal Less fallback Reduced lost circulation	All types

time, thereby saving time and money. The most common cement accelerators are shown in Tables 12–6 and 12–7.

Retarders

For deep wells, cement retarders extend the pumpability of the cement and are normally added to API classes G or H cement for wells deeper than 8,000 ft.

The primary factor that governs the use of additional retarder is the well's temperature, cement volume to be pumped, and well depth. As the temperature increases, the chemical reaction between the cement and water accelerates, which in turn reduces the thickening time and pumpability.

Pressure has some effect, but a temperature increase of 250°F may mean the difference between an unsuccessful and a successful cementing job. Temperature is a critical factor in deep wells because of the longer displacement time. When cementing in deep, high-temperature wells (those 8,000 ft or deeper), materials must be selected carefully and pilot tests conducted using actual job materials under temperature and pressure conditions similar to those in the well.

In high-temperature environments, cement retrogression may become a problem with all API classes of cement. For this reason, 35–40% silica flour should be added to overcome strength retrogression.[15] High temperatures also increase cement permeability. Any additive with a high water requirement, such as bentonite, is not recommended with cement for use in wells where bottom-hole temperatures exceed 230°F because of progressive strength loss over prolonged periods.

Materials commonly used to retard or increase the setting time of cement include calcium lignosulfonate, sodium carboxymethylhydroxyethyl-

■ **TABLE 12–6** Effects of cement accelerators on cement properties (*courtesy Halliburton*)[9]

Cement Additive	Properties			
	Density	**Mixing Water**	**Thickening Time**	**Water Loss**
Calcium Chloride (CaCl$_2$)	No effect	No effect	Decreases (major effect)	Increases (major effect)
Sodium Chloride (NaCl).	Increases (minor effect at high concentration)	No effect	Decreases (minor effect at low concentrations) Increases (minor effect at high concentrations)	Increases (minor effect)
Sea Water	Increases (minor effect)	No effect	Decreases (minor effect)	Increases (minor effect)
Cal-Seal	Decreases (minor effect)	Increases (minor effect)	Decreases (major effect)	No effect

■ **TABLE 12–7** Pressure-temperature thickening-time data, API common cement, API casing tests, hr:min

% Calcium Chloride	0% Bentonite		2% Bentonite		4% Bentonite	
	2,000 ft	4,000 ft	2,000 ft	4,000 ft	2,000 ft	4,000 ft
0	3:36	2:25	3:20	2:25	3:45	2:34
2	1:30	1:04	2:00	1:30	2:41	2:03
4	0:47	0:41	0:56	1:10	1:52	2:00

% Calcium Chloride	Compressive Strength Curing Temperature and Curing Pressure					
	40°F 0 psi	60°F 0 psi	80°F 0 psi	95°F 800 psi	110°F 1,600 psi	140°F 3,000 psi
	8 hr					
0	Not set	45	265	445	730	2,890
2	10	300	1,230	1,250	1,750	3,380
4	Not set	450	1,490	1,650	2,350	2,950
	12 hr					
0	Not set	80	560	800	1,120	2,170
2	64	555	1,675	1,310	1,680	2,545
4	15	705	2,010	2,500	3,725	4,060
	24 hr					
0	30	615	1,905	2,085	2,925	5,050
2	415	1,450	3,125	3,750	5,015	6,110
4	400	1,695	3,080	4,375	4,600	5,410

cellulose derivatives, and blends of lignin materials with organic acids (Tables 12–8 and 12–9).

Lightweight Additives

Additives used to reduce slurry density include bentonite, pozzolans, diatomaceous earth, expanded perlite, and gilsonite (see Table 12–10).[16] Of these, bentonite is the most widely used in cement slurries. Normal concentrations range up to 12% by weight of cement, but 4% represents the average concentration.

Recommended water-cement ratios for given percentages of bentonite for API classes A, B, and G cements are shown in Table 12–11.

For extremely lightweight systems, tiny glass spheres may be added to any API cement to produce densities as low as 9 lb/gal.[17] Other lightweight mechanisms employ nitrogen to produce foam cement, which can

■ **TABLE 12–8** Effects of retarders on cement properties

Cement Additive	Properties			
	Density	Mixing Water	Thickening Time	Water Loss
HR-4	No Effect	No Effect	Increases (major effect)	No Effect
HR-5	No Effect	No Effect	Increases (major effect)	No Effect
HR-6L	No Effect	No Effect	Increases (major effect)	No Effect
HR-7	No Effect	No Effect	Increases (major effect)	Decreases (minor effect)
HR-12	No Effect	No Effect	Increases (major effect)	No Effect
HR-20	No Effect	No Effect	Increases (major effect)	No Effect
Sodium Chloride (NaCl)	Increases (minor effect)	No Effect	Decreases at low concentrations (minor effect) Increases at high concentrations (minor effect)	No Effect

■ **TABLE 12–9** Pressure-temperature thickening-time data, API class G or H cement

Well Depth, ft	Temperature, °F		Lignin Retarder, %	Approximate Thickening Time, hr
	Static	Circulation		
Casing Cementing (Primary)				
0– 6,000	80–170	80–113	0.0	2–4
6,000–10,000	170–230	113–144	0.0–0.5	2–4
10,000–14,000	230–290	144–206	0.5–0.7	3–4
14,000–18,000	290–350	206–300	0.7–1.5	*
Squeeze Cementing				
0– 4,000	80–140	80–116	0.0	2–4
4,000– 8,000	140–200	116–159	0.0–0.5	2–4
8,000–12,000	200–260	159–213	0.5–0.7	2–4
Below 12,000	260 plus	213 plus	0.7–1.5	*

* Requires special laboratory tests. Other retarders may be recommended for these conditions.

produce densities as low as 8.5 lb/gal. Both mechanisms have application to low fracture gradient formations (see Figs. 12–15 and 12–16).

Heavyweight Additives

These additivies are normally added to cement when abnormally high pressures are to be encountered during cementing. Many materials are available to increase slurry density; however, some of the additives affect other properties of the cement in deep wells and should not be used. The specific gravities of the most common heavyweight additives range from

■ **TABLE 12–10** Effects of lightweight additives on cement properties

Cement Additive	Properties			
	Density	**Mixing Water**	**Thickening Time**	**Water Loss**
Halliburton Gel (Bentonite)	Decreases (major effect)	Increases (major effect)	Decreases (major effect)	No Effect
Pozzolans	Decreases (major effect)	Increases (minor effect)	Increases (minor effect)	No Effect
SPHERELITE	Decreases (major effect)	No Effect	No Effect	Decreases (major effect)
Nitrogen	Decreases (major effect)	No Effect	No Effect	Decreases (major effect)
Gilsonite	Decreases (major effect)	Increases (minor effect)	No Effect	No Effect

■ **TABLE 12–11** Recommended water-cement ratio for bentonite cements

Bentonite, %	Water, gal/sack	Slurry Weight, lb/gal	Slurry Volume, cu ft/sack
0	5.2	15.60	1.18
2	6.5	14.70	1.36
4	7.8	14.10	1.53
6	9.1	13.65	1.69
8	10.4	13.10	1.92
*10	11.1	12.95	2.02
*12	12.3	12.60	2.19

* With dispersant.

2.65 to 6.98. The most common materials are hematite, barite, and sand. However, combining friction-reducing additives with hematite is the most common technique of increasing cement slurry density to 22 lb/gal (see Table 12–12).

Lost-Circulation Additives

During drilling, the problem of lost returns or circulation is fairly common. In most instances additives are used with the drilling mud; however, under certain circumstances cement containing lost-circulation materials must be used to maintain circulation. Many materials are available to help restore circulation or prevent lost-circulation difficulties. These materials are classified as fibrous, granular, lamellated, and semisolids (see Table 12–13).

■ **FIG. 12–15** Microspheres used to produce lightweight cement (*courtesy Halliburton Services, OGJ, ref. 18*)

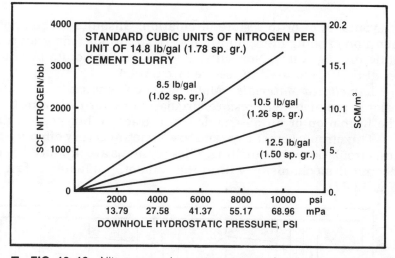

■ **FIG. 12–16** Nitrogen requirements to prepare foam cement (*courtesy Halliburton Services, ref. 19*)

Fibrous Materials Fibrous additives include shredded wood, bark, and sawdust. However, some of these materials are unsuitable for cement slurries because they contain tannins that either retard or in some instances prevent the cement slurry from setting. Nylon fibers (⅛–¼ lb/sack of cement) are sometimes used since they provide a high impact resistant cement when compared with neat cement.

Granular Materials The materials commonly used are gilsonite, nut shells, plastics, and perlites. These materials are bridging agents to help

■ **TABLE 12–12**   Effects of heavyweight additives on cement properties

Cement Additive	Properties			
	Density	**Mixing Water**	**Thickening Time**	**Water Loss**
Sand (Ottawa)	Increases (major effect)	No Effect	No Effect	No Effect
Hi-Dense	Increases (major effect)	Increases (minor effect)	No Effect	No Effect
	Increases (major effect)	Decreases (minor effect)	Increases (minor effect)	Decreases (minor effect)

■ **TABLE 12–13**   Effects of lost circulation additives on cement properties

Cement Additive	Properties			
	Density	**Mixing Water**	**Thickening Time**	**Water Loss**
Gilsonite	Decreases (major effect)	Increases (minor effect)	No Effect	No Effect
Flocele	No Effect	No Effect	No Effect	No Effect

prevent loss of or to restore circulation. Gilsonite is the best material. In addition to being inert, it also has a very low specific gravity and requires little mixing water. Normally, 12½–25 lb of gilsonite per sack of cement is all that is required to restore circulation.

Lamellated Materials   Materials such as mica, cellophane flakes, and related products help restore circulation by forming a mat at the face of the formation or an obstruction in a fracture but possess little strength.

Thixotropic cement systems are sometimes very effective in combating lost circulation while drilling. These systems are produced by cross-linking chemical mechanisms within the cement, producing a fast gelation (see Fig. 12–17).

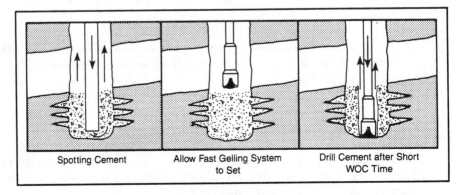

Spotting Cement   Allow Fast Gelling System to Set   Drill Cement after Short WOC Time

■ **FIG. 12–17**   Placement of thixotropic cement system (*courtesy Halliburton Services*)

Low-Fluid-Loss Additives

The application of low-water-loss additives in oilwell cements to reduce filtration rates is similar to that in drilling muds or fracturing fluids.[20] These materials reduce the possibilities of water and/or emulsion blocks, and blocks caused by bentonite clay swelling within formations when penetrated by filtrate from cement slurries. They also help protect water-sensitive shales and reduce the likelihood of premature dehydration of the cement slurry to the formation during high column cementing. Fluid-loss control tends to keep the slurry viscosity low and reduces the possibility of higher circulating pressures (Fig. 12–18).

The ability of the low-fluid-loss additive to aid in reducing the filter cake buildup during squeeze cementing is shown in Fig. 12–19.

While low-water-loss additives were designed primarily for squeeze cementing, they are widely employed today in high column cementing, particularly on deep liners (see Table 12–14).

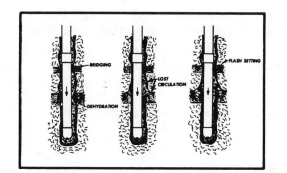

■ **FIG. 12–18** Influence of dehydration during primary cementing (*courtesy Halliburton Services*)

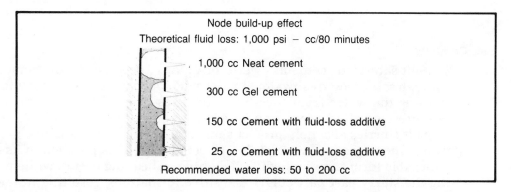

■ **FIG. 12–19** Filtration control in squeeze cementing (*courtesy Halliburton Services*)

■ **TABLE 12–14**   Effects of low-water-loss additives on cement properties

Cement Additive	Properties			
	Density	**Mixing Water**	**Thickening Time**	**Water Loss**
HALAD-9	No Effect	Increases (minor effect)	Increases (minor effect)	Decreases (major effect)
HALAD-22A	No Effect	Increases (minor effect)	Increases (minor effect)	Decreases (major effect)
HALAD 322	No Effect	Increases (minor effect)	Increases (minor effect)	Decreases (major effect)
LATEX (LA-2)	No Effect	No Effect	Increases (minor effect)	Decreases (major effect)
CFR-2 (Dispersant)	Decreases (minor effect)	Decreases (minor effect)	Increases (minor effect)	Decreases (minor effect)

Friction Reducers

Additives or thinners reduce the apparent viscosity of the slurry. The lower viscosity slurries go into turbulent flow at lower pumping rates, reducing circulation rates and allowing cement to be pumped in turbulent flow at less than formation breakdown pressures (see Table 12–15).[21,22]

Cement slurry additives for promoting turbulent flow at low displacement rates include organic dispersants, salt, different types and mixtures of calcium lignosulfonate, and a high-molecular-weight cellulose material in gel cement (Table 12–16).

■ **TABLE 12–15**   Effect of a friction reducer on cement properties

Cement Additive	Properties			
	Density	**Mixing Water**	**Thickening Time**	**Water Loss**
	Decreases (minor effect)	Decreases (minor effect)	Increases (minor effect)	Decreases (minor effect)

Salt Cements

Salt-saturated cements were originally developed for cementing through salt zones; freshwater slurries will not bond properly to salt formations as the water from the slurry dissolves or leaches away the salt at the interface, thus preventing an effective bond (see Table 12–17).

Salt slurries also help protect shale sections that are sensitive to fresh water when used for mixing salt-free cement.[23,24] This problem in its most noticeable form shows up when sloughing or heaving occurs while pumping the slurry past particularly sensitive formations, resulting in the following possible conditions:

■ **TABLE 12–16** Flow rates for turbulence (or plug flow) with and without cement dispersants

Composition	Percent Dispersant	HOLE SIZE (inches)			
		6-3/4	6-3/4	8-3/4	9-7/8
		CASING SIZE OD (inches)			
		2-7/8	4-1/2	5-1/2	7
		FLOW RATE-bbl/min			
API Class H 15.4 lb/gal	0.00	18.18 (3.03)*	13.58 (2.27)*	23.29 (3.88)*	24.93 (4.16)*
	0.50	14.32	11.28	18.66	20.21
	0.75	6.57	5.86	8.91	9.93
API Class A 4% Gel 14.1 lb/gal	0.00	25.17 (5.07)*	17.58 (3.54)*	31.54 (6.35)*	33.26 (6.70)*
	0.50	15.50	11.21	19.65	20.88
	0.75	6.58	5.30	8.65	9.41
API Class A 12% Gel 12.8 lb/gal	0.00	23.55 (4.74)*	16.45 (3.31)*	29.51 (5.94)*	31.12 (6.51)*
	0.75	14.08	10.26	17.89	19.05
	1.00	2.93	2.88	4.10	4.67

*Plug Flow

■ **TABLE 12–17** Sodium chloride

Description of Use	Benefits	Properties	Normal Range of Use in Wells
• Used to improve bonding to water-sensitive shales, clay and salt formations	• Improves the bonding of cement to shales and clay formations	• Specific Gravity: 2.17	• Depth: Surface to 20,000 ft
• Used at low concentrations as an accelerator	• Minimizes damage in zones that are sensitive to fresh water	• How packaged: Bulk and 100-lb sacks	• Circulating Temperature: 50 to 380°F (BHST)
• Used at high concentrations as a mild retarder	• Increases cement expansion at low temperatures	• Bulk Density: 71 lb/cu ft	• Concentration: 18% to saturation
• Also has dispersant properties in some cement slurries		• Water Requirement: None	

- Excessive washouts and channeling behind the pipe
- Lost circulation into the weakened shale structure
- Annular bridging that may prevent slurry circulation

Even shales that are apparently competent in the presence of fresh water may be weakened by continued exposure to freshwater slurry, which contacts the formation before the cement sets.

JOB CONSIDERATIONS

Many factors must be considered in planning the preparation of cement slurry prior to pumping.

Mixing Equipment

The mixing system on any cementing operation proportions and blends the dry cementing composition with the carrier fluid. When this is achieved, a cementing slurry with predictable properties can be supplied to the wellhead.

The most widely used mixing methods are the jet mixer, the recirculating mixer, and the batch system (Fig. 12–20). In the jet system, a stream of water mixes with cement by passing through the mixer bowl, creating a vacuum that pulls the dry cement into the bowl from the hopper immediately above. As the cement enters the jet stream of water, it is thoroughly mixed by the turbulent flow in the discharge pipe. Mixers of this type, when supplied with sufficient water under the optimum mixing pressure and adequate feed rate of cement, can produce a normal slurry at rates up to 50 cu ft/min.

Mixing speed is regulated by the volume of water forced through the jet and by the amount of cement in the hopper while mixing. A bypass line can supply extra water into the bowl discharge line to lower the slurry weight by increasing the water-cement ratio.

Modified jet mixers, called recirculating mixers, are designed to mix a wider range of slurry densities (Fig. 12–20). A recirculating system is basically a pressurized jet mixer with a large tub capacity. It uses recirculated slurry and mixing water to partially mix and discharge the slurry into the tub. Additional shear is provided by the recirculating pump and agitation jets. An in-tub eductor adds extra energy and improves mixing. As a result, more uniform cement slurries having densities as high as 22 lb/gal, pumped as slowly as 0.5 bbl/min, can be achieved with this system.

Batch mixing is also used to blend a large volume of cement slurry at surface conditions before going into the well. A relatively simple mixing method prepares a specific volume of slurry to exacting well requirements. Weight and fluid-loss properties can be controlled before going into the well (Fig. 12–20 and Table 12–18).

Water Sources and Supply

Water for mixing with cement is expensive in many areas and difficult to obtain, while in others an adequate supply is always available. On any job an essential volume with a good margin of safety must be supplied for the cementing operation. Rate of water supply to the mixing and pumping units will affect and often may control the rate of the mixing of the cementing slurry. When the water supply is reduced, the time of pumping may encroach on the time the cement slurry is beginning to set.

The water supply normally available at a drilling well may be anything from reasonably clean and free of soluble chemicals to badly contaminated with varying quantities of silt, organic matter, alkali, salts, or other impurities.

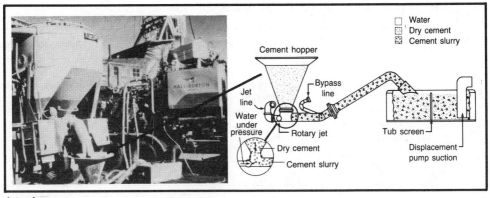

Jet mixer

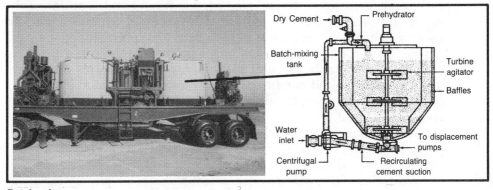

Batch mixer

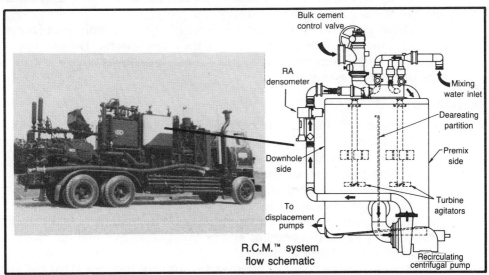

R.C.M.™ system
flow schematic

Recirculating mixer

■ **FIG. 12–20**   Mixing systems (*courtesy Halliburton Services*)

■ **TABLE 12–18** Mixing ranges for different systems

	Jet Mixer		Recirculating Mixer	
	Mixing Rates, bbl/min	Slurry Densities, lb/gal	Mixing Rates, bbl/min	Slurry Densities, lb/gal
Densified and Weighted Slurries	0–5	16–20	0–8	16–22
Neat Slurries	0–8	15–17	0–10	14–18
High-Water-Ratio Slurries	0–14	11–15	0–14	11–15

Water from wells, streams, or ponds often contains sulfates, chlorides, or other dissolved mineral salts that may be detrimental both to drilling muds and cements. Potable water is always recommended when available. Unless a supply of clear water has a noticeable saline or brackish taste, it is usually suitable for use without further questions, although saline waters are useable if conditions are pretested and understood. In the event of questionable water, the purest available supply should be used for mixing with cement.

Water temperature during warm-weather mixing is important. High-temperature mixing water (110°F) may result in a viscous slurry with a shorter pumping time.

Winter conditions must also be considered, primarily because of line freeze-up and because very low water temperatures may significantly increase the setting time of a slurry.

There must be no conflict with other water supply requirements for the drilling rig and possible emergency operations. Do not use a water supply line for cementing that in any way interferes with water in reserve for blowout prevention.

Slurry Density

Cement systems can be designed to cover a density range from 6 to 21 lb/gal (see Fig. 12–21). Slurry density is directly related to the amount of mixing water and additives in the cement and the amount of slurry contamination from drilling mud or other foreign material. In critical low-pressure fracture-gradient wells where lost circulation zones restrict the density of mud and cement, a slurry must be light enough to be supported by the weakest incompetent zone. Where high-pressure gas or salt-water flows exist, the cement slurry must be heavy enough to control the pressure and prevent a blowout.

Cement slurries should always be slightly heavier than the mud, but the mechanisms for achieving light or heavyweight slurries vary greatly.

In field operations, slurry control is customarily maintained by measur-

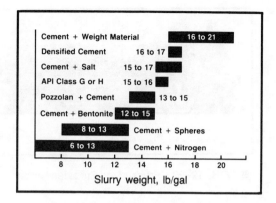

■ **FIG. 12–21** Weight ranges of cementing systems

ing the density with the standard or pressurized mud balance. For accuracy, samples should be selectively taken from the tub and vibrated to remove the finely entrapped air from the jet mixer. Automated weighing devices that fit into the discharge line between the mixing unit and wellhead give a more uniform weight record and are widely used.

A radioactive weighing device is sometimes fitted into the slurry discharge line and provides a strip-chart record of density measurements (Fig. 12–22).

The pressurized density balance is a newer instrument that measures the cement slurry under sufficient pressure to compress the entrained air to a negligible volume. This compression of the slurry under approximately 250 psi allows a more accurate density measurement of the cement when sampled directly from the tub during the mixing process (Fig. 12–23 and Table 12–19).

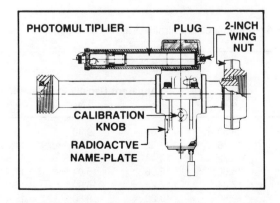

■ **FIG. 12–22** Radioactive density recording system for cement and mud service

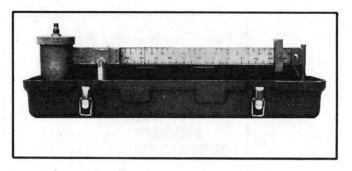

■ **FIG. 12–23** Pressurized fluid density balance

Effects of Drilling Fluids and Additives

A major problem in the cementing of wells is the effective removal of drilling fluids during cementing operations. Cementing systems are influenced by:

- Mud contamination
- Mud dilution
- Mud chemicals
- Bonding to filter cake

Drilling mud additives, organic and inorganic, affect the setting of cement differently. Organic mud additives generally retard the setting of cement, and inorganic additives accelerate the setting of cement (Table 12–20). The degree of retardation or acceleration depends upon the concentration of chemicals in a given system. With some types of cements, mud additives may not conform to a fixed pattern and can cause an erratic gelation or a viscous consistency resembling a set.

Some mud-cement contamination happens during most jobs, but proba-

■ **TABLE 12–19** Effect of entrained air on fluid density

Actual or Absolute Density, lbM/gal	Apparent Density at Indicated Air Content*					
	2%	4%	6%	8%	10%	12%
10.00	9.80	9.60	9.40	9.20	9.00	8.80
12.00	11.76	11.52	11.28	11.04	10.80	10.56
14.00	13.72	13.44	13.16	12.88	12.60	12.32
16.00	15.68	15.36	15.04	14.72	14.40	14.08
18.00	17.64	17.28	16.92	16.56	16.20	15.84
20.00	19.60	19.20	18.80	18.40	18.00	17.60

* Percent air by volume

■ **TABLE 12–20** Mixing water or mud additive contaminants

Source of Contaminant	Type	Effect on Cement Slurry
Mixing Water	Salt (1% to 8% per weight of water)	Accelerates set
	Organic Material (decomposed plant life, waste effluents)	Retards set
	Agricultural Products (farm fertilizer)	Accelerates set
	Sea Water	Accelerates set
Mud Additives	Barium sulfate ($BaSo_4$) (Barite)	Increases density, reduces strength
	Caustics ($NaOH$, Na_2COH_3, etc.)	Accelerates set
	Calcium compounds (CaO, $Ca(OH)_2$, $CaCl_2$, $CaSO$, $2H_2O$)	Accelerates set
	Thinners (tannins, lignosulfonates, quebracho, lignins, etc.)	Retards set
	Fluid-loss control additives (CMC, starch, guar, polyacrylamides, lignosulfonates)	Retards set

bly the most occurs when spotting cement plugs in mud systems highly treated with chemicals. The cement volume in relation to the mud volume is small, and the degree of mud contamination is always unknown. Contamination is evident when soft cement is observed while drilling out the plug.

Oil or inverted emulsion muds having powerful surfactants and sodium or calcium chloride in their formulation have been found to produce a rapid gelation of cement slurries under high temperatures or pressures. For such systems, a generous volume of a selected spacer and/or flush is recommended to prevent interfacial gelation.

The most satisfactory means of combating the effect of drilling mud additives on oilwell cement has been to minimize the contamination of the cement with treated drilling fluids during a cement operation. This is accomplished by using wiper plugs ahead of the cement in casings and by using buffer solutions or washes ahead of the cement, or both.

The wiper plug prevents any contact and resulting contamination between the drilling fluid and the cement slurry inside the casing, and a buffer wash helps eliminate contamination in the annular space between the outside of the casing and the formation.[25]

Cement Volume

API cements, based on depth and temperature requirements, may be purchased in most oil producing areas of the world. While much emphasis is placed on deep, hot wells, the larger volume of cement used in wells is at depths of less than 8,000 ft. API class G and class H cements are designed to fit these conditions where bottom-hole static temperatures are less than 200°F.[26] For deeper wells, other classes of API cements may be purchased, or class G or H cements can be modified with additives to fit individual well conditions.

The cement volume required for a specific fillup on a casing job should be based on field experience and regulatory requirements. In the absence of specific guidelines, a volume equal to 1.5 times the caliper survey volume should be used.

When a good mud program is used, assume that the hole size is the same as the bit size. Make allowances to take care of hole irregularities. Caliper measurements generally show that such allowances are inadequate. Caliper logs may be necessary to determine the proper location of centralizers or scratchers to obtain the maximum efficiency; however, model studies indicate that scratchers, centralizers, or any casing device, whether in a gauged or washed-out section, will aid in inducing turbulence, thereby providing better mud removal and consequently better fillup.

While regulatory rules and hole conditions may dictate the fillup necessary for a given cementing operation, you should have a minimum of 300–500 ft of fillup behind the intermediate and/or production string of casing. Always use too much, rather than too little cement, especially when there is possible contamination or dilution by the mud system.

CASING EQUIPMENT

Casing equipment selected for running a cement job must be strong, versatile, and controllable.

Cementing Heads

Cementing heads or plug containers provide the union for connecting the cementing lines from the service unit to the casing. These heads hold one or more cementing plugs. The dual-plug heads are the most widely used. This type of head allows circulation of the mud in a normal manner, release of the bottom plug, mixing and pumping down of the cement, release of the top plug, and displacing the cement without making or breaking any connections (Figs. 12–1, 12–24, and 12–25).

Subsurface Equipment

Subsurface equipment commonly used includes casing guide shoe, float collars, wiper plugs, casing centralizers, scratchers, and stage tools.[27] Al-

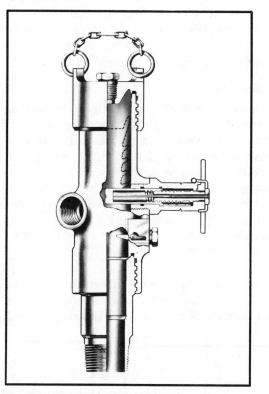

■ **FIG. 12–24**   Cementing head

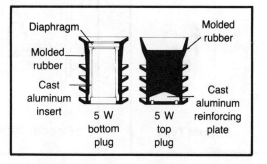

■ **FIG. 12–25**   Cementing plugs

though there are many different types of subsurface equipment, the examples below are representative.

The function of the plain open-ended guide shoe is to guide the casing past any obstruction in the hole. The conventional combination float guide shoe and the float collar provide (1) a means of floating or partially floating the casing into the hole, (2) a back-pressure valve to help prevent the backflow of cement after it has been placed, and (3) a safety valve to help prevent a blowout when going into the hole.

This equipment's essential element is a check valve. Flow of fluids can be directed down the casing; however, the valve prevents flow into and up the casing (Figs. 12–1, 12–26, and 12–27).

Fillup Floating Equipment

Because of lost returns, a frequent cementing problem, plus the extra time required to fill the casing, modified types of float collars have been introduced. This modified float permits the casing to partially fill from the bottom as the casing is run. Fluid entry into the casing is controlled by a differential valve or an orifice.

The back-pressure valve effect is activated by pump pressure, so the same advantages of preventing cement backflow are realized with this equipment as with conventional floats. However, there is no protection against blowouts.

Wiper Plugs

Wiper plugs are equipped with rubber-cupped fins that wipe mud from the walls of the casing ahead of the cement and clean the walls of cement behind the slurry. See Figs. 12–28 and 12–29.[28] The top plug also serves

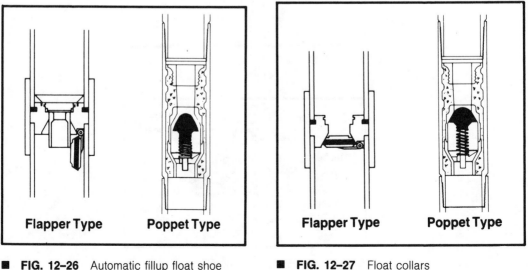

Flapper Type   Poppet Type   Flapper Type   Poppet Type

■ **FIG. 12–26** Automatic fillup float shoe   ■ **FIG. 12–27** Float collars

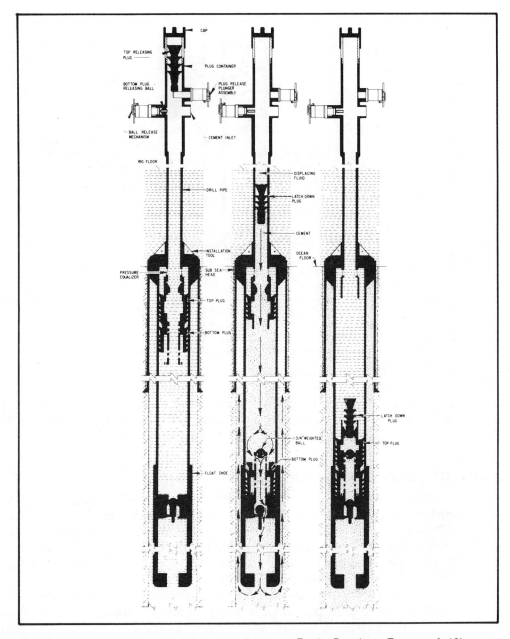

■ **FIG. 12–28** Subsea cementing plugs (*courtesy Farris, Petroleum Trans., ref. 12*)

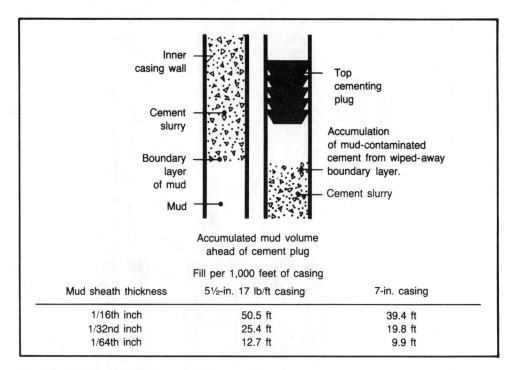

Fill per 1,000 feet of casing

Mud sheath thickness	5½-in. 17 lb/ft casing	7-in. casing
1/16th inch	50.5 ft	39.4 ft
1/32nd inch	25.4 ft	19.8 ft
1/64th inch	12.7 ft	9.9 ft

■ **FIG. 12–29** Need for top and bottom plugs

as a means of determining when cement has been completely displaced. When this plug contacts the bottom plug or hits the plug seat in a float collar, surface pressure increases instantly.

PRIMARY CEMENTING—THE CRITICAL PERIOD[29,30,31]

The cementing of casing is one of the critical operations during the drilling and completion of an oil well. Preparation of the hole, assembly of the necessary surface and subsurface equipment, rigging up, and running the casing are all preliminary to the most important stage, which is that period between the running of the last few joints of casing and the final displacement of the cement.

During this critical period, the success or failure of the entire operation is likely to be determined. A successful casing cementing job is one in which the casing is landed at the exact specific depth and the annular space between the casing and formation is sealed with set cement. This cement allows any one fluid or gas to be produced to the exclusion of any other fluid or gas.

Most primary cement jobs are performed by pumping the slurry down the casing and up the annulus; however, there are modified techniques for special situations (Fig. 12–30).

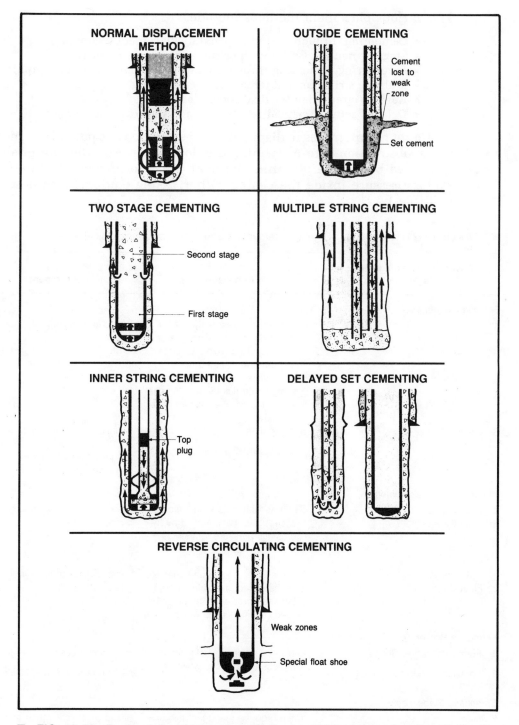

FIG. 12–30 Types of displacement used in the cementing process (*courtesy Halliburton Services*)

The various placement methods include:

- Cementing through casing (normal displacement technique)
- Stage cementing (wells with critical fracture gradients)
- Inner string cementing through drillpipe (for large-diameter pipe)
- Multiple string cement (small-diameter tubing)
- Reverse circulation (critical formations)
- Delayed setting (critical formations and improved placement)

Count pump strokes in displacing cement, since it is possible to encounter an obstruction before reaching the float collar. Also the pumps should be slowed before the plug hits the float collar to prevent the buildup of excess pressure inside the casing. Other casing equipment is described in Table 12–21.

■ **TABLE 12–21** Summary— Digest of cementing equipment and mechanical aids

Cementing Equipment & Types	Application	Placement
Floating Equipment		
1. Guide Shoes	Guides casing past ledges. Minimizes sidewall caving and aids in safely passing hard shoulders and through deviated holes	First joint of casing
2. Float Shoes	Guides casing in hole. Minimizes strain on derrick by allowing casing to float in. Prevents cement backflow	First joint of casing
3. Float Collars	Same as float shoes Catches cementing plugs	1 joint above shoe in wells less than 6,000 ft 2–3 joints above shoe in wells greater than 6,000 ft
Automatic Fillup Equipment		
1. Float Shoes 2. Float Collars	Same as float collars and shoes except fillup is controlled by hydrostatic pressure in annulus	Same as float shoes and float collars
Differential Fill Equipment		
1. Float Shoes 2. Float Collars	Same as float collars and float shoes except maintains a predetermined differential pressure between casing ID and well annulus	Same as float shoes and float collars
Formation Packer Tools		
1. Formation Packer Shoes	Packer expands against wellbore to protect lower zones while cementing	First joint of casing
2. Formation Packer Collars		As hole requirements dictate
Cementing Stage Tools		
2 Stage	When required to cement two or more sections in separate stages	Based on critical zones and formation fracture gradients
3 Stage		

■ **TABLE 12–21** *(Continued)*

Cementing Equipment & Types	Application	Placement
Plug Containers		
1. Quick Opening	To hold cementing plugs in casing string until released	Top joint of casing at surface of well
2. Continuous Cementing Heads		
Cementing Plugs		
1. Top and Bottom Wiper Plugs	Wipes casing ID and provides mechanical spacer between mud and cement (bottom plug) and cement and displacement fluid (top plug)	Between well fluids and cement
2. Omega Plugs		
3. Latch-Down Plugs		
Casing Centralizers		
Variable Types	Center casing in hole or provide minimum standoff to improve distribution of cement in annulus. Prevent differential sticking	Dependent on hole conditions and hole deviation
Scratchers or Wall Cleaners		
1. Rotating	Remove mud cake and circulatable mud from wellbore. Aid in creating turbulence	Place through producing formations and 50–100 ft above. Rotate pipe 15–20 rpm
2. Reciprocating	Improve cement bond	Same as rotating: reciprocate pipe 10–15 ft off bottom
Squeeze Cementing Tools		
1. Drillable	Where pressure is required to be held on cement after squeeze job	Purchased and left in well as permanent plug or drilled out
2. Retrievable	Perform multiple squeeze jobs and then retrieve	Can be moved up or down hole to squeeze as many zones as may be required
Liner Hanger Tools		
1. Mechanical	To hang liner in tension while cementing and left in well after cementing	Set at specific location in last casing string (usually 200 ft overlap in last casing). Set by drillpipe rotation and by reducing casing weight
2. Hydraulic	Same as mechanical	Same as mechanical except set by differential pressure
Bridge Plugs		
1. Wire Line	For permanent or temporary plugging in open hole or casing	May be placed in well on wire line, on tubing, or below retrievable squeeze packers
2. Tubing		
Cementing Baskets and External Packers	For casing or liner where mechanical support is necessary until the cement column sets	Below stage tools or where weak formations occur downhole

Suggested procedures for a good primary cement job follow:

- Condition the hole by reaming, if necessary, to remove tight places.
- Condition the mud. Circulate over a screen until the mud is mostly free of cuttings. Keep viscosity and gel strength low. Select a water-loss additive comparable to that used to drill the lower portion of the hole.
- Install a guide shoe and a float collar. The float collar should be about 30 ft above the guide shoe. This helps prevent overdisplacement of cement and helps obtain a good cement slurry around the casing shoe.
- Install scratchers spaced according to the locations of permeable zones. Use only those needed to obtain adequate coverage.
- Use casing centralizers. Utilize logs to determine the location to place centralizers on casing as well as routing spacing (60–90 ft apart).
- Use a cement slurry as heavy as or heavier than the drilling fluid. This will help prevent overdisplacement. The cement selected depends on temperature, pressure, and hole conditions.
- If a caliper log is available, use it to calculate cement volumes. If none is available, increase volumes according to knowledge of the area. Generally, volumes are increased 50–100%.
- Use a top and a bottom cementing plug.
- Rotate or reciprocate the casing until the top wiper plug has hit the float collar. Premature cessation of movement permits the filter cake to reform.
- After the cement is in place, watch the annulus fluid level. Keep the annulus full of fluid.
- Maintain tension in the entire casing string while the cement is setting. Setting time can be altered to fit most well conditions.
- Before drilling the cement plug or perforating for production, pressure test the cement job. Use a maximum pressure of 80% of the minimum yield of the weakest point.

To determine the height of the cement column in the annulus, a temperature log is often run from 12 to 24 hours after placement of the cement. Because cement generates heat when it hydrates, the top of the cement can be pinpointed by the anomaly in the temperature log. A good approximation of the location of the cement top may be obtained using Eq. 12.1:

$$H = \frac{P_{sd} - P_f}{0.052(W_c - W_m)} \qquad (12.1)$$

where:

$$H = \text{height of cement, ft}$$
$$W_c = \text{weight of cement slurry, lb/gal}$$
$$W_m = \text{weight of mud, lb/gal}$$
$$P_{sd} = \text{surface displacement pressure, psi}$$
$$P_f = \text{friction pressure, psi}$$

To use this formula, the pump should be slowed to the point that it is barely moving the fluid. At this rate, friction losses can probably be neglected. If friction is neglected, remember that the height calculated is probably above the actual cement column.

CONSIDERATIONS DURING CEMENTING

After surface preparations are made and casing equipment is ready, the most critical aspect of a cementing job is the actual conduct of the mixing/pumping operation. See Table 12–22.

Cement Mixing

Job success is correlated with the quality of the cement mixing operation. Slurry density is used to control the relative amounts of water and cement used and should be monitored and recorded to ensure maintenance of the correct water-solids ratio. To avoid the effect of aeration, samples for weighing should be obtained from a special manifold on the discharge side of the displacement pump rather than from the mixing tub.

The last volume of cement mixed is placed around the shoe; thus, take particular care to ensure that the cement meets desired weight specifications. In most instances, cement slurries are mixed with a lower water requirement toward the end of the job to provide greater strength around the shoe joint.

When densified or heavyweight slurries are used (17.0 lb/gal and heavier) and pumped or displaced at rates of less than 5 bbl/min, a recirculating mixer improves the uniformity of slurry over the standard jet mixer.

For critical cementing jobs, batch mixing provides much greater slurry precision and permits measuring and adjusting rheological properties, fluid-loss control, or other specific properties of the slurry before pumping it into the well.

Displacement

The predominant cause of cementing failure is channels of gelled mud remaining in the annulus after the cement is in place[32,33,34,35,36,37,38] (Table 12–23). If mud channels are eliminated, a variety of cementing compositions will provide an effective seal. Should mud channels remain after cementing, regardless of the quality of the cement, there will not be an effective seal between formations. Under dynamic conditions of pressure and temperature, water-based muds and gel deposit a thick filter cake, making mud removal more difficult. Model research supported by field practices indicates the forces listed below aid mud displacement in the annulus.[40,41,42]

- Cement exhibits drag stresses upon mud at different flow rates and flow properties. A well-conditioned mud is much more effectively removed than a thick, viscous mud when using a thin cement in turbulent flow (see Fig. 12–31).
- Drag stress of pipe upon mud and cement due to pipe motion (either rotation or reciprocation). Any pipe movement minimizes static mud pockets in the annulus (see Fig. 12–32).
- Buoyant forces due to the density difference between the mud and the cement aid in mud removal.

■ **TABLE 12–22** Summary of primary cementing

	Type of Job and General Conditions		
	Conductor	**Surface**	**Intermediate and/or Production String**
Casing Size, in.	20–30	7–20	7–11¾
Setting Depth, ft	30–1,500	40–4,500	1,000–15,000
Hole Conditions	Probably enlarged	Probably enlarged	Probably enlarged (particularly in salt)
Muds	Native	Native	Native, water-based or oil emulsion
Properties	Viscous—thick cake	Viscous-thick cake	Controlled viscosity and fluid loss
Cemented To	Surface	Surface	May or may not be circulated to surface pipe
Cement, API Class	A-C-G-H	A-G-H with	A-C-G-H with
Additives	2–3% CaCl	Bentonite or pozzolan	High gel, filter, or pozzolan bentonite
			Dispersant + retarder, if needed
			Salt saturated through salt sections for bonding
Tail-In Slurry	Same—densified (Ready-Mix concrete may be dumped in annulus)	Densified for high strength (deep well may use high-strength slurry for entire job)	Densified for high strength over lower 500–1,000 ft
Technique	Through drillpipe using small plug & sealing sleeve or Down casing with large plugs or Cement pumped or dumped in annulus	Same as conductor	Down casing with plugs (top and bottom) or Stage cementing depending on frac gradient String very heavy, may be set on bottom and cement through ports
	Float collar may or may not be used	Same as conductor	Float collar—guide shoe required
Placement Time	Generally short—less than 30 min.	Usually short—less than 45 min.	Variable, depending on cement volume 45 min.–2½ hr
	Low or high displacement rates employed	High displacement rates	High displacement rates, 6–12 bbl/min.
W.O.C. Time, hr	6–8	6–12	6–12
Mud-Cement Spacers	Plugs and water flush	Plugs and water or thin cement spacer	Plugs and thin cement or spacer compatible with mud
Cementing Hazards	Casing can be pumped out of the hole	Should have competent casing seat	May cover both weak- and high-pressure zones requiring variable weight cement slurries

■ **TABLE 12–22** *(Continued)*

	Type of Job and General Conditions		
	Conductor	**Surface**	**Intermediate and/or Production String**
	Cement may fall back downhole after circulating to the surface	Same as conductor	Prolonged drilling may damage casing
		Lower joints sometimes lost downhole with deeper drilling	Bottomhole temperature necessary for design in hot wells
		Casing can be easily stuck	Centralize in critical areas
		Centralize lower casing	

* See Regulatory Rules

■ **TABLE 12–23** Factors that contribute to cementing failures[39]

Factors	Results
Improper water ratio	
Improper temperature assumption	
Mechanical failures	
Wrong cement or additives	
Hot mixing water	Flash setting of cement
Static too long before movement	
Improper mud-cement spacers	
Contaminates in mixing water	
Dehydration of cement	
Insufficient retarder	
Plug did not leave head	Failure to bump plug
Did not allow for compression	
Mechanical failure	
Insufficient water or pressure	Could not finish mixing
Failure of offshore bulk systems	
Pipe lying against the formation	
Poor mud properties (high yield point and plastic viscosity)	Channeling
Failure to move pipe	
Low displacement rates	
Insufficient hydrostatic head	
Cement mud gelation	Gas leakage in annulus
Cement did not cover gas sands	

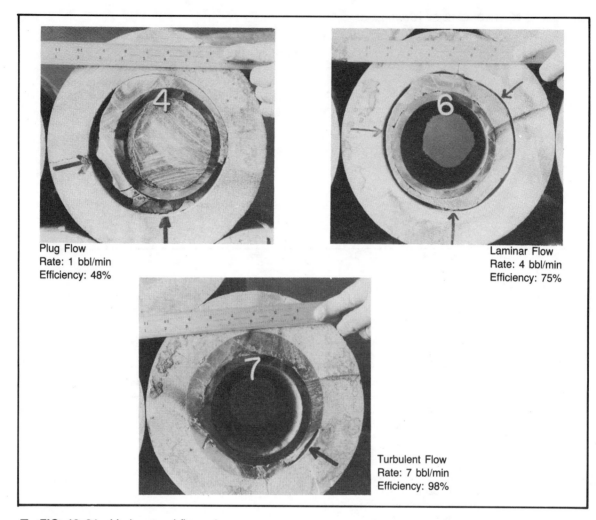

Plug Flow
Rate: 1 bbl/min
Efficiency: 48%

Laminar Flow
Rate: 4 bbl/min
Efficiency: 75%

Turbulent Flow
Rate: 7 bbl/min
Efficiency: 98%

■ **FIG. 12–31** Mud removal flow rate

In evaluating factors affecting displacement of mud, consider the flow pattern in an eccentric annulus condition with the pipe closer to one side of the hole than the other. Flow velocity in an eccentric annulus is not uniform. The highest flow rate occurs on the side of the hole with the largest clearance, as shown in Figs. 12–33 and 12–34.

The length of time cement moves past a point in the annulus in turbulent flow is important.[43] Thus, if a mud channel is put in motion, even though its velocity is much lower than that of the cement flowing on the wide side of the annulus, given enough time, the mud channel may move above the critical productive zone. Contact time is an unimportant factor

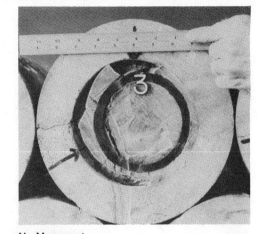

No Movement
Efficiency: 65%

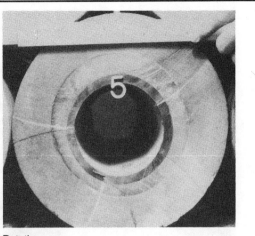

Rotation
Efficiency: 99%

■ **FIG. 12–32** Mud removal rotation

because the cement is in laminar flow, so the cement does not exert sufficient drag stress on the mud to start the mud channel moving.

At a given displacement rate, contact time is directly proportional to the volume of cement.

The same factors as contribute to the success of primary cementing contribute during this critical period. Each of the following factors is aimed at removing mud from the annular cross section:[44]

- Pipe centralization significantly aids mud displacement.
- Pipe movement, either rotation or reciprocation, is a major driving force for mud removal. Pipe motion with scratchers substantially improves mud displacement in areas of hole enlargement.
- A well-conditioned mud (low plastic viscosity and yield point) greatly increases mud displacement efficiency.
- High displacement rates promote mud removal. At equal displacement rates, a thin cement slurry in turbulent flow is more effective than a thick slurry in laminar flow.
- Contact time (cement volume) aids in mud removal if cement is in turbulent flow in some part of the annulus.
- Buoyant force due to density difference between cement and mud is a relatively minor factor in mud removal.

Gas Leakage

In gas well completions the leakage problem is critical, particularly in deep holes, causing a pressure buildup in the annuli of the production and intermediate casings or liners.[45,46,47]

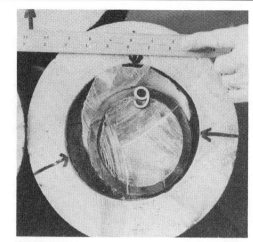

Standoff: 17%
Efficiency: 45%

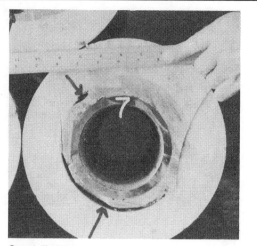

Standoff: 60%
Efficiency: 88%

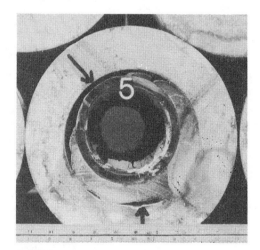

Standoff: 35%
Efficiency: 77%

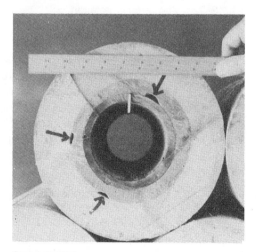

Standoff: 72%
Efficiency: 97%

■ **FIG. 12–33** Mud removal vs centralization

One cubic foot of gas migrating from 12,000 ft to the surface under a given set of conditions will expand to 316 cu ft.

Conditions at 12,000 ft	Conditions at Surface
Pressure: 5,366 psig (5,380 psia) (8.6 lb/gal)	14.7 psi
Temperature: 160°F (620°R)	70°F (530°R)
z: 0.99	1

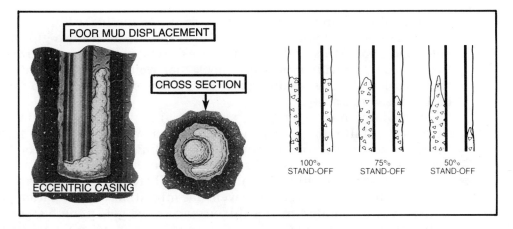

■ **FIG. 12–34** Channeling vs pipe stand-off

$$V_2 = \frac{V_1 z_2 T_2 P_1}{z_1 T_1 P_2}$$
(12.2)

where:

V_1 = volume of gas, cu ft

z_1 = compressibility factor under pressure in formation, dimensionless

T_1 = temperature of the formation, °F

P_1 = pressure of the formation, psi

z_2 = compressibility factor at the surface

T_2 = temperature at the surface, °F

P_2 = pressure at the surface, psi

V_2 = volume of gas at the surface, cu ft

Gas migration is generally attributed to the loss of hydrostatic pressure before the cement slurry has achieved sufficient gel strength to resist gas flow. An arbitrary value used to control most gas leakage is normally around 500 lb/100 sq ft.[48]

The development of this static gel strength, i.e., the internal rigidity within the cement matrix, usually begins shortly after pumping has ceased and continues to increase until the cement develops a set. As gel strengthens, the cement column begins to partially support itself. Hydrostatic pressure of the column decreases as the column becomes more self-supporting. This pressure is directly proportional to any volume changes that may occur when filtrate is lost from the matrix of the cement slurry to the formations or to any small hydration volume reduction as the cement sets.

The slurry properties that affect the cement's ability to maintain hydrostatic pressure are fluid loss, free water, static gel strength, and compressibility. Any filtrate loss from the cement slurry will correspond to a decrease in hydrostatic pressure. Even if this volume reduction does not correspond to an immediate pressure drop, a significant fluid loss could cause a bridge of dehydrated cement to form across from any highly permeable zone. Should this bridge form before the end of the cement's transition time, it could provide a pressure block that would minimize hydrostatic transmission of pressure down the hole, and gas could flow into the area of decreased pressure.

In deviated hole cementing, excessive free water may eliminate the slurry's capability to control fluid, causing a water channel to form as the water separates from the slurry. Gas and fluids under these conditions may migrate into the water channel and up the cement column.

Cement slurries without additives normally require 1–1½ hr to achieve sufficient static gel strength to prevent gas migration. (This gelation period is known as transition time.) A thixotropic or fast gel-strength system has a shorter transition time and may be used to promote rapid gel-strength development to help control gas migration.

The addition of a stable gas to the cement slurry increases the compressibility of the column of cement, which minimizes the effect a volume decrease may have on hydrostatic pressure. Gas is commonly added via a gas-generating additive to provide this property in the system.

Well Conditions that Prevent Gas Migration

The gas flow potential can be reduced by increasing the *overbalanced pressure.* This overbalance is the difference between the hydrostatic pressure at the potential point of flow and the pore pressure of the formation at that depth. When there is an overbalance, the pressure differential keeps the gas in the formation while the cement is developing gel strength. The greater the differential, the more time the cement has to safely pass through the transition phase. Thus a measure of control can be obtained by taking actions to maintain an overbalance through the transition time by increasing slurry density and applying pump pressure to the annulus at the surface. Applying the proper annular pump pressure immediately after pumping the slurry and maintaining this pressure for 1–3 hr minimizes any overbalance condition until the cement has reached 500 lb/100 sq ft gel strength.

Prior to cementing, a gas well should be circulated for an adequate time to condition the mud and remove any trapped gas bubbles located in the annulus. After circulation and before cementing, the pumps should be shut down for a short period and then started again to help remove any microscopic gas bubbles trapped in the washout areas or adhering to the walls of the hole.

This gas is a potential second gas kick, and, if not removed prior to cement placement, may result in a lowering of the fluid column density. During the cementing process, the cement slurry should be kept as heavy as possible to restrain any gas from cutting the slurry.

The following factors reduce gas migration into an annulus during and after primary cementing:[49]

- Centralization of casing string in the hole
- Increased flow rates during displacement
- Pipe movement during displacement
- Scratchers employed across washout sections
- Mud or cement density (positive hydrostatic head)
- Cement filtration control
- Cement setting or changing from a slurry to a solid in a minimum time after placement

Deep Well Cementing

The procedures used in cementing deep casing or liner strings are generally the same as those used for shallower depths, yet the physical factors encountered in shallow wells are of minor importance when compared to those in deep holes (Fig. 12–35).[50,51,52] The magnitude of these factors increases with depth and includes such conditions as:

- Higher temperatures and pressures with corrosive waters or hydrogen sulfide gas zones in some areas (see Fig. 12–36)
- The increased length and reduced clearances of the casing-casing annulus
- Greater mechanical loads imposed on the casing string and drilling rig

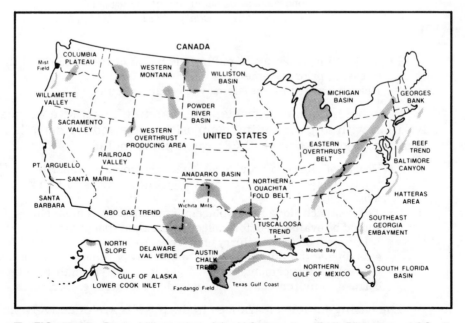

■ **FIG. 12–35** Deep drilling areas of the U.S. (*courtesy Fertl, Pilkington, and Scott, JPT, ref. 53*)

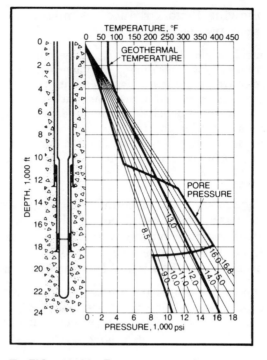

■ **FIG. 12–36** Temperature and pressure conditions encountered with deep drilling in the Delaware Basin (*courtesy API Bulletin D17, ref. 54*)

- Longer time intervals of pulling the bit off of the bottom and the running of casing prior to cementing
- Heavily treated mud systems with high densities

Deep holes demand considerable planning to achieve adequate zonal isolation with any cementing composition selected for deep, hot hole cementing (Figs. 12–37 and 12–38). Such plans include:

- Design of slurry to provide adequate rheological properties and pumping time. Always pretest with rig water and cements to be used on specific jobs.
- The best technique for displacement of mud by cement slurry. Field mud should be evaluated before selecting the mud spacer to precede the cement.
- Attainment of desired slurry properties (weight, viscosity, fluid-loss control, etc.) during the mixing process to effectively remove drilling mud.
- Control setting cements to resist gas leakage, strength retrogression and corrosive environments encountered at elevated temperatures.

The casing-to-hole clearances in deep wells are effectively smaller because of the greater lengths of open hole involved. This results in greater lengths of permeable zones and consequently thicker filter cake. The

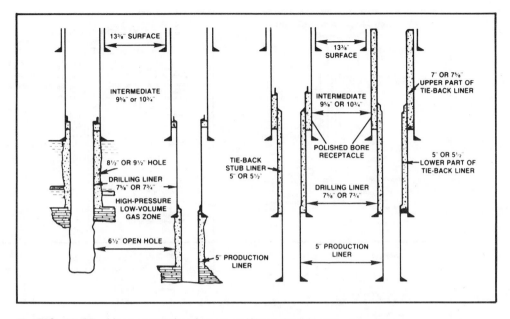

■ **FIG. 12–37** Liner cementing (*courtesy World Oil, ref. 50*)

higher temperatures, higher pressures, and exposure time all contribute to gelation and filter-cake deposition across these permeable sections.[55,56] This factor alone complicates mud removal, as cement has a tendency to channel through gelled mud and to follow the path of least resistance. Fig. 12–39 shows cement channeling through Ferrochrome mud under dynamic test conditions without pipe movement.[57]

Lengthy open-hole exposure causes shales to loosen and accumulate unless oil or invert mud systems are used. Higher mud densities also cause increased hydrostatic pressure differentials over some of the intervening formation pressures, tending to crowd the casing further into the mud cake on the contact side of the hole. This results in differential sticking or frictional drag should the casing remain stationary for any length of time.

When the geometry of the pipe and borehole prohibit turbulent rates, the next most effective mud-removal system is to place the cement in plug flow. Plug flow depends on the same criteria as turbulence except that a different Reynolds number is used.[58,59] The pertinent information is derived from the cement slurry by the use of a Fann VG meter in the laboratory.

These data are then utilized in a graphical log-log plot to derive a flow behavior index (n') and a consistency factor (K') for a particular slurry. These parameters, along with a Reynolds number of 100, are utilized in the equations shown below to calculate the plug flow velocity. This readily converts to a maximum pump rate in order to remain in plug

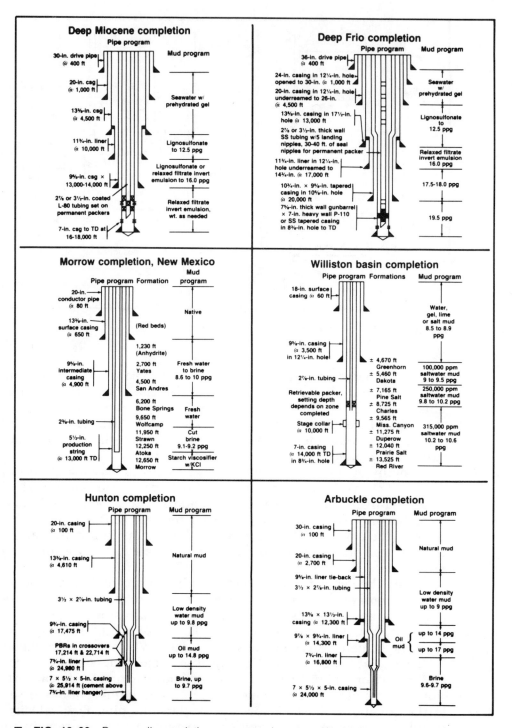

■ **FIG. 12-38** Deep-well completion programs (*courtesy World Oil, ref. 50*)

■ **FIG. 12–39** Simulated liner cementing (*courtesy Graham, 18 December 1972, OGJ, ref. 34*)

flow. In general, these rates correspond well with the 90 ft/min rule normally used for nominal annuli.

1. *Velocity at some specific Reynolds number*
   For generalized calculations:
   N_{Re} for plug flow = 100 (maximum)
   N_{Re} for beginning of turbulence = 3,000 (minimum)

$$V^{2-n'} = \frac{N_{Re} \, K' \, (96/D_i)^{n'}}{1.86\rho} \quad V = \left[\frac{N_{Re} \, K' \, (96/D_i)^{n'}}{1.86\rho} \right]^{1/(2-n')} \tag{12.3}$$

where:

V = velocity, ft/sec
K' = consistency index, lb-sec/sq ft
n' = flow behavior index, dimensionless
ρ = slurry density, lb/gal
D_i = inside diameter of pipe, in.
N_{Re} = specified Reynolds number, dimensionless

For annulus,

$$D_i = D_o - D_I \tag{12.4}$$

where:

D_o = outer pipe inside diameter or hole size, in.
D_I = inner pipe outside diameter, in.

2. *Displacement rate*

$$Q_b = \frac{VD_i^2}{17.15} \quad \text{or} \quad Q_{cf} = \frac{VD_i^2}{3.057} \tag{12.5}$$

where:

Q_b = pumping rate, bbl/min
Q_{cf} = pumping rate, cu ft/min

For annulus,

$$D_i^2 = D_o^2 - D_I^2 \tag{12.6}$$

The desirable cement flow characteristics are a plastic viscosity and a yield point in excess of the mud system. If possible, cement Fann readings in the range of twice those of the mud are preferable. Quite often, at higher densities, this is impractical. The mud system must meet borehole requirements that may make it impractical to design for this ratio, in which case the two systems should approach the PV and YP of the mud system.

Linear Cementing

Liners are commonly used to seal the open hole below a long intermediate casing string to:

- Case off the open hole to allow deeper drilling
- Control water or gas production or hold back unconsolidated or sloughing formations
- Case off zones or lost circulation and/or zones of high pressure encountered during drilling operations

Once the liner has been set, cement is circulated down the drillpipe, out the liner shoe and up the outside of the liner. Plugs are used to prevent mud-cement mixing in the liner, just as they are in a full casing string during primary cementing. At the completion of the cementing operation, the excess cement is reversed out, or allowed to set above the liner and drilled out after setting (Fig. 12–40).

Tie-back liners and/or shorter scab liners are run and cemented to protect the intermediate casing against corrosion under high pressures.[60] They also serve the functions of:

- Reinforcing the intermediate casing worn by drilling

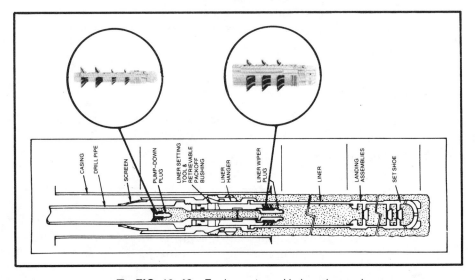

■ **FIG. 12–40** Equipment used in hanging and cementing liners (*courtesy Lindsey and Bateman, ref. 60*)

- Providing greater resistance to collapse stress from abnormal pressures
- Providing corrosion protection
- Sealing an existing liner that may be leaking gas

The tie-back liner may or may not extend back to the surface but does cover the top of an existing liner. Should it extend from the top of a production linear to a point some distance up the hole, it is called a *stub liner.* This short liner may be set with its entire weight on the production liner or hung up the hole. When either liner is set as a tie-back section, it connects to the liner in the well through a polished or honed receptacle 3–6 ft long. The receptacle usually has a setting thread for holding and releasing the liner to permit insertion of a tie-back sealing nipple before, during, or after placement of the cement slurry around the tie-back liner. The diameter of the tie-back sealing nipple should not be less than that of the liner set below it (Fig. 12–41).

Liner Operational Procedure[61]

_____

(Name and location of well)

1. Run drill pipe in the hole and circulate to condition hole for running the liner. Temperature subs should be run on this trip if bottom hole circulating temperatures are not known. Drop hollow rabbit (drift) to check the drill pipe inside diameter for proper pump down plug clearance. On the trip out of the hole, accurately measure and isolate the drill pipe to be used to run the liner. Tie off the remaining drill pipe on the other side of the pipe racking board.

2. Run _____ feet of _____ liner with the float shoe and float collar spaced _____ joints apart. Run the liner plug landing collar _____ joints above the float collar. The volume between float shoe and plug landing collar is _____ barrels. Sandblast joints comprising the lower 1,000 feet and upper 1,000 feet of the liner. Run thread locking compound on the float equipment and bottom eight joints of the liner. Pump through the bottom eight joints to be certain that the float equipment is working.

3. Fill each 1,000 feet of the liner while running, if automatic fill up type equipment is not used.

4. Install liner hanger and liner setting tool assembly. Fill the dead space (if pack-off bushing is used in lieu of liner setting cups) between liner setting tool and liner hanger assembly with inert gel to prevent solids from settling around the setting tool.

5. Run liner on _____ drill pipe with

(size, type connection, weight, and grade)

_____ pounds minimum over pull rating. Run in the hole at 1-2 minutes per stand in the casing and 2-3 minutes per stand in open hole. Circulate last joint to bottom with cement manifold installed. Shut down pump. Hang the liner five feet off bottom. Release liner setting tool and leave 10,000 pounds of drill pipe weight on setting tool and liner top.

6. Circulate bottoms up with _____ barrels per minute to achieve _____ feet per minute annular velocity (approximately equal to previous annular velocities during drilling operations).

7. Cement the liner as follows: _____

_____

_____

8. If unable to continue circulation while cementing, due to plugging or bridging in the liner and hole wall annular area, pump on annulus between drill pipe and liner to maximum _____ psi and attempt to remove bridge. Do not overpressure and frac the formation. If unable to regain circulation, pull out of liner and reverse out any cement remaining in the drill pipe.

9. Slow down pump rate just before pump down plug reaches the liner wiper plug. _____ bbls is capacity of the drill pipe. Watch for plug shear indication, recalculate or correct cement displacement, and continue plug displacement plus _____ bbls maximum over displacement.

10. If no definite indication of plug shear is apparent, pump calculated displacement volume plus _____ bbls (100% + 1 to 3%).

11. Pull out 8-10 drill pipe stands or above the top of the cement, whichever is greatest. Hold pressure on top of the cement to prevent gas migration until cement sets.

12. Trip out of the hole.

13. Wait-on-cement _____ hours.

14. Run _____ in. OD bit and drill cement to top of liner. Test liner overlap with differential test, if possible. Pull out of the hole.

15. Run _____ in. OD bit or mill and drill out cement inside the liner as necessary. Displace the hole for further drilling. Spot perforating fluid (if in production liner) or other conditioning procedures as desired.

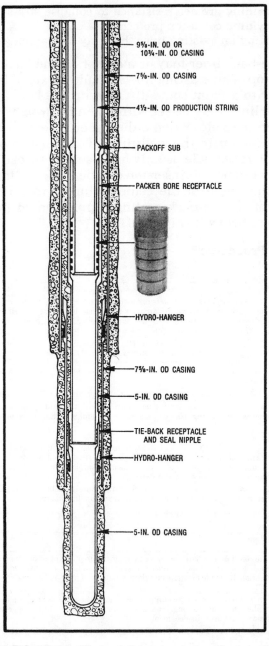

FIG. 12–41 Tie-back liner showing sealing nipple

Regulatory Rules[62,63,64]

Practically all rules (state and federal) cover the setting of specific sections of casing and methods for plugging wells. Conductor pipe is specifically emphasized in offshore federal regulations since it is used to start the hole in a vertical direction and to seal off badly sloughing surface material or water. Conductor pipe also serves as a connection for blowout prevention equipment when drilling shallow gas zones. Although not always done, cement should be circulated to the ocean or Gulf floor level in setting the conductor string.

Most state rules are very specific on the setting of surface casing to protect freshwater sands from contamination and to form a good, solid anchor for blowout-prevention equipment while drilling the well. States with mineral deposits require cementing of the surface pipe below such deposits to prevent any migration of fluids from the lower oil-and-gas producing horizons into the mineral deposits should they be mined at some later date.

Years ago, regulations were almost nonexistent, or applicable rules frequently were not enforced. In the absence of cementing regulations, operators have casing cementing practices today that usually exceed the minimum requirements of state or federal regulatory bodies.

Regulatory rules were not written by the same groups nor were they prepared during the same period. For these reasons they are expressed in a variety of ways.[65]

Many states have appointed industry representatives to assist in writing laws applying to the drilling and cementing of wells. Kansas, New Mexico, Oklahoma, Texas, and Wyoming have followed this practice. The Texas Railroad Commission Rules 13 and 14 (January 1983), dealing with primary cementing and the plugging of wells, are the latest state-industry cooperative efforts in preparing workable regulatory rules (see Fig. 12–

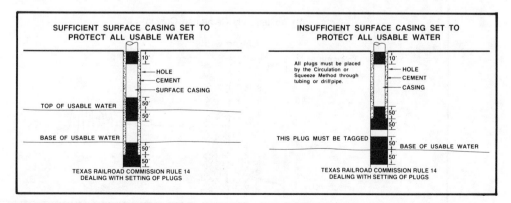

■ **FIG. 12–42** Regulatory rules covering the setting of casing and plugging of wells (*courtesy Texas Railroad Commission, ref. 64*)

42). Much of the reference literature guiding these committees dates back to Farris[66] and others who have published data on the minimum compressive or bond strength required to adequately support pipe.

Technology changes and operating environments can often be attributed to the wide divergence in state regulations, particularly with regard to the volume of cement that should be used and the time allowed for strength development. There are also variations in cementing practices in fields of comparable characteristics in the same state.

ACOUSTIC LOGGING[67,68,69,70]

With the introduction of acoustic logging to the oil industry as a means of evaluating a primary cementing job, more emphasis has been placed on good cementing practices. Most of the available bond logging instruments record only a portion of the acoustic signal transmitted. Therefore, only a small part of the available information is received, and these measurements may be influenced by factors unrelated to bonding. For example,

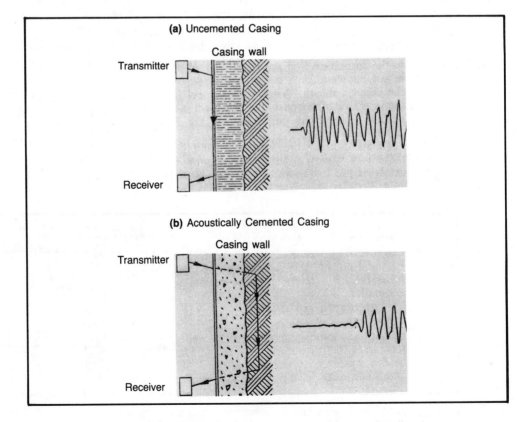

■ **FIG. 12–43** Acoustic energy travel in cased wells

the speed of sound is faster in certain formations than in the steel casing.

The variable-density recording and an alternate, the full-wave-train presentation, alleviate some of the problems of interpretation. Because these presentations record the entire acoustic signal, they are not subject to many of the interpretation limitations of the single-curve "bond log." Interpretations based on either the full-wave-train or the variable-density film recording provide qualitative information of the condition of the cement sheath surrounding the pipe.

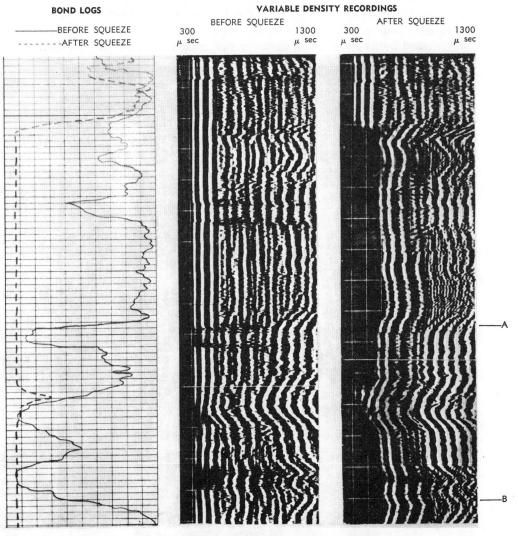

EFFECTS OF SQUEEZE CEMENTING

■ **FIG. 12–44** Effects of squeeze cementing

As shown in Fig. 12–43(A), the first arriving pipe-borne portion of the acoustic signal will have high amplitude if the pipe is not cemented. When the cement is acoustically bonded to both the pipe and formation, the wave travels through the formation without the appearance of a pipe signal. This is illustrated in Fig. 12–43(B). The exact appearance of the signal is affected by the type of cement and its ability to provide the acoustical bonding necessary to transmit the signal and attenuate the pipe-borne signal, e.g., class H vs lightweight cement.

Two factors that affect the interpretation of the amplitude curve on the bond log are calibration and centralization. A properly calibrated tool provides comparative evaluation based upon the known response of a calibration sleeve and not on the uncertainty of "unbonded pipe," which may or may not have some acoustical attenuation. Second, a tool that is decentralized by as little as ⅛ in. does not receive the proper acoustical signal.

ENERGY TRANSMISSION COMPARISON

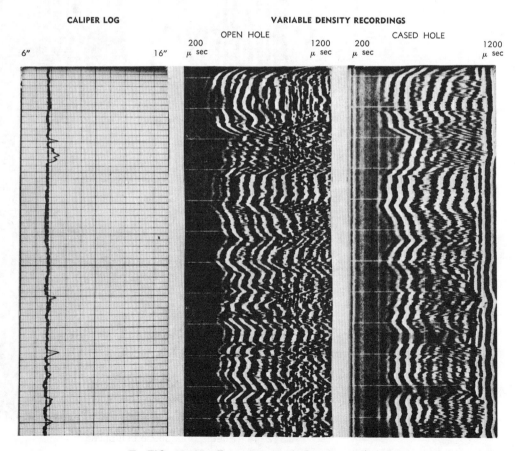

■ **FIG. 12–45** Energy transmission comparison

A transit-time curve is often used to indicate the decentralization and assist in the determination of the adequacy of the acoustic signal for evaluation.

Where channeling exists in the cement sheath, both the casing and formation paths may be evident, as illustrated by Fig. 12–44. Notice that the cement became acoustically bonded after squeezing. Points A and B are locations of perforations where cement was squeezed into the annulus.

Fig. 12–45 shows a section of a log made in the open hole and the same section made in the cemented casing in the same well. Acoustic energy transmission through the formation is shown by the absence of pipe arrivals and the correlation of the formation arrivals on the open- and cased-hole logs. This correlation could not exist if the pipe, cement, and formation were not adequately coupled. This example shows that this log may be used in acoustically cemented wells for many correlation purposes.

Table 12–24 gives a summary of bonding, logging, and perforating.

■ **TABLE 12–24**   Summary—Considerations after cementing bonding, logging, perforating

		Remarks
Bonding	1. Cement to Pipe	Cement properly placed around centralized pipe provides good mechanical bond. Bond can be influenced by heat of hydration, perforation, or stimulation.
	2. Cement to Formation	Most critical area of bonding. Against a clean formation cement will produce good hydraulic, shear, or gas bond. Cement bonds poorly to mud cake in highly treated mud systems. For best bond centralize pipe, use scratchers, and move casing.
Location of Cement	Temperature Survey	Recording thermometer measures heat of hydration of cement. Not as quantitative cement evaluation as bond log. Measurements should be taken 12 to 24 hours after cementing for best results.
	Radioactive Tracer	Iodine 131 or Zirconium — Niobium 95 added to mixing water for tracer purposes. Gamma log run before and after cement job provides cement location in annulus.
	Bond Log	Widely used for location of cement behind pipe. Better method of identification than temperature or radioactive tracer logs. New logs provide both pipe and formation bond identification.
Location of Cement	Bond Log	Harder cements or longer WOC times provide best bond logs. Sensitive to interpretation.
Perforating	Jets	Most widely used. Hollow carrier jets do less damage to pipe or cement. Expendable charges do some damage to small pipe. Hard cements (2,000 psi plus) perforate best with less damage.
	Bullets	Not used as much as jets. Perforate soft cement (500 psi) best with very little damage. In hard cements or formations less penetration occurs.
	Hydraulic Cutting	Does little damage to cement or pipe. Frequently used to initiate fracture.
	Mechanical Cutters	Used in pipe recovery. Does very little damage to cement or pipe.
	Permeaters	Attachments put on casing before going in well. Extended with pressure before cement sets. Opened with chemicals. Not widely used.

■ **BIBLIOGRAPHY**

1. Union Oil Co. of California: "On Tour," November–December 1952.
2. Shyrock, S.H. and Slagle, K.A.: "Problems Related to Squeeze Cementing," *Journal of Petroleum Technology (JPT)*, August 1968.
3. Portland Cement Assoc.: "Design and Control of Concrete Mixtures," 1983.
4. Aspdin, Joseph: "An Improvement in the Modes of Producing Artificial Stone," British patent no. 5022 (1824).
5. Portland Cement Assoc., op. cit.
6. Ibid.
7. Lea, F.M. and Desch, C.H.: *Chemistry of Cement and Concrete,* Arnold and Co., London, 1935, reprinted 1937 and 1940.
8. Caveny, William J.: "Practical Oil-Well Cement Microscopy," paper presented at International Cement Microscopy Assoc. meeting in Fort Worth, TX, March 1985.
9. Halliburton Technical Literature—Cementing.
10. After API Specifications 10, 2d ed., 15 June 1984.
11. Ibid.
12. Farris, R. Floyd: "Method for Determining Minimum Waiting-on-Cement Time," Petroleum Transactions, AIME v165, (1946) 178–188.
13. Bearden, W.G. and Lane, R.D.: "Engineered Cementing Operations to Eliminate WOC Time," Midcontinent API district meeting, 5–7 April 1961, Tulsa, OK.
14. Carter, Fred and Smith, Dwight K.: "Properties of Cementing Compositions at Elevated Temperatures and Pressures," AIME paper 892–G, 32d annual fall meeting, Dallas, TX, 6–9 October 1957.
15. SPE Monograph series: "Cementing," Society of Petroleum Engineers (SPE), Dallas, TX, 1976–1986.
16. Murphy, W.C. and Smith, Dwight: "A Critique of Filler Cements," SPE preprint, March 1967.
17. Smith, R.C., Powers, C.A., and Dobkins, T.A.: "A New Ultralightweight Cement with Super Strength," *JPT,* August 1980.
18. Harms, W.M. and Lingenfelter, J.T.: "Microspheres Cut Density of Cement Slurry," *Oil & Gas Journal (OGJ)*, 2 February 1981, 59–66.
19. Harms, W.M. and Febus, J.S.: "Cementing of Fragile-Formation Wells with Foamed Cement Slurries," SPE paper 12755 presented at the 1984 California regional meeting in Long Beach, 11–13 April.
20. Beach, H.J., O'Brien, T.B., and Goins, W.C., Jr.: "The Role of Filtration in Cement Squeezing," spring meeting of API southern district, div. of production, Shreveport, LA, 1961.
21. Parker, P.N., Ladd, B.J., and Wahl, W.W.: "An Evaluation of a Primary Cementing Technique Using Low Displacement Rates," API 1234 paper presented at SPE 40th annual fall meeting of AIME, Denver, CO, 3–6 October 1965.
22. Slagle, Knox A.: "Rheological Design of Cementing Operations," *JPT,* March 1962, 323–328.
23. Slagle, K.A. and Smith, D.K.: "Salt Cement for Shale and Bentonitic Sands," *JPT,* February 1963, 187–194.
24. Hewitt, Charles H.: "Analytical Techniques for Recognizing Water-Sensitive Reservoir Rocks," *JPT,* August 1963, 813–818.
25. Owsley, W.D.: "Improved Casing Cementing Practices in the United States,"

paper presented at AIME meeting, San Antonio, TX, 6 October 1949 and in Los Angeles, CA, 20 October 1949, *OGJ* v48, n32, 15 December 1949, 76, 78.

26. API Spec. 10, op. cit.
27. API Spec. for Casing Centralizers, API Spec. 10D, 2d ed., February 1973, and Supplement 1, March 1976.
28. API Spec. 10, op. cit.
29. Clark, E.H. and Murray, A.S.: "A Study of Primary Cementing," API spring meeting, Los Angeles, CA, 22–23 May 1958.
30. Jones, P.H. and Berdine, D.: "Oil-Well Cementing—Factors Influencing Bond between Cement and Formation," *Drill. & Prod. Prac.,* API (1940) 45.
31. Teplitz, A.J. and Hasebroek, W.E.: "An Investigation of Oil-Well Cementing," *Drill. & Prod. Prac.,* API (1946) 76.
32. Brice, J.W., Jr. and Holmes, R.C.: "Engineering Casing Cementing Programs Using Turbulent Flow Techniques," *JPT,* May 1964, 503.
33. McLean, R.H., Manry, C.W., and Whitaker, W.W.: "Displacement Mechanics in Primary Cementing," SPE Santa Barbara, CA, 1966, SPE paper 1488.
34. Graham, Harold L.: "Rheology-Balanced Cementing Improves Primary Success," *OGJ,* 18 December 1972.
35. Carter, L.G., Cook, C., and Snelson, L.: "Cementing Research in Directional Gas Well Completions," 2d annual SPE of AIME Europe meeting, preprint SPE 4313.
36. Carter, Greg and Slagle, Knox A.: "A Study of Completion Practices to Minimize Gas Communication," Amarillo, TX regional SPE meeting, November 1970, SPE paper 3164.
37. Hoch, R.D.: "Cementing Techniques Used for High-Angle, S-Type, Directional Wells," *OGJ,* 22 June 1970, 88–93.
38. Christian, W.W. and Stone, W.H.: "The Inability of Unset Cement to Control Formation Pressure," SPE of AIME Formation Damage Meeting, preprint SPE 4783 (1974), 147–140.
39. Clark, and Murray, op. cit.
40. Brice, J.W., Jr. and Holmes, R.C.: "Engineering Casing Cementing Programs Using Turbulent Flow Techniques," *JPT,* May 1964, 503.
41. McLean, Manry, and Whitaker, op. cit.
42. Clark, Charles R. and Carter, Greg L.: "Mud Displacement with Cement Slurries," October 1972, San Antonio, TX annual SPE meeting, SPE paper 4090.
43. Brice and Holmes, op. cit.
44. Haut, R.C. and Crook, R.J.: "Primary Cementing: The Mud Displacement Process," SPE paper 8253 presented at the SPE 54th Annual Technical Conference and Exhibition, Las Vegas, Nevada, 23–26 September 1979.
45. Carter and Slagle, op. cit.
46. Christian and Stone, op. cit.
47. Sabins, Fred L. and Sutton, David L.: "The Relationship of Thickening Time, Gel Strength, and Compressive Strengths of Oil-Well Cements," SPE paper 11205 presented at the 57th Annual Fall Technical Conference and Exhibition of the SPE of AIME, New Orleans, LA, 26–29 September, 1982.
48. Ibid.
49. Carter, Cook, and Snelson, op. cit.
50. "Deep Well Drilling and Completion Report," *World Oil,* May 1982.
51. Kirk, W.L.: "Deep Drilling Practices in Mississippi," *JPT,* June 1972, 633.

52. Bowman, Glenn: "Proper Liner Cementing Plans Will Prevent Later Problems," *World Oil,* November 1982–January 1983.
53. Fertl, W.H., Pilkington, P.E., and Scott, J.B.: "A Look at Cement Bond Logs," *JPT,* June 1974.
54. "Running and Cementing Liners in the Delaware Basin, Texas," API Bulletin D17, 2d ed., March 1983.
55. Gibbs, Max A.: "Delaware Basin Cementing—Problems and Solutions," *JPT,* October 1966.
56. Lindsey, H. Ed, Jr., and Bateman, S.J.: "Liner Cementing in High-Pressure Gas Zones," 1973.
57. Carter, L.G. and Clark, C.R.: "Mud Displacement with Cement Slurries," 47th Annual SPE of AIME Fall Meeting, preprint SPE 4090 (1972).
58. Bearden and Lane, op. cit.
59. Slagle, op. cit.
60. Lindsey and Bateman, op. cit.
61. API Bulletin D17, op. cit.
62. Carter and Smith, op. cit.
63. McRee, Boyd C.: "Cementing Regulations Applied to Oil and Gas Wells," *Oil Well Cementing Practices in the United States,* chapter 20, API, Johnston Printing Co., Dallas, TX (1959).
64. Texas Railroad Commission Rules 13 and 14, 1 November 1982 and effective 1 January 1983.
65. Herndon, J. and Smith, D.K.: "Setting Downhole Plugs: A State of the Art," *Petroleum Engineer,* April 1978, 56–71.
66. Farris, op. cit.
67. Anderson, T.O., Winn, R.H., and Walker, Terry: "A Qualitative Cement-Bond Evaluation Method," API paper 875–18–A, spring meeting of the Rocky Mountain district, div. of production, Billings, Montana, 20–22 April 1964.
68. Pickett, G.R.: "Acoustic Character Logs and their Applications in Formation Evaluation," *JPT,* June 1963, v15, n6.
69. Pilkington, P.E. and Scott, J.B.: "Comparing Cement Bonds after Ten-Plus Years," *Petroleum Engineer,* April 1976, 52–62.
70. Fertl, Pilkington, and Scott, op. cit.

CHAPTER 13

PREDICTION OF PORE PRESSURES AND FRACTURE GRADIENTS

Accurate prediction of pore pressures and fracture gradients has become almost essential to drilling deep wells with higher-than-normal pore pressures. Costs and drilling problems can be reduced substantially by recognizing abnormally high pore pressures early. The accurate determination of fracture gradients defines the need for protective casing and limits that must be placed on total annulus pressures during drilling.

Every method used to recognize or predict changes in pore pressure is based on formation compaction or the lack of it. Compaction occurs primarily in shales; thus, most prediction methods are based on what happens when drilling or logging shale sections. Normally, shale becomes denser as depth increases. If this does not happen, then suspect abnormally high pore pressures.

Some primary methods used to detect pore pressure changes follow:

- Seismic data
- Drilling rates
- Sloughing shales
- Shale densities
- Gas units in mud
- Chloride increases
- Mud properties
- Temperature measurements
- Bentonite content in shales
- Paleo information
- Wireline logs

SEISMIC DATA

Pennebaker explains in detail the use of seismic field data for locating the existence and depth of formations with abnormally high pressures.[1] The technique mainly relies on normal formation compaction with depth. When this normal compaction trend is not followed, the velocity of sound waves is reduced. These velocity changes can be detected and converted to the degree of abnormal pore pressures.

To be accurate, precise sound velocity data are needed. These may determine the successful application of this method. In coastal areas, seismic data have been used to predict the tops of abnormal pressure formations within a few hundred feet. In some new areas, seismic data have given erroneous results. Actually, seismic data have more utility in exploratory areas. Specific interpretations of seismic data are difficult, and experienced geophysicists are necessary for meaningful results. Any method of this type improves as field data are collected. Therefore, drilling personnel and geophysicists can profit by working closely with each other.

DRILLING RATE*

Experienced drilling personnel, both technical and nontechnical, often refer to the fact that a well talks. The language of the well is international. If we can understand what the well is saying and react accordingly, many of the serious problems encountered in drilling can be avoided.

The secret to understanding the well is to ascertain that any talking by the well is clear. The drilling practices program needs to be fully understood and the results recorded accurately.

The well talks primarily through changes in monitoring devices, which include:

- pumping pressure and pump strokes
- torque gauges
- pit level indicators
- mud properties
- drilling rate recorders
- weight indicators.

What the well says cannot be properly interpreted unless the cuttings below the bit are removed as they are generated. The significance of good bottom-hole cleaning is shown in Fig. 13-1.

If cuttings below the bit are allowed to accumulate, the results are similar to listening to someone with a gag in his mouth. We know he is trying to say something, but the words are unintelligible. Without understanding the reply, our actions may be directed toward solving the wrong problem. By the time we recognize our mistake, it may be too late to use the obvious solution.

---

* Most of this section is adapted from "How to Predict Pore Pressure" by Preston L. Moore in the March 1982 *Petroleum Engineer International,* pp. 144, 146, 148, 152.

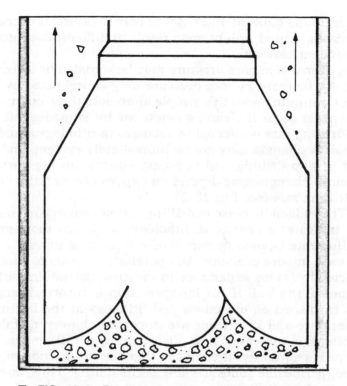

■ **FIG. 13–1** Formation cuttings accumulating below the bit (*courtesy Petroleum Engineer International*)

For this reason, the hydraulics program plays the most important role in the total drilling practices program. For years, cleaning below the bit was considered mainly for economic reasons. Now cleaning below the bit may be the key to survival in deep, high-pressure wells.

All of the well talk is affected if cuttings are allowed to accumulate below the bit. This section is limited primarily to recognizing changes in lithology and pore pressure. Lithology changes are important both to exploration and drilling personnel and often are indicated by a change in drilling rate. Changes in pore pressure also are related to changes in drilling rate.

Thus, the way to better understand the well is to determine whether a change in drilling rate signals a variance in lithology or a change in pore pressure. For this reason exploration personnel must work closely with drilling personnel. If doubt exists and the operator must know if the lithology or pore pressure has changed, then cuttings should be circulated to the surface before drilling continues.

Recognizing changes in pore pressure while drilling is important so mud weight can be altered before problems develop. Decreases in pore pressure may signal lower fracture gradients and a high vulnerability to underground blowout. Increases in pore pressure, if not contained by quick increases in mud weight, may result in difficult-to-control well kicks and subsequent blowouts.

A decrease in pore pressure may be signaled by a reduction in drilling rate. An increase in pore pressure may be indicated by a rise in drilling rate. Continuing with this simple approach, these changes in drilling rate may never occur if cleaning below the bit is inadequate.

Drilling rate is affected by changes in lithology and drilling practices. Lithology changes may not be immediately evident, and it may be necessary to stop drilling and circulate cuttings to the surface. Recognizing lithology changes may depend on experience and also on abrupt changes in drilling rate (see Fig. 13–2).

The sudden increase in drilling rate shown in part A of Fig. 13–2 generally indicates a change in lithology, while the more gradual change in drilling rate (shown in part B of Fig. 13–2) generally is indicative of an increase in pore pressure. An operator's reaction to such changes will be affected by (1) his experience in the area, (2) the limitations of his ability to control the well if his interpretation is incorrect, and (3) the data he was furnished on objectives and lithology at the beginning of the well. Experience and knowledge are important; however, if doubt exists when drilling rate increases, flow checks are generally made, and the operator proceeds cautiously. Remember that the top portion of many high-pressured, permeable zones has low permeability. Thus, one flow check after 5 ft may be insufficient.

Stories are told about three flow checks being made, 5 ft apart, and then, after drilling 100 ft, the well kicks and serious problems develop. There are always gray zones and stories that prove any set of practices inadequate. If you are unsure, never take a chance when the alternative

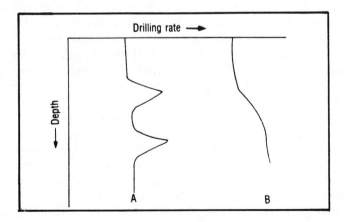

■ **FIG. 13–2** Drilling vs depth (*courtesy Petroleum Engineer International*)

results introduce the possibility of exceeding your control equipment's limitations.

The rest of this section is directed toward a method for determining the change in pore pressure as a result of a variance in drilling rate. In most cases, the operator's ability to use drilling rate as an indicator of a pore pressure change is limited to drilling changes in shales. However, some predictions are possible while drilling in limestone. Because data are limited for drilling in limestone, the following techniques should be applied only to drilling in shale. Note, too, that the techniques apply to both soft and hard shales.

CONTROLLING KICKS

A specific problem facing an operator, particularly in pressure transition zones, is controlling any type of kick. Part of the problem is shown in the example below.

☐   **EXAMPLE 13.1**

Assume:

Surface casing set at 4,000 ft
Fracture gradient just below casing seat at 4,000 ft = 0.74 psi/ft
Drilling in soft shale in pressure transition zone at 12,000 ft
Mud weight = 13.0 lb/gal

Determine: Control limitations

SOLUTION:

$$\text{Mud gradient} = \frac{13.00}{19.25} = 0.675 \text{ psi/ft}$$

Maximum pressure that can be held at the surface without losing circulation at the casing seat:

$$(0.740 - 0.675)\, 4,000 = 260 \text{ psi}$$

Permissible underbalance on mud weight at 12,000 ft:

$$\frac{(260)(19.25)}{12,000} = 0.42 \text{ lb/gal}$$

The solutions for Example 13.1 show that the maximum shut-in pressure at the surface is 260 psi. At 12,000 ft a differential pressure of 260 psi is equal to a mud weight of 0.42 lb/gal. Thus, in the situation shown in this example, if a permeable formation is encountered in the pressure transition zone, the mud weight cannot be more than 0.42 lb/gal less than the equivalent pore pressure without losing control of the well. For this reason very little mud underbalance is allowed in the pressure transition zone.

To minimize mud underbalance in the pressure transition zone, drilling rate should be controlled and mud weight increased to keep the drilling rate below a normal drilling rate. Assuming good bottom-hole cleaning, drilling rate increases at the top of the pressure transition zone. Drilling rate can be reduced by decreasing bit weight or rotary speed or by increasing mud weight. A normal procedure is to lessen bit weight or rotary speed to reduce drilling rate immediately and then increase mud weight. This procedure makes it more difficult to accurately determine the required mud weight increase. Fig. 13–3 illustrates the basic problem in the pressure transition zone.

From Fig. 13–3, the drilling rate in the normal pressure region is extrapolated into the abnormal pressure region. With no change in bit weight or rotary speed, the mud weight increase required to reduce the drilling rate to the normal line should equal the pore pressure increase. A simply way to determine the pore pressure is simply to keep the drilling rate on the normal line by increasing mud weight. In most cases the problem is compounded because of reductions in bit weight or rotary speed. Drilling

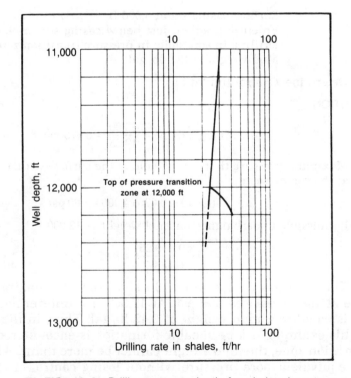

■ **FIG. 13–3** Drilling rate vs depth for shales (*courtesy Petroleum Engineer International*)

rate can be normalized for changes in bit weight and rotary speed by using Eq. 13.1*:

$$R = K_d W N^2 \tag{13.1}$$

where:

R = drilling rate, fph
K_d = drillability constant
W = bit weight, lb/in.
N = rotary speed, rpm
a = experimentally determined exponent

In soft shale formations $a = 1.0$, and in hard shale formations a can be determined by rewriting Eq. 13.1 as shown in Eq. 13.2:

$$\log R/W = \log K_d + a \log N \tag{13.2}$$

Thus, in field operations, R/W is plotted versus N on log paper. The line R/W versus N plot should be a straight line with a slope equal to a. One way to avoid the determination of a is to always reduce bit weight to lower the drilling rate in the pressure transition zone and to keep the rotary speed constant.

The procedure for controlling drilling rate and pore pressure in the pressure transition zone when changes in bit weight and rotary speed occur is shown in this next example.

☐   **EXAMPLE 13.2**

Assume:
Well depth at the top of the pressure transition zone = 12,000 ft

R = 25 fph at 12,000 ft
Bit weight = 5,000 lb/in.
Rotary speed = 120 rpm
Experience has shown $R = K_d W N^{0.8}$
Bit weight is reduced to 2,500 lb/in.
Rotary speed is reduced to 80 rpm
R (at 12,200 ft and new bit weight and rotary speed) = 15 fph

SOLUTION:
Fig. 13–4 shows a plot of the log of drilling rate versus depth for the well.
At 12,200 ft, the normal drilling rate would have been 24 fph. With no change in bit weight and rotary speed, the normal drilling rate should be as follows:

$$R = 24 \ (2,500/5,000)(80/120)^{0.8} = 8.7 \text{ fph}$$

This shows the mud weight should be increased enough to reduce the drilling rate from 15 to 8.7 fph. To be safe, raise the mud weight enough

---

* Eq. 13.1 assumes no bit dulling, which is generally true with insert bits but may not be true for milltooth-type bits. In any event, the only way bit dulling could be included would be with milltooth bits, where the drilling rate as a function of dulling is predictable. There is no way to do this for insert bits.

to reduce the drilling rate to less than 8.7 fph. The mud weight increase required to reduce the drilling rate to 8.7 fph represents the actual increase in pore pressure. The drilling rate should be kept at or below the new normal drilling rate line with no further changes in bit weight and rotary speed shown on Fig. 13–4 until the mud weight is increased to no more than 0.5 lb/gal below the equivalent fracture gradient at the last casing seat.

From a practical standpoint, the above procedure applies to any type of shale zone where pore pressures are increasing. Also, the same procedure can be used if the pore pressure is below normal. Mistakes are not uncommon. If the mud weight is increased more than necessary, the drilling rates will be too low. A decrease in mud weight will raise the drilling rate to the permitted level. The success of this procedure depends on good communication with the well.

Predicting Pore Pressure

In addition to the practical method discussed above, pore pressure can be determined from a change in drilling rate. One general method used to predict pore pressure from drilling rate is the d exponent. In addition, there are other variations based on Eq. 13.1. All of these methods, including the d exponent, fail to consider any change in shale compaction and for this reason have limited applications. A new way to predict pore pressure from drilling rate is shown in Eq. 13.3.

$$\rho_1{}^c \log R_1 = \rho_2{}^c \log R_2 \tag{13.3}$$

where:

ρ = mud weight, lb/gal
c = experimentally determined shale compactibility constant

The exponent c depends on the shale compactibility and must be determined experimentally. After c is determined, Eq. 13.3 can calculate pore pressure from increases in drilling rate. Eq. 13.3 can be used for any formation with a predictable compactibility constant. The determination of the compactibility constant, c, is demonstrated in the following example:

☐ **EXAMPLE 13.3**

Assume:

Well depth = 10,000 ft
Mud weight = 9.0 lb/gal
Normal pore pressure gradient = 0.43 psi/ft
Drilling rate = 40 fph
Mud weight is increased to 9.8 lb/gal
Drilling rate = 32 fph

Determine: The shale compaction coefficient c

SOLUTION:

Using Eq. 13.3,

$$\rho_1 = 9.0 \text{ lb/gal and } R_1 = 40 \text{ fph}$$
$$\rho_2 = 9.8 \text{ lb/gal and } R_2 = 32 \text{ fph}$$
$$9.0^c \log 40 = 9.8^c \log 32$$
$$c = 0.74$$

The value of c can be redetermined from other changes in mud weight and drilling rate. Also, the value of c can be checked after running logs to determine pore pressure. Remember that c can be calculated on the basis of differential pressure.

For example, if the pore pressure is assumed equal to an equivalent mud weight of 9.6 lb/gal, and 10.1 lb/gal mud is being used, the differential is 0.5 lb/gal. If later logs or tests show the formation pore pressure is 9.0 lb/gal, the actual differential pressure was equivalent to 1.1 lb/gal, and the value of c should be recalculated. This recalculation of c can be done simply by making the respective mud weights equivalent to the actual pressure differentials. Use some caution with this procedure at low mud weights because c is very sensitive and changes considerably with small variances in pore pressure. Example 13.4 illustrates the effect on the value of c of a small change in pore pressure.

☐ **EXAMPLE 13.4**

Assume: Same data as Example 13.3 except that the pore pressure gradient rose to 0.44 psi/ft when the mud weight was increased to 9.8 lb/gal

Determine: A new value for c

SOLUTION:

The increase in pore pressure is equivalent to a change in mud weight of 0.1925 lb/gal. This surge in effective pore pressure reduces the pressure differential by the same amount as the change in mud weight. For this reason the effective mud weight is reduced from 9.8 lb/gal to about 9.6075 lb/gal. Then c is recalculated as follows:

$$9.0^c \log 40 = 9.6075^c \log 32$$
$$c = 0.95$$

At higher mud weights and greater pressure differentials the value of c is less sensitive.

When the value of c is determined, pore pressure can be estimated using Eq. 13.3 and changes in drilling rate. Example 13.5 shows the suggested procedure for the determination of pore pressure.

☐ **EXAMPLE 13.5**

Assume:

Drilling at 10,000 ft in shale
Mud weight = 9.6 lb/gal

Drilling rate = 30 fph
Drilling rate increases to 50 fph
c = 1.1

Determine: The increase in pore pressure

SOLUTION:

Use Eq. 13.3; however, in this case determine the mud weight increase that is necessary to reduce the drilling rate from 50 to 30 fph

$$9.6^{1.1} \log 50 = \rho_2^{1.1} \log 30$$

$$\left(\frac{\rho_2}{9.6}\right)^{1.1} = \frac{\log 50}{\log 30} = \frac{1.699}{1.477} = 1.15$$

$$\rho_2 = 9.6(1.15)^{0.909} = 10.9 \text{ lb/gal}$$

The equivalent rise in pore pressure is $10.9 - 9.6 = 1.3$ lb/gal. At 10,000 ft the pore pressure increase is

$$\frac{(1.3)(10,000)}{19.25} = 675 \text{ psi}$$

One problem encountered with this procedure is that an increase in drilling rate may be immediately reduced by reducing bit weight or rotary speed. If bit weight or rotary speed is reduced, the drilling rate must be corrected as shown in Example 13.1. Using the corrected drilling rate, pore pressure is determined as shown in Example 13.5.

D EXPONENT

Jorden and Shirley developed a useful method of evaluating drilling rate, known as the d exponent.[2] The theoretical base for this exponent is derived from a standard drilling rate equation shown in Eq. 13.4:

$$R = K_d N \left(\frac{W}{D_b}\right)^d \tag{13.4}$$

where:

R = drilling rate, fph
K_d = drillability constant
N = rotary speed, rpm
W = bit weight, lb/in.
D_b = bit diameter, in.
d = d exponent

With no changes in rotary speed, bit weight, or lithology, the only factor in Eq. 13.4 that changes the d exponent is drilling rate. Thus, under these conditions drilling rate should be used instead of the d exponent. Note that drilling rate is shown as directly proportional to rotary speed. This is true in soft formations such as the soft shales in coastal areas but is

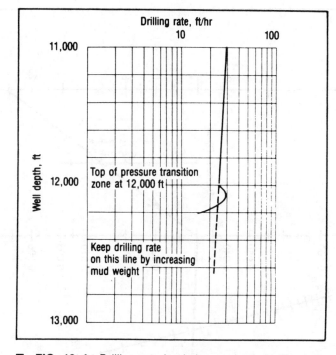

■ **FIG. 13–4** Drilling rate in shales vs depth for Example
13.2 (*courtesy Petroleum Engineer International*)

not true in hard rock areas. Consequently, this correlation is limited to
soft, compactible rocks. If the *d* exponent is used in hard rock areas, the
corrections for rotary speed are invalid and should be made as shown
for drilling rate in Example 13.2.

In the Gulf Coast areas of the United States, the *d* exponent has proved
to be a valuable tool in general field applications. Therefore, the nomograph solution offered by Jorden and Shirley is included as Fig. 13–5. This
nomograph is a solution to Eq. 13.5.

$$d = \frac{\log \dfrac{R}{60N}}{\log \dfrac{12W}{10^6 D_b}} \tag{13.5}$$

Note that the nomograph solution for the *d* exponent does not include
the mud weight correction. Mud weight corrections depend on the normal
pore pressure for a specific area and the actual shale compactibility as
shown just for the drilling rate case. In Gulf Coast areas, the normal pore
pressure is considered equivalent to a mud weight of 9.0 lb/gal, and corrections for mud weight increases are made on that basis. In hard rock areas
the normal pore pressure is generally equivalent to the weight of water,

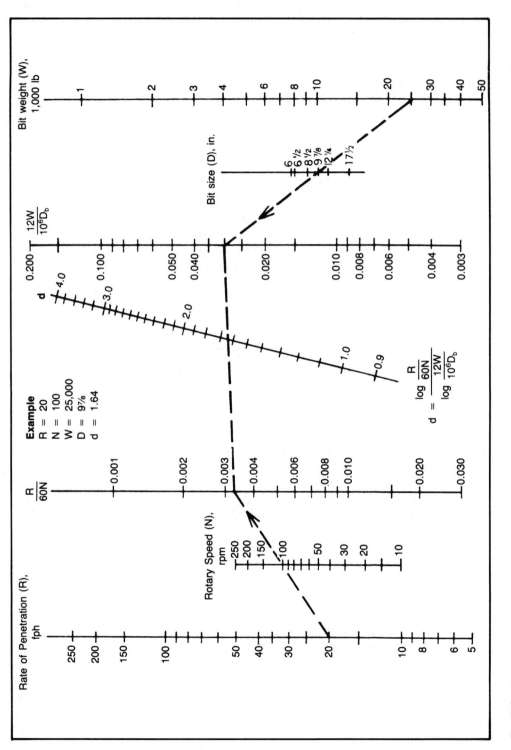

■ **FIG. 13–5** Nomogram for *d*-exponent determination (from Jorden and Shirley, "Application of Drilling Performance Data to Overpressure Detection," courtesy *Journal of Petroleum Technology*, November 1966)

8.33 lb/gal. Shale compactibility must be determined as shown for the drilling rate prediction method.

Example 13.6 illustrates the mud weight correction that should be made for the d exponent.

☐ **EXAMPLE 13.6**

Assume:

Normal pore pressure = 9.0 lb/gal equivalent
Mud weight in use = 12.0 lb/gal
Compaction coefficient = 0.8
d exponent from nomograph = 1.65

Determine: d exponent corrected for an increase in mud weight

SOLUTION:

$$d_c = 1.65 \left(\frac{9.0}{12.0} \right)^{0.8} = 1.31$$

where:

d_c represents the d exponent corrected for mud weight and shale compactibility

If the normal pore pressure is equivalent to a mud weight of 8.33 lb/gal, the solution to Example 13.6 would be to simply substitute 8.33 for 9.0 lb/gal mud. Historically d exponents have been determined ignoring any shale compaction effect. Obviously this is incorrect; even in soft rock areas shale compaction changes with deeper burial depths.

SLOUGHING SHALE

Sloughing shale may result from the following hole conditions:

1. Formation fluid pressures exceeding the hydrostatic pressure
2. Hydration or swelling of shale
3. Erosion caused by fluid circulation, surge pressures, or pipe movement

In some cases the sloughing problem may be a combination of more than one of the above causes. For this reason operators should always try to diagnose the causes. In the coastal areas of the U.S. items 1 and 2 are the main reasons for sloughing shale, with most of the serious sloughing problems in coastal areas caused by item 1.

Thus, this symptom is watched carefully when penetration rate begins to increase. If penetration rate increases and sloughing shales are noted about the same time, an operator is alerted to the problem of increasing pore pressures. Again, this is a phenomenon associated primarily with soft, compactible shales. In the midcontinent, Permian Basin, or Rocky Mountain areas, shale sloughing is associated mainly with items 1 and 3.

SHALE DENSITY

The normal trend is for the density of compactible shales to increase with depth. If this trend is reversed, as shown in Fig. 13–6 by the dashed lines, assume that a pore pressure increase within the shales has prevented compaction. In theory, this is an excellent tool; however, its actual field value is clouded by the difficulties of making precise measurements and selecting the shale that represents drill cuttings.

Two methods of measuring shale density are commonly used in the field: (1) the variable-density liquid column and (2) the mud balance density. The variable-density liquid column theoretically is an accurate procedure if all the shale cuttings represent bottom-hole cuttings from the depth assumed and if the shale cuttings are prepared exactly the same way each time. The mud balance method is based on Eq. 13.6:

$$\text{Specific gravity} = \frac{8.33}{16.66 - W_s} \qquad (13.6)$$

where:

W_s = weight of shale + weight of water, lb/gal

The mud cup is filled with shale until the weight is 8.33 lb/gal; then it is filled with water. The shale weight plus the added water is W_s. Again,

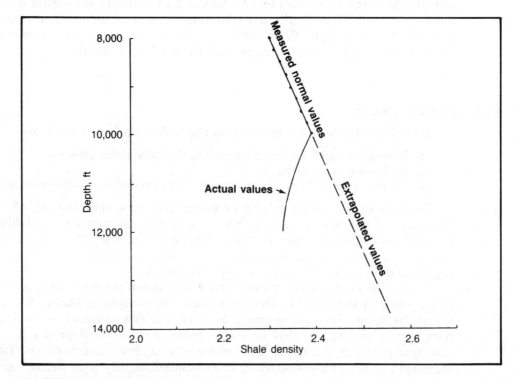

■ **FIG. 13–6** Shale density vs depth

the problem is to equate depth with the shale cuttings and to prepare the cuttings in the same manner each time.

Determining the source depth for the shale cuttings is not easy. The velocity profile of the mud in laminar flow is not flat, and shale cuttings from the same depth may arrive at the surface at time intervals that vary as much as several hours. This variation may be minimized by close observation of the sizes and shapes of cuttings used in the analysis, which means expertise is a significant part of the procedures.

Washing and drying the shale samples is important. The water content influences the results of any measurement method. Precise methods of shale preparation are described in service-company manuals. In general, preparation methods call for washing carefully and towel-drying the cuttings.

GAS IN MUD

Gas-cut mud is always a warning signal but not necessarily a serious problem. There are multiple gas sources; gas may enter the mud as a result of the following:

- Gas in shales that form a base line for a continuous gas unit level
- Gas from sands that may cause sudden changes in the gas concentration level
- Connection gas resulting from swabbing on connections
- Trip gas associated with swabbing following round trips with the drill-string
- Gas that enters the mud because of insufficient mud weight to control formation fluids

Because it is a compressible material, gas often gives the appearance of being a more serious problem at the surface than it actually is. Many shales contain gas in the pore spaces and furnish a continuous level of gas in the mud. This continuous gas level forms a base-line reference. For given areas, it is predictable. Little attention is paid to this source of gas.

Some gas sands increase the gas in the mud substantially and can result in severe reductions in surface mud weights. This always concerns an operator and may cause trouble if it comes unexpectedly. To illustrate the effect of this gas, Example 13.7 shows an assumed set of conditions.

☐   **EXAMPLE 13.7**

Well depth = 15,000 ft
Hole size = 7⅞ in.
Drillpipe size = 4½ in.
Mud weight = 15 lb/gal
Drilling rate = 20 fph
Sand: gas saturation = 70%
      porosity = 25%
Circulation rate = 7.0 bbl/min

z_s, gas compressibility factor at surface = 1.0
z_b, gas compressibility factor at bottom of hole = 1.35
T_b, bottom-hole temperature = 250°F
T_s, temperature at surface ≈ 100°F
Bottom-hole ratio* of mud volume to gas volume:

$$\frac{\text{mud}}{\text{gas}} = \frac{(7)(60)}{(0.062)(20)(0.25)(0.70)} = 1{,}935$$

Ratio of surface volume of gas to bottom-hole volume of gas:

$$\frac{V_s}{V_b} = \frac{P_{bh}}{P_s} \frac{z_s}{z_b} \frac{T_s}{T_b} = \frac{11{,}700(1.0)(560)}{14.7(1.35)(700)} = 466$$

where:

P_{bh} = bottom-hole pressure, psi
P_s = surface pressure, psi

$$\frac{1{,}935}{466} = 4.15 \frac{\text{mud vol}}{\text{gas vol}} \text{ at surface}$$

Mud weight:

$$\frac{4.15(15)}{4.15+1} = \frac{62.3}{5.15} = 12.1 \text{ lb/gal}$$

A reduction in mud weight from 15 to 12.1 lb/gal would be frightening if the operator was unaware of the source. In actual practice, all the gas probably would not be retained in the mud when it was weighed, and, most likely, the observed reduction in mud weight would be less than that shown in this example. In any event, any mud weight reduction is basis for concern, and the symptom needs to be associated with others in diagnosing the problem and the course of action.

For this specific example, the reduction in mud weight in the annulus would not be very high if the gas occupied the entire annulus. The reduction in hydrostatic pressure in Example 13.7 can be estimated by using Eq. 13.7:

$$\rho_m D_w - P_h = \frac{c_f P_s Z_a T_a}{(100 - c_f) z_s T_s} \ln\left(\frac{P_h + P_s}{P_s}\right) \tag{13.7}$$

where:

ρ_m = mud gradient, psi/ft
D_w = well depth, ft
P_h = hydrostatic pressure exerted by a column of mud and gas, psi
c_f = % of total fluid at the surface that is gas
Z_a = average gas compressibility factor
T_a = average temperature, R

---

* Use same units for gas and mud volumes. Here the units are barrels.

$$\rho_m D_w = 0.78 \frac{psi}{ft} (15,000 \text{ ft}) = 11,700 \text{ psi}$$

$$c_f = \frac{(1)(100)}{1 + 4.15} = 19.4$$

Assume $P_h = \rho_m D_w$ and solve the right-hand side of Eq. 13.7:

$$\frac{c_f P_s z_a T_a}{(100 - c_f) z_s T_s} \ln\left(\frac{P_h + P_s}{P_s}\right) = \frac{(19.4)(14.7)(1.18)(635)}{(80.6)(1)(560)} \ln\left(\frac{11,700 + 14.7}{14.7}\right)$$

$$\rho_m D_w - P_h = 4.74 \ln 798 = (4.74)(6.7) = 31.8 \text{ psi}$$

$$P_h = 11,700 - 31.8 = 11,668.2 \text{ psi}$$

Thus, in this 15,000 ft-well, a continuous column of mud, cut back in surface weight as shown, results in only a reduction of about 32 psi in hydrostatic pressure. With this type of gas cut, the emphasis is placed on recognizing the source. Sudden increases in mud weight to control the influx of gas might result in excessive hydrostatic pressure, cause a loss of circulation, and result in substantially more hole trouble.

On the other hand, failure to recognize a well kick could be disastrous. In a 15,000-ft well, if there are no other indications of a well kick or blowout by the time the gas is seen at the surface, then the problem can be associated with gas entry due to drilling the sand or a high-pressure, low-productivity gas zone. In either case an operator generally has time to determine the problem's source.

Again, the need to be familiar with the area's lithology is emphasized. In a shallow well, little decision time is available, regardless of the source. In new areas, the operator may need to close in the well and observe bottom-hole pressures. If doubt continues, he should stop drilling and circulate under controlled conditions until the gas source is determined. If the gas disappears after two or three circulations, it is probably gas from the sand. If it continues, the problem is most likely a low-productivity gas sand.

Connection gas and trip gas are introduced into the mud by swabbing or just by the reduction in total annulus back pressure when the pump is stopped. Any increasing trend in either connection or trip gas should be watched carefully. Both gas sources are affected by the practices of specific drillers.

Thus, when conditions are critical, standardize practices of making connections and pulling pipe on round trips. With standard practices, an increasing trend in connection or trip gas may be a warning signal that pore pressures are increasing.

CHLORIDE TRENDS

Chloride trends in the mud are not as easily recognizable as changes in gas concentration. Methods of measurement make it more difficult to obtain information on chloride changes. Also, in many cases, the water used in mud is a brackish or seawater type with a high chloride level.

However, the operator cannot afford to ignore the symptom and should regularly check the chloride content of the mud both going into and coming out of the hole. A comparison of trends may provide a warning or confirmation signal of increasing pore pressures.

An alternative to measuring the chloride content of the filtrate is a continuous method of measuring the mud resistivities into and out of the hole. Such resistivity measurements may be a part of a regular mud logging service or could be added at a small additional cost.

MUD PROPERTIES

The measurement of mud properties into and out of the hole may provide the first signal of gas or chloride changes. A sudden increase in yield point or a sudden decrease in the n value signals potential flocculation of the mud. While the flocculation may be caused by excessive solids or high temperatures, contamination could also cause it. In any event, the operator should check immediately for the cause.

This places a premium on the regular measurement of mud properties into and out of the hole and emphasizes the need to record the precise time that each measurement is made. A drilling mud is normally not a homogeneous mixture. When comparing mud properties, know what part of the mud system is being checked. A plot of mud property trends should be maintained and any addition of additives noted.

TEMPERATURE MEASUREMENTS

A continuous recording of flow-line temperature measurements may or may not aid in detecting pore pressure increases. Fig. 13–7 illustrates a typical flow-line temperature profile, and point A shows a temperature increase above the normal trend. This rise may be caused by an increase in circulation rate, a change in solids content of the mud, a change in drilling practices, an increase in bit torque, or an increase in pore pressure. Thus, the temperature curve alone is not definitive.

However, in many areas such as the Anadarko Basin, the temperature log may be the most effective tool available. In coastal areas, where drilling rates and other symptoms of increasing pore pressure are more definitive, the use of flow-line temperature plays a secondary role. Many specific examples can be cited when the flow-line temperature was very useful in coastal operations. When equipment permits, an accurate record of the flow-line temperature should be maintained.

BENTONITE CONTENT IN SHALE CUTTINGS

The bentonite content in shales normally decreases with depth. If the bentonite content does not decrease or if it increases in deeper shales,

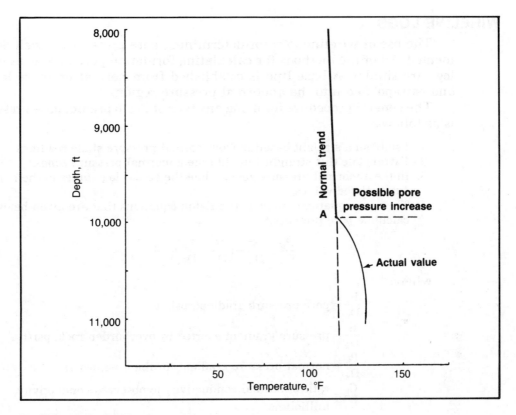

■ **FIG. 13–7** Depth vs temperature

this is a signal that normal shale compaction trends are not followed. The accuracy of this method is hampered by the available measurement methods.

PALEO INFORMATION

Abnormally high pore pressures are frequently related to certain environmental conditions within a given geologic period. For example, in the coastal areas of south Louisiana, if deposition occurs in water deeper than 1,500 ft, the shale-to-sand ratio increases, introducing permeability barriers to sand continuity. This isolation of sands results in traps for entrained fluids, and the later overburden deposition results in part of the overburden being supported by the trapped fluids.

The water depth during deposition is marked by certain fossils. If an examination of cuttings reveals these fossils, then the operator should be alerted to the potential problem of encountering abnormally high pore pressures.

WIRELINE LOGS

The use of wireline logs for determining pore pressures is well documented. All of the methods for calculating formation pore pressures from logs are similar. A base line is established from normal pressure levels and extrapolated into the abnormal pressure region.

The general procedure for using any type of log to predict pore pressure is as follows:

1. Establish a straight baseline from normal pressure shale readings.
2. Extrapolate this straight line into the abnormal pressure zones.
3. In the abnormal pressure zone, relate the actual log values to the normal extrapolated values.
4. Then use equations, such as the Eaton equations that are given below, to calculate pore pressure.[4]

$$\frac{P_p}{D_w} = \frac{S}{D_w} - \left(\frac{S}{D_w} - \frac{P_n}{D_w}\right)\left(\frac{C_n}{C_o}\right)^{1.2} \tag{13.8}$$

where:

$\dfrac{P_p}{D_w}$ = pore pressure gradient, psi/ft

$\dfrac{S}{D_w}$ = pressure gradient exerted by overburden rock, psi/ft

$\dfrac{P_n}{D_w}$ = normal pressure gradient for the area, psi/ft

$\dfrac{C_n}{C_o}$ = ratio of normal conductivity to observed conductivity, milliohms

$$\frac{P_p}{D_w} = \frac{S}{D_w} - \left(\frac{S}{D_w} - \frac{P_n}{D_w}\right)\left(\frac{R_o}{R_n}\right)^{1.2} \tag{13.9}$$

where:

$\dfrac{R_o}{R_n}$ = ratio of observed resistivity to normal resistivity, ohm-m

$$\frac{P_p}{D_w} = \frac{S}{D_w} - \left(\frac{S}{D_w} - \frac{P_n}{D_w}\right)\left(\frac{\Delta t_n}{\Delta t_o}\right)^{3} \tag{13.10}$$

where:

$\dfrac{\Delta t_n}{\Delta t_o}$ = ratio of normal travel time to observed travel time, μsec

$$\frac{P_p}{D_w} = \frac{S}{D_w} - \left(\frac{S}{D_w} - \frac{P_n}{D_w}\right)\left(\frac{\Delta v_o}{\Delta v_n}\right)^{3} \tag{13.11}$$

where:

$\dfrac{\Delta v_o}{\Delta v_n}$ = ratio observed velocity to normal velocity, fps

$$\frac{P_p}{D_w} = \frac{S}{D_w}\left(\frac{S}{D_w} - \frac{P_n}{D_w}\right)\left(\frac{d_{c_n}}{d_{c_o}}\right)^{1.2} \tag{13.12}$$

where:

$$\frac{d_{c_n}}{d_{c_o}} = \text{ratio of normal modified } d \text{ exponent to the observed modified } d \text{ exponent}$$

The accuracy of these equations depends on shale compactibility and the net effective overburden stress gradient. As shale compactibility is reduced, because of an increased burial depth or an older geologic age, the exponent in Eqs. 13.6–13.10 should be reduced. The amount of this reduction must be determined experimentally. The net effective overburden stress gradient, S/D_w, in soft rock areas is generally assumed to be the total pressure exerted, in psi/ft, of the overlying rock. In any area with a hard zone that has the matrix capacity to support part of the overburden, it is incorrect to use an S/D_w for the total overlying rock. For this reason, if a hard zone is present it is necessary to determine the effective overburden stress gradient that will support part of the overlying rock.

Actually, the presence of hard rock zones (when the rock matrix is strong enough to support the overburden rock) is the precise reason why pore pressures may be less or return to normal at depths below higher pore pressure zones. Even in soft rock areas a reduction of pore pressure at greater depths may occur. A warning signal that pore pressure may be reduced is a hard zone such as a dense limestone or dolomite in a lithology sequence of sands and shales. In hard rock areas, pore pressure increases and decreases are commonly found at greater drilling depths.

There are no precise methods for calculating the net effective overburden stress. Experience in an area helps, and some information can be developed from sonic logs. It is possible to obtain cores and check the matrix strengths of the rocks under simulated downhole conditions. Using this information, the potential magnitude of the overburden stress gradient may possibly be determined. For example, if 100 ft of rock is strong enough to support the entire overburden weight, then expect the pore pressure below that rock to be no more than normal.

Efforts should continue to predict pore pressures in hard rock areas more accurately. The same forces that affect pore pressures in soft rock areas are present in hard rock areas. These forces are simply harder to define in the hard rock areas, and the magnitude of their effect is minimized by rocks that are not very compressible.

Another problem common to both soft and hard rock areas is the construction of a straight line in the normal pressure zone. It is normal practice to use a log scale for the log readings and a regular coordinate scale for depth. This plotting method became a standard procedure because the plotted data generally formed a straight line; obviously, straight lines can be extrapolated with more accuracy than curves. Obtaining a straight line on log paper in normal pressure zones may be difficult, for the following reasons:

- Clean shale sections may be difficult to define from lithology logs
- Logs affected by fluid content may give false readings because of changes in water salinity

Fig. 13–8 shows a classic example of how abnormal pressure zones affect both conductivity and sonic logs. The departure of the actual log data from the normal pressure straight line shows a reduction in shale compaction, indicating an increase in pore pressure. The increase in formation conductivity shown for the conductivity log indicates a higher porosity with more fluid in the pore spaces. The increase in travel time shown for the sonic log indicates a reduction in shale compaction, a very good indication of abnormal pressure. Example 13.8 illustrates the use of Eaton's equations to calculate the pore pressure at 12,000 ft using both logs in Fig. 13–8.

☐   **EXAMPLE 13.8**

For conductivity log use Eq. 13.8:
From log data in Fig. 13–8,

$$C_0 = 1,600 \text{ and } C_n = 330$$

$$\frac{P_n}{D_w} = 0.465 \text{ psi/ft for South Texas}$$

$$\frac{S}{D_w} = 0.96, \text{ from Fig. 13–9}$$

Then,

$$\frac{P_p}{D_w} = 0.96 - (0.96 - 0.465)\left(\frac{330}{1,600}\right)^{1.2} = 0.89 \text{ psi/ft}$$

For sonic log use Eq. 13.10:

$$t_n = 53$$
$$t_o = 100$$

Then,

$$\frac{P_p}{D_w} = 0.96 - (0.96 - 0.465)\left(\frac{53}{100}\right)^{3} = 0.89 \text{ psi/ft}$$

Later tests showed the pore pressure gradient in the well illustrated by the log data in Fig. 13–8 to be 0.86 psi/ft rather than the calculated gradient of 0.89 psi/ft. A reduction in the exponent from 1.2 to 1.0 in Eq. 13.8 makes the calculated data match the measured data of 0.86 psi/ft. Likewise, a reduction in the exponent from 3.0 to 2.72 in Eq. 13.10 makes the calculated data match the measured data of 0.86 psi/ft. It is necessary to reduce the exponent on log data as shale zones become more compacted.

Recognizing the symptoms of increasing pore pressures, estimating the magnitude of pore pressures, knowing the fracture gradients of exposed formations, and maintaining the drilling practices within controllable lim-

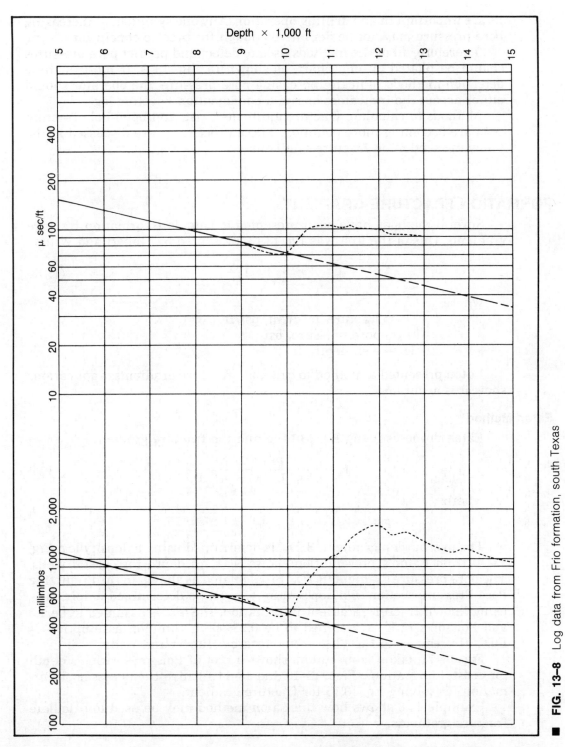

■ **FIG. 13–8** Log data from Frio formation, south Texas

its are important in any drilling operation. Any one symptom of increasing pore pressure may not be definitive enough for precise conclusions.

Therefore, all of the methods used to detect and predict pore pressures should be plotted on one chart, as shown in Fig. 13–9. If most of these detection methods indicate increasing pore pressure, the operator should either accept the indicators or run a confirmation wireline log.

Methods to calculate fracture gradients exist, and open-hole fracture tests are becoming more common. These procedures are discussed in the following section on fracture gradients.

FORMATION FRACTURE GRADIENT

Several methods have been presented to calculate formation fracture gradients. One of the first was Eq. 13.13, presented by Hubert and Willis.[5]

$$F_g = \frac{1}{3}\left(1 + 2\frac{P_p}{D_w}\right) \tag{13.13}$$

where:

F_g = fracture gradient, psi/ft
P_p = pore pressure, psi
D_w = well depth, ft

Eaton presented a method to calculate fracture gradients; applications show it is accurate.[6]

Eaton Method

Eaton proposes using Eq. 13.14 to find the fracture gradient.

$$F_g = \left[\frac{S}{D_w} - \frac{P_p}{D_w}\right]\left(\frac{\gamma}{1-\gamma}\right) + \frac{P_p}{D_w} \tag{13.14}$$

where:

γ = Poisson's ratio

The overburden gradient, S/D_w, is determined using a density log; Fig. 13–10 from Eaton's presentation shows S/D_w vs depth for coastal wells. The pore pressure gradient, P_p/D_w, is calculated as shown in the Predicting Pore Pressure section. Poisson's ratio, γ, must be determined empirically using known fracture gradient information. If S/D_w is assumed to be 1.0 and Poisson's ratio is assumed to be 0.25—both common assumptions—Eq. 13.14 reduces to Eq. 13.13 introduced by Hubert and Willis.

Fig. 13–11, taken from Eaton, shows a plot of Poisson's ratio vs depth for coastal formations. Figs. 13–10 and 13–11 furnish the necessary information for solving Eq. 13.15 for fracture gradient.

Example 13.8 shows how the Eaton method may be used to calculate fracture gradients.

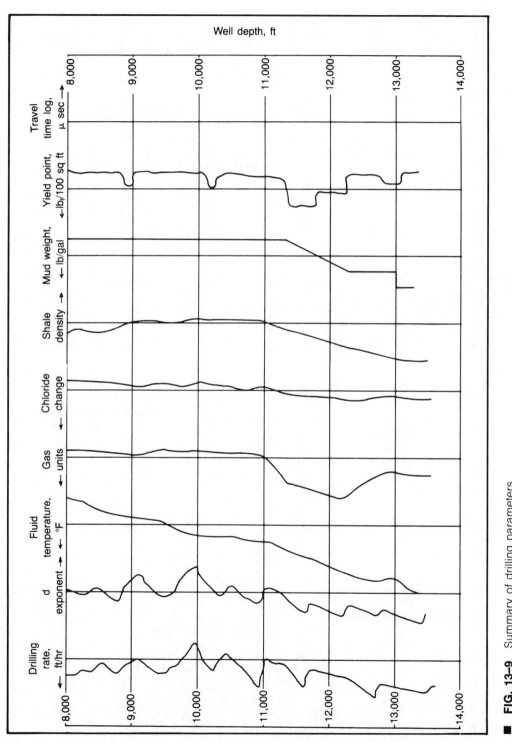

■ **FIG. 13–9** Summary of drilling parameters

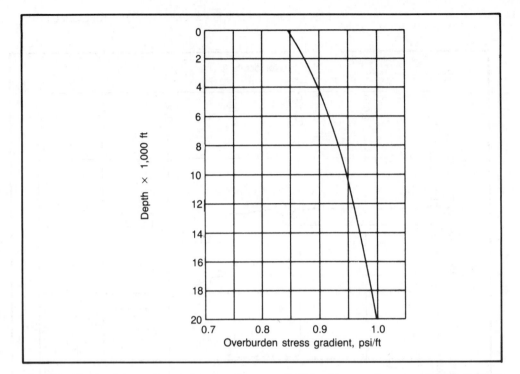

■ **FIG. 13–10** Composite overburden stress gradient for all normally compacted Gulf Coast formations

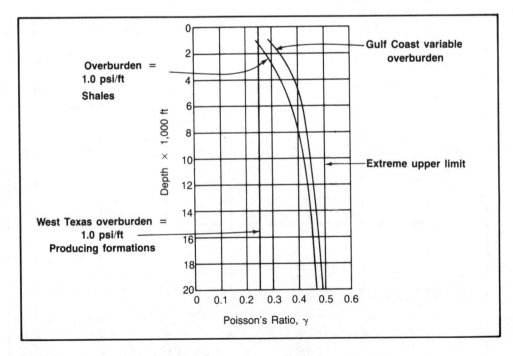

■ **FIG. 13–11** Variations of Poisson's ratio was depth

☐   **EXAMPLE 13.8**

Find:
Fracture gradient at 11,600 ft in the Frio formation of south Texas. Use the information shown in Fig. 13–8.

EATON SOLUTION:

$$P_p = 9,876 \text{ psi from sonic log data}$$

$$\frac{S}{D_w} = 0.97$$

$$\gamma = 0.46 \text{ (from Fig. 13–10)}$$

$$F_g = (0.97 - 0.86)\frac{0.46}{0.54} + 0.86$$

$$F_g = 0.0937 + 0.86 = 0.954 \text{ psi/ft}$$

Note: Using Eq. 13.13:

$$F_g = \left\{\frac{1}{3}[1 + 2(0.86)]\right\} = 0.907 \text{ psi/ft}$$

The Eaton procedure is commonly used; however, the most definitive procedure is to pressure test just below the casing seat and open-hole sections. The procedure for these tests is outlined in the following discussion.

Pressure Testing

Pressure testing below the casing seat is performed for two reasons: (1) to test the cement job and (2) to determine the fracture gradient in the first formation below the casing shoe.

The general procedure for pressure testing below the casing seat is to close in the annulus and start pumping slowly at a rate of 0.3–0.5 bbl/min. Record the pressure regularly. The maximum pressure may be at a preselected level, the leakoff pressure, or the rupture pressure. If for some reason the operator anticipates a need for a total hydrostatic pressure greater than the test pressure, he should either retest with the drillstring back in the last casing string or set a liner before drilling ahead.

Testing initially to leakoff is an arbitrary decision that depends on the operator's objectives. If experience shows the hydrostatic pressure plus circulating pressures will not exceed 16.0 lb/gal, then there is no need to increase the pressure to the leakoff point that may reach 18.0 lb/gal. On the other hand, if the next casing string depends on what can be contained with the last casing string, then maximum advantage can be obtained from a specific knowledge of the maximum permissible wellbore pressure that can be imposed on the formation. Thus, an operator can profit from knowing the maximum leakoff and rupture pressures.

Some examples of leakoff tests are shown in Figs. 13–12, 13–13, and 13–14. Fig. 13–12 shows a leakoff test below the surface casing. Conse-

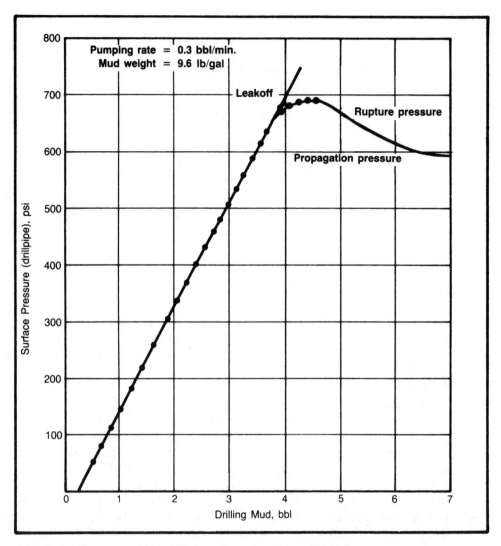

■ **FIG. 13–12** Casing seat leakoff test—casing seat at 3,000 ft

quently, this is primarily a test of the cement job. In this example the leakoff occurs at 660 psi, the rupture pressure is 690 psi, and the propagation pressure is 595 psi. Considering the leakoff pressure as the maximum permissible results in a maximum mud weight of 13.8 lb/gal as evidenced below:

$$9.6 + \frac{660}{(3,000)(0.052)} = 13.8 \text{ lb/gal}$$

This calculation does not prove that all formations below the surface casing will hold 13.8-lb/gal mud. It does show that no more than 13.8-

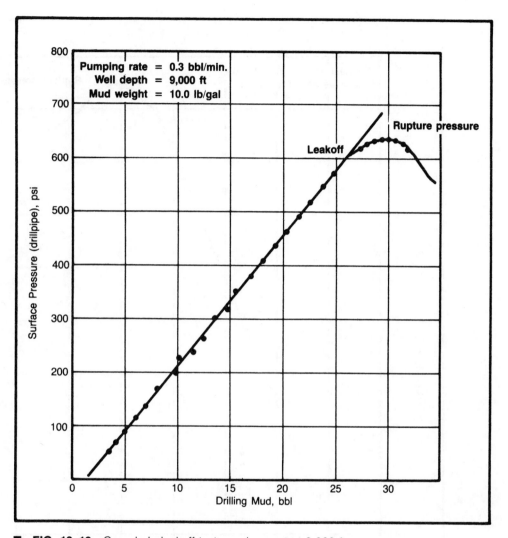

■ **FIG. 13–13**   Open-hole leakoff test—casing seat at 3,000 ft

lb/gal mud should be used unless a retest shows an increase in the fracture resistance. Often an increase in formation strength is noted after several days of drilling. A zone that held only 13.8 lb/gal (as shown in Fig. 13–12) frequently holds 15.0 lb/gal later.

A knowledge of this strength increase might be very helpful when drilling into a pressure transition zone. The strength increase, when it occurs, is probably due to plugging of pore spaces by drilled solids. It should be emphasized that this strength increase may or may not occur; it is not something an operator can assume will happen.

Fig. 13–13 is a leakoff test taken when the drillstring was back in the surface casing but with 6,000 ft of open hole. Note that the leakoff occurred

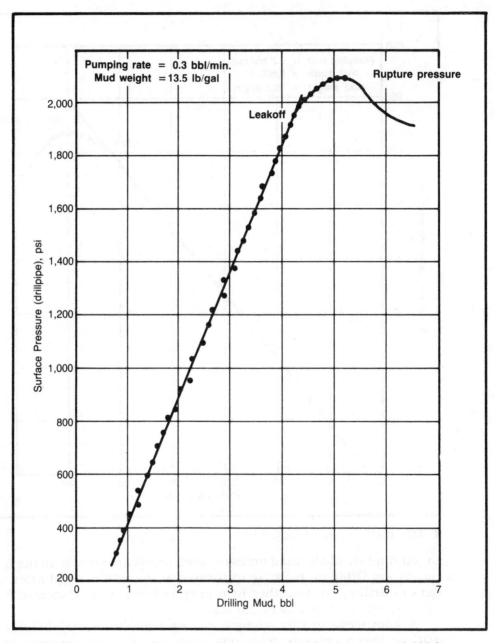

■ **FIG. 13–14** Leakoff test at first sand below protective casing seat at 10,000 ft

at about 600 psi. Also notice that 26 bbl of mud are required to reach this point, while in Fig. 13–12 only 3.8 bbl of mud are required for the casing seat test.

The primary difference is the amount of open hole. In Fig. 13–12 only 20 ft of hole has been opened below the casing seat. In Fig. 13–13, 6,000 ft of hole have been opened. The additional mud is required because of filtration and mud loss to very permeable sands.

The leakoff pressure of 600 psi in Fig. 13–13 shows the formation just below the casing seat will hold a 13.8-lb/gal mud. Again, this does not ensure that all of the open formations below 3,000 ft will hold 13.8-lb/gal mud because 600 psi imposed at, say, 6,000 ft would represent only a 1.9-lb/gal increase in mud weight. Thus, the 6,000-ft formation has been tested to only 11.9 lb/gal in the test shown in Fig. 13–13. However, in young sediments normally associated with most offshore and coastal formations, the leakoff test results taken just below the casing shoe are generally indicative of the maximum mud weight that can be used.

Fig. 13–14 is a leakoff test for the first sand below protective casing. Notice that the leakoff occurs when the surface pressure increase reaches 1,950 psi. This happened with a 13.5-lb/gal mud in the hole. Surface pressure plus mud weight represents a formation fracture gradient equal to a mud weight of 17.25 lb/gal, calculated as follows:

$$13.5 + \frac{1,950}{(10,000)(0.052)} = 17.25 \text{ lb/gal}$$

After several days of drilling, this formation resistance may increase; however, field results show that in most cases subsequent tests remain in the same range as the initial test when testing below protective casing.

Special Considerations

Special considerations in running leakoff tests include:

- Pumping rate
- Decision to test to a leakoff pressure
- Which pressure to use if there is a difference in drillpipe and annulus pressures
- Changes in line slope during the test
- Frequency of testing and the effect on formation resistance
- Maximum mud weight relative to that shown on a leakoff test

The pumping rate should be kept low, such as 0.25–0.5 bbl/min. The tests in Figs. 13–12, 13–13 and 13–14 were run at 0.3 bbl/min. This means the normal rig pump generally should not be used. Exceptions include plunger-type pumps with which the suggested low volumes can be attained. A cementing unit with pump and volume tank is usually preferred.

If pumping rates are too high, the leakoff test may follow the pattern in Fig. 13–15. In this graph there is no indicated leakoff pressure, the forma-

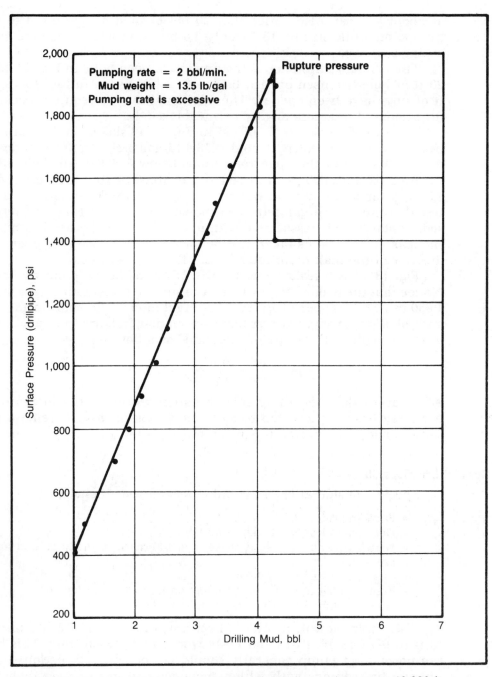

■ **FIG. 13–15** Leakoff test at first sand below protective casing seat at 10,000 ft

tion suddenly ruptures, and whole mud is lost quickly. Even this type of test probably has no long-range detrimental effects. The primary problem is that the objective of determining the leakoff pressure has not been attained.

Some operators are repelled by the concept of increasing surface pressure until some mud is lost to the formation. Others think that their drilling conditions do not justify such tests. If all the drilling is to be in formations with a normal pore pressure, leakoff tests are unnecessary. The well-accepted drilling belief that once the formation is tested to leakoff it will never again hold that much pressure is outdated.

Several hundred leakoff tests were conducted during a three-year period in U.S. coastal areas; the tests have not been detrimental to subsequent formation resistance. There may, however, be valid reasons for not testing to a leakoff pressure. At times, surface equipment may not permit the surface pressure necessary to reach the leakoff point. If the maximum leakoff pressure is desirable under these conditions, a retest may be performed after the mud weight has been increased.

Another reason for not testing to leakoff is that an operator knows, based on offset well data or geologic information, that future mud weights will not be high enough to justify a test to leakoff.

Encountering differences between the drillpipe and annular pressure during leakoff tests is rare. If an operator is pumping only through the drillstring, then he should use the drillpipe pressure. One method to overcome the problem is to pump slowly into both the drillstring and the annulus. This has been done on a limited basis.

In some leakoff tests, the pressure may not increase in a perfectly straight line (see Figs. 13–12, 13–13, and 13–14). Fig. 13–16 demonstrates what has been observed in some past tests. The pressure leaked off at 600 psi and then continued to increase to 900 psi, where it leaked off again and then continued to increase to the true leakoff. The question that arises immediately is, "How does an operator know the first leveling of pressure buildup is not the true leakoff?"

Actually, it may be the true leakoff; however, if the pressure at which this occurs is substantially below that anticipated, pumping should be continued to the rupture pressure because remedial action will most likely be necessary. The cause of this phenomenon is unknown. Gas or air pockets may be in the annulus, or some mud may be lost where there is a substantial amount of open hole, as illustrated in Fig. 13–16, and the pressure buildup may continue.

The frequency of testing is normally not a problem because standard retests are not run immediately. However, a limited number of field tests show that a leakoff test may be repeated immediately when the pressure is bled off. Thus, retesting is not generally a problem from the standpoint of hole conditions.

To determine the maximum permissible mud weight after a definitive

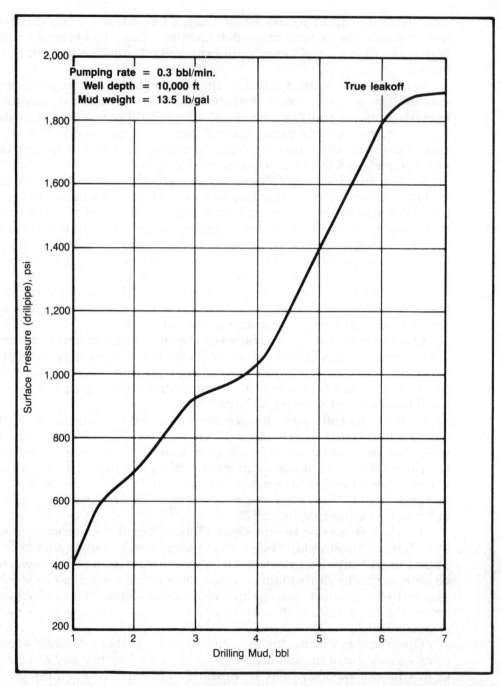

■ **FIG. 13–16** Leakoff test with 3,000 ft of open sand below protective casing seat at 10,000 ft

leakoff test requires a knowledge of mud properties, circulating pressure losses in the annulus, and an operator's requirements for safety factors. Specifying a mud weight 0.5 lb/gal below the equivalent mud weight at leakoff below surface casing and 1.0 lb/gal below the equivalent mud weight at leakoff below protective casing is not unusual.

These numbers are general. Circumstances may demand more precise limits. Circulating pressures can be calculated or estimated, then converted to mud weight, and added to the actual mud weight. After this, an operator may add 100 or 200 psi as a safety factor. Actually, the risk involved determines the operator's decision.

If there is a risk of an underground blowout, then an operator probably takes less of a chance on losing circulation. One fact is obvious: if an operator exceeds the leakoff pressure from his last test, he is inviting problems. In all probability, they will materialize.

SUMMARY

Pressure testing below the casing seat has application all over the world. An operator flirts with disaster if he uses a mud weight exceeding what test pressure shows to be safe. Testing to leakoff has mainly been applied in coastal areas where sediments are young geologically and the open formations are mostly shales and sands. Limestone and dolomite formations introduce a different problem.

If these formations are ruptured, the fracture may or may not heal. At this point, there are not enough data to justify any routine practice of testing to leakoff in limestone or dolomite. The same may hold for geologically old shale zones, although recent data indicate compaction trends occur in these shales. If this is generally true, then leakoff tests in these geologically old shales should have no permanent detrimental effect.

At this time, data are too limited to recommend leakoff tests when drilling the so-called hard rock. In any event, pressure testing is still recommended when an operator is unsure of fracture gradients and when mud weights must be increased to control high-pore-pressure zones.

Pressure testing and fracturing a zone is better than raising the mud weight and fracturing the zone with a permeable high-pore-pressure zone open that requires the higher-weight mud. The fracture during the pressure test may result in costly remedial action. A fracture with the high-weight mud required to contain the fluid in a high-pore-pressure zone may result in a disastrous underground blowout and eventually an uncontrolled surface blowout.

■ REFERENCES

1. Pennebaker, E.S., "Detection of Abnormal Pressure Formations from Seismic Field Data," API paper no. 926–13–C, 6–8 March 1968, San Antonio, Texas.
2. Jorden, J.R., and Shirley, O.J., "Application of Drilling Performance Data to

Overpressure Detection," *Journal of Petroleum Technology,* November 1966.

3. Combs, George D., "Prediction of Pore Pressure from Penetration Rate," SPE of AIME paper no. 2162, September 1968.

4. Eaton, Ben A., "Graphical Method Predicts Geopressures Worldwide," *World Oil,* July 1976.

5. Hubert, M. King and Willis, D.G., "Mechanics of Hydraulic Fracturing," SPE of AIME paper no. 210, October 1969.

6. Eaton, Ben A., "Fracture Gradient Prediction and its Application in Oilfield Operations," *SPE Journal,* October 1969, p. 1353.

14

PRESSURE. CONTROL

During the 1960s, pressure control received more attention than any other phase of drilling practices. One reason for this emphasis was increasing awareness of the problems of deep drilling and the extremely high drilling costs associated with abnormally pressured formations. In addition, ecology became an important part of the American scene; several blowouts with the associated problems of pollution gained international attention. Politicians thrived on any problem related to the industry.

In the meantime, engineers and scientists were working hard to further develop and put into practice technology that would prevent a blowout. They also were working on methods to bring the blowout under control if it occurred. In addition, methods were developed to handle oil spills on open waters. No expense has been spared to develop methods of prevention and cure.

One of the most important parts of any operation is the planning stage. With geological information and experience in the same or similar areas, wells are planned to reach total depth with no problems. Casing programs are designed to maximize safety, mud programs are selected carefully, and control equipment is designed to handle potential problems.

Even with this planning, exploratory or new areas present problems. Specific knowledge of the new area is unavailable; however, geophysical exploration can estimate formation pressures and give insight into formation characteristics. Continued development of these tools will aid substantially future exploratory drilling.

In development or exploratory areas related to known areas, an operator can be more precise. He can calculate formation pressures fairly accurately. He can predict formation fracture gradients closely and can test predictions while drilling.

Therefore, in the well planning stage, the operator can design the best casing program and prepare for problems that may or may not occur. In the actual drilling he can test using imposed wellhead pressures before raising mud weights to levels that might fracture some of the formations open to the wellbore. An operator should never fight lost causes, situations in which the available technology indicates little chance for success.

The objectives of this discussion on pressure control are to (1) show pressure relationships that exist in a borehole and (2) illustrate methods used to control well kicks and blowouts.

BOREHOLE PRESSURE RELATIONSHIPS

An operator planning a well needs some knowledge of overburden and formation fluid pressures in order to select the necessary hydrostatic or drilling-fluid pressure. Industry generally accepts a theoretical overburden pressure of 1.0 psi/ft, which is calculated as follows:

$$\text{Assume rock specific gravity} = 2.5$$
$$\text{Average void or pore space} = 10\%$$

$$\text{Overburden gradient} = (0.433 \text{ psi/ft})(2.5)(0.90) + (0.433 \text{ psi/ft})(0.10) = 1.018 \text{ psi/ft}$$

Some rocks like dense limestone have a specific gravity greater than 3.0 and little void space. Other rocks, such as surface shales in young formations, have a pore space of more than 40% of the bulk rock volume. This means the pressure exerted by overburden rock may vary from 0.75 to more than 1.3 psi/ft. Using 1.0 psi/ft as the theoretical overburden weight places an upper limit on wellbore pressures.

With this upper boundary in mind, operators try to prevent any combination of surface pressure plus hydrostatic pressure from exceeding 1.0 psi/ft. In actual operations formation fracture gradients are generally less than this upper boundary. However, with only surface casing set, there is a danger of the fluid penetrating through the earth to the surface if the 1.0 psi/ft is exceeded. This penetration is obviously much worse than simply fracturing an underground formation and losing circulation. With the communication of fluids to the surface, an operator loses complete control of the well.

Formation fluid or pore pressures are frequently categorized as subnormal, normal, and abnormal. *Subnormal pore pressures* are those below what is considered normal in a specific area. In the Gulf coastal areas of the U.S., the *normal pore pressure gradient* is 0.465 psi/ft.

In the north Texas and Midcontinent areas, subnormal pore pressures result when formations have been produced for many years or when formation outcrops are below a well's surface location, and direct fluid communication is established with the outcrop. The latter condition is normal in several Rocky Mountain drilling locations.

Abnormal pore pressures are those above normal for a certain area. Generally, abnormal pore pressures are associated with fluids trapped within the pore spaces of rocks by a permeability barrier. These permeability barriers may be faults, folds, salt domes, or permeability pinchouts.

A common cause of permeability pinchouts in coastal areas is a reduction in the sand:shale ratio associated with the deepwater deposition of sediments. This geologic feature may be associated with specific fossils of the period and permit the early recognition of environmental conditions conducive to abnormal pore pressure.

In addition to the permeability barriers, there must be another reason for the increasing fluid pressure within the pore spaces. In soft rocks, a normal compaction (with depth) trend in shale zones exists. Abnormal pore pressures result when this trend is reversed by the partial support of the overburden rock by fluids within the pore spaces. In this environment, the upper limit on pore pressure is the pressure exerted by the overburden rock.

In hard, high-compressive-strength rocks, there may be no evidence of compaction with depth. Yet in many areas abnormal pressures are frequently encountered. This typically occurs in the deep, high-pore-pressure zones in west Texas and the shallower zones in southern Iran. Because compaction is not normal with depth in these areas, deposition occurred within the pore spaces and thus forced the same quantity of fluid to occupy a smaller volume.

The basic cause of abnormal pressure appreciably affects the detection methods and drilling programs. Compaction can be detected by different logging methods. When compaction is common, the abnormal pore pressures usually increase with depth. In the hard rock areas, abnormal pressures may be detected; however, the logging methods are less definitive. For these areas, abnormal pressure zones frequently lie above normal pressure zones.

Compactible formations are typically found in coastal areas. The normal pressure gradient for these areas is 0.465 psi/ft. The top of the so-called pressure transition zone is the point at which the pore pressure increases above normal. No specific level of pore pressure signals the bottom of the pressure transition zone.

Generally the requirement to set protective casing is based on the length of the surface casing, the fracture gradients or normal pressure formations, and the attitude and experience of the operator. Efforts to minimize casing requirements are made by planned drilling programs.

A significant observation related to abnormal pore pressure is that many of the permeable zones are small. For developmental drilling, high fluid pressures may be bled off quickly or in fact may be nonexistent in an offset well. Consequently, well correlations must be used carefully. This can work both ways. An operator may encounter no abnormal pressure and suddenly find it in his next well. In coastal areas, compaction may continue as hydrocarbons are removed from permeable zones; this may result in premeability shutoffs in the region of the wellbore, where pressure drawdowns are highest.

Another potential danger and a cause of abnormal pressure is thick gas sands. Consider this example.

☐ **EXAMPLE 14.1**

Assume:

Gas sand top = 2,000 ft
Effective sand thickness = 600 ft
Normal pressure gradient for the area = 0.465 psi/ft

SOLUTION:

Bottom of the sand = 2,600 ft
Fluid pressure at 2,600 ft = (0.465)(2,600) = 1,210 psi
Weight of the gas is negligible; thus pressure at 2,000 ft is also about 1,210 psi.
At 2,000 ft, 1,210 psi gives a pressure gradient = 1,210/2,000 = 0.605 psi/ft
Mud weight required to balance = 11.65 lb/gal

In new or exploratory areas thick gas sands represent a potential hazard. Drilling with normal mud weights into a zone like this may result in well kicks that are difficult to control properly. This problem becomes significant for shallow gas zones because, by the time a pit level increase is noted, gas is at the surface, which complicates the drilling and casing programs in new areas.

Hydrostatic pressure is a pressure imposed by a static column of drilling fluid. Most mud balances provide a direct reading of this pressure in psi/ft. If only the weight of the fluid in pounds per gallon or pounds per cubic foot is given, the pressure gradient is calculated by converting the units, as shown in Example 14.2.

☐ **EXAMPLE 14.2**

$$\frac{\text{Mud weight, lb/gal}}{19.25 \text{ sq. in.-ft/gal}} = \text{psi/ft}$$

$$\frac{\text{Mud weight, lb/cu ft}}{144 \text{ sq in./sq ft}} = \text{psi/ft}$$

The selection of mud weight is related to formation pressures, swabbing, formation fracture gradients, penetration rate, and general knowledge of an area. In coastal regions of the U.S., a mud weight of 9.0 lb/gal is required to balance most normal pressure formations; thus, the selected mud weight is at least 9.2 lb/gal. In the Rocky Mountain and Midcontinent areas, the required mud weight for normal pressure formations is less than 9.0 lb/gal.

Well Kick and Blowout

Defining a well kick and a blowout may be difficult. An uncontrolled influx of formation fluids into the wellbore is considered a *well kick*. If this uncontrolled influx of formation fluids is not controlled by surface equipment so the fluids can be displaced without losing control, the kick turns into a blowout.

There are two types of blowouts. A *surface blowout* occurs when formation fluids are flowing uncontrolled at the surface. An *underground blowout* occurs when the formation fluids are flowing uncontrolled into underground zones. Note that a simple influx of formation fluids into the wellbore is not considered a well kick unless surface equipment must be closed to prevent further entry of formation fluids.

CAUSES AND INDICATION OF WELL KICKS AND BLOWOUTS

A well kick or a well blowout may occur for the following reasons:

- Insufficient mud weight
- Failure to keep the hole full of fluid
- Swabbing
- Lost circulation (see chapter 7)
- Mud cut by gas or water

Early warning signals are listed below:

- Increase in fluid volume at the surface, commonly termed a "pit level increase"
- Sudden increase in drilling rate
- Mud cut by gas or water
- Reduction in pump pressure
- Reduction in drillpipe weight

Insufficient Mud Weight

Drilling with minimum mud weights to maximize drilling rates has been emphasized. More operators are drilling closer to balanced pressure conditions, when the hydrostatic pressure is barely sufficient to contain the formation pore pressure. In many cases, well kicks are taken to determine the specific pore pressure to help plan future operations. In certain areas, such as the Delaware Basin of west Texas, an operator may drill a section of the hole underbalanced, when the pore pressure is greater than the hydrostatic pressure. In these cases, upper formations are stable, and the productivities of permeable formations are low.

Mud weight requirements are not always known. Operators may encounter shallow abnormal pressure zones. Shallow gas sands present a problem, as noted in the discussion on abnormal pressures (see this chapter's Borehole Pressure Relationships section). Correlations around salt domes are always difficult. Any new or exploratory area presents a problem in selecting the correct mud weight as well.

Failing to Keep the Hole Full of Fluid and Swabbing

The problem of keeping the hole full of fluid has been emphasized almost as long as fluid has been used to control subsurface pressures. Surveys made on the causes of well kicks and blowouts show that about 25% of the time these problems occur when pulling the drillstring. Keeping the hole full should be no problem if the crews follow volume tables.

Data are abundant on pipe displacement volumes. Special precautions must be followed with drill collars. A drill collar that weighs 100 lb/ft displaces five times as much fluid as 5-in. drillpipe. Thus, one stand of drill collars this weight equals five stands of 5-in. drillpipe. This emphasizes precautions that should be taken.

One problem in filling the hole is measuring the fluid that goes into the hole when pulling the pipe. Reading flowmeters, counting pump strokes, and other measuring methods are employed to ascertain that the hole is taking a volume of mud equal to the pipe removed. Accurate measurement is very important. A small-volume trip tank is desirable to determine actual mud volumes.

In addition, a rig supervisor should keep a record of the mud required to fill the hole on each round trip with the drillstring. Also, the amount of the water loss should be recorded to explain the variations in mud volumes required to fill the hole.

One problem is a result of the use of heavy mud slugs just before pulling the drillstring. The heavy slug of mud causes the mud inside the drillstring to drop and the mud level in the annulus to rise. For this reason the hole may not take the proper amount of mud initially. The hole's failure to take the right amount of mud may confuse the operator as to the specific cause. Thus, ten stands of pipe are often pulled before going back to the bottom if the hole is not taking the correct amount of mud.

Mud Cut by Gas or Water

Gas-cut mud has always been a warning signal but not necessarily a serious problem. There are multiple gas sources. Gas may enter the mud as a result of the following.

- Gas in shales that form a base line for continuous gas unit level
- Gas from sands that may cause sudden changes in the mud's gas concentration
- Connection gas that comes from swabbing on connections
- Trip gas associated with swabbing following round trips with the drillstring
- Gas that enters the mud because of insufficient mud weight to control formation fluids

Because gas is a compressible material, it often gives the appearance of being a more serious problem at the surface than actually exists. Many shales contain gas in the pore spaces and furnish a continuous level of gas in the mud. This continuous gas level forms a base-line reference, and for given areas it is predictable. Little attention is paid to this gas source.

Some gas sands increase the gas in the mud substantially, thus severely reducing surface mud weights. (These effects are covered in Chapter 13.)

Connection and trip gas are both introduced into the mud by swabbing or just by the reduction in hydrostatic head when the pump is stopped. The plot of gas from this source should be maintained to determine trends. If there is a general increase of gas in the mud attributable to this source, use it as a warning signal. Check other indicators to see if the mud weight should be increased. If these indicators are negative, watch the well closely; be prepared to act quickly if the well kicks.

Gas that enters when the mud weight is too low forms the basis for current well control technology.

CONTROL OF WELL KICKS AND BLOWOUTS

Starting a discussion on controlling well kicks and blowouts is difficult because everything should be said first. However, two points are essential in well control technology:

1. The surface equipment for proper control must be available.
2. The operator must understand the compression and expansion characteristics of gas.

From this beginning, the specific control procedures are almost classic. The drillpipe pressure method of control has become standard practice. Application methods are refined, and efforts to make procedures more understandable continue. In addition, certain control problems, casing setting depths, and drilling practices are examined constantly to ensure a minimum of trouble.

Surface Equipment

The required surface equipment is based on the type of well being drilled. Some of the equipment used is listed below:

- Hydraulic-adjustable choke
- Two hand-adjustable chokes
- Hydraulic control valve
- Method for removing gas from the mud
- Back-pressure valve
- Full-opening positive shutoff valve for the drillstring
- At least one bag-type blowout preventer
- At least two ram-type preventers
- Closing units for preventers
- Accurate pressure gauges, mounted close to the driller's position

A possible arrangement for surface control equipment on a low-risk well is shown in Fig 14–1. An arrangement for surface control equipment

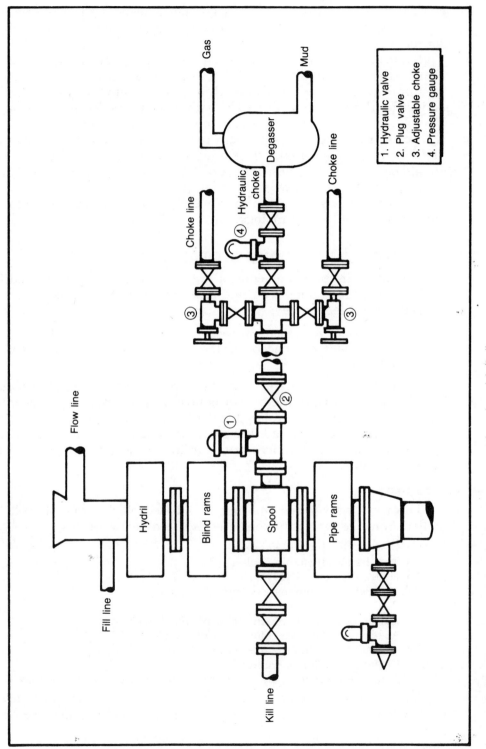

■ **FIG. 14–1** Pressure control equipment arrangement for a low-risk well

on a high-risk well is illustrated in Fig. 14–2. There is no specific standard arrangement for surface control equipment that is accepted by all operators. An operator is generally influenced by experience and the related experiences of others. In offshore areas, a second annular preventer that is run on the bottom of the riser pipe is commonly utilized.

Several types of hydraulic chokes are marketed. Low-pressure hydraulic chokes (below 3,000 psi) frequently have rubber closing mechanisms and are subject to severe erosion if used for extended periods. Most of the high-pressure hydraulic chokes (above 3,000 psi) use metal seals and are much more resistant to erosion. The hand-adjustable chokes are backups to the hydraulic chokes and bleed off small volumes of fluid to

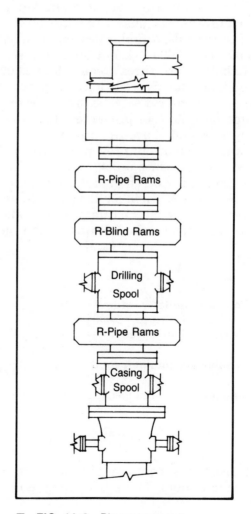

■ **FIG. 14–2** Blowout preventer

control wellbore pressures. Some operators also use a full-opening line to bleed off pressure if the chokes should become plugged or inoperable.

A hydraulic shutoff valve on the flow line upstream from the chokes permits leaving one of the adjustable chokes open while drilling. A separator or degasser is needed to remove gas from the mud. A series of baffles may be sufficient for low-risk wells, but for areas where trouble is likely, arrangements need to be made for a degassing unit.

The full-opening positive shutoff valve needs to be available for stabbing into the drillstring if the well kicks while pulling the pipe. As mentioned, 25% of the well kicks occur when pulling pipe. The regular back-pressure valve offers a flow restriction. If fluid flow is through the drillstring, the full-opening valve is needed.

Blowout preventers are, of course, the key to successful well control. The annular preventer is generally required for all wells other than low-risk, low-pressure wells. Annular preventers may be rated at working pressures of 5,000 and 10,000 psi, but operators seldom rely on the annular preventers to hold at pressures greater than 2,000 psi. Thus, when displacing formation fluids during well kicks, the pipe rams should be used instead of the annular preventer if surface pressures exceed 2,000 psi. Blind rams or shear rams in a blowout preventer stack are generally just below the annular preventer and above the pipe rams.

Closing units for blowout preventers are an important part of the total control system. The closing units, or accumulators, as they are normally called, are sized to provide enough fluid volume to open and close all preventers in the stack. Remember that usable accumulator fluid volumes depend on total volume, precharge pressure, and pressure maintained on the fluid bottles. Consider the following example:

☐ **EXAMPLE 14.3**

Given:

> Total accumulator volume $= 160$ gal
> Precharge pressure $= 1,000$ psi
> Maximum bottle pressure $= 3,000$ psi

$$\text{Usable accumulator fluid volume} = \frac{2,000}{3,000}(160) = 107 \text{ gal}$$

If maximum pressure is reduced to 2,500 psi, the maximum usable fluid volume is

$$\frac{1,500}{3,000}(160) = 80 \text{ gal}$$

On high-risk wells, common practice is to use air volume tanks as a backup to accumulators. These backup systems should be checked regularly.

Emphasis is placed on the availability of accurate pressure gauges.

The spread between the control of entering formation fluids and losing circulation may be small. Quick-responding gauges are also needed when controlling the well.

BEHAVIOR OF GAS IN DRILLING FLUIDS

The emphasis on well control became necessary because of the behavior of gas. When displacing a gas kick, the gas must be permitted to expand or most of the formation pressure is transmitted directly to the surface. Excessive expansion may result in additional influx magnifying the problems. All well control procedures are dedicated to keeping the bottomhole pressure constant during displacement. With liquids, this same principle applies; however, fewer problems are encountered because the liquids are essentially incompressible.

This section on gas behavior defines the general behavior of gas, the migration of gas in drilling fluids, and the general procedures that should be used to compare actual with predicted behavior.

General Behavior of Gas

The accuracy required in predicting gas behavior depends on the objectives. If gas is being bought or sold, precise measurements are required. In drilling operations, precise measurements or predictions are not necessary. However, generalized predictions of gas behavior are useful in making decisions to minimize the likelihood of major problems. Some aspects of gas behavior are predictable. Others, such as the difference between a gas bubble and the same volume of gas strung out through the mud, are not.

Gas behaves ideally only at very low pressures. Gas behavior in drilling fluids never follows ideal behavior patterns. To illustrate this boundary condition, Eqs. 14.1 and 14.2 relate to ideal gas behavior.

$$PV = nRT \tag{14.1}$$

where:

P = pressure, psi
V = gas volume, cu ft
n = gas, lb-mols
R = units conversion constant
T = temperature, °R

$$\frac{P_1 V_1}{T_1} = \frac{P_2 V_2}{T_2} = \text{Constant} \tag{14.2}$$

Absolute temperatures are used in Eqs. 14.1 and 14.2. Eqs. 14.3 and 14.4 show how absolute temperature is obtained.

$$°R = 460 + °F \tag{14.3}$$

$$°K = 273 + °C \tag{14.4}$$

In Eq. 14.3, the absolute temperature is in degrees Rankine; in Eq. 14.4, the absolute temperature is in Kelvin. Degrees Rankine is used with the pound-foot-second system of units and Kelvin is used with the centimeter-gram-second unit system. In Eq. 14.2, the units of absolute temperature may be either degrees Rankine or Kelvin. Other useful conversions are shown in Eqs. 14.5 and 14.6.

$$°F = \frac{9}{5}°C + 32 \tag{14.5}$$

$$°R = 1.8°K \tag{14.6}$$

Temperature is sometimes omitted when considering maximum surface pressures, as shown by Eq. 14.7.

$$P_1V_1 = P_2V_2 = \text{Constant} \tag{14.7}$$

Under many conditions, Eq. 14.7 may suffice as an acceptable estimate of gas behavior, but remember that this equation is less accurate than Eq. 14.2 and that Eq. 14.2 is not very accurate. Even in drilling operations, some effort may be necessary to give a better representation of gas behavior. One such procedure is shown in the equations below:

$$PV = znRT \tag{14.8}$$

where:

z = gas compressibility factor

$$\frac{P_1V_1}{z_1T_1} = \frac{P_2V_2}{z_2T_2} = \text{Constant} \tag{14.9}$$

In Eqs. 14.8 and 14.9, the compressibility factor, z, is the deviation from ideal behavior. The z factor is determined empirically for specific gases and for gas mixtures. This discussion relates only to the generalized behavior of gas mixtures. In most cases, the type of gas that enters the wellbore is unknown, so assumptions must be made. The best estimate is to assume some specific gravity for the gas and then determine the generalized z factor. The specific procedures used to determine the z factor are listed here:

1. Assume a gas specific gravity—a good assumption is usually 0.6 to 0.7.
2. After assuming the gas specific gravity, determine the pseudocritical pressure and pseudocritical temperature from Fig. 14–1.
3. Use the pseudocritical temperature and pseudocritical pressure to calculate the reduced pressure and reduced temperature, as shown in Eqs. 14.10 and 14.11.

$$T_r = \frac{T}{T_{pc}} \tag{14.10}$$

where:

T_r = reduced temperature

$$T = \text{temperature, } °F$$
$$T_{pc} = \text{pseudocritical temperature, } °R$$

$$P_r = \frac{P}{P_{pc}} \qquad (14.11)$$

where:

$$P_r = \text{reduced pressure}$$
$$P = \text{pressure, psi}$$
$$P_{pc} = \text{pseudocritical pressure, psi}$$

4. Use Fig. 14–2 to determine the z factor. The procedure is illustrated in Example 14.4.

□   **EXAMPLE 14.4**

Assume:

$$\text{Temperature} = 250°F$$
$$\text{Pressure} = 10,000 \text{ psi}$$
$$\text{Gas specific gravity} = 0.6$$

Determine: z factor

SOLUTION:

From Fig. 14–3,
$$T_{pc} = 359°R$$
$$P_{pc} = 671 \text{ psia}$$
$$T = 250 + 460 = 710°R$$
$$T_r = \frac{710}{359} = 1.978°R$$
$$P_r = \frac{10,000}{671} = 14.903 \text{ psi}$$

From Fig. 14–4,

$$z = 1.40$$

To illustrate the effects of simplifications, Example 14.5 shows predicted surface pressures and volumes using Eqs. 14.2, 14.7, and 14.9.

□   **EXAMPLE 14.5**

Assume:

$$\text{Well depth} = 15,000 \text{ ft}$$
$$\text{Bottom-hole pressure} = 10,000 \text{ psia}$$
$$\text{Gas specific gravity} = 0.6$$
$$\text{Kick volume} = 20 \text{ bbl}$$
$$\text{Bottom-hole temperature} = 250°F$$

Determine:
(a) Surface pressure, assuming no gas expansion
(b) Surface volume, assuming atmospheric conditions of pressure and temperature

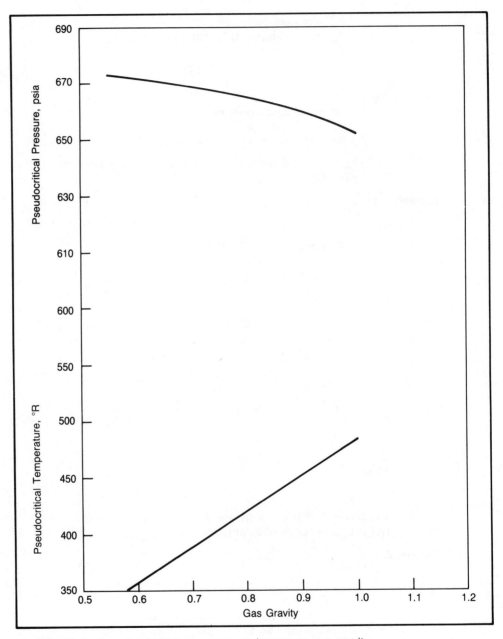

■ **FIG. 14–3** Pseudocritical temperature and pressure gas gravity

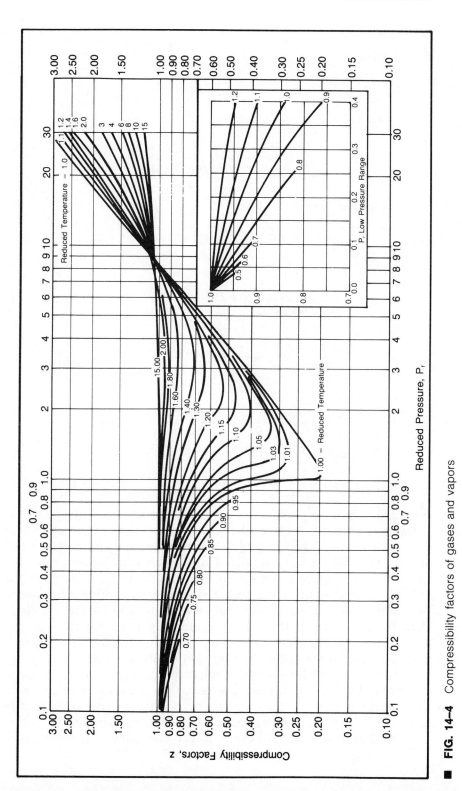

■ **FIG. 14–4** Compressibility factors of gases and vapors

SOLUTION:

(a) Using Eq. 14.2:

$$\frac{P_s V_s}{T_s} = \frac{P_{bh} V_{bh}}{T_{bh}}$$

where:

$$s = surface$$
$$bh = bottom\ hole$$

$$P_s = \frac{P_{bh} V_{bh} T_s}{V_s T_{bh}} = \frac{(10{,}000)(1)(540)}{710} = 7{,}606\ psia$$

Using Eq. 14.7:

$$P_s = \frac{V_{bh} P_{bh}}{V_s} = (1)(10{,}000) = 10{,}000\ psia$$

Using Eq. 14.9:

$$P_s = \frac{V_{bh}}{V_s} \frac{z_s}{z_{bh}} \frac{T_s P_{bh}}{T_{bh}} = \frac{(1)(1.0)(540)(10{,}000)}{(1.4)(710)} = 5{,}433\ psia$$

(b) Using Eq. 14.2:

$$V_s = \frac{P_{bh}}{P_s} \frac{T_s}{T_{bh}} V_{bh} = \frac{(10{,}000)(540)(20)}{(14.7)(710)} = 10{,}348\ bbl$$

Using Eq. 14.7:

$$V_s = \frac{P_{bh} V_{bh}}{P_s{}^2} = \frac{(10{,}000)(20)}{(14.7)} = 13{,}605\ bbl$$

Using Eq. 14.9:

$$V_s = \frac{P_{bh}}{P_s} \frac{z_s}{z_{bh}} \frac{T_s}{T_{bh}} V_{bh} = \frac{(10{,}000)(1.0)(540)}{(14.7)(1.4)(710)}(20) = 7{,}391\ bbl$$

Note from Example 14.5 that Eq. 14.7 introduces almost a 50% error as compared with Eq. 14.9. Drilling personnel often dismiss this difference by simply saying it represents a safety factor. Unfortunately, an additional safety factor may also be required, and a drilling operator may have penalized himself relative to the size of blowout preventer equipment and other surface control equipment required to control a high-pressure well. The best approach is always to calculate results as accurately as possible and then to use a reasonable safety factor. By ignoring the compressibility factor and, in some cases, the changes in temperature, the drilling industry places an additional predicted burden on themselves.

In actual fact, the gas will probably migrate as gas stringers through the mud, and the predicted gas behavior may be only theoretical. The importance of the calculation is that it represents the worst possible condition and is important when considering the next course of action.

Gas that enters a wellbore is never one pure gas. Most gases contain methane; however, others that may be present include offensive gases such as carbon dioxide and hydrogen sulfide. Methane alone is always in a gaseous state when it enters the wellbore and remains in a gaseous state as it moves up the annulus. This may not be true if other gases are present. A typical pressure-temperature diagram is shown in Fig. 14–5.

Anywhere inside the envelope in Fig. 14–5 part of the fluid is liquid and part is gas. Above the critical temperature, all the fluid is gas. The critical pressure is the pressure at the critical temperature. The fluid above the critical pressure and below the critical temperature may be either gas or liquid.

For pure gases, a typical pressure-temperature diagram is shown in Fig. 14–6.

A comparison of Figs. 14–5 and 14–6 shows that the behavior of a specific pure gas is affected by other gases present. The question of when a mixture of gases turns to liquid often arises. Actually, the only way this question can be answered is to prepare a phase diagram from a specific bottom-hole sample of the formation fluid. This can be done in a development area.

One reason for this gas behavior discussion is that there is a general belief that some fluids may enter the wellbore as liquids and turn to gas during displacement or enter the wellbore as gas and turn to liquid during displacement. Fig. 14–5 illustrates that this may in fact occur. One scary part of gas displacement is when carbon dioxide and hydrogen sulfide are present. Table 14–1 shows the critical temperature and pressure for water, carbon dioxide, and hydrogen sulfide.

Thus, considering pure forms of carbon dioxide and hydrogen sulfide, they are generally gases entering the wellbore in deep wells because the temperature is above the critical temperature. Again, in actual situations, fluid mixtures substantially affect the behavior of all pure materials, and in the wellbore there is no way to determine specific fluid behavior.

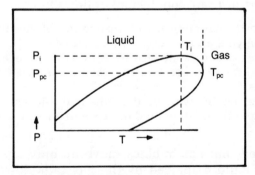

■ **FIG. 14–5** Pressure vs temperature for a mixture of gases

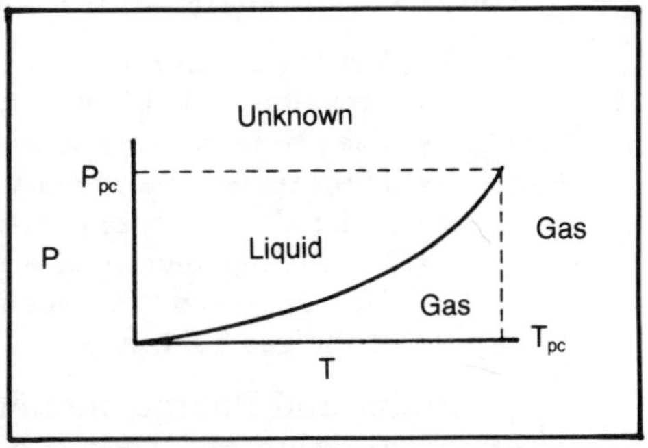

■ **FIG. 14–6** Pressure vs temperature, pure gas

■ **TABLE 14–1** Critical temperature and pressure

	T_{pc}, °F	P_{pc}, psi
Water	703	3,302
Carbon Dioxide	88	1,073
Hydrogen Sulfide	213	1,306

Both hydrogen sulfide and carbon dioxide are corrosive gases. If hydrogen sulfide is in a gaseous form, the hydrogen ion invades steel readily and failures occur. A combination of carbon dioxide and hydrogen sulfide is dangerous because, in water-based muds, hydrogen sulfide can be kept out of solution by maintaining the pH of the mud above 10. When carbon dioxide is present, the pH of the water-based mud may be reduced to substantially below 10 in one cycle of the mud.

This theoretical discussion says we do not know what will happen, and, unless a specific sample of the fluid has been analyzed, that is correct. However, our failure to accurately predict emphasizes the need to carefully observe everything during the displacement of formation fluids.

Most of this discussion was directed toward having a specific volume of fluid in bubble form and illustrates what might be expected from the gas present. All fluids, particularly gas, remaining in bubble form is unlikely, even if they were swabbed into the borehole. Certainly, when a well kicks during the drilling operation, the formation fluids are mixed into the drilling fluid. The migration of fluid through drilling mud depends on many conditions other than the composition of the formation fluid.

Theoretically, fluids segregate according to densities. Of equal concern is the time required for the segregation of different fluids, particularly when the formation fluid is gas.

Factors affecting the rise of a gas bubble were discussed by D.W. Radar and A.T. Bourgoyne, and a part of their conclusions follows:[1]

- A rising gas bubble in a vertical annulus travels up one side of the annulus with liquid backflow occupying an area opposite the bubble.
- The rising velocity of a gas bubble is affected by annular clearances.
- The greater the density difference between the gas and the liquid, the faster the gas bubble rises.
- The rising velocity of a gas bubble is reduced as the mud thickens.
- An increase in the velocity of the drilling fluid increases the rising velocity of the gas bubble.

Radar and Bourgoyne noted that gas bubbles move up only one side of the annulus and that the liquid displaced by the gas falls down the remaining annulus area. Based on this liquid fallback and the other factors listed above, they developed a mathematical approach to calculate the predicted rate of rise of the gas bubble. Their mathematical model is not presented, but it is a useful means for determining the magnitude of the effect of certain parameters on gas bubble rising velocity.

The rate of rise of the gas bubble can easily be determined in the field after a well kicks. A procedure for establishing the gas bubble's location and the rate it is rising is shown in the following example:

☐ **EXAMPLE 14.6**

Assume:

Shut-in annular pressure = 300 psi

Mud weight = 9.6 lb/gal = 0.50 psi/ft

Closed-in annular pressure increases to 400 psi in ½ hour with the well closed in

$$\text{Gas bubble has risen} \frac{(400 - 300)\ \text{psi ft}}{0.5\ \text{psi}} = 200\ \text{ft}$$

Rate of rise of gas bubble = 200 ft × 2/1 hr = 400 ft/hr

Radar and Burgoyne's paper pointed out that both the mud thickness and velocity contributed to gas bubble velocity. As the mud thickness increased, the rate of rise for the gas bubble was substantially reduced. As mud velocity increased, gas velocity increased. This increase in gas velocity is probably due to shear thinning of the mud.

Theory provides a means for evaluating the effects of certain parameters on gas behavior. By fitting theory to observed results in the field, practices that minimize the potential danger of disaster in field operations can be developed.

WELL CONTROL PRACTICES

Well control is divided into the following groupings:

- Controlling a well kick when the drillstring is on the bottom of the hole
- Controlling a well kick when the drillstring is off the bottom of the hole
- Specialized well control practices

Controlling a Well Kick When the Drillstring Is on the Bottom of the Hole

Two methods generally are used to control a well kick with the drillstring on the bottom. The driller's method, in which mud weight is unchanged during displacement, requires few mathematical calculations. The wait-and-weight method increases mud weight and takes more care during displacement.

With either displacement method, the basic theory is the same: keep constant pressure against the formation to prevent the further entry of formation fluids. Maintaining constant pressure against the formation is simplified by controlling the pressure on the drillpipe. A well is like a big U-tube. When control is achieved, the pressure on the inside of the drillstring equals the pressure in the annulus surrounding the drillstring.

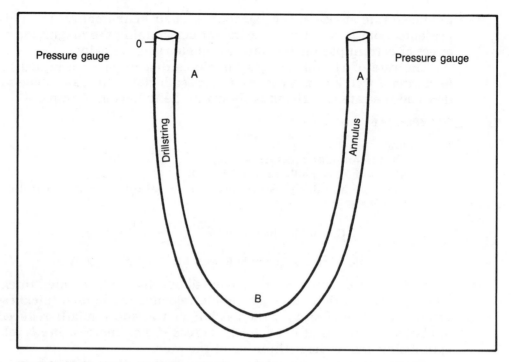

■ **FIG. 14–7** U-tube illustrating drillstring and annulus

To explain, note Fig. 14–7, which is a U-tube diagram representing the inside of the drillstring and the annulus.

If fluid of the same density is put into the U-tube, it will reach a constant level shown as *A*. This simply implies that what happens in one-half of the U-tube will be reflected in the other half. If the U-tube is filled with the same density fluid and pressure is applied, the gauge readings on both the drillstring and annulus will be identical.

If a constant-density fluid is put in the drillstring portion of the U-tube, two or more fluids of different densities are allowed to enter the annulus. Pressure is applied at point B; then the pressure gauges on the annulus and drillstring show different values. However, if the entire U-tube is shut in, the same total pressure is exerted at point B by the combinations of pressure and fluid on each side of the U-tube.

On the drillstring side, there is a fluid of known density and a pressure that can be read at the surface. The addition of the drillstring pressure reading and the pressure exerted by the known-density fluid in the drillstring give the pressure being exerted at point B.

In like manner, the addition of the annulus pressure to the pressure exerted by the fluids in the annulus will give the pressure being exerted at point B. Using the annulus side for this purpose is complicated by:

1. The operator does not know accurately the total fluid that has entered the annulus
2. Fluid type is difficult to recognize so the formation fluid might be a combination of gas, oil, and water
3. The hole may not be to gauge, thus the height of the formation fluid could not be determined even if an accurate measurement of fluid influx were possible

(This explains why the drillstring side of the U-tube is used.)

Eq. 14.12 shows how the pressure in each side of the U-tube can be equated:

$$P_d + P_{md} = P_a + P_{ma} + P_{ff} \qquad (14.12)$$

where:

P_d = shutin drillpipe pressure, psi
P_{md} = pressure exerted by mud inside drillstring, psi
P_a = pressure on annulus gauge at the surface, psi
P_{ma} = pressure exerted by mud in the annulus, psi
P_{ff} = pressure exerted by mass of formation fluid, psi

If the well kicks while drilling, the formation fluid follows the direction of fluid movement, which is up the annulus. If the well kicks when pulling the drillstring, the formation fluid enters the entire wellbore below the drillstring. When stripping the pipe back into the hole, the drillstring must be equipped with a back-pressure valve that prevents the fluid's entry into the drillstring. One exception might occur if, when pulling the pipe with no back-pressure valve, flow continued long enough for the formation fluids to enter both the drillstring and the annulus.

This exception would be recognized and would force the operator to guess the type of fluid that entered and the formation pressure. Common differentiations have been introduced for control methods. The simpler method is called the driller's method. Several names have been applied to the other method, but in this discussion it is called the engineer's method.

Driller's Method

The driller's method of well control is outlined below:

- Pull the kelly out of the blowout preventers
- Shut down the mud pump
- Close the blowout preventers
- Read and record shutin drillpipe and annular pressures
- Add the shutin drillpipe pressure to the circulating pressure required to overcome friction at the kill rate
- Pump at a constant pressure and circulation rate

Two common procedures are used to shut in a well: the soft shutin and the hard shutin. The *soft shutin* refers to a procedure in which the choke lines are open when the blowout preventers are closed. The *hard shutin* is a method in which the choke lines are closed when the blowout preventers are closed.

If there is less than 2,000 ft of casing set when the well kicks, then the soft shutin is generally implemented. With more than 2,000 ft of casing set, either shutin procedure may be used.

The soft shutin minimizes any impact force against the formation and preventers when the well is shut in. The hard shutin minimizes the size of the kick by shutting in the well quickly. With less than 2,000 ft of casing set, the soft shutin may be utilized until it is determined whether the well can be shut in without losing returns below the casing.

Some fear that the loss of returns below the casing may result in fluid flowing to the surface around the casing, which could turn into an uncontrollable disaster. When more than 2,000 ft of casing is set, any fluids flowing underground should not flow to the surface.

Shutin pressures can be read quickly in areas with a high formation permeability. In hard-rock areas, the shutin pressures may build slowly with time. At the same time, if the formation fluid that has entered the wellbore is gas, the gas may start rising in the annulus. The operator may have difficulty distinguishing between a rising gas bubble and a slow buildup of formation pressure. One procedure is shown in Fig. 14–8.

The estimated shutin pressures are assumed to be at the point the rapid pressure buildup becomes gradual. This may or may not be the correct shutin pressure to use to determine pore pressure, but it is probably as close as you can get. In any event, the control procedures are the same, regardless of the pore pressure estimate. Anticipate volumetric expansion of gas within reasonable limits. If the control pressures are inadequate, then increase the control pressures until further entry of formation fluids has been stopped.

The pressure required to overcome friction at the circulation rate selected to control the well may be determined in the following two ways:

1. Running the pump at the anticipated circulation rate that will be used

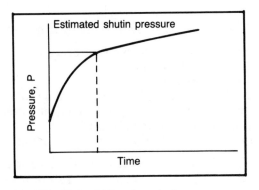

■ **FIG. 14–8** Estimating shutin pressure

to control a well kick and recording the pumping pressure is common practice. This low-circulating-rate pump pressure often is recorded every drilling tour; however, there are no specific requirements for this type of test.

2. Another procedure for determining the pumping pressure required to overcome friction at the kill rate is to conduct the measurement after a well kick occurs. This type of measurement is illustrated by Fig. 14–9.

Fig. 14–9 illustrates an approximate annular pressure profile for the displacement of a typical gas kick. The dropoff in annular pressure to the right of point A represents a shortening of the gas bubble when the gas enters the annulus around the drillpipe from the annulus around the drill collars or heavyweight drillpipe.

When the annular pressure starts to increase as shown, hold the annular pressure constant and establish the pumping pressure. This should take only two or three minutes. After the pumping pressure is established at the selected circulation rate, keep the pumping pressure constant and let the annular pressure continue to increase.

The pumping pressure required to overcome friction is the pumping pressure minus the shutin drillpipe pressure. The same procedure for establishing pumping pressure can be used if the operator wants to change circulating rates.

Using the driller's method, the circulating drillpipe pressure is held steady at a constant circulation rate. The circulation rate can be changed during the displacement of a gas kick. The procedure is simply to hold the annular pressure constant for two or three minutes while changing the circulation rate. Then keep the new circulation rate and new pumping pressure constant.

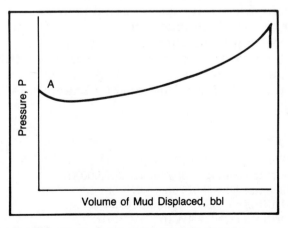

■ **FIG. 14–9** Pressure vs mud displaced

Wait-and-Weight Control Method

The first four steps of the driller's method are the same for the wait-and-weight method. In addition, the actions listed below are necessary.

- From shutin drillpipe pressure, determine the mud weight increase required to balance the pore pressure
- Raise the mud weight in the suction tank to balance the pore pressure and calculate the rate of weighting material additions necessary to increase the mud weight while pumping
- Determine the pump pressure required to overcome friction at the selected pumping rates for both the unweighted and weighted mud slurries
- Prepare a pumping schedule to reduce pump pressure as weighted mud enters the drillstring
- After weighted mud reaches the bit, keep the pumping pressure constant at the selected pumping rate until formation fluids have been displaced and the mud weight of the total mud system has been raised to the same level

Increasing mud weight just enough to balance the pore pressure is normal because another complete pumping cycle of the mud generally is required to condition the mud before drilling is continued. If drilling is to continue, the mud weight will be increased enough to continue drilling on the second pumping cycle. In actual fact, the precise procedure depends on the operator's preference.

The pump pressure required to overcome friction is determined as previously demonstrated. If the pumping rate is to be changed and a new pumping pressure established, do these after the weighted mud reaches the bit.

When pumping starts with the weighted mud, the pump pressure is reduced as weighted mud goes down the drillstring because the hydrostatic pressure is increased. The total reduction in pump pressure should equal the increase in hydrostatic pressure minus the additional pressure required to overcome friction for the weighted mud.

After the weighted mud reaches the bit, the pumping pressure should remain constant if the pumping rate stays the same. This is true because there will be no further change in the drillstring side of the well.

The pumping procedures for both the driller's method and the wait-and-weight method are illustrated by Example 14.7.

□   **EXAMPLE 14.7**

Assume:

> Well depth = 10,000 ft
> Surface casing = 9⅝ in. set at 2,500 ft
> Casing I.D. = 8.921 in. (0.077 bbl/ft)
> Drilling in 8½-in. hole at 10,000 ft
> Drillpipe size = 4½ in., 16.6 lb/ft
> Drill collar size and length = 6¼-in. O.D. and 625 ft

Mud weight = 9.6 lb/gal
Fracture gradient at 2,500 ft = 0.80 psi/ft

The well kicks, and the following information is recorded:

Pit level increase = 20 bbl
Shutin drillpipe pressure = 260 psi
Shutin annular pressure = 500 psi
Volume of 8½-in. hole = 0.07 bbl/ft
Volume of 4½-in. drillpipe − 8¼-in. hole annulus = 0.05 bbl/ft
Volume of 6¼-in. drill collar − 8½-in. hole annulus = 0.032 bbl/ft
Volume of 4½-in. drillpipe − 9⅝-in. casing annulus = 0.057 bbl/ft
Internal volume of 4½-in. drillpipe = 0.014 bbl/ft
Internal volume of 6¼-in. drill collars = 0.007 bbl/ft
Temperature gradient = 70°F + 1.2°F/100 ft

General Information:

Pore pressure = 0.5 psi/ft(10,000 ft) + 260 psi = 5,260 psi
Pressure required to overcome friction at a kill rate of
3.0 bbl/mi = 500 psi
Kick volume = 20 bbl

Height of the formation fluids in the annulus $= \dfrac{20 \text{ bbl ft}}{0.032 \text{ bbl}} = 625$ ft

$$5{,}260 \text{ psi} = 500 \text{ psi} + (10{,}000 - 625) \text{ ft} \times 0.5 \text{ psi/ft} + 625 \text{ ft} \times \rho_f \text{ psi/ft}$$

$$625 \, \rho_f = 72 \text{ psi}$$

$$\rho_f = 0.115 \text{ psi/ft}$$

This is a gas gradient, and the entire 20 bbl of influx will be considered gas.

Driller's method

Pumping pressure = 260 + 500 = 760 psi
Keep pumping rate constant and pumping pressure constant at 760 psi
until the gas that entered the wellbore has been displaced.
The annular profile for this procedure is shown in Fig. 14–10.

Wait-and-Weight Method

Mud weight required to control well $= \dfrac{(5{,}260)(19.25)}{10{,}000} = 10.13$ lb/gal

Use mud weight = 10.2 lb/gal
Pressure gradient of weighted mud $= \dfrac{10.2}{19.25} = 0.53$ psi/ft

Pressure required to overcome friction with 10.2 lb/gal mud:

$$500 \left(\frac{10.2}{9.6} \right)^{0.8} = 525 \text{ psi}$$

Internal volume of drillstring = 136 bbl
Pumping rate = 3 bbl/min at 50 strokes/min (spm)

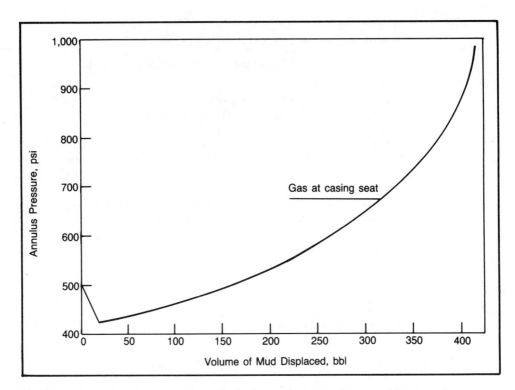

■ FIG. 14–10 Annular pressure profile for Example 14.7 using unweighted mud

Reduce pump pressure until weighted mud reaches the bit = 760 − 525 = 235 psi
Pumping time for weighted mud to reach the bit = 136/3 = 44 min or 2,200 pump strokes
The pumping schedule is shown in Table 14–2.

■ TABLE 14–2 Pumping schedule

Time, min	Number of Pump Strokes	Pump Pressure, psi
0	0	760
10	500	707
20	1,000	654
30	1,500	601
40	2,000	548
44	2,200	525

Fig. 14–11 illustrates the annular pressure profile for this example.

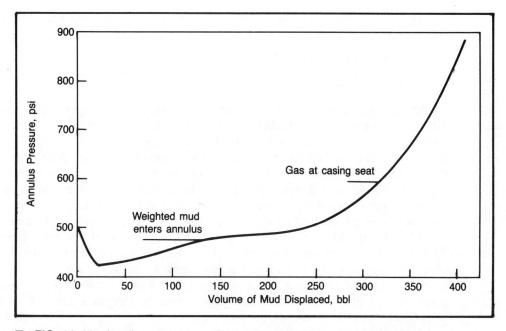

■ **FIG. 14–11** Annulus pressure profile for Example 14.7 using weighted mud

CONTROLLING A WELL KICK WHEN THE DRILLSTRING IS OFF THE BOTTOM OF THE HOLE

The control of well kicks is complicated when the pipe is off the bottom of the hole. Procedures that may be used depend on the potential problems. If the pipe is off the bottom in an open hole with permeable zones exposed, strip the pipe back to bottom. If the pipe is in a cased hole or in an open hole with essentially no permeable formations exposed, an operator has more options. Remember that some rigs and some operators are not prepared to strip pipe back into the hole; therefore, other choices must be made. Procedures that may be used to maintain control of the well, when the pipe is off the bottom, include the following:

1. Strip or snub the pipe back to the bottom.
2. Push the formation fluid that entered the wellbore back into the formation.
3. Control the well by adjusting surface pressure without circulation.
4. Control the well by increasing mud weight and displacing the weighted mud with the pipe in place.

Strip or Snub the Pipe Back to the Bottom

Pipe is stripped through the blowout preventers if the weight of the pipe is sufficient to overcome the support given by wellbore pressure. If the pipe weight is insufficient to pull the pipe into the hole, then the pipe may have to be snubbed or pushed into the hole.

The drillstring may be stripped or snubbed into the hole using the following preventers:

- Annular preventer only
- Combination annular preventer and ram preventer
- Two ram preventers

In general, the annular preventer is used for stripping pipe into or out of the hole. The annular preventer allows the use of one preventer and permits the tool joints to pass through the packing element. To minimize wear, bleed a small amount of fluid around the pipe during stripping operations. The escape of fluid can be observed if the preventer can be seen during stripping operations; however, the problem is more complicated if the preventer is below water. Many of the annular preventers use wellbore pressure to help keep the preventer closed. Thus, to bleed fluid by the packing element might require the use of opening pressure. This pressure must be adjusted as the wellbore pressure changes. In general, the singular use of annular preventers is limited to a maximum wellbore pressure of 1,500 psi. This is an arbitrary maximum that is routinely used, and conditions require that pipe could be stripped through annular preventers at much higher pressures.

A combination annular and ram-type preventer may be used to help prolong the life of the packing element in the annular preventer. With this combination, each preventer is opened to allow the passage of the tool joint. Again, this type of operation is simplified if the blowout preventers can be observed. Below water the operator must know the pipe measurements and tool joint location.

The use of two ram-type preventers for stripping operations is generally confined to conditions in which excessive wear has been experienced with annular preventers or at wellbore pressures above 1,500 psi. The stripping procedure is one in which one preventer is closed and the other preventer is opened to pass the tool joint. This is a slow procedure and, if implemented, there should always be a ram-type preventer below those used for stripping. The ram-type preventer not utilized will permit shutting-in of the well if excessive wear occurs on the packing elements of the preventers used in the stripping operation.

Regardless of the method for stripping pipe into the hole, measure all of the fluid that comes out of the wellbore. Formation fluid that enters the well-bore may be gas; during the stripping operation, an operator needs to watch for gas migration. If there is no migration of fluid, the volume of mud released from the wellbore should equal the pipe displacement. If fluid migrates, let the gas expand to prevent excessive surface pressures.

If the fluid released is measured, it is possible to determine if fluid is migrating. For example, if there is no fluid migration, the surface pressure should remain the same until the pipe reaches the formation fluid, providing the fluid volume released equals the pipe volume. When the pipe reaches the formation fluid that entered the wellbore, the surface pressure increases if the formation fluid weighs less than the drilling fluid. The surface pressure increases because the formation fluid cannot enter the drillstring, and the height of the formation fluid increases. If there is fluid migration, the surface pressure increases, even though the volume of drilling fluid released at the surface is exactly equal to the pipe displacement. In the case of fluid migration, precautions must be taken to prevent excessive surface pressures.

If problems occur while drilling, decisions to solve the problems must be made quickly and the chosen procedures must be implemented as soon as possible. If pipe stripping or snubbing is an alternative when a well kicks with the pipe off bottom, everything should be arranged to implement the stripping operation quickly. This means there should be a way to accurately measure any mud volumes released, accurate gauges should be available, pressure regulators on blowout preventers should be in good working order, and the equipment and procedures to be used should be known and practiced.

To illustrate a procedure for stripping pipe, consider Example 14.8. The first procedure illustrates the stripping operation with no migration, and the second procedure illustrates the stripping operation if fluid migration occurs.

☐   **EXAMPLE 14.8**

Well depth = 14,000 ft
9⅝-in. protective casing set at 11,000 ft
Fracture gradient at 11,000 ft = 0.90 psi/ft
8½-in. hole to 14,000 ft; volume of hole = 0.07 bbl/ft
Temperature gradient = 70°F + 1.2°F/100 ft
Drillstring: 4½-in. drill collars
                    270 ft of 7-in. drill collars
                    900 ft of 5-in. heavyweight drillpipe
Drilling fluid: Weight = 16.0 lb/gal
Well kicks while pipe is being pulled at 12,000 ft
Estimated pit level increase = 20 bbl
Shutin drillpipe and annulus pressure = 200 psi

Determine: The type of fluid that entered the wellbore.
Show: (a) The stripping procedure if no migration of formation fluid occurs
         (b) The stripping procedure if the formation fluid migrates

SOLUTION:

Determine the type of fluid that entered the wellbore. First, assume that the formation pore pressure at 14,000 ft is slightly under the hydrostatic pressure of the mud column. The gradient of gas is determined as follows:

$$\rho_f = \frac{SP_p}{53.3zT_R}$$

where:

$$S = \text{specific gravity of gas, dimensionless}$$
$$P = \text{pressure at any point, psi}$$
$$T_R = \text{temperature, °R}$$

Assume:

P_p = 11,648 psi, which is equal to the gradient of 16-lb/gal mud
T_R = 70°F + (1.2)(140) = 238°F = 698°R
Gas specific gravity = 0.7

Then

T_{pc} = 668 psia
T_{pc} = 390°R

$$P_r = \frac{11,648}{668} = 17.4 \text{ psi}$$

$$T_r = \frac{698}{390} = 1.79°R$$

z = 1.75

$$\rho_f = \frac{(0.7)(11,648)}{(53.3)(1.75)(698)} = 0.125 \text{ psi/ft}$$

Length of formation fluid column $= \dfrac{20 \text{ bbl/ft}}{0.07 \text{ bbl}} = 286 \text{ ft}$

Pressure exerted by mass of formation fluid = (286)(0.125) = 36 psi

Based on shutin pressures, the pressure exerted by the formation fluid can be determined as follows:

$$11,648 = (14,000 - 286)0.832 + 200 + P_{ff}$$

$$P_{ff} = 11,648 - 11,610 = 38 \text{ psi}$$

This shows the formation fluid to be gas, as assumed.

(a) *Show procedure for stripping pipe if no gas migration occurs:*
In this case, assume that no migration occurs. The shutin pressure and the amount of fluid released must be carefully observed as the pipe is stripped into the hole. If no gas migration occurs, the surface pressure should remain constant if the amount of mud released is equal to the pipe displacement. When the pipe enters the gas, the surface pressure increases because the pipe displacement increases the length of the gas column. The annular volumes for the different pipe sizes are as follows:

4½-in. hole = 0.05 bbl/ft
Heavyweight hole = 0.041 bbl/ft
Drill collar hole = 0.022 bbl/ft

Pipe displacements for the various sizes of pipe follow:

Drill collars = 0.048 bbl/ft
Heavyweight pipe = 0.029 bbl/ft
Drillpipe = 0.020 bbl/ft

270 ft $\times$ 0.048 bbl/ft = 13 bbl
$\times$ 0.029 bbl/ft = 0.5 bbl

Total fluid displaced up the annulus by pipe
= 13 + 0.5 = 13.5 bbl $\dfrac{13.5 \text{ bbl ft}}{0.041 \text{ bbl}}$ = 329 ft
329 ft $\times$ (0.832 − 0.100) psi/ft = 241 psi
Note: 0.100 is the assumed gradient of the gas in psi/ft

The 241 psi is the potential increase in surface pressure caused by increasing the length of the formation fluid. During the stripping operations, the shutin annular pressure increases to 441 psi. In actual operations, the exact increase in annular pressure may vary because the formation fluids may mix with mud during the stripping operations. An operator must know about this potential increase in annular pressure. Otherwise, he may suggest corrections that could be hazardous. The actual increase in annular pressure due to the fluid displacement of the pipe can be determined by measuring the fluid displaced. The actual fluid displaced gives a record of the increase in annular length of the formation fluid. From the increase in the annular length of the formation fluid, determining the increase in annular pressure is simple.

(b) *The stripping procedure if the formation fluid migrates*
The gas will probably migrate during the stripping operation. If the pipe is stripped into the hole holding annular pressure constant and releasing a volume of mud equal to the pipe volume and gas migration is occurring, gas will enter the hole. If a volume of mud equal to the pipe volume is released and the annular pressure increases before the pipe reaches the gas, gas migration is indicated. Corrections can be made for gas migration, and the stripping procedure can be modified to allow the necessary gas expansion. To permit accurate modifications, it is necessary to maintain control of the mud released. For Example 14.8, assume that the pore pressure at 14,000 ft is 11,648 psi. In fact, the pore pressure is slightly less than 11,648 psi because the formation gas entered the wellbore after the pipe was pulled.

The fracture gradient at 11,000 ft is 0.91 psi/ft. Because the pressure gradient exerted by the 16.0 lb/gal mud is equal to 0.832 psi/ft, the total annulus pressure that can be maintained without losing circulation is 858 psi:

$$(0.91 - 0.832) \, 11,000 = 858 \text{ psi}$$

A knowledge of the maximum permissible pressure emphasizes the need to be careful; however, it may be necessary in some cases to exceed surface pressure limitations. In this instance, assume that the pipe is being stripped into the hole while controlling the annulus pressure and measuring the volume of mud released. Table 14.5 shows one procedure that may be used. Remember that, without gas migration, the annulus pressure increased to 441 psi because of the increase in the length of the gas column.

To begin the stripping operations, the surface pressure increases to 500 psi before releasing any mud in excess of the pipe volume. When mud in excess of the pipe volume is released, determining the exact location of the gas may be difficult. One procedure is determining the effective rise by the increase in surface pressure. This procedure is shown in Table 14–3:

■ **Table 14–3** Stripping procedure, assuming gas migration

Pore Pressure, psi	Shutin Annular Pressure, psi	Pipe in Hole, ft	Pipe in Hole, bbl	Mud Released, bbl	Length of Gas, ft	Top of Gas Above Bottom, ft	Reduction in Hydrostatic Pressure, psi	Borehole Pressure, psi
11,648	200	0	0.0	0.0	286	286	0	11,648
11,648	400	300	5.9	5.9	286	526	0	11,848
11,648	500	400	7.9	7.9	286	646	0	11,948
11,648	700	500	9.8	9.8	286	886	0	12,148

Reduce borehole pressure by bleeding off the mud and letting the gas expand. Bleed off 0.75 bbl of mud.

11,648	300	500	9.8	10.55	297	897	9	11,739
11,648	600	742	14.7	15.45	297	1,258	9	12,039

The bit has reached the top of the gas at this point. Further stripping will result in a more rapid rise in annular pressure. Bleed off another 0.5 bbl of mud.

11,648	300	742	14.7	15.95	320	1,281	25	11,722
11,648	750	1,012	19.9	21.15	613	1,597	248	11,965
11,648	800	1,312	25.8	27.05	513	1,774	160	12,085

Bleed off 0.5 bbl of mud or the amount required to reduce the annular pressure to 550 psi.

11,648	550	1,312	25.8	27.55	523	1,784	167	11,803
11,648	650	2,000	39.3	41.05	523	1,904	167	11,903

Several significant points are noted from Table 14–3:

- Only 1.75 bbl of mud had to be bled off to get the drillstring to the bottom of the hole.
- Increasing the length of the gas bubble when stripping the pipe into the hole could cause an underground blowout.
- Changes in annulus size make accurately assessing the effect of letting the gas bubble expand difficult.
- This procedure is hard to understand.

The choke line should discharge into a small tank so the volumes of mud released can be measured precisely. Discharging mud into a regular mud tank where 1 in. of height represents 4 bbl of mud makes it almost impossible to strip the drillstring into the hole properly.

The small amount of mud released to allow gas expansion can easily be a problem. If too much mud is released, the borehole pressure may drop enough to allow additional formation fluids to enter the wellbore. The example illustrated in Table 14–3 does not require narrow limits on annular pressure levels; however, there may be cases when there is little tolerance between additional fluid entry and lost returns.

When pipe enters the gas while stripping, the increased length of the gas bubble reduces hydrostatic pressure in the borehole. This reduction in hydrostatic pressure forces the annular pressure to rise to prevent further entry of formation fluids. The increased annular pressure with a full hydrostatic head of mud may cause a fracture in a formation above the gas. As a result, determine with a gas kick if it is safe to strip the pipe back into the hole. If it is unsafe to strip in the pipe, then the gas bubble should be allowed to rise without this stripping in.

The changes in annular size increase the problems of pressure control with a rising gas bubble. That is the primary reason for the column showing the top of the gas bubble. In field operations, the gas can possibly be located by the change in pressure when the mud is released. In deep high-pressure wells, recognizing the gas bubble location by changes in surface pressure when the mud is released will be difficult because of the small mud volumes involved.

Unquestionably, the precise procedure for stripping pipe with a rising gas bubble is difficult in field operations. The problem is simplified if the range of pressures between fracturing a formation and further fluid entry is several hundred pounds per square inch.

Pushing the gas bubble back into the formation should be considered first if presence of hydrogen sulfide gas is indicated. In Example 14.7 the maximum surface pressure that can be exerted without fracturing just below the casing seat is 858 psi. The procedure for pushing gas back into the formation is as follows:

1. Pump mud slowly into the drillstring until the pressure reaches 500 psi or any pressure below the 500-psi level that is required to push the gas back into the formation.
2. If the pressure is still increasing at 500 psi, stop pumping and observe. If gas is going back into the formation, the pressure will decrease.
3. Measure the mud displaced into the hole. After the annulus pressure stabilizes, repeat the procedure.
4. Twenty barrels of mud must be pumped into the hole to replace the 20 bbl of gas influx. Even if only 15 bbl of mud were displaced into the hole, the surface pressure would be reduced and potential problems minimized.

Another alternative is to simply control the well without circulation or stripping. While this may not be a long-range solution to the problem, this procedure may have to be used while decisions are made on the next course of action. A well should never be shut in without some plan for controlling wellbore pressure until some other action is taken. In Example 14.8, the procedure for controlling the well is outlined as follows:

1. Watch the annular pressure, and time the increase in annular pressure. For example, if the annular pressure increases from 200 psi to 300 psi in 30 min, the rate of rise of the gas can be determined as follows:

$$\frac{(300 - 200)\text{ psi ft}}{0.832\text{ psi}} \times \frac{2}{1}\text{ hr} = 240\text{ fph}$$

2. Next, determine the pressure increase that would occur if the gas migrates to 12,000 ft with no expansion permitted:

$$2,000\text{ ft} \times 0.832\,\frac{\text{psi}}{\text{ft}} = 1,664\text{ psi}$$

This calculation shows that, if the well were shut in and no adjustments were made in shutin pressure during the shut-down time, the annular pressure could increase by 1,664 psi. Based on the maximum permissible annular pressure of 858 psi, an underground blowout would occur before the gas reached 12,000 ft.

3. One problem in releasing mud is knowing the reduction in hydrostatic pressure per barrel of mud released. This is no problem in the open hole below the pipe but is a problem when the gas reaches the pipe at 12,000 ft and starts to migrate up the annulus. Two procedures can be used to determine when the gas has reached the pipe at 12,000 ft:

   (a) If there is no back-pressure valve in the drillstring, the annular pressure should exceed the drillpipe pressure after the gas passes the pipe at 12,000 ft. This assumes mud is being released in the annulus.
   (b) The top of the gas can be determined by keeping a record of the feet the gas bubble has risen and the feet the gas bubble has expanded.

The migration of gas under static conditions is easy to follow if it is possible to read the drillpipe pressure. If drillpipe pressure cannot be read, then the gas must be followed as shown in Table 14-4. As shown, there were only 1.5 bbl of mud in excess of the pipe volume released when the gas reached the bottom of the drillstring. Also, the reduction in hydrostatic pressure was only 18 psi. Thus, deep wells can have a 50–100 psi safety factor, and the reduction in hydrostatic pressure can be forgotten until the gas gets closer to the surface.

This simply means that the controls shown in Table 14-4 need not be very precise, and specific volume measurements may be unnecessary. The need to control surface pressures by allowing the gas to expand

■ **TABLE 14–4**   Gas migration under static conditions

Pore Pressure, psi	Shutin Annular Pressure, psi	Mud Released, bbl	Length of Gas, ft	Top of Gas Above Bottom, ft	Reduction in Hydrostatic Pressure, psi	Borehole Pressure, psi
11,648	200	0.00	286	286	0	11,648
11,648	700	0.00	286	887	0	12,148

Bleed off 0.75 bbl of mud.

Pore Pressure, psi	Shutin Annular Pressure, psi	Mud Released, bbl	Length of Gas, ft	Top of Gas Above Bottom, ft	Reduction in Hydrostatic Pressure, psi	Borehole Pressure, psi
11,648	270	0.75	297	898	9	11,709
11,648	750	0.75	297	1,475	9	12,189

Bleed off 0.75 bbl of mud.

Pore Pressure, psi	Shutin Annular Pressure, psi	Mud Released, bbl	Length of Gas, ft	Top of Gas Above Bottom, ft	Reduction in Hydrostatic Pressure, psi	Borehole Pressure, psi
11,648	330	1.50	308	1,486	18	11,760
11,648	758	1.50	308	2,000	18	12,188

Gas is at the bottom of the drillstring.
Bleed off 0.75 bbl of mud.

Pore Pressure, psi	Shutin Annular Pressure, psi	Mud Released, bbl	Length of Gas, ft	Top of Gas Above Bottom, ft	Reduction in Hydrostatic Pressure, psi	Borehole Pressure, psi
11,648	376	2.25	342	2,034	46	11,777
11,648	800	2.25	342	2,544	46	12,000

Bleed off 0.50 bbl of mud.

Pore Pressure, psi	Shutin Annular Pressure, psi	Mud Released, bbl	Length of Gas, ft	Top of Gas Above Bottom, ft	Reduction in Hydrostatic Pressure, psi	Borehole Pressure, psi
11,648	300	2.75	365	2,567	62	11,740
11,648	800	2.75	365	3,168	62	12,210

Bleed off 0.75 bbl of mud.

Pore Pressure, psi	Shutin Annular Pressure, psi	Mud Released, bbl	Length of Gas, ft	Top of Gas Above Bottom, ft	Reduction in Hydrostatic Pressure, psi	Borehole Pressure, psi
11,648	426	3.50	383	3,186	77	11,820

is self-evident. Although, as shown, the gas expansion in deep wells is minimal, the surface pressures may rise rapidly if the gas is not permitted to expand.

The final alternative suggested for the problem presented in Example 14.8 was to increase mud weight by circulating a heavier mud into the hole. All kinds of problems exist with this procedure, particularly if the gas bubble is rising during the displacement of the heavy mud. Mud weight has been increased in similar situations; during displacement of the heavy mud, the gas rose and no adjustments in annular pressure were made.

In all probability, an underground blowout will result before all of the heavy mud is displaced. When this happens, the operator must shoulder some of the responsibility because circulation should never be commenced without some way of determining if the gas bubble is rising and the course of action if it is.

The procedure for increasing the mud weight to kill the pressure without moving the pipe is shown below:

1. First, the increase in mud weight required must be determined. The mud weight increase ought to be equal to a hydrostatic pressure increase, in this particular case, of 300 psi.

$$\text{Mud weight} = \frac{300}{(12{,}000)(0.052)} = 0.48 \text{ lb/gal} \approx 0.5 \text{ lb/gal}$$

2. While increasing the mud weight, watch the surface pressures. If the surface pressure increases, the gas bubble is rising and this procedure should not be used.
3. If there is no increase in annular pressure while increasing mud weight, the gas bubble is not rising, and operations can proceed as shown in step 4.
4. After increasing the mud weight in the surface mud tank to 16.5 lb/gal, begin displacement holding the annular pressure constant at 200 psi.
5. The annular pressure should remain at 200 psi at a constant choke size and circulation rate until the 16.5 lb/gal mud reaches the bit. If the annular pressure starts to increase with the constant choke size and circulation rate, the gas bubble is rising. Then the circulation should be stopped and adjustments made to allow expansion of the gas bubble.
6. When the 16.5 lb/gal mud enters the annulus, the 200 psi on the annulus starts down at the constant choke size and flow rate. The annular pressure should be zero when the mud fills two-thirds of the annulus. The annular volume is 684 bbl, 0.057 × 12,000. When 456 bbl of mud have been displaced after the weighted mud enters the annulus, the annular pressure should be zero. If the annular pressure does not reduce to zero when 456 bbl of 16.5 lb/gal mud have been displaced into the annulus, the gas bubble is rising. If the gas bubble starts rising at this point, the 16.5 lb/gal mud should be displaced completely; the gas bubble should be allowed to expand enough to maintain a constant borehole pressure. The procedure for this is the same as that shown in Tables 14–3 and 14–4.

The remark, "My company would never do that" is not an uncommon one. Remember that many alternative solutions to a problem may be necessary and that a final decision on the course of action may be based on the least undesirable alternative rather than the most desirable one. In any event, never wait until a potentially hazardous condition exists to choose contingency plans. In most cases, the order of application of any group of plans is determined after the problem occurs.

SPECIALIZED WELL CONTROL PRACTICES

The practices discussed in this section include the following:

- Methods for determining annular pressures while displacing gas kicks
- Methods for controlling a gas kick when the pipe is out of the hole
- Constant annulus-constant drillpipe control methods
- Variable circulation rates while displacing a gas kick
- Blowout through the drillstring
- Momentum control procedure if the drillstring is off the bottom of the hole and the well is flowing

Methods for Determining Annular Pressures While Displacing Gas Kicks

Eq. 14.13 can be used to calculate annular pressures while displacing a gas kick with unweighted mud.

$$P = \frac{A}{2} + \left[\frac{A^2}{4} + \frac{\rho_m \, P_b \, z T_{ab} h_b}{z_b \, T_b} \right]^{1/2} \qquad (14.13)$$

where:

P = pressure at the top of gas column, psi
$A = P_p - (D_g - X)\rho_m - P_{ff}$
P_p = pore pressure, psi
D_g = depth at which gas entered the wellbore, ft
X = distance from the surface to the top of the gas bubble, ft
ρ_m = pressure gradient of unweighted mud, psi/ft
P_{ff} = pressure exerted by mass of formation gas that entered the wellbore, psi
T_{ab} = temperature in absolute units at the top of the gas
T_b = bottom-hole temperature in absolute units at the point gas entered the wellbore
h_b = height of gas at the bottom of the hole, ft

Note: Eq. 14.13 does not allow for a change in annulus size. Thus, when changing annulus size, h_b should be changed to conform with the new annulus size. Also, P_{ff} must be altered to conform to a change in annulus size.

The use of Eq. 14.13 is illustrated by Example 14.9.

☐   **EXAMPLE 14.9**

Assumed conditions:

Well depth = 11,000 ft
Surface casing = 9⅝ in. at 3,000 ft
Casing I.D. = 8.92 in.
Hole size = 8½ in.
Drillstring assembly = 700 ft of 6¼-in. drill collars
                = 10,300 ft of 4½-in. drillpipe
Temperature gradient = 70 + 1.2°F/100 ft
Fracture gradient = 0.74 psi/ft just below casing seat at 3,000 ft
Mud weight = 11.0 lb/gal

Well kicks:

Shutin drillpipe pressure = 200 psi
Shutin annulus pressure = 450 psi
Pit level increase = 20 bbl
Assume gas specific gravity = 0.6

Determine:

(a) The maximum pressure at 3,000 ft when the gas first reaches that point
(b) The maximum surface pressure

SOLUTION:

(a) Hydrostatic pressure = (0.571)(11,000) = 6,285 psi
Pore pressure = 6,285 + 200 = 6,485 psi
Annulus volume around drill collars = 0.032 bbl/ft
$$h_b \text{ around drill collars} = \frac{20 \text{ bbl}}{0.032 \text{ bbl/ft}} = 625 \text{ ft}$$
$6,485 = (11,000 - 625)(0.571) + 350 + P_{ff}$
$6,481 = 5,924 + 450 + P_{ff}$
$P_{ff} = 107$ psi around drill collars
Annulus volume around drillpipe = 0.05 bbl/ft
$$h_b \text{ around drillpipe} = 20 \frac{\text{bbl}}{0.05 \text{ bbl/ft}} = 400 \text{ ft}$$
$$P_{ff} = 107 \frac{0.032}{0.050} = 64 \text{ psi}$$
$A = P_p - (D - X) \rho_m - P_{ff} = 6,485 - (11,000 - 3,000)(0.571) - 64 = 1,853$ psi

Eq. 14.13 can determine the pressure at 3,000 ft when the top of the gas column first reaches that point.

$$T_{3,000} = 70 + \frac{(1.2)(3,000)}{100} = 106°F + 460 = 566°R$$

$$T_{11,000} = 70 + \frac{(1.2)(11,000)}{100} = 202°F + 460 = 662°R$$

From example 14.8:

$T_{pc} = 359°R$
$P_{pc} = 671$ psia

At 3,000 ft:
$$T_r = \frac{566}{359} = 1.577$$

Estimate $P_{3,000} = 2,200$ psi
$$P_r = \frac{2,200}{671} = 3.28$$
$z = 0.8$
At 11,000 ft:

$$T_r = \frac{662}{359} = 1.844$$

$$P_r = \frac{6,485}{671} = 9.665$$

$$z_b = 1.15$$

$$P_{3,000} = \frac{1,853}{2} + \left[\frac{(1,853)^2}{4} + \frac{(0.571)(6,485)(0.80)(566)(400)}{(1.15)(662)}\right]^{1/2}$$

$$P_{3,000} = 927 + 1,319 = 2,246 \text{ psi}$$

$$\text{Pressure gradient at 3,000 ft} = \frac{2,246 \text{ psi}}{3,000 \text{ ft}} = 0.749 \frac{\text{psi}}{\text{ft}}$$

$$h = \frac{(6,485)(0.80)(566)(400)}{(2,246)(1.15)(662)} = 687 \text{ ft}$$

This is an expansion of $(687 - 400) = 287$ ft and equals a pit level increase of $(287)(0.05) = 14.4$ bbl

(b) Maximum surface pressure, $X = 0$:

$$A = 6,485 - (11,000 - 0)0.571 - 64 = 140 \text{ psi}$$

$$P_{\text{surface}} = \frac{140}{2} + \left[\frac{140^2}{2} + \frac{(0.571)(6,485)(1.0)(530)(400)}{(1.15)(662)}\right]^{1/2}$$

$$P_{\text{surface}} = 70 + 1,018 = 1,088 \text{ psi}$$

$$h \text{ at surface} = \frac{P_p z T_{ab} h_b}{P z_b T_b} = \frac{(6,485)(1)(530)(400)}{(1,088)(1.15)(662)} = 1,600 \text{ ft}$$

$$3,000 - 1,660 = 1,340 \text{ ft of mud above the casing seat}$$

Total pressure at casing seat:

$$(1,088) + (1,340)(0.571) + 44 = 1,897 \text{ psi}$$

These calculations show the theoretical conditions if the gas is allowed to migrate to the surface or if it is displaced holding the borehole pressure constant. Assuming these conditions actually exist during displacement, circulation would be lost before the gas reached the surface casing shoe. If this situation were real, the operator would have to decide on the next course of action. Some of his alternatives are listed below:

1. Displace with the unweighted mud and hope the formation gas does not expand as much as the calculations show or hope the formation fracture gradient is higher than indicated
2. Increase the mud weight enough to control the formation pore pressure before beginning displacement of the formation gas or increase the mud weight more than enough to control the formation pore pressure
3. Displace the formation gas and let the pressure against the formation drop by an amount equal to or just below the circulating pressure

The first alternative is a gamble and should be taken only if repeated experience shows the calculated pressures to be higher than those experienced. At times the fracture gradient below the casing shoe also increases; however, unless this is positively the case, the operator is gambling.

The second alternative of increasing the mud weight to control the formation pore pressure is probably the most desirable option. The pressure at any point in the annulus using the weighted mud can be calculated with Eq. 14.14.

$$P = \frac{A_1}{2} + \left[\frac{A_1{}^2}{4} + \frac{P_p \rho_{m_1} z\, T_{ab} h_b}{z_b T_b} \right]^{1/2} \qquad (14.14)$$

where:

$$A_1 = P_p - (D_g - X)\, \rho_{m_1} + D_{g_1} (\rho_{m_1} - \rho_m) - P_{ff}$$

Two new terms are introduced in Eq. 14.14. ρ_{m_1} is the pressure gradient of the weighted mud in psi/ft, and D_{g_1} represents the annular height of the low-weight mud inside the drillstring. Example 14.9 will now be considered using a mud weight high enough to control the formation pore pressure.

☐   **EXAMPLE 14.9 continued:**

Mud weight required to control the pore pressure:

$$\frac{(6,485)(19.35)}{(11,000)} = 11.35\ \text{lb/gal}$$

Use mud weight = 11.4 lb/gal

Pressure gradient of 11.4 lb/gal mud $= \dfrac{11.4}{19.25} = 0.592$ psi/ft

Use the same z factors.

Determine D_{g_1}:

Volume of unweighted mud inside pipe = $(0.014)(10,300) + (700)(0.007) = 149$ bbl

$$D_{g_1} = \frac{149\ \text{bbl}}{0.05\ \text{bbl/ft}} = 2,980\ \text{ft}$$
$$A_1 = 6,485 - (11,000 - 3,000)(0.592) + 2,980(0.592 - 0.571) - 64$$
$$A_1 = 6,485 - 7,736 + 63 - 64 = 1,748$$
$$P_{3,000} = \frac{1,748}{2} + \frac{1,748^2}{4} + \left[\frac{(6,485)(0.592)(0.80)(566)(400)}{(1.15)(662)} \right]^{1/2}$$
$$= 874 + 1,295 = 2,169\ \text{psia}$$

Pressure gradient at 3,000 ft $= \dfrac{2,169}{3,000} = 0.723$ psi/ft

$$h_{3,000} = \frac{6,485\,(0.80)(566)(400)}{2,169(1.15)(662)} = 711\ \text{ft}$$

This is an expansion of 311 ft, 711 − 400, and represents a pit level increase of 15.6 bbl, (311 × 0.05).

When the mud weight is increased enough to control the formation pore pressure, circulation is not lost. This example demonstrates that the common practice of using a safety factor results in an underground blowout in both cases.

For this example, when the mud weight was increased enough to control the pore pressure, circulation was not lost. However, in some cases the mud weight may have to be increased more than enough to control pore pressure to prevent the loss of circulation. The effect of further increasing mud weight can be determined easily by Eq. 14.14. The only danger in this procedure is if circulation is stopped when the weighted mud reaches the bit where the extra pressure exerted by the weighted mud is U-tubed into the annulus. To prevent this, the operator would need to let the weighted mud U-tube freely into the annulus until the pressure was balanced.

The third suggested alternative procedure, the low-choke method, is to use the circulating pressure as a part of the total back pressure. Again, considering Example 14.9, assume the annular circulating pressure is 180 psi.

In this case, the maximum pressure at 3,000 ft can be 2,220 psi, (3,000 × 0.74). If the annulus pressure is reduced 100 psi during the circulation by adjusting the annulus choke, the maximum pressure at 3,000 ft when using unweighted mud would be 2,146 psi, (2,246 + 37 − 100). This prevents the loss of circulation and an underground blowout. This reduction in annulus pressure of 100 psi can be regained before circulation is stopped without any danger of losing circulation. Note that in the case of displacing with an unweighted mud, the maximum pressure at 3,000 ft when the gas first reaches the surface is 1,962 psi, (1,918 + 44), which is below the breakdown pressure of 2,220 psi.

Method for Controlling a Well When the Pipe Is Out of the Hole

If a well kicks while the pipe is out of the hole, the operator has the options of controlling the well with the pipe out of the hole or of snubbing the pipe back into the hole.

Controlling the well with pipe out of the hole depends on the drilling fluid used and the type of formation fluid that has entered the wellbore. It would be difficult to list all of the situations that might exist. One typical situation is with water for the drilling fluid and with gas for the formation fluid. The potential solution to the problem is shown in this example:

□ **EXAMPLE 14.10**

> Well depth = 10,000 ft
> Drilling fluid = Salty water that weighs 8.7 lb/gal
> Casing setting depth = 2,500 ft of 9⅝-in. O.D.
> Drillpipe = 4½-in. O.D. and 3.82-in. I.D.
> Drillpipe length = 9,400 ft
> Drill collars = 6¼-in. O.D. and 2¾-in. I.D.
> Drill collar length = 600 ft
> Hole size = 7⅞ in.

The well kicked while the pipe was out of the hole. When it was finally closed in, the shutin pressure was 140 psi.

The fracture gradient at the casing seat = 0.70 psi/ft
The noted pit level increase was about 20 bbl
Temperature gradient = 70°F + 1.2°F/100 ft

Determine: (a) The type of fluid that has entered the wellbore
(b) The procedure required to control the well with the pipe out of the hole

SOLUTION:

(a) The internal volume of the casing and hole are estimated as follows:

Hole volume = 0.06 bbl/ft
Casing volume = 0.078 bbl/ft

$$10 \text{ bbl of influx} = \frac{20 \text{ bbl}}{0.06 \text{ bbl/ft}} = 333 \text{ ft of fluid}$$

Assume: Pore pressure = hydrostatic pressure of 8.7 lb/gal water

$$A = 4{,}520 - 3{,}390 - 11 = 1{,}119$$

$$P_{2,500} = \frac{1{,}119}{2} + \left[\frac{(1{,}119)^2}{4} + \frac{(4{,}520)(0.452)(1)(560)(333)}{(1)(650)} \right]^{1/2}$$

$$P = 560 + 948 = 1{,}508 \text{ psi}$$

The pressure of 1,508 psi is well below the indicated fracture pressure of 1,750 psi. Unless serious mistakes are made, the gas should be permitted to migrate to the surface with no particular problems. When the gas reaches the surface, the operator should replace any gas bled-off with water. If salt water is bullheaded* into the well, an underground blowout is likely. The proper way to control the well is to accurately measure the water pumped into the well and to be sure the pressure reduction allowed at the surface by bleeding off gas is just under the increase in hydrostatic pressure. This procedure is illustrated in Table 14–5.

With gas to the surface, 93 bbl of water have been released. To kill the well, the gas must be replaced with water. The procedure for pumping water into the well and releasing gas is shown in Table 14–6.

This procedure permits the gas to reach the surface by releasing water. Next water is pumped into the hole, gas is allowed to migrate through the water, and then the annulus pressure is reduced by bleeding off gas. The method is repeated until all of the gas is removed, and the well is under control.

These practical procedures can control gas under the conditions shown in the examples. Many other possibilities that may arise are not covered by these examples. As a result, an operator must become familiar with the expansion and compression characteristics of nonideal gases. In field operations, some of the calculations shown in this discussion may be un-

---

* When a substance is pumped into a well, with no way for it to circulate; a dead end.

■ **TABLE 14–5**   Gas migration with the drillstring out of the hole

P_p, psi	Shutin Annular Pressure, psi	Water Released, bbl	Length of Gas, ft	Top of the Gas from the Bottom, ft	Hydrostatic Pressure, psi	Reduction in Hydrostatic Pressure, psi	P_{ff}, psi	Borehole Pressure, psi
4,520	140	0	333	333	4,369	0	11	4,520
4,520	500	0	333	1,133	4,369	0	11	4,880
4,520	166	0.75	346	1,146	4,363	6	11	4,540
4,520	500	0.75	346	1,885	4,363	6	11	4,874
4,520	188	1.50	346	2,624	4,357	12	11	4,556
4,520	392	12	860	7,500	4,131	240	11	4,534
4,520	649	93.0	1,454	10,000	3,863	507	8	4,520

necessary. However, keeping an accurate record of the volume of fluid released and the resulting effect on hydrostatic pressure will always be necessary. Most guesswork can be eliminated if an operator fully understands the mechanisms at work.

Constant Annulus-Constant Drillpipe Control Procedure

After a well kick, the basic philosophy of control is the same, regardless of the procedure used. The total pressure against the kicking formation is maintained at a level high enough to prevent further fluid entry and below the level that would fracture another exposed formation. Simplicity is emphasized because decisions must be made quickly when a well kicks, and many people may be involved in the control procedure.

With the inception of current well control methods, displacing formation fluids with the unweighted mud in use when the well kicked was emphasized. In recent years, it has become apparent that such a generalized procedure, although simple, may result in an underground blowout that could have been prevented.

■ **TABLE 14–6**   Controlling a well by replacing the gas with water

P_p, psi	Shutin Annular Pressure, psi	Water into Well, bbl	Increase in Hydrostatic Pressure, psi	Hydrostatic Pressure, psi	P_{ff}, psi	Borehole Pressure, psi
4,520	649	0	0	3,863	8	4,520
4,520	789	20	116	3,979	8	4,776
4,520	600	20	116	3,979	6	4,585
4,520	764	40	232	4,095	6	4,865
4,520	0	113	657	4,520	0	4,520

Some operators prefer to raise the mud weight before displacement or during displacement of formation fluids. This increase in mud weight requires adjusting the drillpipe pressure during displacement of the weighted mud down the drillstring. As a result, calculations are necessary to determine the levels of downward pressure adjustments required to maintain a constant level of back pressure against the formation.

To minimize the required calculations, another procedure has become common. This one calls for keeping the casing pressure constant until the weighted mud reaches the bit and then keeping the drillpipe pressure constant until the formation fluid that entered the wellbore has been displaced.

The geometry of the drillstring makes the constant casing pressure-constant drillpipe pressure method acceptable. If the annulus size did not change, the procedure would introduce the possibility of further influx of formation fluids unless additional back pressures were held on the annulus.

The effect of the larger O.D. pipe on the bottom of the hole is shown in Fig. 14–12.

If the weighted mud reaches the bit before point B on the pressure profile is reached, the constant annulus-constant drillpipe pressure control method is acceptable. An operator can determine whether the constant annulus-constant drillpipe pressure control method can be used based on the drillstring geometry and the kick size.

Variable Circulation Rates While Displacing a Gas Kick

A gas kick is frequently displaced with a constant circulation rate. There are advantages to displacing some gas kicks with more than one circulation rate. Each time the circulation rate is changed, a new pumping pressure must be determined (see Fig. 14–13).

The annular pressure profile in Fig. 14–13 changes slowly until the

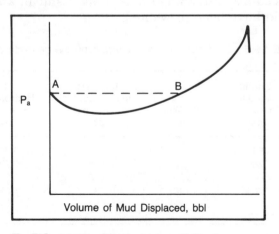

■ **FIG. 14–12** Effect of larger O.D. pipe

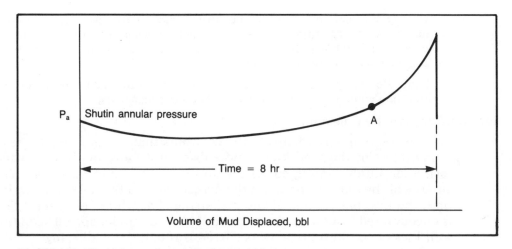

■ FIG. 14–13   Volume of mud displaced, bbl

gas is close to the surface. Thus the circulation rate can be altered by holding annular pressure constant until a new circulation rate is established. The pumping pressure at the new circulation rate is then kept constant until the gas kick is displaced. If there are to be other circulation rate changes, make these before reaching point A on Fig. 14–13. As shown, the annular pressure changes rapidly after point A is reached; using a new circulation rate and obtaining an accurate pumping pressure might be difficult.

Also note that any change in circulation rate should be made after all of the gas is above the bottom-hole drill collars or heavyweight pipe. If weighted mud is being used, any circulation rate changes must be made after the weighted mud reaches the bit and enters the annulus.

Changing the circulation rate may offer several advantages:

- By utilizing a higher circulation rate during a portion of the gas displacement, the total time required to displace the gas kick is reduced.
- When drilling from drillships, reducing the circulation rate when the gas is close to the surface may be desirable.
- Knowing that the circulation rate may be changed is important in the event of a mechanical problem that requires a change in pumps or at least a shutdown period.

The specific procedure for changing the circulation rate is shown in the following example.

☐   **EXAMPLE 14.11**

Assume:

> Circulation rate = 3.0 bbl/min
> Pumping pressure = 700 psi
> Annular pressure = 200 psi

Change circulation rate to 6.0 bbl/min. Hold the annular pressure constant at 200 psi. New pumping pressure = 2,440 psi. Then hold the circulation rate constant at 6.0 bbl/min and pumping pressure constant at 2,440 psi. Annular pressure will continue to change.

Note: Remember that a change in circulation rate should be made during the time when the annular pressure is changing slowly.

Blowout Through the Drillstring

This section only relates to the blowouts that occur through the drillstring. Most of the well kicks and subsequent blowouts have started by formation fluids coming out of the annulus. The main reason is that the volume of the annulus between the drillstring and the hole is much larger than the internal volume of the drillstring. Therefore, less back pressure is encountered by flow up the annulus. However, under specific conditions, well kicks and blowouts have occurred through the drillstring.

For a well kick to start through the drillstring, the total back pressure against the formation must be less in that direction than through the annulus. This may happen when pulling the drillstring because of external equipment on the drillstring. Drillstring stabilizers, a necessity in many locations, may become packed with clays and shales and form an annular swab. The barrel of mud swabbed out of the annulus may be replaced, at least partially, by mud from inside the drillstring. The barrel of mud removed from the annulus may reduce the length of the hydrostatic head, say 10 ft, while a barrel of mud removed from inside the drillstring may reduce the length of the hydrostatic head by 100 ft. This excessive reduction in hydrostatic pressure inside the drillstring may send the flow in that direction when pulling pipe rather than through the annulus. The potential dangers of this process are illustrated by an assumed set of conditions shown in the following example.

☐ **EXAMPLE 14.12**

Well depth = 5,000 ft
Surface casing set at 2,000 ft
Surface casing size = 13⅜ in.
Hole size = 12¼ in.
Drillstring size = 4½-in. O.D., 3.78-in. I.D.
Drill collars = 600 ft of 7-in. O.D., 3-in. I.D.
Stabilizers are used on the drill collars.
Gas sand at 5,000 ft
Pore pressure gradient = 0.45 psi/ft
Mud weight = 9.2 lb/gal = 0.478 psi/ft
Annulus volume = 0.126 bbl/ft
Drillpipe volume = 0.014 bbl/ft
Drill collar volume = 0.0087 bbl/ft

Determine:
    (a) The drop in hydrostatic head necessary for the well to kick and while pulling pipe

(b) The number of barrels of mud required to prevent a well kick through the annulus and through the drillpipe

SOLUTION:

(a) Hydrostatic pressure of mud at 5,000 ft = (0.478)(5,000) = 2,390 psi
Pore pressure = (0.45)(5,000) = 2,250 psi
The drop in hydrostatic pressure necessary for a well kick = 140 psi, 2,390 − 2,250
(b) Number of barrels of mud that must be swabbed out of the annulus to produce a well kick is (140)(0.126)/(0.478) = 36.9 bbl
Number of barrels of mud that must be removed from the drillpipe to produce a well kick is (140)(0.014)/(0.478) = 4.1 bbl

The mud level inside the drillstring can be dropped by swabbing in the annulus. One such condition is the so-called balling of the stabilizers by accumulations of shale and clay. If this happens, an effective swab can be built around the drill collars so the hole below the swab is evacuated and the mud in the annulus is replaced by mud out of the drillstring.

To remove 4.1 bbl of mud from the annulus requires a complete swabbing of only 32.5 ft of hole, 4.1 ÷ 0.126. It is possible to swab the well in through the drillpipe while pulling only one stand of drill collars. Under these conditions, recognizing the swabbing as a danger is difficult until the well starts flowing through the drillstring.

Most pit-level indicators activate an alarm with a pit-level change of 10 bbl or more; a change of only 4.1 bbl would not activate the alarm. If the pit-level indicator was watched closely, an operator would, of course, see the pit level increase. The real danger is that, while pulling only one stand of pipe, most people would be unaware of the instantaneous problem illustrated in Example 14.12.

There would be warning signals if considerable swabbing was occurring in the annulus. First, the drillstring would be pulling tight, and the pulling weight would increase appreciably. Second, a close observation of the annulus would show the swabbing.

One way to prevent the problem of swabbing through the drillstring is to use a back pressure valve in the drillstring at all times. There are arguments for and against the routine use of a drillstring back pressure valve. In actual practice, any specific cure for a problem may also affect other problems; therefore, particular problems must be examined in total. The purpose of this discussion is to show that the problem exists and that caution should be exercised.

Momentum Control Procedure if the Drillstring is Off the Bottom of the Hole and the Well is Flowing

Controlling a blowing well when the drillstring is off the bottom of the hole may be difficult for the following reasons:

- Fluid pumped into the well may not exert enough downward force to stop the flow.

- Pumping heavy mud at high rates may rupture the casing before control is achieved, thus worsening the problem.

The momentum force required to control the flow before pumping into the well can be estimated. Thus, the operator can determine, based on the limitations of this equipment, whether the procedure is feasible. Eq. 14.15 shows how to calculate the momentum force exerted by flowing gas.

$$F_{mo} = \frac{S\,V_r^2}{1{,}715(10^{10})\,z\,D_c^2} \tag{14.15}$$

where:

F_{mo} = momentum force, lb force
S = gas specific gravity (assume 0.7)
V_r = volume rate of gas flow, scf/day
z = average gas compressibility factor
D_c = diameter of flow chamber, in.

Eq. 14.16 shows how to determine the momentum force exerted by flowing liquid.

$$F_{mo} = \frac{\rho Q^2}{2.6785\,D_c^2} \tag{14.16}$$

where:

ρ = mud weight, lb mass/gal
Q = liquid flow rate, bbl/min

The use of Eqs. 14.15 and 14.16 is illustrated by Example 14.11.

☐   **EXAMPLE 14.11**

Assume:

A gas flow rate = 20,000,000 scf/day
Casing size = 5½-in. O.D., 5.0-in. I.D.
Gas specific gravity = 0.7
z factor = 1.0

$$F_{mo} = \frac{(0.7)(20^2)(10^{12})}{1.715(10^{10})(5^2)} = 653 \text{ lb force}$$

The well is to be killed using a 20-lb/gal mud.

$$Q = \left[\frac{(653)(2.6785)(5^2)}{20}\right]^{1/2} = 46.76 \text{ bbl/min}$$

Assume:

A liquid saltwater flow rate = 40,000 bbl/day
Casing size = 9⅝-in. O.D., 9-in. I.D.
Saltwater weight = 10 lb/gal

$$F_{mo} = \frac{(10)(27.7778)^2}{(2.6785)(9^2)} = 35.56 \text{ lb force}$$

The well is to be killed using a 20-lb/gal mud.

$$Q = \left[\frac{(2.6785)(9^2)(35.56)}{20}\right]^{1/2} 19.64 \text{ bbl/min}$$

The required flow rates to kill a well are directly proportional to the internal diameter of the pipe the kill-weight mud is pumped through. Thus, if 4½-in. drillpipe is in the hole, the required flow rate of the kill mud to control the saltwater flow is shown as follows:

$$Q = \frac{(19.64)(3.82)}{9} = 8.34 \text{ bbl/min}$$

where:

3.82 in. is the I.D. of the 4½-in. drillpipe

There are many potential problems involved in controlling a well in this manner:

- High flow rates and high mud weights necessitate high pumping pressures.
- The pressures required may exceed the burst strength of the pipe.
- Estimating the flow rate is difficult under free-flow conditions.

Momentum calculations are certainly an improvement over pumping at high rates on a trial-and-error basis. Thus, a substantial portion of the guesswork is removed from the kill procedures during blowouts for all conditions.

SUMMARY

This discussion on pressure control, by necessity, is abbreviated. The subject of well control is broad. Covering all the potential problems in one chapter or even in one book is impossible.

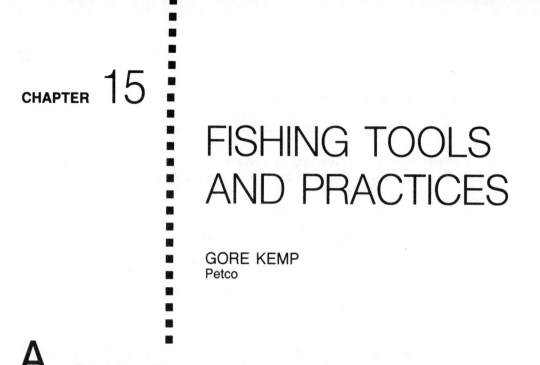

CHAPTER 15

FISHING TOOLS AND PRACTICES

GORE KEMP
Petco

Any object that obstructs the hole or impedes further drilling or running of pipe is called a *fish,* and removing this obstruction or correcting the problem is called "fishing." Since fishing is required on up to 20% of all drilling jobs and 80% of all workover jobs, the skills and tools required form a sizable service industry. Most operating companies find it more economical to rely on service companies to furnish the tools and fishing tool operators or supervisors than to provide the service themselves.

Fishing is more art than science.[1] Though the available tools and practices have improved tremendously, there is an art in guiding a fishing tool to its target and causing it to perform the most exacting function. Fishing tool operation is usually limited to right-hand rotation, limited left-hand rotation, reciprocation of the pipe, and pump operation, or a combination of these functions.

Planning a fishing job and deciding how to attack the problem are most important. Everyone concerned should contribute input. Fishing tool specialists recommend the proper tools for the job and supervise the running of the tools. The superintendent or the person responsible for the well should consult the drilling supervisor, tool pusher, petroleum engineer, and mud engineer. All should be consulted for their knowledge of the hole conditions, zones that require special attention, and possible causes of the fishing problem.

Fishing must be an economical solution. Obviously, a shallow hole with little rig time and equipment invested can justify only the cheapest fishing. When there is a large investment in the hole and substantial capital equipment to be recovered, more time and expense can be committed feasibly. Studies, papers, formulas, and models help make the economic deci-

This chapter is adapted from *Oilfield Fishing Operations: Tools and Techniques* by Gore Kemp. Published by Gulf Publishing Company, P.O. Box 2608, Houston, Texas 77001.

sion of "to fish or not to fish, and if so, for how long?" All have merit, but so many factors affect the decision that converting them into a standard formula or pattern is almost impossible. Good judgment, a careful analysis of the problem, and then skilled application of the decision insofar as the rig and tools are concerned is the best solution.

STUCK PIPE

Pipe sticking can be caused by several factors.[2] The problem can be easier to solve if the cause is known. Sticking can result from failure to remove the cuttings from the hole sufficiently, hole sloughing, lost circulation, keyseating, and differential sticking. Each situation may require a different approach, so any information regarding the hole conditions and symptoms prior to sticking is very valuable.

Differentially stuck pipe and reasons for its sticking need to be thoroughly understood in order to select the best method for correcting the problem. Differential sticking is caused by the hydrostatic mud pressure in the wellbore exceeding the formation pore pressure in the interval where the pipe is stuck. This formation is usually a sand, limestone, or dolomite with high permeability. Normally a thick filter cake is across this formation; the pipe or drill collars have been left stationary, such as in making a connection, and the pipe is in contact with the formation. Ordinarily, the mud may be circulated freely around the stuck zone.

Fig. 15–1 shows the forces acting against the pipe. Assume a reasonable set of values to calculate a hypothetical situation. If the formation pressure is 5,000 psi at the stuck point of 11,000 ft and a mud weight of 11.0 lb/gal produces a hydrostatic pressure of 6,292 psi, there would be a differential pressure of 1,292 psi. Assuming that 90 ft of drill collars are in contact with the formation in an area 3 in. wide, there would be a total force of 348,840 lb required to pull the pipe loose. If this force is added to the normal *hook load* (the calculated air weight of the drillstring down to the stuck point), the pipe's tensile strength may be exceeded.

To free the pipe, certain steps should be taken. The sequence for these steps is determined by how quickly they can be put into operation, their usual success ratio, and economic factors. Usually the sticking increases with time.

First, the pumps should be shut down, and the pipe worked without circulation. In *working the pipe*, it should be reciprocated with approximately 10% slack-off and no pull on the pipe. It should also be torqued to the right during the working. The pump pressure adds to the hydrostatic pressure in the wellbore; sometimes this small reduction is sufficient to free the pipe. It is ordinarily better to slack off weight on the stuck portion rather than to pull tension, as tension tends to pull the pipe into the formation and increase the stuck area. As the example above demonstrates, it is impossible to pull the pipe free.

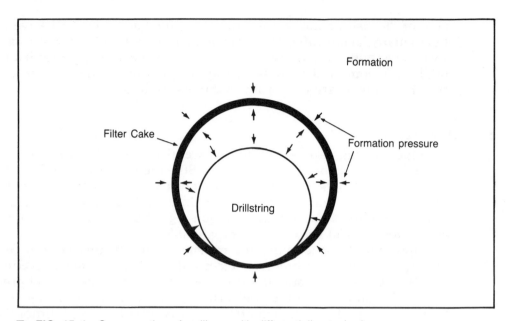

■ FIG. 15–1 Cross section of wellbore with differentially stuck pipe

The second step recommended is the *U-tube technique* in which gas or diesel or some other fluid lighter than the mud is used to reduce the hydrostatic pressure.[3] This procedure must be planned carefully to arrive at a previously determined hydrostatic head in the annulus. The U-tube method can be used only when there is no check or float valve in the drilling string. The lighter liquid is pumped down the drillpipe in a precalculated volume and then allowed to flow back due to the heavier fluid in the annulus. As the fluid level in the annulus goes down, the effective hydrostatic pressure is reduced, and the pipe may be "blown free" by the pore pressure at the stuck point. This method induces a well kick of barely sufficient pressure to free the pipe. Be careful to control the induced kick.

Another method of approaching the problem is to *spot a fluid* opposite the stuck interval that will soften or dissolve the mud filter cake. This fluid may be weighted to approximately the same weight as the drilling fluid to minimize migration as well as control the pressure. Unweighted or semiweighted fluids may be used to obtain the added advantage of decreasing the hydrostatic head against the pipe. The disadvantage is the migration that can be expected of the lighter fluid up the hole. Any fluid that is spotted should contain the proper chemicals and surfactants to penetrate the filter cake. Before abandoning this procedure, spend at least 8 hr spotting the fluid and working the pipe, as this is the average time required. Apply reasonable torque to the pipe according to the size and weight, and alternate approximately 10,000 lb of weight with the static weight of the drillstring. Pulling on the pipe is counterproductive. In areas where sticking may be prevalent, make advance arrangements for immediate action in case of sticking. The necessary pipe-freeing chemicals, fluid, oil, or surfactants should be stored on the location. The quicker corrective procedures can be put into operation, the higher the chances for success.

If the pipe cannot be freed by one of the above methods, then the drillstring must be *parted* so fishing tools can be placed in the string to free it. The first step in this procedure is to determine accurately the top of the stuck pipe. This requires a free-point instrument run on a conductor line. The instrument measures a flow of current through the pipe when the instrument is relaxed and when it is stressed either by pulling or torquing. If the pipe is free at that point, a change takes place in the pipe, and there is a different flow of current. If the pipe is stuck, it does not move or change when stress is applied.

The most popular method of parting the pipe is to "back it off" with a string shot. Left-hand torque is applied to the pipe at the surface, and the torque is then worked down the string by reciprocating it until the torque is evenly distributed. A string shot consists of a firing cap and a length of explosive line called primacord. When the primacord is fired across the tool joint, the concussion momentarily relieves the stress, and the pipe unscrews. There has been much discussion about the weight on the pipe when the back-off is made.

Common practice is to establish the neutral point at the free (stuck) point by holding buoyed weight of the string to the stuck point.[4] Contrary to popular belief, this is not an actual neutral point because there is no effective buoyant force acting on the stuck pipe. Therefore, the pipe at that point is in axial compressive load. To create a zero-axial-load neutral point at the stuck point, air weight of the string to that point would have to be held. Although in theory the pipe would unscrew more readily if air weight were held, as soon as the pipe is loosened added compression from hydrostatic pressure acting on the end could bind the threads. Conversely, when hydrostatic force acts on a string held with only buoyed weight, the string does not shorten, and the pipe readily unscrews.

The back-off leaves a tool joint looking up on the stuck portion of a string. Normally, the work string with the fishing tools is screwed into this tool joint if jars (impact tools) are being used.

Free pipe should always be left below the point selected for back-off or cutting. Ordinarily, a full joint of free pipe is left if the back-off method is used. At least half a joint is left if the pipe is cut. Some operators prefer an added joint to the above in case the string shot should damage the tool joint or some other unforeseen damage should occur. The free pipe acts as a guide for tools that may be run to go over or to catch the pipe. Many times solids in the mud settle, and fill decreases the length of the guide.

In some wells, such as highly deviated holes, torque cannot be applied to the stuck point, and the pipe must be cut. In such cases, the chemical cutter or the jet cut may be selected. Sometimes the pipe must be cut mechanically by running an internal cutter on a small-diameter string of pipe.

The chemical cutter, run on a conductor cable, cuts the pipe by burning holes in the pipe with a fluoride chemical forced under extreme pressure by a propellant. This method has the advantage of not swelling the pipe, which permits it to be caught either inside or outside without milling. The cutter must be held in place in the pipe by hold-down slips actuated by the propellant. In coated pipe, the slips do not always hold and the cutter is forced up the hole, fouling the electric line. The chemical cutter cannot be run in mud that has a high volume of solids.

The jet cut is a shaped or "beehive" charge of plastic explosive in which the face toward the pipe to be cut is shaped in a parabola or a modified parabola. When the charge is fired, the majority of force is exerted at a point, and the pipe is cut. Since an explosive force cuts the pipe, the pipe swells with a burr both inside and outside. This burr can usually be trimmed off with a mill control in the overshot or a carbide mill container to "dress" the top of the fish.

In some areas an *open-hole packer* or drillstem test tool successfully reduces the hydrostatic pressure against the stuck pipe or drill collars. The pipe is backed off above the stuck area into a known formation that will provide a suitable packer seat in an interval where the hole is in gauge. The open-hole packer is run with an appropriate jar assembly and with a perforated nipple near the bottom of the string in case the fish is plugged. When the packer string is screwed back into the fish and the packer is set, the hydrostatic head holding the drillstring against the wall is immediately reduced to such a degree that the formation pressure blows the pipe or drill collars free. The drillstring can be worked immediately and the pipe pulled from the hole. This method of freeing the stuck pipe is probably the quickest solution available, but hole conditions limit it to certain areas and wells.

Washover operations are another method of freeing the stuck pipe. Large-diameter heavy-wall pipe is run into the hole and used to cut and circulate out whatever is causing the sticking. Washpipe size must be selected so that it goes over the fish with clearance and has clearance outside between the borehole and the washpipe. Usually relatively short strings of pipe are run, but there is no rule that may be applied. Hole conditions dictate the safe length of the string.

Washpipe is run with a connection at the top back to the work string, either a safety joint or a bushing. An appropriate rotary shoe is on the bottom that will cut the formation or the fill. If the fish is stuck off of the bottom, then consider running a washpipe spear. This will catch the stuck string when it is freed and prevent its falling to the bottom of the hole. Factors to consider when running the spear are the weight of the fish, a safety factor to allow for impact, thread strength of the washpipe, and the distance off the bottom.

Pipe that parts when the bit is near the bottom of the hole always causes substantially more damage than parting that occurs when the string is substantially off of the bottom. When pipe is suspended in the borehole, it is in tension and is actually stretched. When it parts and falls to the bottom, the pipe vacillates from tension to compression. If the distance off the bottom is small, this movement is still taking place in the pipe when it strikes bottom; the next compressive cycle drives the bit downward and tends to corkscrew the pipe. If the distance off the bottom is substantial, all of the longitudinal forces are worked out of the pipe before it touches the bottom.

Jarring is a popular method of freeing stuck pipe and is quite effective unless the pipe is differentially stuck. In this case, the forces are so great that the cause must be corrected.

Jars may be divided into two general categories, hydraulic and mechanical. Hydraulic or oil jars are favored as fishing jars since the impact may be varied to almost any desired blow. Mechanical jars are subject to a pre-set stress necessary in order to be tripped. This cannot be varied substantially downhole. In order to jar the fish in a downward direction as well as upward, the oil jar is run with a bumper jar or a bumper sub below. With the two jars assembled in this manner, the fish can be jarred in either direction. Drill collars are run immediately above the oil jar to afford a greater impact.

Above the drill collars, an accelerator or intensifier may be run, greatly increasing the effectiveness of the jarring action (see Fig. 15-2). Without the accelerator, as the oil jar strokes, the work string must absorb the travel, and it is sometimes pushed up the hole and into a compressive stress. Friction of the pipe being pushed up the hole acts as a shock absorber to the jarring action. This absorbing of the force decreases the effectiveness of the impact.

An accelerator is a "fluid spring" that stores up energy in a compressed fluid. When released, the accelerator speeds the movement of the drill collars upward. The free stroke in the accelerator also offsets the stroke in the oil jar, thereby eliminating the loss due to the friction of moving the drillpipe or tubing up the hole. Fewer drill collars may be run when an accelerator is added to the jarring string. The complete jarring string should be specified by an experienced fishing tool operator or supervisor, as several criteria affect it.

Mechanical jars are used in fishing at shallow depths when there is not enough stretch in the drillpipe to create impact with the oil jar. These are also used in extreme temperatures in which the oil jar would be ineffective.

Drilling jars frequently are run in the drillstring so they are readily

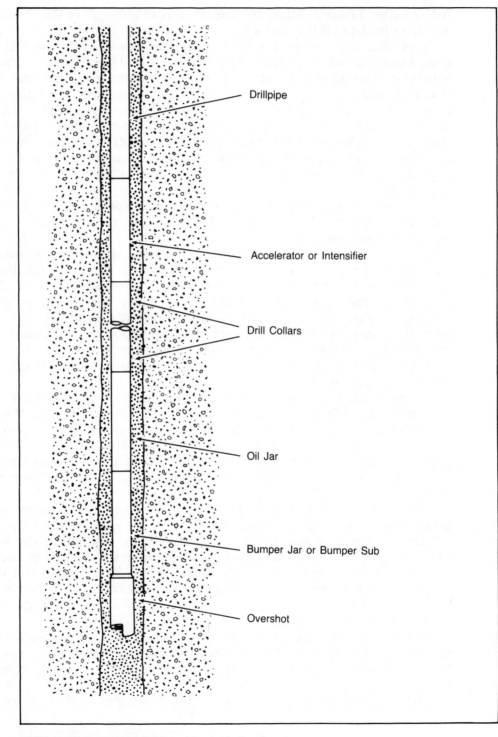

■ FIG. 15–2 Typical fishing string including jars

available when the need occurs. Both hydraulic and mechanical types are used. These differ from the fishing jars only in that they are much heavier and sturdier in order to withstand the rigors of continuous drilling.

CATCHING THE PIPE

When the pipe has been parted by any means except the string shot back-off, then the pipe must be caught with an appropriate tool. This includes parting such as twist-offs, jet cutting, chemical cutting, and mechanical cutting. Overshots and spears are usually used to catch the pipe.

Overshots

The overshot is the basic tool used to catch the pipe from the outside. Modern overshots are very reliable, can usually be released, and will pack off around the fish so you can pump through the fish and have a full opening.

In ordering an overshot, give the hole or casing size, as the tool is selected with an appropriate clearance in the hole. List the sizes that you wish to catch, such as the pipe body size, tool joint size, and drill collar size. Last, the fishing tool company needs to know what threads are on the work string. Appropriate grapples will be selected to catch the various sizes (see Fig. 15–3). Since a grapple must be selected rather accurately, you must know the exact size of the catch. In open-hole sizes, overshot grapples usually catch about $1/8$-in. larger and $1/16$-in. smaller than the size marked on the grapple.

Overshots are furnished as full strength (F.S.) and slim hole (S.H.). The full-strength overshots have at least the capacity of the pipe that is being caught. The slim-hole models are turned down for close tolerances and will not withstand a pull equal to the tensile strength of the pipe. Overstressing an overshot causes the bowl to either swell or split. When running a slim-hole tool, consult the strength charts to be certain you do not exceed the normal capacity.

Spears

A spear (Fig. 15–4) is made to catch the inside of tubular material, and by necessity it has a very small hole in the mandrel. Because of this, run an overshot when a choice can be made, as the overshot is full opening and allows any tools or instruments to be run through it. The spear is difficult to pack off so that pump pressure can be exerted through the fish. The advantage of the spear is its strength. It is usually stronger than any pipe that it is run on or that is being caught.

LOOSE JUNK

Loose junk, such as bit cones and hand tools, is frequently lost in the hole and impedes drilling until it is eliminated. Knowing what is lost in

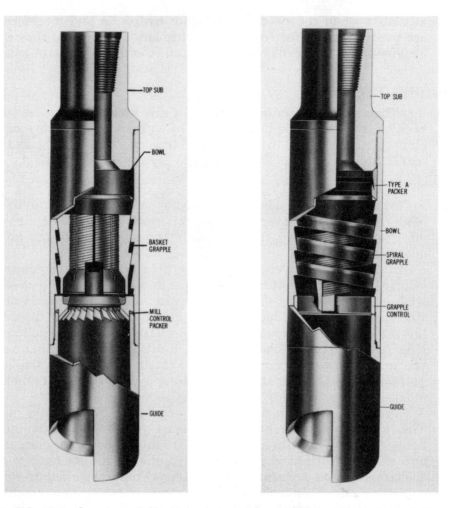

■ FIG. 15–3 Overshots: (left) with basket grapple and (right) with spiral grapple (*courtesy Bowen Tools*)

the hole is always helpful, as it may be possible to duplicate the situation on the surface and actually try the tools being considered. Specific tools used are magnets, junk baskets, carbide mills, or some tool specifically designed for the problem.

Magnets

Permanent magnets are very effective if there is enough surface on the fish for good contact. Magnets attract only ferrous metal objects and do not pick up carbide buttons, brass, bronze, or aluminum. Magnets are made with circulating ports opening around the outer edge of the tool so they can be washed down the hole. They should always be run on pipe

■ **FIG. 15–4** Releasing spear (*courtesy Bowen Tools*)

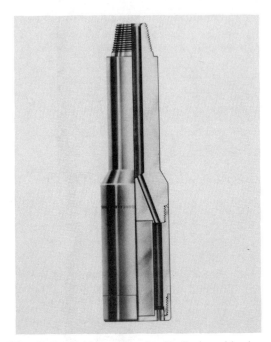

■ **FIG. 15–5** Fishing magnet with flush guide (*courtesy Bowen Tools*)

so that the fill on top of the fish can be washed away and actual contact made. The tools are usually furnished with guides (see Fig. 15–5) that extend below the magnet surface and thus prevent the fish's being pulled off while retrieving. When fishing with any tool, the pipe should be pulled using the spinning chain. Breaking out with the rotary will spin the fish and probably lose it.

Core-Type Junk Baskets

The core is the original finger-catcher type of junk basket, installed with two sets of catchers. The upper cage is dressed with short fingers and the lower cage with alternating long and short fingers. The junk basket is designed to go over the fish and cut a core from the formation. The short fingers help to break the core and retrieve the junk above the core. The finger cages must rotate freely in the shoe or body so the catchers are not torn out by the junk. The inside diameter of the cage is the limiting dimension of the junk that can be caught.

Reverse-Circulation Junk Basket

The core-type junk basket has one fault: the circulation of the fluid down the drillpipe and up the annulus tends to kick the junk away from the catchers. Current designs (Fig. 15–6) of catcher-type junk baskets incor-

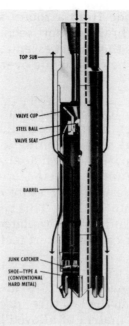

TOP SUB

VALVE CUP
STEEL BALL
VALVE SEAT

BARREL

JUNK CATCHER

SHOE—TYPE A
(CONVENTIONAL
HARD METAL)

■ **FIG. 15–6** Reverse-circulation junk basket shown with ball in place after reverse circulation has been established (*courtesy Bowen Tools*)

porate a method of reverse circulating the wellbore in the immediate area of the tool. This tends to kick the junk into the catcher instead of away from it. In operation, the hole is cleaned of any fill by normal circulation. When this is accomplished, a ball is pumped down. The fluid is diverted to the annulus and up inside the bowl of the junk basket before flowing into the annulus again. This reversing action is much more efficient than other methods.

Poor-Boy Junk Basket

The poor-boy basket is a shop-made device that has the advantage of quite a large opening, enabling it to catch large irregular shapes. The material should always be of very low-grade pipe, as the fingers must bend without breaking. A welder fashions teeth with a curvature on the leading edge and a gap between the teeth. When running, the basket should be rotated carefully and circulated without weight on the fingers until sufficient hole has been made to completely cover the fish. When this has been done, the running should be smooth, indicating that the fish is above the gaps cut in the pipe. As rotation continues, weight should be applied gradually, and the teeth will close up or "orange peel" around the junk.

Many junk-basket designs may be fashioned for a particular job. One popular device is made from a rotary shoe by cutting a series of holes around the shoe and brazing short sections of wire rope in the basket as a catcher. This leaves the basket essentially full opening, but it also limits the rotation that can be done while over a fish.

Boot Basket

Boot baskets (Fig. 15–7) must be run in conjunction with some other tool such as a magnet, a reverse-circulation basket, or a mill. You must circulate the long way when running a boot basket for it to be effective. The boot is large compared to the hole size, thereby creating a high velocity of fluid coming up the annulus. At the top of the boot is a suddenly increased size annulus with a resultant pressure drop that creates turbulence; the heavy particles fall into the boot. This type of basket can be most effective when mill cuttings cannot be circulated to the surface or when fishing for junk like cones or carbide insert teeth.

Carbide Mills

Tungsten carbide mills (Fig. 15–8) are one of the most innovative items added to fishing tools in the last three decades. The sharp tungsten carbide particles are set in a matrix of alloyed bronze. A welder applies them to a steel mill body by brazing. Quality products must be utilized, the steel body must be absolutely clean, and the proper temperature must be used

■ **FIG. 15–7** Boot basket. (Note that gussets have been welded onto the mandrel by the manufacturer and then stress relieved. Caution: Do not field weld on the mandrel.)

■ **FIG. 15–8** Tungsten carbide junk mills (*courtesy Petco Fishing and Rental Tools*)

applying the material to the body. A quality mill, when properly run, cuts a tremendous amount of steel. It should be operated in the same manner that a carbide cutting tool is run in a machine shop. Normally, it is turned fast, perhaps one-third to one-half faster than a similar sized rock bit. Light weight must be used on the mill, as carbide as well as the bronze matrix will not support a heavy load. As the mill turns, the effect should be a cutting action similar to a lathe tool; all grinding should be eliminated.

Carbide mills run with excess weight will be pulled and will have a pattern of wear that appears to be polished. In planning a milling job, consider removing the metal chips that have been cut. A viscous fluid will be required. Usually circulation the long way is necessary, as reverse circulation tends to plug. At the surface the mill cuttings must be screened out and perhaps a ditch magnet used in the return line to pick up any cuttings that might be going back through the pump and cause damage.

Carbide mills are made in many configurations. There are flat-bottomed mills for milling junk, tapered mills to open up the tops of liners, and pilot mills with a guide to keep the mill cutting on the top of the pipe. In open-hole work, mills should be dressed with the carbide on the sides as well as the bottom of the mill body. Inside the pipe, however, be careful to avoid cutting the casing. Carbide on the outside of the body must be ground smooth. Sometimes it is good practice to braze a pure brass or bronze bushing on the outside to keep the cutting surfaces away from the casing and also to reduce friction.

Junk Shots

Junk shots are used occasionally to break up a large piece of junk that cannot be retrieved. The shot is a large shaped charge made to direct the major force downward. Since all of the forces cannot be directed down, junk shots should *never* be run inside casing as the explosion will split and destroy the pipe.

Junk shots are run on both drillpipe and conductor lines. If the shot is run on an electric line, then a trip should be made to the top of the fish with a bit, as shaped charges must be spaced accurately with the target. These shots are usually quite effective in breaking the junk into pieces small enough to be fished with a magnet or a junk basket.

WIRELINE FISHING

If a tool or instrument becomes hung, do not pull the line in two. As long as the line is intact, the location of the instrument is known. The line furnishes a guide to direct a recovery tool to it. Common practice is to pull out of the rope socket, if possible, but many times, the line is parted before the line pulls out of the rope socket.

Cable Guide

When the line is intact, the cable guide method is the safest and most nearly certain of all possible solutions. After a clamp is placed on the line at the surface, the line is cut at a convenient distance above the floor. Rope sockets are placed on each end of the line. On the top part, which goes to the reel unit, a sinker bar and quick-action catching tool is made up. On the bottom part, a spear point that fits the catching tool is fitted.

An overshot dressed to catch the rope socket or body of the tool or instrument is made up on the first stand of pipe. Then the pipe is stripped into the hole over the line. The operation requires good voice contact and coordination between the reel operator, derrickman, and floorman. Once the fish is caught in the overshot, the line may be pulled out of the rope socket. In critical jobs, some operators prefer to clamp off the line and cut it as each stand of pipe is pulled. This ensures that the fish will not be lost coming out of the hole. This method is also known as "cut and strip" or "cut and thread."

Side Door Overshot

Side door overshots may be used inside pipe and at very shallow depths; they can create problems in a deep open hole. The line is outside the drillpipe or tubing and may easily become *keyseated* (imbedded in the formation or filter cake). The advantage of side door overshots is that the line does not need to be cut.

Rope Spear

The rope spear or center spear is preferred as the tool to catch a parted line. The most frequent mistake in fishing for the line is assuming that it has fallen far down the hole. Actually, considering the size of the line and the hole size, it does not fall far but stands up in the hole. In using the rope spear, the parted line must be caught near the top, as it tends to ball up and stick the work string if it is caught far below the top.

■ **REFERENCES**
1. E.W. Porter, "Fishing is More Art than Science," *OGJ,* (21 September 1970) 95–96.
2. Mike Brouse, "How to Handle Stuck Pipe and Fishing Problems," *World Oil* (November 1982) 103–110, 124.
3. Ibid.
4. W.E. Goins, "Better Understanding Prevents Tubular Buckling Problems," *World Oil* (1 February 1980) 35–40.

APPENDIX

TABLE 1
MINIMUM PERFORMANCE PROPERTIES OF CASING

1	2	3	4	5	6	7	8	9	10	11	12
					Threaded and Coupled			Extreme Line			
Size: Outside Diameter in.	Nominal Weight, Threads and Coupling lb. per ft.	Grade	Wall Thickness	Inside Diameter	Drift Diameter	Outside Diameter of Coupling	Outside Diameter Special Clearance Coupling	Drift Diameter	Outside Diameter of Box Power-tight	Collapse Resistance	Pipe Body Yield Strength
			in.	in.	in.	in.	in.	in.	in.		
D			t	d	W	W	W_c		M	psi	1000 lbs.
4½	9.50	H-40	0.205	4.090	3.965	5.000	—	—	—	2,760	111
	9.50	J-55	0.205	4.090	3.965	5.000	—	—	—	3,310	152
	10.50	J-55	0.224	4.052	3.927	5.000	4.875	—	—	4,010	165
	11.60	J-55	0.250	4.000	3.875	5.000	4.875	—	—	4,960	184
	9.50	K-55	0.205	4.090	3.965	5.000	—	—	—	3,310	152
	10.50	K-55	0.224	4.052	3.927	5.000	4.875	—	—	4,010	165
	11.60	K-55	0.250	4.000	3.875	5.000	4.875	—	—	4,960	184
	11.60	C-75	0.250	4.000	3.875	5.000	4.875	—	—	6,100	250
	13.50	C-75	0.290	3.920	3.795	5.000	4.875	—	—	8,140	288
	11.60	L-80	0.250	4.000	3.875	5.000	4.875	—	—	6,350	267
	13.50	L-80	0.290	3.920	3.795	5.000	4.875	—	—	8,540	307
	11.60	N-80	0.250	4.000	3.875	5.000	4.875	—	—	6,350	267
	13.50	N-80	0.290	3.920	3.795	5.000	4.875	—	—	8,540	307
	11.60	C-90	0.250	4.000	3.875	5.000	4.875	—	—	6,820	300
	13.50	C-90	0.290	3.920	3.795	5.000	4.875	—	—	9,300	345
	11.60	C-95	0.250	4.000	3.875	5.000	4.875	—	—	7,030	317
	13.50	C-95	0.290	3.920	3.795	5.000	4.875	—	—	9,660	364
	11.60	P-110	0.250	4.000	3.875	5.000	4.875	—	—	7,580	367
	13.50	P-110	0.290	3.920	3.795	5.000	4.875	—	—	10,680	422
	15.10	P-110	0.337	3.826	3.701	5.000	4.875	—	—	14,350	485
5	11.50	J-55	0.220	4.560	4.435	5.563	—	—	—	3,060	182
	13.00	J-55	0.253	4.494	4.369	5.563	5.375	—	—	4,140	208
	15.00	J-55	0.296	4.408	4.283	5.563	5.375	4.151	5.360	5,560	241
	11.50	K-55	0.220	4.560	4.435	5.563	—	—	—	3,060	182
	13.00	K-55	0.253	4.494	4.369	5.563	5.375	—	—	4,140	208
	15.00	K-55	0.296	4.408	4.283	5.563	5.375	4.151	5.360	5,560	241
	15.00	C-75	0.296	4.408	4.283	5.563	5.375	4.151	5.360	6,940	328
	18.00	C-75	0.362	4.276	4.151	5.563	5.375	4.151	5.360	9,960	396
	21.40	C-75	0.437	4.126	4.001	5.563	5.375	—	—	11,970	470
	23.20	C-75	0.478	4.044	3.919	5.563	5.375	—	—	12,970	509
	24.10	C-75	0.500	4.000	3.875	5.563	5.375	—	—	13,500	530
	15.00	L-80	0.296	4.408	4.283	5.563	5.375	4.151	5.360	7,250	350
	18.00	L-80	0.362	4.276	4.151	5.563	5.375	4.151	5.360	10,500	422
	21.40	L-80	0.437	4.126	4.001	5.563	5.375	—	—	12,760	501
	23.20	L-80	0.478	4.044	3.919	5.563	5.375	—	—	13,830	543
	24.10	L-80	0.500	4.000	3.875	5.563	5.375	—	—	14,400	566
	15.00	N-80	0.296	4.408	4.283	5.563	5.375	4.151	5.360	7,250	350
	18.00	N-80	0.362	4.276	4.151	5.563	5.375	4.151	5.360	10,500	422
	21.40	N-80	0.437	4.126	4.001	5.563	5.375	—	—	12,760	501
	23.20	N-80	0.478	4.044	3.919	5.563	5.375	—	—	13,830	543
	24.10	N-80	0.500	4.000	3.875	5.563	5.375	—	—	14,400	566
	15.00	C-90	0.296	4.408	4.283	5.563	5.375	4.151	5.366	7,840	394
	18.00	C-90	0.362	4.276	4.151	5.563	5.375	4.151	5.366	11,530	475
	21.40	C-90	0.437	4.126	4.001	5.563	5.375	—	—	14,360	564
	23.20	C-90	0.478	4.044	3.919	5.563	5.375	—	—	15,560	611
	24.10	C-90	0.500	4.000	3.875	5.563	5.375	—	—	16,200	636
	15.00	C-95	0.296	4.408	4.283	5.563	5.375	4.151	5.360	8,110	416
	18.00	C-95	0.362	4.276	4.151	5.563	5.375	4.151	5.360	12,030	501
	21.40	C-95	0.437	4.126	4.001	5.563	5.375	—	—	15,160	595
	23.20	C-95	0.478	4.044	3.919	5.563	5.375	—	—	16,430	645
	24.10	C-95	0.500	4.000	3.875	5.563	5.375	—	—	17,100	672
	15.00	P-110	0.296	4.408	4.283	5.563	5.375	4.151	5.360	8,850	481
	18.00	P-110	0.362	4.276	4.151	5.563	5.375	4.151	5.360	13,470	580
	21.40	P-110	0.437	4.126	4.001	5.563	5.375	—	—	17,550	689
	23.20	P-110	0.478	4.044	3.919	5.563	5.375	—	—	19,020	747
	24.10	P-110	0.500	4.000	3.875	5.563	5.375	—	—	19,800	778

	†Internal Pressure Resistance, psi						★Joint Strength — 1,000 lbs. (Threaded and Coupled)						Extreme Line	
Plain End or Extreme Line	Round Thread		Buttress Thread				Round Thread		Buttress Thread				Standard Joint	Optional Joint
	Short	Long	Regular Coupling Same Grade	Regular Coupling Higher Grade	Special Clearance Same Grade	Special Clearance Higher Grade	Short	Long	Regular Coupling	Regular Coupling Higher Grade‡	Special Clearance Coupling	Special Clearance Coupling Higher Grade‡		
13	14	15	16	17	18	19	20	21	22	23	24	25	26	27
3,190	3,190	—	—	—	—	—	77	—	—	—	—	—	—	—
4,380	4,380	—	—	—	—	—	101	—	203	203	203	203	—	—
4,790	4,790	—	4,790	4,790	4,790	4,790	132	—	225	225	225	225	—	—
5,350	5,350	5,350	5,350	5,350	5,350	5,350	154	162	—	—	—	—	—	—
4,380	4,380	—	—	—	—	—	112	—	249	249	249	249	—	—
4,790	4,790	—	4,790	4,790	4,790	4,790	146	—	277	277	277	277	—	—
5,350	5,350	5,350	5,350	5,350	5,350	5,350	170	180	—	—	—	—	—	—
7,290	—	7,290	7,290	—	7,290	—	—	212	288	—	288	—	—	—
8,460	—	8,460	8,460	—	7,490	—	—	257	331	—	320	—	—	—
7,780	—	7,780	7,780	7,780	7,780	7,780	—	212	291	—	291	—	—	—
9,020	—	9,020	9,020	9,020	7,990	9,020	—	257	334	—	320	—	—	—
7,780	—	7,780	7,780	7,780	7,780	7,780	—	223	304	304	304	304	—	—
9,020	—	9,020	9,020	9,020	7,990	9,020	—	270	349	349	349	349	—	—
8,750	—	8,750	8,750	—	8,750	—	—	223	309	—	309	—	—	—
10,150	—	10,150	10,150	—	9,000	—	—	270	355	—	337	—	—	—
9,240	—	9,240	9,240	—	9,240	—	—	234	325	325	325	—	—	—
10,710	—	10,710	10,710	—	9,490	—	—	284	374	374	353	—	—	—
10,690	—	10,690	10,690	10,690	10,690	10,690	—	279	385	385	385	385	—	—
12,410	—	12,410	12,410	12,410	10,990	12,410	—	338	443	443	421	443	—	—
14,420	—	14,420	13,460	14,420	10,990	13,910	—	406	509	509	421	509	—	—
4,240	4,240	—	—	—	—	—	133	—	252	252	252	252	—	—
4,870	4,870	4,870	4,870	4,870	4,870	4,870	169	182	293	293	287	293	328	—
5,700	5,700	5,700	5,700	5,700	5,130	5,700	207	223	—	—	—	—	—	—
4,240	4,240	—	—	—	—	—	147	—	309	309	309	309	—	—
4,870	4,870	4,870	4,870	4,870	4,870	4,870	186	201	359	359	359	359	416	—
5,700	5,700	5,700	5,700	5,700	5,130	5,700	228	246	—	—	—	—	—	—
7,770	—	7,770	7,770	—	6,990	—	—	295	375	—	364	—	416	—
9,500	—	9,500	9,290	—	6,990	—	—	376	452	—	364	—	446	—
11,470	—	10,140	9,290	—	6,990	—	—	466	510	—	364	—	—	—
12,550	—	10,140	9,290	—	7,000	—	—	513	510	—	364	—	—	—
13,130	—	10,140	9,290	—	6,990	—	—	538	510	—	364	—	—	—
8,290	—	8,290	8,290	8,290	7,460	8,290	—	295	379	—	364	—	416	—
10,140	—	10,140	9,910	10,140	7,460	10,140	—	376	457	—	364	—	446	—
12,240	—	10,810	9,910	—	7,460	—	—	466	510	—	364	—	—	—
13,380	—	10,820	9,910	—	7,460	—	—	513	510	—	364	—	—	—
14,000	—	10,810	9,910	—	7,460	—	—	538	510	—	364	—	—	—
8,290	—	8,290	8,290	8,290	7,460	8,290	—	311	396	396	383	396	437	—
10,140	—	10,140	9,910	10,140	7,460	10,140	—	396	477	477	383	477	469	—
12,240	—	10,810	9,910	12,240	7,460	10,250	—	490	537	566	383	479	—	—
13,380	—	10,820	9,910	13,380	7,460	10,260	—	540	537	614	383	479	—	—
14,000	—	10,810	9,910	13,620	7,460	10,250	—	567	537	639	383	479	—	—
9,320	—	9,320	9,320	—	8,400	—	—	311	404	—	383	—	430	—
11,400	—	11,400	11,150	—	8,400	—	—	396	487	—	383	—	469	—
13,770	—	12,170	11,150	—	8,400	—	—	490	537	—	383	—	—	—
15,060	—	12,170	11,150	—	8,400	—	—	540	537	—	383	—	—	—
15,750	—	12,170	11,150	—	8,400	—	—	567	537	—	383	—	—	—
9,840	—	9,840	9,840	—	8,850	—	—	326	424	—	402	—	459	—
12,040	—	12,040	11,770	—	8,850	—	—	416	512	—	402	—	493	—
14,530	—	12,840	11,770	—	8,850	—	—	515	563	—	402	—	—	—
15,890	—	12,850	11,770	—	8,860	—	—	567	563	—	402	—	—	—
16,630	—	12,850	11,770	—	8,850	—	—	595	563	—	402	—	—	—
11,400	—	11,400	11,400	11,400	10,250	11,400	—	388	503	503	479	503	547	—
13,940	—	13,940	13,620	13,940	10,250	13,940	—	495	606	606	479	606	587	—
16,820	—	14,870	13,620	16,820	10,250	13,980	—	613	671	720	479	613	—	—
18,400	—	14,880	13,630	18,400	10,260	13,990	—	675	671	780	479	613	—	—
19,250	—	14,870	13,620	18,580	10,250	13,980	—	708	671	812	479	613	—	—

TABLE 1
MINIMUM PERFORMANCE PROPERTIES OF CASING

1	2	3	4	5	6	7	8	9	10	11	12
					Threaded and Coupled			Extreme Line			
Size: Outside Diameter in. D	Nominal Weight, Threads and Coupling lb. per ft.	Grade	Wall Thickness in. t	Inside Diameter in. d	Drift Diameter in.	Outside Diameter of Coupling in. W	Outside Diameter Special Clearance Coupling in. W_r	Drift Diameter in.	Outside Diameter of Box Power-tight in. M	Collapse Resistance psi	Pipe Body Yield Strength 1000 lbs.
5½	14.00	H-40	0.244	5.012	4.887	6.050	—	—	—	2,620	161
	14.00	J-55	0.244	5.012	4.887	6.050	—	—	—	3,120	222
	15.50	J-55	0.275	4.950	4.825	6.050	5.875	4.653	5.860	4,040	248
	17.00	J-55	0.304	4.892	4.767	6.050	5.875	4.653	5.860	4,910	273
	14.00	K-55	0.244	5.012	4.887	6.050	—	—	—	3,120	222
	15.50	K-55	0.275	4.950	4.825	6.050	5.875	4.653	5.860	4,040	248
	17.00	K-55	0.304	4.892	4.767	6.050	5.875	4.653	5.860	4,910	273
	17.00	C-75	0.304	4.892	4.767	6.050	5.875	4.653	5.860	6,040	372
	20.00	C-75	0.361	4.778	4.653	6.050	5.875	4.653	5.860	8,410	437
	23.00	C-75	0.415	4.670	4.545	6.050	5.875	4.545	5.860	10,470	497
	17.00	L-80	0.304	4.892	4.767	6.050	5.875	4.653	5.860	6,280	397
	20.00	L-80	0.361	4.778	4.653	6.050	5.875	4.653	5.860	8,830	466
	23.00	L-80	0.415	4.670	4.545	6.050	5.875	4.545	5.860	11,160	530
	17.00	N-80	0.304	4.892	4.767	6.050	5.875	4.653	5.860	6,280	397
	20.00	N-80	0.361	4.778	4.653	6.050	5.875	4.653	5.860	8,830	466
	23.00	N-80	0.415	4.670	4.545	6.050	5.875	4.545	5.860	11,160	530
	17.00	C-90	0.304	4.892	4.767	6.050	5.875	4.653	5.860	6,740	447
	20.00	C-90	0.361	4.778	4.653	6.050	5.875	4.653	5.860	9,630	525
	23.00	C-90	0.415	4.670	4.545	6.050	5.875	4.545	5.860	12,380	597
	26.00	C-90	0.476	4.548	4.423	6.050	5.875	—	—	14,240	676
	35.00	C-90	0.650	4.200	4.075	6.050	5.875	—	—	18,760	891
	17.00	C-95	0.304	4.892	4.767	6.050	5.875	4.653	5.860	6,940	471
	20.00	C-95	0.361	4.778	4.653	6.050	5.875	4.653	5.860	10,010	554
	23.00	C-95	0.415	4.670	4.545	6.050	5.875	4.545	5.860	12,940	630
	17.00	P-110	0.304	4.892	4.767	6.050	5.875	4.653	5.860	7,480	546
	20.00	P-110	0.361	4.778	4.653	6.050	5.875	4.653	5.860	11,100	641
	23.00	P-110	0.415	4.670	4.545	6.050	5.875	4.545	5.860	14,540	729
6⅝	20.00	H-40	0.288	6.049	5.924	7.390	—	—	—	2,520	229
	20.00	J-55	0.288	6.049	5.924	7.390	7.000	—	—	2,970	315
	24.00	J-55	0.352	5.921	5.796	7.390	7.000	5.730	7.000	4,560	382
	20.00	K-55	0.288	6.049	5.924	7.390	7.000	—	—	2,970	315
	24.00	K-55	0.352	5.921	5.796	7.390	7.000	5.730	7.000	4,560	382
	24.00	C-75	0.352	5.921	5.796	7.390	7.000	5.730	7.000	5,550	520
	28.00	C-75	0.417	5.791	5.666	7.390	7.000	5.666	7.000	7,790	610
	32.00	C-75	0.475	5.675	5.550	7.390	7.000	5.550	7.000	9,800	688
	24.00	L-80	0.352	5.921	5.796	7.390	7.000	5.730	7.000	5,760	555
	28.00	L-80	0.417	5.791	5.666	7.390	7.000	5.666	7.000	8,170	651
	32.00	L-80	0.475	5.675	5.550	7.390	7.000	5.550	7.000	10,320	734
	24.00	N-80	0.352	5.921	5.796	7.390	7.000	5.730	7.000	5,760	555
	28.00	N-80	0.417	5.791	5.666	7.390	7.000	5.666	7.000	8,170	651
	32.00	N-80	0.475	5.675	5.550	7.390	7.000	5.550	7.000	10,320	734
	24.00	C-90	0.352	5.921	5.796	7.390	7.000	5.730	7.000	6,140	624
	28.00	C-90	0.417	5.791	5.666	7.390	7.000	5.666	7.000	8,880	732
	32.00	C-90	0.475	5.675	5.550	7.390	7.000	5.550	7.000	11,330	826
	24.00	C-95	0.352	5.921	5.796	7.390	7.000	5.730	7.000	6,310	659
	28.00	C-95	0.417	5.791	5.666	7.390	7.000	5.666	7.000	9,220	773
	32.00	C-95	0.475	5.675	5.550	7.390	7.000	5.550	7.000	11,810	872
	24.00	P-110	0.352	5.921	5.796	7.390	7.000	5.730	7.000	6,730	763
	28.00	P-110	0.417	5.791	5.666	7.390	7.000	5.666	7.000	10,160	895
	32.00	P-110	0.475	5.675	5.550	7.390	7.000	5.550	7.000	13,220	1009

13	14	15	16	17	18	19	20	21	22	23	24	25	26	27
†Internal Pressure Resistance, psi							★Joint Strength — 1,000 lbs. Threaded and Coupled						Extreme Line	
Plain End or Extreme Line	Round Thread		Buttress Thread				Round Thread		Buttress Thread				Standard Joint	Optional Joint
	Short	Long	Regular Coupling		Special Clearance Coupling		Short	Long	Regular Coupling	Regular Coupling Higher Grade‡	Special Clearance Coupling	Special Clearance Coupling Higher Grade‡		
			Same Grade	Higher Grade	Same Grade	Higher Grade								
3,110	3,110	—	—	—	—	—	130	—	—	—	—	—	—	—
4,270	4,270	—					172	—	—	—	—	—	—	—
4,810	4,810	4,810	4,810	4,810	4,730	4,810	202	217	300	300	300	300	339	339
5,320	5,320	5,320	5,320	5,320	4,730	5,320	229	247	329	329	318	329	372	372
4,270	4,270	—					189	—	—	—	—	—	—	—
4,810	4,810	4,810	4,810	4,810	4,730	4,810	222	239	366	366	366	366	429	429
5,320	5,320	5,320	5,320	5,320	4,730	5,320	252	272	402	402	402	402	471	471
7,250	—	7,250	7,250	—	6,450	—	—	327	423	—	403	—	471	471
8,610	—	8,610	8,430	—	6,450	—	—	403	497	—	403	—	497	479
9,900	—	9,260	8,430	—	6,450	—	—	473	550	—	403	—	549	479
7,740	—	7,740	7,740	7,740	6,880	7,740	—	338	428	—	403	—	471	471
9,190	—	9,190	8,990	9,190	6,880	9,190	—	416	503	—	403	—	497	479
10,560	—	9,880	8,990	10,560	6,880	9,460	—	489	550	—	403	—	549	479
7,740	—	7,740	7,740	7,740	6,880	7,740	—	348	446	446	424	446	496	496
9,190	—	9,190	8,990	9,190	6,880	9,190	—	428	524	524	424	524	523	504
10,560	—	9,880	8,990	10,560	6,880	9,460	—	502	579	596	424	530	577	504
8,710	—	8,710	8,710	—	7,740	—	—	356	456	—	424	—	496	496
10,340	—	10,340	10,120	—	7,740	—	—	438	536	—	424	—	523	504
11,880	—	11,110	10,120	—	7,740	—	—	514	580	—	424	—	577	504
13,630	—	11,110	10,120	—	7,740	—	—	598	580	—	424	—	—	—
18,610	—	11,110	10,120	—	7,740	—	—	614	580	—	424	—	—	—
9,190	—	9,190	9,190	—	8,170	—	—	374	480	—	445	—	521	521
10,910	—	10,910	10,680	—	8,170	—	—	460	563	—	445	—	549	530
12,540	—	11,730	10,680	—	8,170	—	—	540	608	—	445	—	606	530
10,640	—	10,640	10,640	10,640	9,460	10,640	—	445	568	568	530	568	620	620
12,640	—	12,640	12,360	12,640	9,460	11,880	—	548	667	667	530	667	654	630
14,520	—	13,160	12,360	14,520	9,460	11,880	—	643	724	759	530	668	772	630
3,040	3,040	—	—	—	—	—	184	—	—	—	—	—	—	—
4,180	4,180	4,180	4,180	4,180	4,060	4,180	245	266	374	374	374	374	—	—
5,110	5,110	5,110	5,110	5,110	4,060	5,110	314	340	453	453	390	453	477	477
4,180	4,180	4,180	4,180	4,180	4,060	4,180	267	290	453	453	453	453	—	—
5,110	5,110	5,110	5,110	5,110	4,060	5,110	342	372	548	548	494	520	605	605
6,970	—	6,970	6,970	—	5,540	—	—	453	583	—	494	—	605	605
8,260	—	8,260	8,260	—	5,540	—	—	552	683	—	494	—	648	644
9,410	—	9,410	9,200	—	5,540	—	—	638	771	—	494	—	717	644
7,440	—	7,440	7,440	—	5,910	—	—	473	592	—	494	—	605	605
8,810	—	8,810	8,810	—	5,910	—	—	576	693	—	494	—	648	644
10,040	—	10,040	9,820	—	5,910	—	—	666	783	—	494	—	717	644
7,440	—	7,440	7,440	7,440	5,910	7,440	—	481	615	615	520	615	637	637
8,810	—	8,810	8,810	8,810	5,910	8,120	—	586	721	721	520	650	682	678
10,040	—	10,040	9,820	10,040	5,910	8,120	—	677	814	814	520	650	755	678
8,370	—	8,370	8,370	—	6,650	—	—	520	633	—	520	—	637	637
9,910	—	9,910	9,910	—	6,650	—	—	633	742	—	520	—	682	678
11,290	—	11,290	11,050	—	6,650	—	—	732	837	—	520	—	755	678
8,830	—	8,830	8,830	—	7,020	—	—	546	665	—	546	—	668	668
10,460	—	10,460	10,460	—	7,020	—	—	665	780	—	546	—	716	712
11,920	—	11,830	11,660	—	7,020	—	—	769	880	—	546	—	793	712
10,230	—	10,230	10,230	10,230	8,120	8,310	—	641	786	786	650	786	796	796
12,120	—	11,830	12,120	12,120	8,120	8,310	—	781	922	922	650	832	852	848
13,800	—	11,830	13,500	13,800	8,120	8,310	—	904	1,040	1,040	650	832	944	848

TABLE 1
MINIMUM PERFORMANCE PROPERTIES OF CASING

1	2	3	4	5	6	7	8	9	10	11	12
					Threaded and Coupled			Extreme Line			
Size: Outside Diameter in. D	Nominal Weight, Threads and Coupling lb. per ft.	Grade	Wall Thickness in. t	Inside Diameter in. d	Drift Diameter in.	Outside Diameter of Coupling in. W	Outside Diameter Special Clearance Coupling in. W_c	Drift Diameter in.	Outside Diameter of Box Power-tight in. M	Collapse Resistance psi	Pipe Body Yield Strength 1000 lbs.
7	17.00	H-40	0.231	6.538	6.413	7.656	—	—	—	1,420	196
	20.00	H-40	0.272	6.456	6.331	7.656	—	—	—	1,970	230
	20.00	J-55	0.272	6.456	6.331	7.656	—	—	—	2,270	316
	23.00	J-55	0.317	6.366	6.241	7.656	7.375	6.151	7.390	3,270	366
	26.00	J-55	0.362	6.276	6.151	7.656	7.375	6.151	7.390	4,320	415
	20.00	K-55	0.272	6.456	6.331	7.656	—	—	—	2,270	316
	23.00	K-55	0.317	6.366	6.241	7.656	7.375	6.151	7.390	3,270	366
	26.00	K-55	0.362	6.276	6.151	7.656	7.375	6.151	7.390	4,320	415
	23.00	C-75	0.317	6.366	6.241	7.656	7.375	6.151	7.390	3,750	499
	26.00	C-75	0.362	6.276	6.151	7.656	7.375	6.151	7.390	5,220	566
	29.00	C-75	0.408	6.184	6.059	7.656	7.375	6.059	7.390	6,730	634
	32.00	C-75	0.453	6.094	5.969	7.656	7.375	5.969	7.390	8,200	699
	35.00	C-75	0.498	6.004	5.879	7.656	7.375	5.879	7.530	9,670	763
	38.00	C-75	0.540	5.920	5.795	7.656	7.375	5.795	7.530	10,680	822
	23.00	L-80	0.317	6.366	6.241	7.656	7.375	6.151	7.390	3,830	532
	26.00	L-80	0.362	6.276	6.151	7.656	7.375	6.151	7.390	5,410	604
	29.00	L-80	0.408	6.184	6.059	7.656	7.375	6.059	7.390	7,020	676
	32.00	L-80	0.453	6.094	5.969	7.656	7.375	5.969	7.390	8,610	745
	35.00	L-80	0.498	6.004	5.879	7.656	7.375	5.879	7.530	10,180	814
	38.00	L-80	0.540	5.920	5.795	7.656	7.375	5.795	7.530	11,390	877
	23.00	N-80	0.317	6.366	6.241	7.656	7.375	6.151	7.390	3,830	532
	26.00	N-80	0.362	6.276	6.151	7.656	7.375	6.151	7.390	5,410	604
	29.00	N-80	0.408	6.184	6.059	7.656	7.375	6.059	7.390	7,020	676
	32.00	N-80	0.453	6.094	5.969	7.656	7.375	5.969	7.390	8,610	745
	35.00	N-80	0.498	6.004	5.879	7.656	7.375	5.879	7.530	10,180	814
	38.00	N-80	0.540	5.920	5.795	7.656	7.375	5.795	7.530	11,390	877
	23.00	C-90	0.317	6.366	6.241	7.656	7.375	6.151	7.390	4,030	599
	26.00	C-90	0.362	6.276	6.151	7.656	7.375	6.151	7.390	5,740	679
	29.00	C-90	0.408	6.184	6.059	7.656	7.375	6.059	7.390	7,580	760
	32.00	C-90	0.453	6.094	5.969	7.656	7.375	5.969	7.390	9,380	839
	35.00	C-90	0.498	6.004	5.879	7.656	7.375	5.879	7.530	11,170	915
	38.00	C-90	0.540	5.920	5.795	7.656	7.375	5.795	7.530	12,820	986
	23.00	C-95	0.317	6.366	6.241	7.656	7.375	6.151	7.390	4,140	632
	26.00	C-95	0.362	6.276	6.151	7.656	7.375	6.151	7.390	5,880	717
	29.00	C-95	0.408	6.184	6.059	7.656	7.375	6.059	7.390	7,830	803
	32.00	C-95	0.453	6.094	5.969	7.656	7.375	5.969	7.390	9,750	885
	35.00	C-95	0.498	6.004	5.879	7.656	7.375	5.879	7.530	11,650	966
	38.00	C-95	0.540	5.920	5.795	7.656	7.375	5.795	7.530	13,440	1041
	26.00	P-110	0.362	6.276	6.151	7.656	7.375	6.151	7.390	6,230	830
	29.00	P-110	0.408	6.184	6.059	7.656	7.375	6.059	7.390	8,530	929
	32.00	P-110	0.453	6.094	5.969	7.656	7.375	5.969	7.390	10,780	1025
	35.00	P-110	0.498	6.004	5.879	7.656	7.375	5.879	7.530	13,020	1119
	38.00	P-110	0.540	5.920	5.795	7.656	7.375	5.795	7.530	15,140	1205
7⅝	24.00	H-40	0.300	7.025	6.900	8.500	—	—	—	2,030	276
	26.40	J-55	0.328	6.969	6.844	8.500	8.125	6.750	8.010	2,890	414
	26.40	K-55	0.328	6.969	6.844	8.500	8.125	6.750	8.010	2,890	414
	26.40	C-75	0.328	6.969	6.844	8.500	8.125	6.750	8.010	3,280	564
	29.70	C-75	0.375	6.875	6.750	8.500	8.125	6.750	8.010	4,650	641
	33.70	C-75	0.430	6.765	6.640	8.500	8.125	6.640	8.010	6,300	729
	39.00	C-75	0.500	6.625	6.500	8.500	8.125	6.500	8.010	8,400	839
	42.80	C-75	0.562	6.501	6.376	8.500	8.125	—	—	10,240	935
	45.30	C-75	0.595	6.435	6.310	8.500	8.125	—	—	10,790	986
	47.10	C-75	0.625	6.375	6.250	8.500	8.125	—	—	11,290	1031

13	14	15	16	17	18	19	20	21	22	23	24	25	26	27
							★Joint Strength — 1,000 lbs.							
†Internal Pressure Resistance, psi							Threaded and Coupled						Extreme Line	
Plain End or Extreme Line	Round Thread		Buttress Thread				Round Thread		Buttress Thread				Standard Joint	Optional Joint
	Short	Long	Regular Coupling		Special Clearance Coupling		Short	Long	Regular Coupling	Regular Coupling Higher Grade‡	Special Clearance Coupling	Special Clearance Coupling Higher Grade‡		
			Same Grade	Higher Grade	Same Grade	Higher Grade								
2,310	2,310	—	—	—	—	—	122	—	—	—	—	—	—	—
2,720	2,720	—	—	—	—	—	176	—	—	—	—	—	—	—
3,740	3,740	—	—	—	—	—	234	—	—	—	—	—	—	—
4,360	4,360	4,360	4,360	4,360	3,950	4,360	284	313	432	432	421	432	499	499
4,980	4,980	4,980	4,980	4,980	3,950	4,980	334	367	490	490	421	490	506	506
3,740	3,740	—	—	—	—	—	254	—	—	—	—	—	—	—
4,360	4,360	4,360	4,360	4,360	3,950	4,360	309	341	522	522	522	522	632	632
4,980	4,980	4,980	4,980	4,980	3,950	4,980	364	401	592	592	533	561	641	641
5,940	—	5,940	5,940	—	5,380	—	—	416	557	—	533	—	632	632
6,790	—	6,790	6,790	—	5,380	—	—	489	631	—	533	—	641	641
7,650	—	7,650	7,650	—	5,380	—	—	562	707	—	533	—	685	674
8,490	—	8,490	7,930	—	5,380	—	—	633	779	—	533	—	761	674
9,340	—	8,660	7,930	—	5,380	—	—	703	833	—	533	—	850	761
10,120	—	8,660	7,930	—	5,380	—	—	767	833	—	533	—	917	761
6,340	—	6,340	6,340	6,340	5,740	6,340	—	435	565	—	533	—	632	632
7,240	—	7,240	7,240	7,240	5,740	7,240	—	511	641	—	533	—	641	641
8,160	—	8,160	8,160	8,160	5,740	7,890	—	587	718	—	533	—	685	674
9,060	—	9,060	8,460	9,060	5,740	7,890	—	661	791	—	533	—	761	674
9,960	—	9,240	8,460	9,960	5,740	7,890	—	734	833	—	533	—	850	761
10,800	—	9,240	8,460	10,800	5,740	7,890	—	801	833	—	533	—	917	761
6,340	—	6,340	6,340	6,340	5,740	6,340	—	442	588	588	561	588	666	666
7,240	—	7,240	7,240	7,240	5,740	7,240	—	519	667	667	561	667	675	675
8,160	—	8,160	8,160	8,160	5,740	7,890	—	597	746	746	561	702	721	709
9,060	—	9,060	8,460	9,060	5,740	7,890	—	672	823	823	561	702	801	709
9,960	—	9,240	8,460	9,960	5,740	7,890	—	746	876	898	561	702	895	801
10,800	—	9,240	8,460	10,800	5,740	7,890	—	814	876	968	561	702	965	801
7,130	—	7,130	7,130	—	6,450	—	—	447	605	—	561	—	666	666
8,150	—	8,150	8,150	—	6,450	—	—	563	687	—	561	—	675	675
9,180	—	9,180	9,180	—	6,450	—	—	648	768	—	561	—	721	709
10,190	—	9,520	9,520	—	6,450	—	—	729	847	—	561	—	801	709
11,210	—	9,520	9,520	—	6,450	—	—	809	876	—	561	—	895	801
12,150	—	9,520	9,520	—	6,450	—	—	883	876	—	561	—	965	80
7,530	—	7,530	7,530	—	6,810	—	—	505	636	—	589	—	699	699
8,600	—	8,600	8,600	—	6,810	—	—	593	722	—	589	—	709	709
9,690	—	9,520	9,690	—	6,810	—	—	683	808	—	589	—	757	744
10,760	—	9,520	10,050	—	6,810	—	—	768	891	—	589	—	841	744
11,830	—	9,520	10,050	—	6,810	—	—	853	920	—	589	—	940	841
12,820	—	9,520	10,050	—	6,810	—	—	931	920	—	589	—	1,013	841
9,960	—	9,520	9,960	9,960	7,480	7,480	—	693	853	853	702	853	844	844
11,220	—	9,520	11,220	11,220	7,480	7,480	—	797	955	955	702	898	902	886
12,460	—	9,520	11,640	11,790	7,480	7,480	—	897	1,053	1,053	702	898	1,002	886
13,700	—	9,520	11,640	11,790	7,480	7,480	—	996	1,096	1,150	702	898	1,118	1,002
14,850	—	9,520	11,640	11,790	7,480	7,480	—	1,087	1,096	1,239	702	898	1,207	1,002
2,750	2,750	—	—	—	—	—	212	—	—	—	—	—	—	—
4,140	4,140	4,140	4,140	4,140	4,140	4,140	315	346	483	483	483	483	553	553
4,140	4,140	4,140	4,140	4,140	4,140	4,140	342	377	581	581	581	581	700	700
5,650	—	5,650	5,650	—	5,650	—	—	461	624	—	624	—	700	700
6,450	—	6,450	6,450	—	6,140	—	—	542	709	—	709	—	700	700
7,400	—	7,400	7,400	—	6,140	—	—	635	806	—	735	—	766	744
8,610	—	8,610	8,610	—	6,140	—	—	751	929	—	735	—	851	744
9,670	—	9,670	9,190	—	6,140	—	—	852	1,035	—	735	—	—	—
10,240	—	9,840	9,180	—	6,140	—	—	905	1,090	—	764	—	—	—
10,760	—	9,840	9,190	—	6,140	—	—	953	1,140	—	735	—	—	—

TABLE 1
MINIMUM PERFORMANCE PROPERTIES OF CASING

1	2	3	4	5	6	7	8	9	10	11	12
					Threaded and Coupled			Extreme Line			
Size: Outside Diameter in. D	Nominal Weight, Threads and Coupling lb. per ft.	Grade	Wall Thickness in. t	Inside Diameter in. d	Drift Diameter in.	Outside Diameter of Coupling in. W	Outside Diameter Special Clearance Coupling in. W_c	Drift Diameter in.	Outside Diameter of Box Power-tight in. M	Collapse Resistance psi	Pipe Body Yield Strength 1000 lbs.
7⅝	26.40	L-80	0.328	6.969	6.844	8.500	8.125	6.750	8.010	3,400	602
	29.70	L-80	0.375	6.875	6.750	8.500	8.125	6.750	8.010	4,790	683
	33.70	L-80	0.430	6.765	6.640	8.500	8.125	6.640	8.010	6,560	778
	39.00	L-80	0.500	6.625	6.500	8.500	8.125	6.500	8.010	8,820	895
	42.80	L-80	0.562	6.501	6.376	8.500	8.125	—	—	10,810	998
	45.30	L-80	0.595	6.435	6.310	8.500	8.125	—	—	11,510	1051
	47.10	L-80	0.625	6.375	6.250	8.500	8.125	—	—	12,040	1100
	26.40	N-80	0.328	6.969	6.844	8.500	8.125	6.750	8.010	3,400	602
	29.70	N-80	0.375	6.875	6.750	8.500	8.125	6.750	8.010	4,790	683
	33.70	N-80	0.430	6.765	6.640	8.500	8.125	6.640	8.010	6,560	778
	39.00	N-80	0.500	6.625	6.500	8.500	8.125	6.500	8.010	8,820	895
	42.80	N-80	0.562	6.501	6.376	8.500	8.125	—	—	10,810	998
	45.30	N-80	0.595	6.435	6.310	8.500	8.125	—	—	11,510	1051
	47.10	N-80	0.625	6.375	6.250	8.500	8.125	—	—	12,040	1100
	26.40	C-90	0.328	6.969	6.844	8.500	8.125	6.750	8.010	3,610	677
	29.70	C-90	0.375	6.875	6.750	8.500	8.125	6.750	8.010	5,040	769
	33.70	C-90	0.430	6.765	6.640	8.500	8.125	6.640	8.010	7,050	875
	39.00	C-90	0.500	6.625	6.500	8.500	8.125	6.500	8.010	9,620	1007
	42.80	C-90	0.562	6.501	6.376	8.500	8.125	—	—	11,890	1122
	45.30	C-90	0.595	6.435	6.310	8.500	8.125	—	—	12,950	1183
	47.10	C-90	0.625	6.375	6.250	8.500	8.125	—	—	13,540	1237
	26.40	C-95	0.328	6.969	6.844	8.500	8.125	6.750	8.010	3,710	714
	29.70	C-95	0.375	6.875	6.750	8.500	8.125	6.750	8.010	5,140	811
	33.70	C-95	0.430	6.765	6.640	8.500	8.125	6.640	8.010	7,280	923
	39.00	C-95	0.500	6.625	6.500	8.500	8.125	6.500	8.010	10,000	1063
	42.80	C-95	0.562	6.501	6.376	8.500	8.125	—	—	12,410	1185
	45.30	C-95	0.595	6.435	6.310	8.500	8.125	—	—	13,660	1248
	47.10	C-95	0.625	6.375	6.250	8.500	8.125	—	—	14,300	1306
	29.70	P-110	0.375	6.875	6.750	8.500	8.125	6.750	8.010	5,350	940
	33.70	P-110	0.430	6.765	6.640	8.500	8.125	6.640	8.010	7,870	1069
	39.00	P-110	0.500	6.625	6.500	8.500	8.125	6.500	8.010	11,080	1231
	42.80	P-110	0.562	6.501	6.376	8.500	8.125	—	—	13,920	1372
	45.30	P-110	0.595	6.435	6.310	8.500	8.125	—	—	15,430	1446
	47.10	P-110	0.625	6.375	6.250	8.500	8.125	—	—	16,550	1512
8⅝	28.00	H-40	0.304	8.017	7.892	9.625	—	—	—	1,610	318
	32.00	H-40	0.352	7.921	7.796	9.625	—	—	—	2,200	366
	24.00	J-55	0.264	8.097	7.972	9.625	—	—	—	1,370	381
	32.00	J-55	0.352	7.921	7.796	9.625	9.125	7.700	9.120	2,530	503
	36.00	J-55	0.400	7.825	7.700	9.625	9.125	7.700	9.120	3,450	568
	24.00	K-55	0.264	8.097	7.972	9.625	—	—	—	1,370	381
	32.00	K-55	0.352	7.921	7.796	9.625	9.125	7.700	9.120	2,530	503
	36.00	K-55	0.400	7.825	7.700	9.625	9.125	7.700	9.120	3,450	568
	36.00	C-75	0.400	7.825	7.700	9.625	9.125	7.700	9.120	4,000	775
	40.00	C-75	0.450	7.725	7.600	9.625	9.125	7.600	9.120	5,330	867
	44.00	C-75	0.500	7.625	7.500	9.625	9.125	7.500	9.120	6,660	957
	49.00	C-75	0.557	7.511	7.386	9.625	9.125	7.386	9.120	8,180	1059
	36.00	L-80	0.400	7.825	7.700	9.625	9.125	7.700	9.120	4,100	827
	40.00	L-80	0.450	7.725	7.600	9.625	9.125	7.600	9.120	5,520	925
	44.00	L-80	0.500	7.625	7.500	9.625	9.125	7.500	9.120	6,950	1021
	49.00	L-80	0.557	7.511	7.386	9.625	9.125	7.386	9.120	8,580	1129
	36.00	N-80	0.400	7.825	7.700	9.625	9.125	7.700	9.120	4,100	827
	40.00	N-80	0.450	7.725	7.600	9.625	9.125	7.600	9.120	5,520	925
	44.00	N-80	0.500	7.625	7.500	9.625	9.125	7.500	9.120	6,950	1021
	49.00	N-80	0.557	7.511	7.386	9.625	9.125	7.386	9.120	8,580	1129
	36.00	C-90	0.400	7.825	7.700	9.625	9.125	7.700	9.120	4,250	930
	40.00	C-90	0.450	7.725	7.600	9.625	9.125	7.600	9.120	5,870	1040
	44.00	C-90	0.500	7.625	7.500	9.625	9.125	7.500	9.120	7,490	1149
	49.00	C-90	0.557	7.511	7.386	9.625	9.125	7.386	9.120	9,340	1271

13	14	15	16	17	18	19	20	21	22	23	24	25	26	27
											★Joint Strength — 1,000 lbs.			
		†Internal Pressure Resistance, psi							Threaded and Coupled				Extreme Line	
Plain End or Extreme Line	Round Thread		Buttress Thread				Round Thread		Buttress Thread				Standard Joint	Optional Joint
	Short	Long	Regular Coupling		Special Clearance Coupling		Short	Long	Regular Coupling	Regular Coupling Higher Grade‡	Special Clearance Coupling	Special Clearance Coupling Higher Grade‡		
			Same Grade	Higher Grade	Same Grade	Higher Grade								
6,020	—	6,020	6,020	6,020	6,020	6,020	—	482	635	—	635	—	700	700
6,890	—	6,890	6,890	6,890	6,550	6,890	—	566	721	—	721	—	700	700
7,900	—	7,900	7,900	7,900	6,550	7,900	—	664	820	—	735	—	766	744
9,180	—	9,180	9,180	9,180	6,550	9,000	—	786	945	—	735	—	851	744
10,320	—	10,320	9,790	—	6,550	—	—	892	1,053	—	735	—	—	—
10,920	—	10,500	9,790	—	6,550	—	—	947	1,109	—	764	—	—	—
11,480	—	10,490	9,790	—	6,550	—	—	997	1,160	—	735	—	—	—
6,020	—	6,020	6,020	6,020	6,020	6,020	—	490	659	659	659	659	737	737
6,890	—	6,890	6,890	6,890	6,550	6,890	—	575	749	749	749	749	737	737
7,900	—	7,900	7,900	7,900	6,550	7,900	—	674	852	852	773	852	806	784
9,180	—	9,180	9,180	9,180	6,550	9,000	—	798	981	981	773	967	896	784
10,320	—	10,320	9,790	10,320	6,550	9,000	—	905	1,093	1,093	773	967	—	—
10,920	—	10,500	9,790	10,920	6,550	8,030	—	962	1,152	1,152	804	1,005	—	—
11,480	—	10,490	9,790	11,480	6,550	9,000	—	1,013	1,205	1,204	773	967	—	—
6,780	—	6,780	6,780	—	6,780	—	—	532	681	—	681	—	737	737
7,750	—	7,750	7,750	—	7,370	—	—	625	773	—	773	—	737	737
8,880	—	8,880	8,880	—	7,370	—	—	733	880	—	804	—	806	784
10,330	—	10,330	10,330	—	7,370	—	—	867	1,013	—	804	—	896	784
11,610	—	11,610	11,020	—	7,370	—	—	984	1,129	—	804	—	—	—
12,290	—	11,800	11,020	—	7,370	—	—	1,045	1,189	—	804	—	—	—
12,910	—	11,800	11,020	—	7,370	—	—	1,100	1,239	—	804	—	—	—
7,150	—	7,150	7,150	—	7,150	—	—	560	716	—	716	—	774	774
8,180	—	8,180	8,180	—	7,780	—	—	659	813	—	812	—	774	774
9,380	—	9,380	9,380	—	7,780	—	—	772	925	—	812	—	846	823
10,900	—	10,900	10,900	—	7,780	—	—	914	1,065	—	812	—	941	823
12,250	—	11,800	11,620	—	7,780	—	—	1,037	1,187	—	812	—	—	—
12,970	—	11,800	11,630	—	7,780	—	—	1,101	1,251	—	854	—	—	—
13,630	—	11,800	11,620	—	7,780	—	—	1,159	1,300	—	812	—	—	—
9,470	—	9,470	9,470	9,470	8,030	8,030	—	769	960	960	960	960	922	922
10,860	—	10,860	10,860	10,860	8,030	8,030	—	901	1,093	1,093	967	1,093	1,008	979
12,620	—	11,800	12,620	12,620	8,030	8,030	—	1,066	1,258	1,258	967	1,237	1,120	979
14,190	—	11,800	12,680	12,680	8,030	8,030	—	1,210	1,402	1,402	967	1,237	—	—
15,020	—	11,800	12,680	12,680	8,030	8,030	—	1,285	1,477	1,477	1,005	1,287	—	—
15,780	—	11,800	12,680	12,680	8,030	8,030	—	1,353	1,545	1,545	967	1,237	—	—
2,470	2,470	—	—	—	—	—	233	—	—	—	—	—	—	—
2,860	2,860	—	—	—	—	—	279	—	—	—	—	—	—	—
2,950	2,950	—	—	—	—	—	244	—	—	—	—	—	—	—
3,930	3,930	3,930	3,930	3,930	3,930	3,930	372	417	579	579	579	579	686	686
4,460	4,460	4,460	4,460	4,460	4,060	4,460	434	486	654	654	654	654	688	688
2,950	2,950	—	—	—	—	—	263	—	—	—	—	—	—	—
3,930	3,930	3,930	3,930	3,930	3,930	3,930	402	452	690	690	690	690	869	869
4,460	4,460	4,460	4,460	4,460	4,060	4,460	468	526	780	780	780	780	871	871
6,090	—	6,090	6,090	—	5,530	—	—	648	847	—	839	—	871	871
6,850	—	6,850	6,850	—	5,530	—	—	742	947	—	839	—	942	886
7,610	—	7,610	7,610	—	5,530	—	—	834	1,046	—	839	—	1,007	886
8,480	—	8,480	8,480	—	5,530	—	—	939	1,157	—	839	—	1,007	886
6,490	—	6,490	6,490	6,490	5,900	6,490	—	678	864	—	839	—	871	871
7,300	—	7,300	7,300	7,300	5,900	7,300	—	776	966	—	839	—	942	886
8,120	—	8,120	8,120	8,120	5,900	8,110	—	874	1,066	—	839	—	1,007	886
9,040	—	9,040	9,040	9,040	5,900	8,110	—	983	1,180	—	839	—	1,007	886
6,490	—	6,490	6,490	6,490	5,900	6,340	—	688	895	895	883	895	917	917
7,300	—	7,300	7,300	7,300	5,900	6,340	—	788	1,001	1,001	883	1,001	992	932
8,120	—	8,120	8,120	8,120	5,900	6,340	—	887	1,105	1,105	883	1,103	1,060	932
9,040	—	9,040	9,040	9,040	5,900	6,340	—	997	1,222	1,222	883	1,103	1,060	932
7,300	—	7,300	7,300	—	6,340	—	—	749	928	—	883	—	917	917
8,220	—	8,220	8,220	—	6,340	—	—	858	1,038	—	883	—	992	992
9,130	—	9,130	9,130	—	6,340	—	—	965	1,146	—	883	—	1,060	932
10,170	—	10,170	10,170	—	6,340	—	—	1,085	1,268	—	883	—	1,060	932

TABLE 1
MINIMUM PERFORMANCE PROPERTIES OF CASING

1	2	3	4	5	6	7	8	9	10	11	12
					Threaded and Coupled			Extreme Line			
Size: Outside Diameter in. D	Nominal Weight, Threads and Coupling lb. per ft.	Grade	Wall Thickness in. t	Inside Diameter in. d	Drift Diameter in.	Outside Diameter of Coupling in. W	Outside Diameter Special Clearance Coupling in. W_c	Drift Diameter in.	Outside Diameter of Box Power-tight in. M	Collapse Resistance psi	Pipe Body Yield Strength 1000 lbs.
8⅝	36.00	C-95	0.400	7.825	7.700	9.625	9.125	7.700	9.120	4,350	982
	40.00	C-95	0.450	7.725	7.600	9.625	9.125	7.600	9.120	6,020	1098
	44.00	C-95	0.500	7.625	7.500	9.625	9.125	7.500	9.120	7,740	1212
	49.00	C-95	0.557	7.511	7.386	9.625	9.125	7.386	9.120	9,710	1341
	40.00	P-110	0.450	7.725	7.600	9.625	9.125	7.600	9.120	6,390	1271
	44.00	P-110	0.500	7.625	7.500	9.625	9.125	7.500	9.120	8,420	1404
	49.00	P-110	0.557	7.511	7.386	9.625	9.125	7.386	9.120	10,740	1553
9⅝	32.30	H-40	0.312	9.001	8.845	10.625	—	—	—	1,370	365
	36.00	H-40	0.352	8.921	8.765	10.625	—	—	—	1,720	410
	36.00	J-55	0.352	8.921	8.765	10.625	10.125	—	—	2,020	564
	40.00	J-55	0.395	8.835	8.679	10.625	10.125	8.599	10.100	2,570	630
	36.00	K-55	0.352	8.921	8.765	10.625	10.125	—	—	2,020	564
	40.00	K-55	0.395	8.835	8.679	10.625	10.125	8.599	10.100	2,570	630
	40.00	C-75	0.395	8.835	8.679	10.625	10.125	8.599	10.100	2,990	859
	43.50	C-75	0.435	8.755	8.599	10.625	10.125	8.599	10.100	3,730	942
	47.00	C-75	0.472	8.681	8.525	10.625	10.125	8.525	10.100	4,610	1018
	53.50	C-75	0.545	8.535	8.379	10.625	10.125	8.379	10.100	6,350	1166
	40.00	L-80	0.395	8.835	8.679	10.625	10.125	8.599	10.100	3,090	916
	43.50	L-80	0.435	8.755	8.599	10.625	10.125	8.599	10.100	3,810	1005
	47.00	L-80	0.472	8.681	8.525	10.625	10.125	8.525	10.100	4,760	1086
	53.50	L-80	0.545	8.535	8.379	10.625	10.125	8.379	10.100	6,620	1244
	40.00	N-80	0.395	8.835	8.679	10.625	10.125	8.599	10.100	3,090	916
	43.50	N-80	0.435	8.755	8.599	10.625	10.125	8.599	10.100	3,810	1005
	47.00	N-80	0.472	8.681	8.525	10.625	10.125	8.525	10.100	4,760	1086
	53.50	N-80	0.545	8.535	8.379	10.625	10.125	8.379	10.100	6,620	1244
	40.00	C-90	0.395	8.835	8.679	10.625	10.125	8.599	10.100	3,250	1031
	43.50	C-90	0.435	8.755	8.599	10.625	10.125	8.599	10.100	4,010	1130
	47.00	C-90	0.472	8.681	8.525	10.625	10.125	8.525	10.100	5,000	1221
	53.50	C-90	0.545	8.535	8.379	10.625	10.125	8.379	10.100	7,120	1399
	40.00	C-95	0.395	8.835	8.679	10.625	10.125	8.599	10.100	3,320	1088
	43.50	C-95	0.435	8.755	8.599	10.625	10.125	8.599	10.100	4,120	1193
	47.00	C-95	0.472	8.681	8.525	10.625	10.125	8.525	10.100	5,090	1289
	53.50	C-95	0.545	8.535	8.379	10.625	10.125	8.379	10.100	7,340	1477
	43.50	P-110	0.435	8.755	8.599	10.625	10.125	8.599	10.100	4,420	1381
	47.00	P-110	0.472	8.681	8.525	10.625	10.125	8.525	10.100	5,300	1493
	53.50	P-110	0.545	8.535	8.379	10.625	10.125	8.379	10.100	7,950	1710
10¾	32.75	H-40	0.279	10.192	10.036	11.750	—	—	—	840	367
	40.50	H-40	0.350	10.050	9.894	11.750	—	—	—	1,390	457
	40.50	J-55	0.350	10.050	9.894	11.750	11.250	—	—	1,580	629
	45.50	J-55	0.400	9.950	9.794	11.750	11.250	9.794	11.460	2,090	715
	51.00	J-55	0.450	9.850	9.694	11.750	11.250	9.694	11.460	2,700	801
	40.50	K-55	0.350	10.050	9.894	11.750	11.250	—	—	1,580	629
	45.50	K-55	0.400	9.950	9.794	11.750	11.250	9.794	11.460	2,090	715
	51.00	K-55	0.450	9.850	9.694	11.750	11.250	9.694	11.460	2,700	801
	51.00	C-75	0.450	9.850	9.694	11.750	11.250	9.694	11.460	3,110	1092
	55.50	C-75	0.495	9.760	9.604	11.750	11.250	9.604	11.460	3,920	1196
	51.00	L-80	0.450	9.850	9.694	11.750	11.250	9.694	11.460	3,220	1165
	55.50	L-80	0.495	9.760	9.604	11.750	11.250	9.604	11.460	4,020	1276
	51.00	N-80	0.450	9.850	9.694	11.750	11.250	9.694	11.460	3,220	1165
	55.50	N-80	0.495	9.760	9.604	11.750	11.250	9.604	11.460	4,020	1276

13	14	15	16	17	18	19	20	21	22	23	24	25	26	27
									★Joint Strength — 1,000 lbs.					
	†Internal Pressure Resistance, psi							Threaded and Coupled					Extreme Line	
Plain End or Extreme Line	Round Thread		Buttress Thread				Round Thread		Buttress Thread				Standard Joint	Optional Joint
	Short	Long	Regular Coupling		Special Clearance Coupling		Short	Long	Regular Coupling	Regular Coupling Higher Grade‡	Special Clearance Coupling	Special Clearance Coupling Higher Grade‡		
			Same Grade	Higher Grade	Same Grade	Higher Grade								
7,710	—	7,710	7,710	—	6,340	—	—	789	976	—	927	—	963	963
8,670	—	8,670	8,670	—	6,340	—	—	904	1,092	—	927	—	1,042	979
9,640	—	9,640	9,640	—	6,340	—	—	1,017	1,206	—	927	—	1,113	979
10,740	—	10,380	10,740	—	6,340	—	—	1,144	1,334	—	927	—	1,113	979
10,040	—	10,040	10,040	10,040	6,340	6,340	—	1,055	1,288	1,288	1,103	1,288	1,240	1,165
11,160	—	10,380	11,160	11,160	6,340	6,340	—	1,186	1,423	1,423	1,103	1,412	1,326	1,165
12,430	—	10,380	11,230	11,230	6,340	6,340	—	1,335	1,574	1,574	1,103	1,412	1,326	1,165
2,270	2,270	—	—	—	—	—	254	—	—	—	—	—	—	—
2,560	2,560	—	—	—	—	—	294	—	—	—	—	—	—	—
3,520	3,520	3,520	3,520	3,520	3,520	3,520	394	453	639	639	639	639	—	—
3,950	3,950	3,950	3,950	3,950	3,660	3,950	452	520	714	714	714	714	770	770
3,520	3,520	3,520	3,520	3,520	3,520	3,520	423	489	755	755	755	755	—	—
3,950	3,950	3,950	3,950	3,950	3,660	3,950	486	561	843	843	843	843	975	975
5,390	—	5,390	5,390	—	4,990	—	—	694	926	—	926	—	975	975
5,930	—	5,930	5,930	—	4,990	—	—	776	1,016	—	934	—	975	975
6,440	—	6,440	6,440	—	4,990	—	—	852	1,098	—	934	—	1,032	1,032
7,430	—	7,430	7,430	—	4,990	—	—	999	1,257	—	934	—	1,173	1,053
5,750	—	5,750	5,750	—	5,140	—	—	727	947	—	934	—	975	975
6,330	—	6,330	6,330	—	5,140	—	—	813	1,038	—	934	—	975	975
6,870	—	6,870	6,870	—	5,140	—	—	893	1,122	—	934	—	1,032	1,032
7,930	—	7,930	7,930	—	5,140	—	—	1,047	1,286	—	934	—	1,173	1,053
5,750	—	5,750	5,750	5,750	5,140	5,140	—	737	979	979	979	979	1,027	1,027
6,330	—	6,330	6,330	6,330	5,140	5,140	—	825	1,074	1,074	983	1,074	1,027	1,027
6,870	—	6,870	6,870	6,870·	5,140	5,140	—	905	1,161	1,161	983	1,161	1,086	1,086
7,930	—	7,930	7,930	7,930	5,140	5,140	—	1,062	1,329	1,329	983	1,229	1,235	1,109
6,460	—	6,460	6,460	—	5,140	—	—	804	1,021	—	983	—	1,027	1,027
7,120	—	7,120	7,120	—	5,140	—	—	899	1,119	—	983	—	1,027	1,027
7,720	—	7,720	7,720	—	5,140	—	—	987	1,210	—	983	—	1,086	1,086
8,920	—	8,460	8,920	—	5,140	—	—	1,157	1,386	—	983	—	1,235	1,109
6,820	—	6,820	6,820	—	5,140	—	—	847	1,074	—	1,032	—	1,078	1,078
7,510	—	7,510	7,510	—	5,140	—	—	948	1,178	—	1,032	—	1,078	1,078
8,150	—	8,150	8,150	—	5,140	—	—	1,040	1,273	—	1,032	—	1,141	1,141
9,410	—	8,460	8,460	—	5,140	—	—	1,220	1,458	—	1,032	—	1,297	1,164
8.700	—	8,700	8,700	8,700	5,140	5,140	—	1,106	1,388	1,388	1,229	1,388	1,283	1,283
9,440	—	9,440	9,160	9,160	5,140	5,140	—	1,213	1,500	1,500	1,229	1,500	1,358	1,358
10,900	—	9,670	9,160	9,160	5,140	5,140	—	1,422	1,718	1,718	1,229	1,573	1,544	1,386
1,820	1,820	—	—	—	—	—	205	—	—	—	—	—	—	—
2,280	2,280	—	—	—	—	—	314	—	—	—	—	—	—	—
3,130	3,130	—	3,130	3,130	3,130	3,130	420	—	700	700	700	700	—	—
3,580	3,580	—	3,580	3,580	3,290	3,580	493	—	796	796	796	796	975	—
4,030	4,030	—	4,030	4,030	3,290	4,030	565	—	891	891	822	891	1,092	—
3,130	3,130	—	3,130	3,130	3,130	3,130	450	—	819	819	819	819	—	—
3,580	3,580	—	3,580	3,580	3,290	3,580	528	—	931	931	931	931	1,236	—
4,030	4,030	—	4,030	4,030	3,290	4,030	606	—	1,043	1,043	1,041	1,043	1,383	—
5,490	5,490	—	5,490	—	4,150	—	756	—	1,160	—	1,041	—	1,383	—
6,040	6,040	—	6,040	—	4,150	—	843	—	1,271	—	1,041	—	1,515	—
5,860	5,860	—	5,860	—	4,150	—	794	—	1,190	—	1,041	—	1,383	—
6,450	6,450	—	6,450	—	4,150	—	884	—	1,303	—	1,041	—	1,515	—
5,860	5,860	—	5,860	5,860	4,150	4,150	804	—	1,228	1,228	1,096	1,228	1,456	—
6,450	6,450	—	6,450	6,450	4,150	4,150	895	—	1,345	1,345	1,096	1,345	1,595	—

TABLE 1
MINIMUM PERFORMANCE PROPERTIES OF CASING

1	2	3	4	5	6	7	8	9	10	11	12
					Threaded and Coupled			Extreme Line			
Size: Outside Diameter in. D	Nominal Weight, Threads and Coupling lb. per ft.	Grade	Wall Thickness in. t	Inside Diameter in. d	Drift Diameter in.	Outside Diameter of Coupling in. W	Outside Diameter Special Clearance Coupling in. W_c	Drift Diameter in.	Outside Diameter of Box Power-tight in. M	Collapse Resistance psi	Pipe Body Yield Strength 1000 lbs.
10¾	51.00	C-90	0.450	9.850	9.694	11.750	11.250	9.694	11.460	3,400	1310
	55.50	C-90	0.495	9.760	9.604	11.750	11.250	9.604	11.460	4,160	1435
	51.00	C-95	0.450	9.850	9.694	11.750	11.250	9.694	11.460	3,480	1383
	55.50	C-95	0.495	9.760	9.604	11.750	11.250	9.604	11.460	4,290	1515
	51.00	P-110	0.450	9.850	9.694	11.750	11.250	9.694	11.460	3,660	1602
	55.50	P-110	0.495	9.760	9.604	11.750	11.250	9.604	11.460	4,610	1754
	60.70	P-110	0.545	9.660	9.504	11.750	11.250	9.504	11.460	5,880	1922
	65.70	P-110	0.595	9.560	9.404	11.750	11.250	—	—	7,500	2088
11¾	42.00	H-40	0.333	11.084	10.928	12.750	—	—	—	1,040	478
	47.00	J-55	0.375	11.000	10.844	12.750	—	—	—	1,510	737
	54.00	J-55	0.435	10.880	10.724	12.750	—	—	—	2,070	850
	60.00	J-55	0.489	10.772	10.616	12.750	—	—	—	2,660	952
	47.00	K-55	0.375	11.000	10.844	12.750	—	—	—	1,510	737
	54.00	K-55	0.435	10.880	10.724	12.750	—	—	—	2,070	850
	60.00	K-55	0.489	10.772	10.616	12.750	—	—	—	2,660	952
	60.00	C-75	0.489	10.772	10.616	12.750	—	—	—	3,070	1298
	60.00	L-80	0.489	10.772	10.616	12.750	—	—	—	3,180	1384
	60.00	N-80	0.489	10.772	10.616	12.750	—	—	—	3,180	1384
	60.00	C-90	0.489	10.772	10.616	12.750	—	—	—	3,360	1,557
	60.00	C-95	0.489	10.772	10.616	12.750	—	—	—	3,440	1644
	60.00	P-110	0.489	10.772	10.616	12.750	—	—	—	3,610	1903
13⅜	48.00	H-40	0.330	12.715	12.559	14.375	—	—	—	740	541
	54.50	J-55	0.380	12.615	12.459	14.375	—	—	—	1,130	853
	61.00	J-55	0.430	12.515	12.359	14.375	—	—	—	1,540	962
	68.00	J-55	0.480	12.415	12.259	14.375	—	—	—	1,950	1069
	54.50	K-55	0.380	12.615	12.459	14.375	—	—	—	1,130	853
	61.00	K-55	0.430	12.515	12.359	14.375	—	—	—	1,540	962
	68.00	K-55	0.480	12.415	12.259	14.375	—	—	—	1,950	1069
	68.00	C-75	0.480	12.415	12.259	14.375	—	—	—	2,220	1458
	72.00	C-75	0.514	12.347	12.191	14.375	—	—	—	2,600	1558
	68.00	L-80	0.480	12.415	12.259	14.375	—	—	—	2,260	1556
	72.00	L-80	0.514	12.347	12.191	14.375	—	—	—	2,670	1661
	68.00	N-80	0.480	12.415	12.259	14.375	—	—	—	2,260	1556
	72.00	N-80	0.514	12.347	12.191	14.375	—	—	—	2,670	1661
	68.00	G-90	0.480	12.415	12.259	14.375	—	—	—	2,320	1750
	72.00	G-90	0.514	12.347	12.191	14.375	—	—	—	2,780	1869
	68.00	C-95	0.480	12.415	12.259	14.375	—	—	—	2,330	1847
	72.00	C-95	0.514	12.347	12.191	14.375	—	—	—	2,820	1973
	68.00	P-110	0.480	12.415	12.259	14.375	—	—	—	2,330	2139
	72.00	P-110	0.514	12.347	12.191	14.375	—	—	—	2,890	2284

13	14	15	16	17	18	19	20	21	22	23	24	25	26	27
										★Joint Strength — 1,000 lbs.				
	†Internal Pressure Resistance, psi							Threaded and Coupled					Extreme Line	
Plain End or Extreme Line	Round Thread		Buttress Thread				Round Thread		Buttress Thread				Standard Joint	Optional Joint
	Short	Long	Regular Coupling		Special Clearance Coupling		Short	Long	Regular Coupling	Regular Coupling Higher Grade‡	Special Clearance Coupling	Special Clearance Coupling Higher Grade‡		
			Same Grade	Higher Grade	Same Grade	Higher Grade								
6,590	6,590	—	6,590	—	4,150	—	692	—	1,287	—	1,112	—	1,456	—
7,250	6,880	—	7,250	—	4,150	—	771	—	1,409	—	1,112	—	1,595	—
6,960	6,880	—	6,960	—	4,150	—	927	—	1,354	—	1,151	—	1,529	—
7,660	6,880	—	7,450	—	4,150	—	1,032	—	1,483	—	1,151	—	1,675	—
8,060	7,860	—	7,450	7,450	4,150	4,150	1,080	—	1,594	1,594	1,370	1,594	1,820	—
8,860	7,860	—	7,450	7,450	4,150	4,150	1,203	—	1,745	1,745	1,370	1,745	1,993	—
9,760	7,860	—	7,450	7,450	4,150	4,150	1,338	—	1,912	1,912	1,370	1,754	2,000	—
10,650	7,860	—	7,450	7,450	4,150	4,150	1,472	—	2,077	2,077	1,370	1,754	—	—
1,980	1,980	—	—	—	—	—	307	—	—	—	—	—	—	—
3,070	3,070	—	3,070	3,070	—	—	477	—	807	807	—	—	—	—
3,560	3,560	—	3,560	3,560	—	—	568	—	931	931	—	—	—	—
4,010	4,010	—	4,010	4,010	—	—	649	—	1,042	1,042	—	—	—	—
3,070	3,070	—	3,070	3,070	—	—	509	—	935	935	—	—	—	—
3,560	3,560	—	3,560	3,560	—	—	606	—	1,079	1,079	—	—	—	—
4,010	4,010	—	4,010	4,010	—	—	693	—	1,208	1,208	—	—	—	—
5,460	5,460	—	5,460	—	—	—	869	—	1,361	—	—	—	—	—
5,830	5,820	—	5,830	—	—	—	913	—	1,399	—	—	—	—	—
5,830	5,820	—	5,830	—	—	—	924	—	1,440	1,440	—	—	—	—
6,550	5,820	—	6,300	—	—	—	1,011	—	1,517	—	—	—	—	—
6,920	5,820	—	6,300	—	—	—	1,066	—	1,596	—	—	—	—	—
8,010	5,820	—	6,300	6,300	—	—	1,242	—	1,877	1,877	—	—	—	—
1,730	1,730	—	—	—	—	—	322	—	—	—	—	—	—	—
2,730	2,730	—	2,730	2,730	—	—	514	—	909	909	—	—	—	—
3,090	3,090	—	3,090	3,090	—	—	595	—	1,025	1,025	—	—	—	—
3,450	3,450	—	3,450	3,450	—	—	675	—	1,140	1,140	—	—	—	—
2,730	2,730	—	2,730	2,730	—	—	547	—	1,038	1,038	—	—	—	—
3,090	3,090	—	3,090	3,090	—	—	633	—	1,169	1,169	—	—	—	—
3,450	3,450	—	3,450	3,450	—	—	718	—	1,300	1,300	—	—	—	—
4,710	4,550	—	4,710	—	—	—	905	—	1,496	—	—	—	—	—
5,040	4,550	—	4,930	—	—	—	978	—	1,598	—	—	—	—	—
5,020	4,550	—	4,930	—	—	—	952	—	1,545	—	—	—	—	—
5,380	4,550	—	4,930	—	—	—	1,029	—	1,650	—	—	—	—	—
5,020	4,550	—	4,930	4,930	—	—	963	—	1,585	1,585	—	—	—	—
5,380	4,550	—	4,930	4,930	—	—	1,040	—	1,693	1,693	—	—	—	—
5,650	4,550	—	4,930	—	—	—	1,057	—	1,683	—	—	—	—	—
6,050	4,550	—	4,930	—	—	—	1,142	—	1,797	—	—	—	—	—
5,970	4,550	—	4,930	—	—	—	1,114	—	1,772	—	—	—	—	—
6,390	4,550	—	4,930	—	—	—	1,204	—	1,893	—	—	—	—	—
6,910	4,550	—	4,930	4,930	—	—	1,297	—	2,079	2,079	—	—	—	—
7,400	4,550	—	4,930	4,930	—	—	1,402	—	2,221	2,221	—	—	—	—

TABLE 1
MINIMUM PERFORMANCE PROPERTIES OF CASING

1	2	3	4	5	6	7	8	9	10	11	12
					Threaded and Coupled			Extreme Line			
Size: Outside Diameter in. D	Nominal Weight, Threads and Coupling lb. per ft.	Grade	Wall Thickness in. t	Inside Diameter in. d	Drift Diameter in.	Outside Diameter of Coupling in. W	Outside Diameter Special Clearance Coupling in. W_c	Drift Diameter in.	Outside Diameter of Box Powertight in. M	Collapse Resistance psi	Pipe Body Yield Strength 1000 lbs.
16	65.00	H-40	0.375	15.250	15.062	17.000	—	—	—	630	736
	75.00	J-55	0.438	15.124	14.936	17.000	—	—	—	1,020	1178
	84.00	J-55	0.495	15.010	14.822	17.000	—	—	—	1,410	1326
	75.00	K-55	0.438	15.124	14.936	17.000	—	—	—	1,020	1178
	84.00	K-55	0.495	15.010	14.822	17.000	—	—	—	1,410	1326
18⅝	87.50	H-40	0.435	17.755	17.567	20.000	—	—	—	*630	994
	87.50	J-55	0.435	17.755	17.567	20.000	—	—	—	*630	1367
	87.50	K-55	0.435	17.755	17.567	20.000	—	—	—	*630	1367
20	94.00	H-40	0.438	19.124	18.936	21.000	—	—	—	*520	1077
	94.00	J-55	0.438	19.124	18.936	21.000	—	—	—	*520	1480
	106.50	J-55	0.500	19.000	18.812	21.000	—	—	—	*770	1685
	133.00	J-55	0.635	18.730	18.542	21.000	—	—	—	1,500	2125
	94.00	K-55	0.438	19.124	18.936	21.000	—	—	—	*520	1480
	106.50	K-55	0.500	19.000	18.812	21.000	—	—	—	*770	1685
	133.00	K-55	0.635	18.730	18.542	21.000	—	—	—	1,500	2125

13	14	15	16	17	18	19	20	21	22	23	24	25	26	27
										★Joint Strength — 1,000 lbs.				
	†Internal Pressure Resistance, psi								Threaded and Coupled				Extreme Line	
	Round Thread		Buttress Thread				Round Thread		Regular Coupling	Buttress Thread			Standard Joint	Optional Joint
Plain End or Extreme Line			Regular Coupling		Special Clearance Coupling					Regular Coupling Higher Grade‡	Special Clearance Coupling	Special Clearance Coupling Higher Grade‡		
	Short	Long	Same Grade	Higher Grade	Same Grade	Higher Grade	Short	Long						
1,640	1,640	—	—	—	—	—	439	—	—	—	—	—	—	—
2,630	2,630	—	2,630	2,630	—	—	710	—	1,200	1,200	—	—	—	—
2,980	2,980	—	2,980	2,980	—	—	817	—	1,351	1,351	—	—	—	—
2,630	2,630	—	2,630	2,630	—	—	752	—	1,331	1,331	—	—	—	—
2,980	2,980	—	2,980	2,980	—	—	865	—	1,499	1,499	—	—	—	—
1,630	1,630	—	—	—	—	—	559	—	—	—	—	—	—	—
2,250	2,250	—	2,250	2,250	—	—	754	—	1,329	1,329	—	—	—	—
2,250	2,250	—	2,250	2,250	—	—	794	—	1,427	1,427	—	—	—	—
1,530	1,530	1,530	—	—	—	—	581	—	—	—	—	—	—	—
2,110	2,110	2,110	2,110	2,110	—	—	784	907	1,402	1,402	—	—	—	—
2,410	2,400	2,400	2,320	2,320	—	—	913	1,057	1,596	1,596	—	—	—	—
3,060	2,400	2,400	2,320	2,320	—	—	1,192	1,380	2,012	2,012	—	—	—	—
2,110	2,110	2,110	2,110	2,110	—	—	824	955	1,479	1,479	—	—	—	—
2,410	2,400	2,400	2,320	2,320	—	—	960	1,113	1,683	1,683	—	—	—	—
3,060	2,400	2,400	2,320	2,320	—	—	1,253	1,453	2,123	2,123	—	—	—	—

‡For P-110 casing the next higher grade is 150YS, a non-API steel grade having a minimum yield strength of 150,000 psi.

★Some joint strengths listed in Col. 20 through 27 are greater than the corresponding pipe body yield strength listed in Col. 12.

*Collapse resistance values calculated by elastic formula.

†Internal pressure resistance is the lowest of the internal yield pressure of the pipe, the internal yield pressure of the coupling, or the internal pressure leak resistance at the E_1 or E_7 plane.

INDEX